The Romans were the first great builders of bridges in the western world. Professor O'Connor, a civil engineer and expert in bridge construction, has personally examined a large number of those bridges that still remain, often in daily use, throughout the Roman empire – an area extending from Hadrian's Wall in the north of England to Shushtar in modern Iran. In this book he presents a thorough listing and description of all known bridges, including some 330 stone arch bridges, 34 timber bridges and 94 aqueducts. In many cases the construction of the bridges is illustrated by the author's own photographs and sketches.

Introductory chapters place the bridges in their geographical and historical contexts, with detailed maps of the empire-wide system of Roman roads and discussion of how these came to be constructed, and an investigation of the technology available to the Romans. An important element of Professor O'Connor's purpose is to elucidate the principles used by the Romans in designing their bridges, and in his concluding chapters he examines the proportions of the stone arches and subjects the rules that emerge to modern structural analysis.

This handsome book will be of value to archaeologists and engineers and to all those interested in the history of technology.

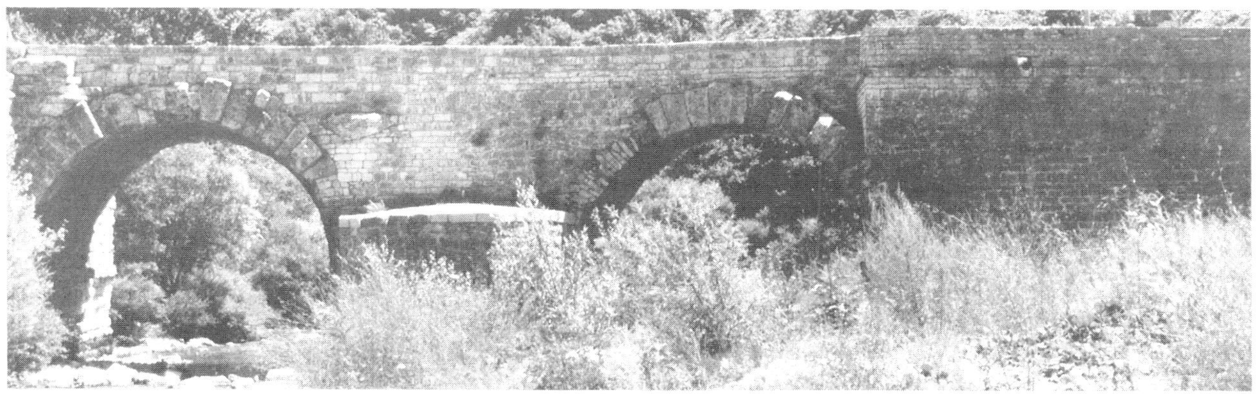

ROMAN BRIDGES

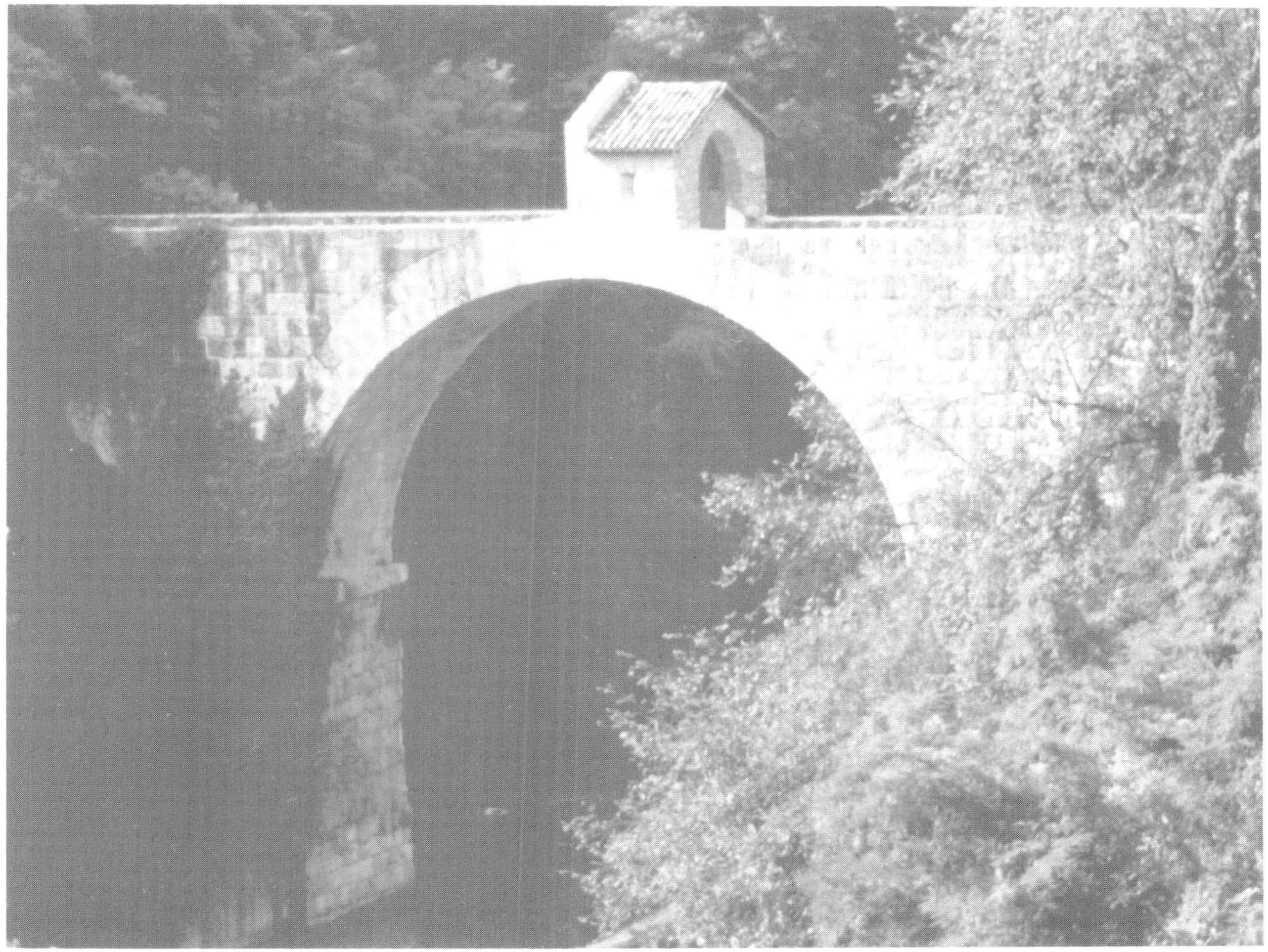

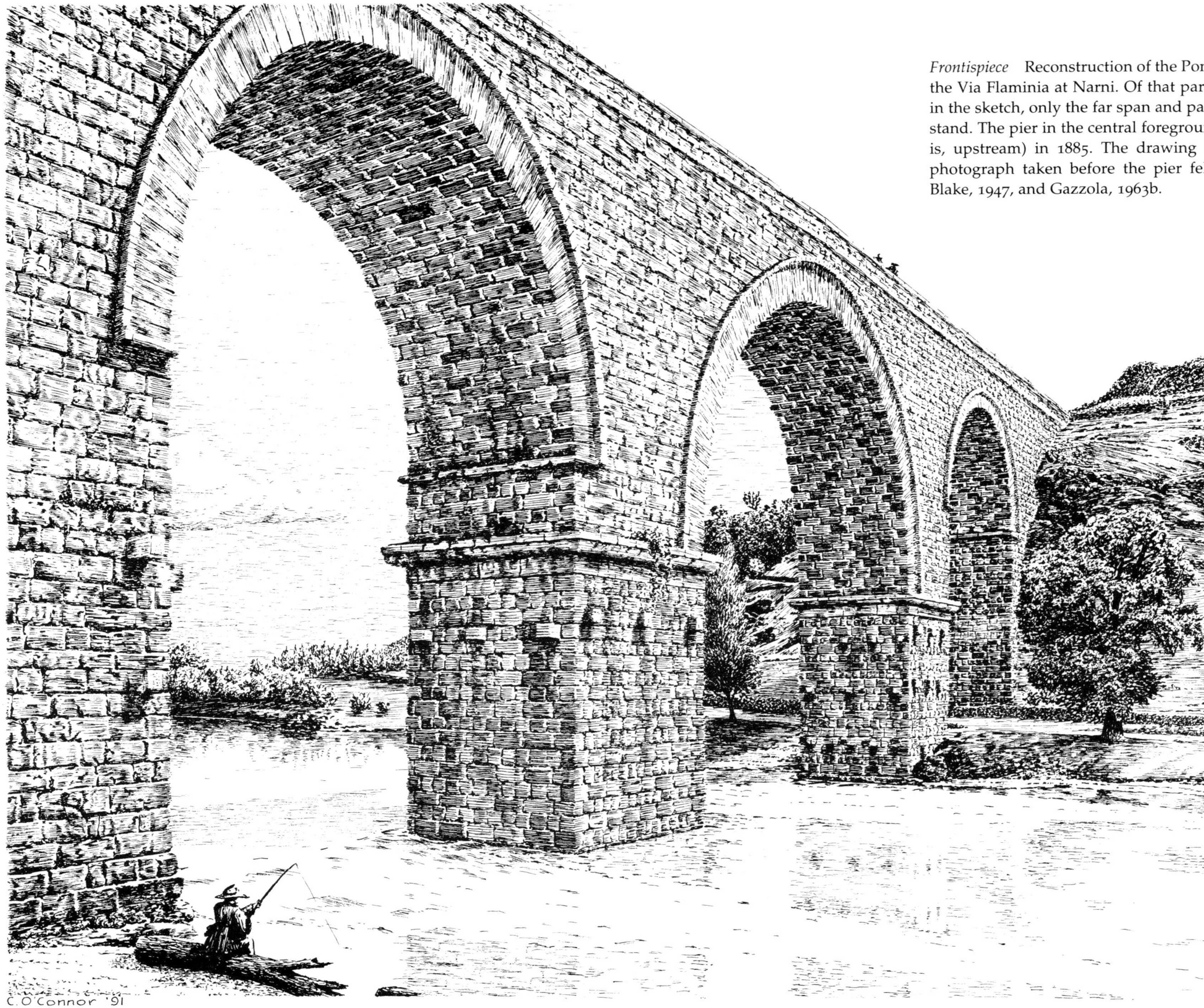

Frontispiece Reconstruction of the Ponte d'Augusto (I87) on the Via Flaminia at Narni. Of that part of the bridge shown in the sketch, only the far span and part of the near pier still stand. The pier in the central foreground fell to the left (that is, upstream) in 1885. The drawing is based on an early photograph taken before the pier fell and reproduced in Blake, 1947, and Gazzola, 1963b.

ROMAN BRIDGES

COLIN O'CONNOR

Emeritus Professor of Civil Engineering, University of Queensland

With photographs, sketches and diagrams by the author

CAMBRIDGE
UNIVERSITY PRESS

Published by the Press Syndicate of the University of Cambridge
The Pitt Building, Trumpington Street, Cambridge CB2 1RP
40 West 20th Street, New York, NY 10011-4211, USA
10 Stamford Road, Oakleigh, Melbourne 3166, Australia

First published 1993

Printed in Great Britain at the University Press, Cambridge

A catalogue record for this book is available from the British Library

Library of Congress cataloguing in publication data

O'Connor, Colin.
Roman bridges / by Colin O'Connor, with photographs, sketches, and diagrams by the author.
p. cm.
Includes bibliographical references.
ISBN 0 521 39326 4
1. Rome – Antiquities. 2. Bridges – Rome. 3. Roads, Roman. 4. Aqueducts – Rome.
I. Title.
DG28.O36 1993
388.1'32'0937 – dc20 92-30900 CIP

ISBN 0 521 39326 4 hardback

CE

To my wife
and the glory of God

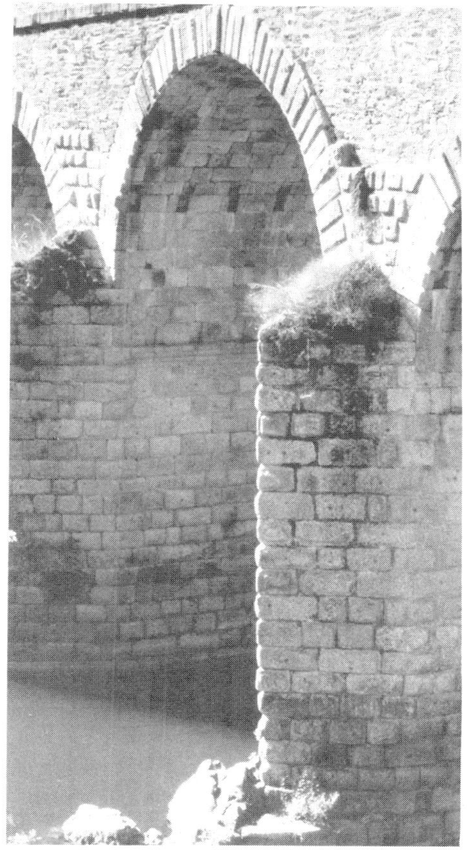

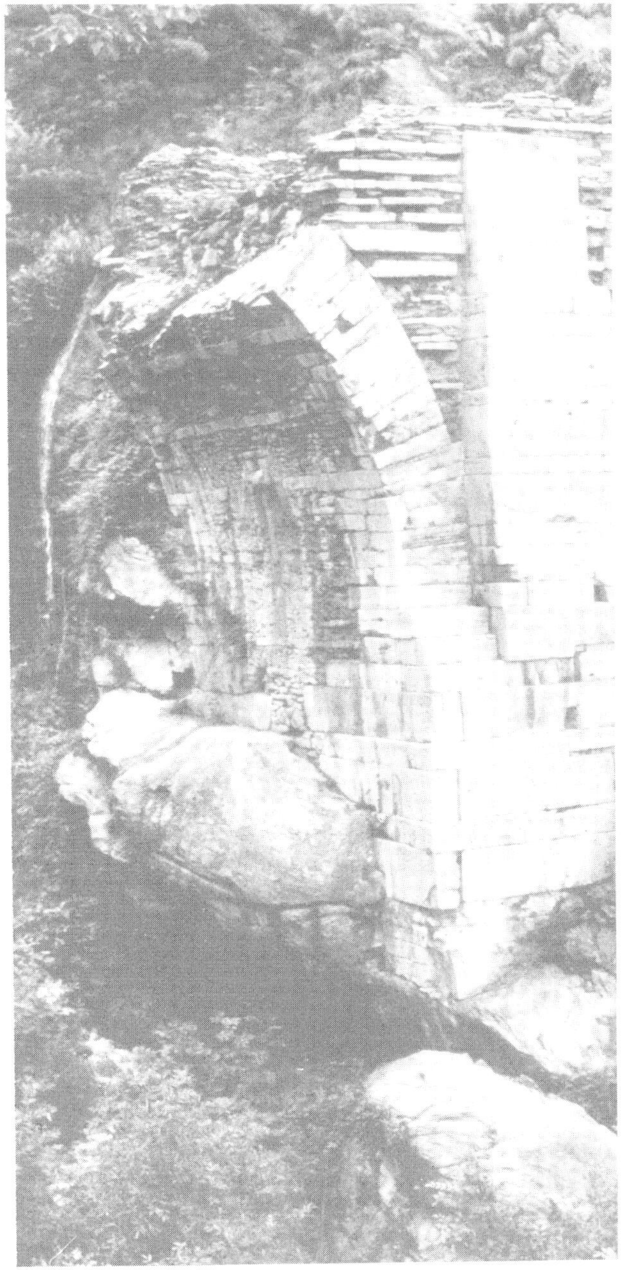

CONTENTS

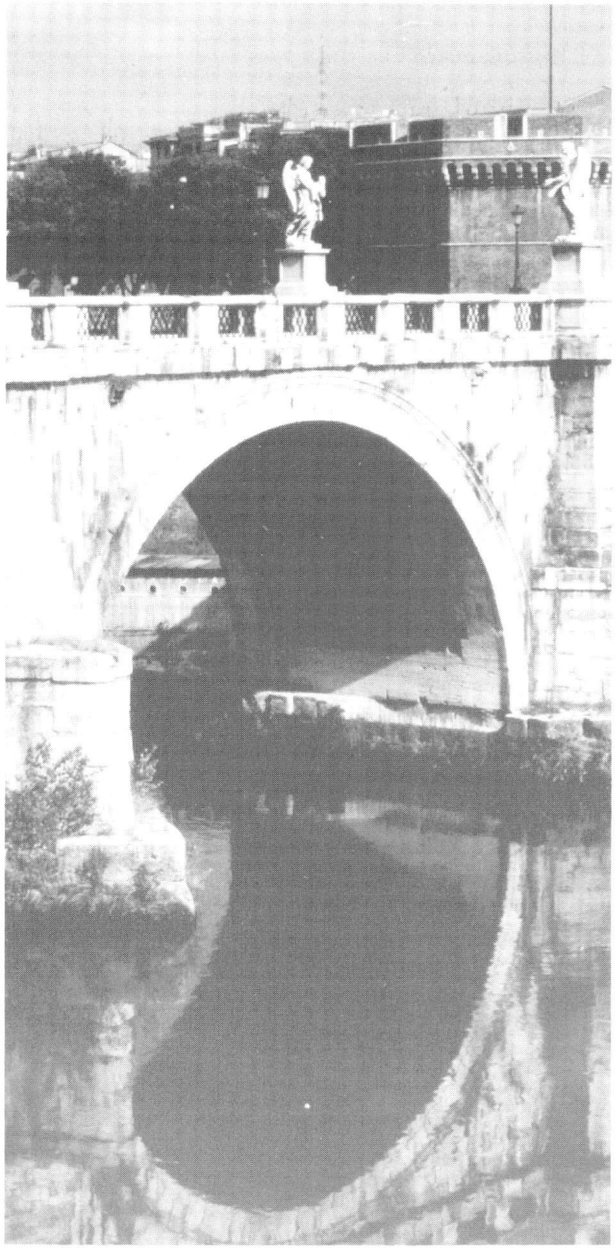

ILLUSTRATIONS

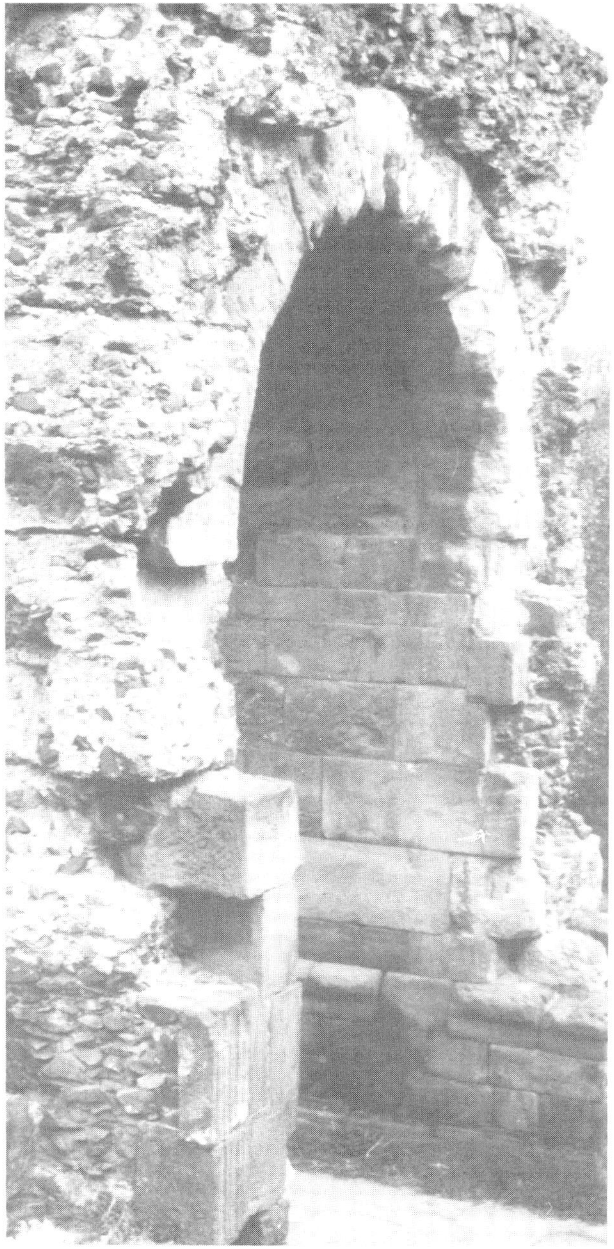

MAPS AND TABLES

MAPS

TABLES

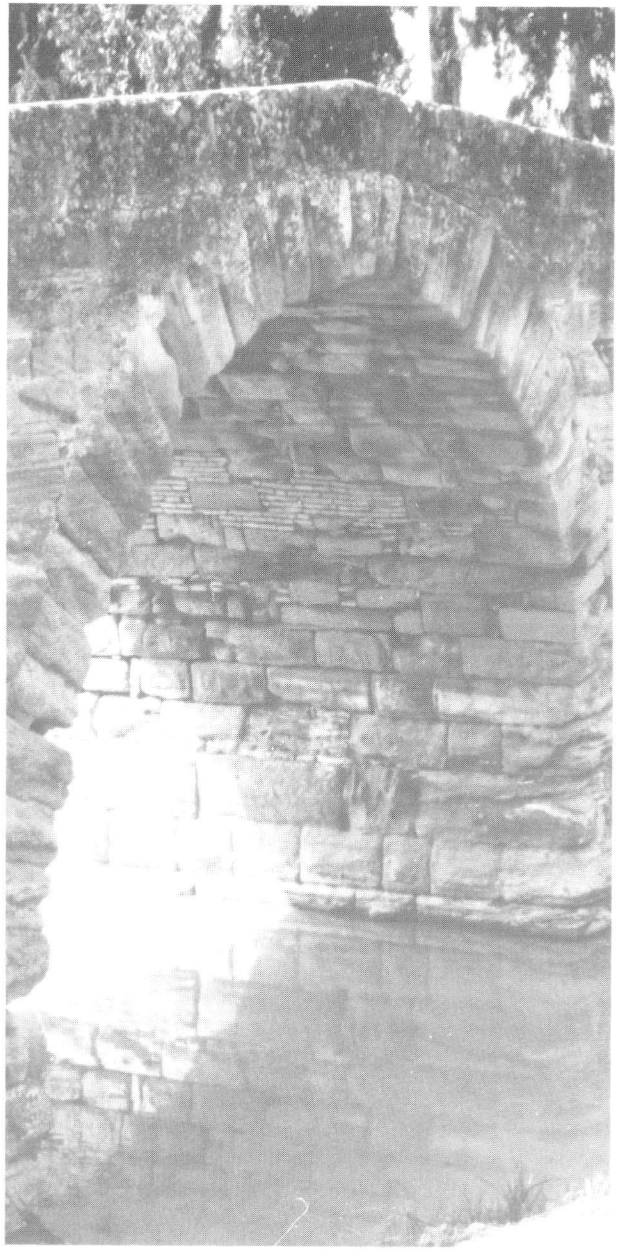

PREFACE

This book began in two ways.

It was written first because I am an engineer with an interest in bridges. In that regard it differs from many published works on Roman structures, produced typically by archaeologists and historians. Following my graduation in 1949 I worked on bridge design; my research since then has largely been related to bridges; and in 1971 I published *Design of Bridge Superstructures*, a technical book for the bridge engineer. After I was appointed professor in 1970, I gave in 1974 a somewhat belated inaugural address. Its title was 'The design of bridges: an historical study', and it was my intention to take data for existing bridges and submit these to a statistical review somewhat like that presented in Chapter 8. It was then that I found that, although many books had been published on the history of bridges, they were for the most part selective and anecdotal (see the Introduction), and there was no systematic study available of the bridges themselves. This was not a criticism of these books, for they were valuable and served a useful purpose, but, nevertheless, the fact remained. I decided then to prepare material for a history of bridges, and where else could one start but with Roman bridges?

In the 1970s I was heavily involved in university administration, but about 1980 I was invited to take part in a programme, sponsored by the Institution of Engineers, Australia, and funded by the Australian Heritage Commission, to study and record Australia's historic bridges. In the next few years I visited bridges in all Australian states; in 1980 I published the booklet, *How to Look at Bridges*; in 1983 the *Register of Australian Historic Bridges*; and in 1985 the book, *Spanning Two Centuries, Historic Bridges of Australia*. I was asked, finally, to identify bridges for inclusion in the Australian Register of the National Estate. I selected 250, of which 85 were already on the Register. It was not, then, until 1987 that I was able to return to Roman bridges.

This book was written secondly because I am a Christian. After my conversion in 1951, I decided in 1960 to become better equipped, and studied for the Bachelor of Divinity degree of the University of London, which I received in 1965. At that time I learnt New Testament Greek, which I still use.

I became impressed some years ago by something Paul wrote to the Galatian churches (Galatians 4:4), concerning the coming of Christ. 'But when the time had fully come, God sent his son' (NIV). He meant to say two things: first, that when Jesus came it was the right time; and second, that he came when certain preliminary conditions had been fulfilled. It was, of course, very much at the right time, for there was then a unified world government and a developed and continuous road system. Paul used this system. From Jerusalem he went to Tarsus, his home town, and then to Antioch or Antakya, that most strategically placed of all Roman towns of the east. It was from here that he travelled with his message of judgement, hope and deliverance, along the roads of Asia, the Via Egnatia in Greece and finally by the Via Appia to Rome. A study of Roman roads is, then, part of the background of

Christian church history and I initially studied them, partly from this perspective.

There are many persons or bodies I should thank for their assistance in the writing of the book. The first is my wife, who has been patient in our travels and constant in her support. I owe her much in so many ways. I am grateful, too, to the University of Queensland, for in 1987 they granted me study leave to travel and prepare material for this book. During that period I spent some time in Cambridge, England, through the kindness of Professor J. Heyman, and worked in the University Library as a visiting fellow of Corpus Christi College. My friend, Professor R. D. Milns, of the Department of Classics and Ancient History of the University of Queensland, has encouraged me in the writing of the book and has read much of the final text. Associate Professor B. Jones has read the book in detail, and I am grateful to them both, and to other members of that department, for their assistance.

The book has been typed for me by Beverly Watson and, to a greater extent, Anne Dombrovskis, both friends for many years. I am deeply grateful to them.

Colin O'Connor
Brisbane
January 1992

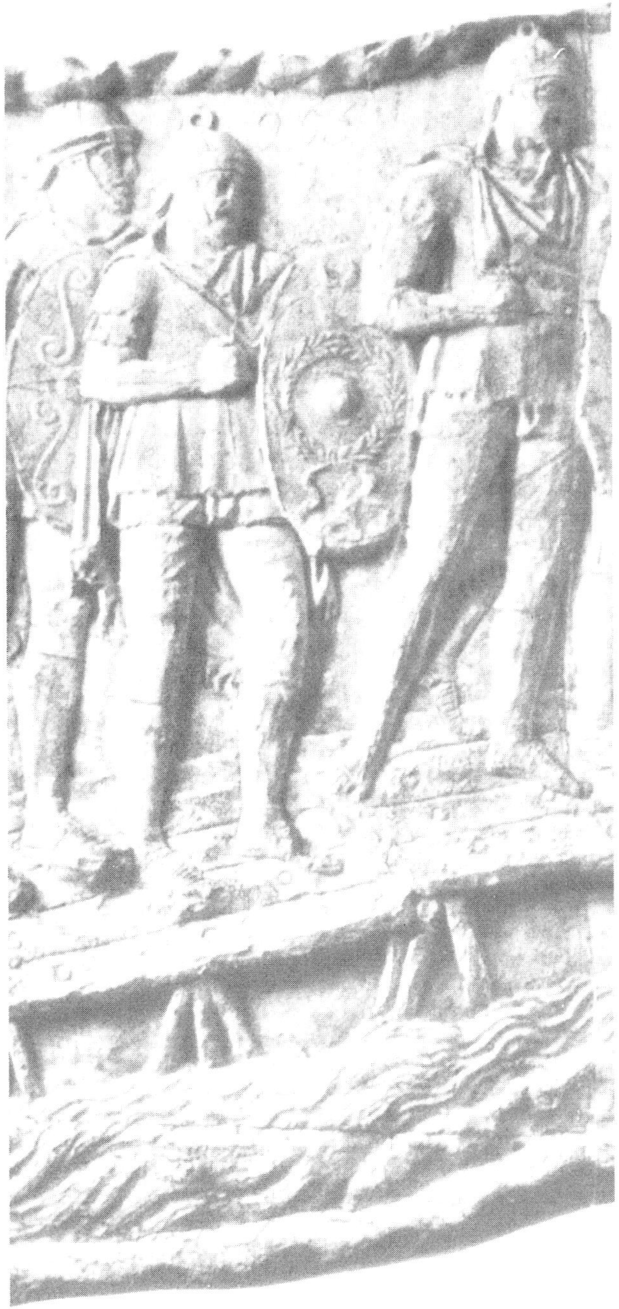

ABBREVIATIONS

ASCE J. Struct. Eng. *American Society of Civil Engineers, Journal of Structural Engineering*
BNI British Naval Intelligence Division
CIG *Corpus Inscriptionum Graecarum*
CIL *Corpus Inscriptionum Latinarum*
HMSO Her Majesty's Stationery Office
ICE Proc. *Institution of Civil Engineers Proceedings*
Int. J. Mech. Sci. *International Journal of Mechanical Sciences*
JHS *Journal of Hellenic Studies*
JRS *Journal of Roman Studies*
PBSR *Papers of the British School at Rome*
Sci. Am. *Scientific American*
Struct. Eng. *Structural Engineer*

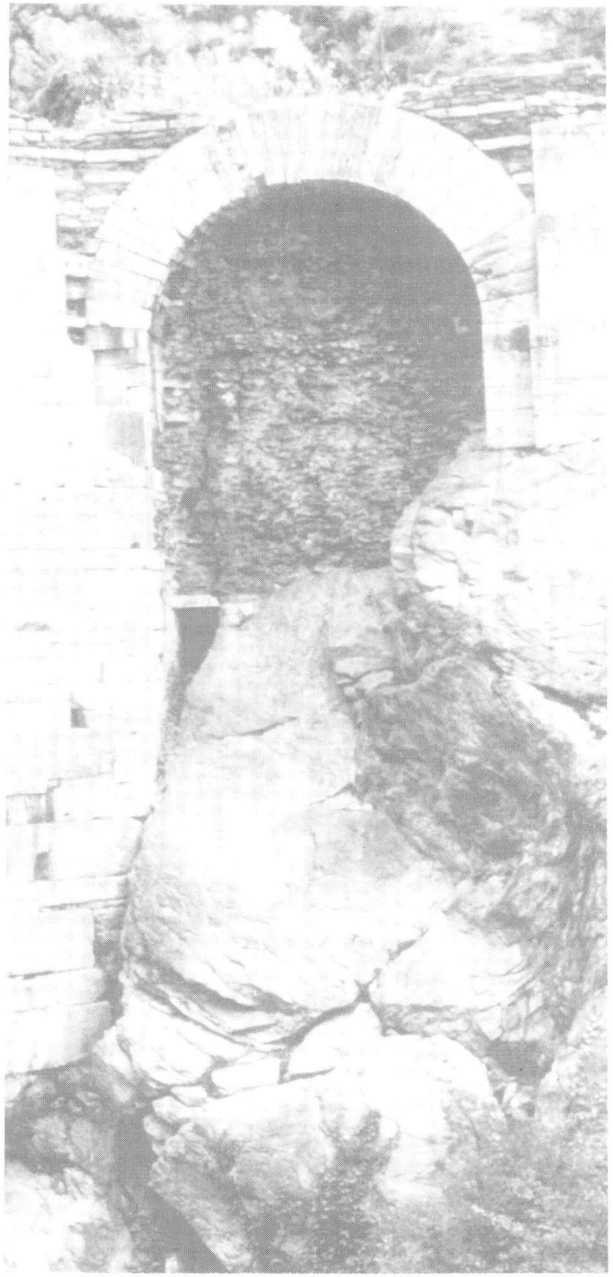

INTRODUCTION

The Romans were not the first bridge builders. To claim that they were would fly against common sense, for there is no doubt that, from time immemorial, men have built structures to cross streams and other obstacles. These may have been primitive in form, such as the early log and vine bridges, but it would have been only a small step to advance from the use of a log, fortuitously fallen across a stream, to the deliberate construction of a timber bridge. There is some evidence that this was so. Caesar, for example, refers to pre-existent timber bridges, and the oldest of all recorded road bridges was built by Nabopolassar over the Euphrates at Babylon in the seventh century BC, with a timber deck on piers of brick and stone (Section 6.3). Even earlier was Sennacherib's aqueduct (690 BC), which at Jerwan crossed a ravine, with five small arches of corbelled stone (Section 7.6).

There is evidence, too, that the Greeks built bridges. Tyrrell, for example (1911: 20), has a drawing of a bridge at Assos with stone slabs on stone piers. Bean (1971: 149) refers to a bridge at Kumyer, between Cnidus Vetus and Cnidus Nova in Turkey as 'among the more notable specimens of ancient Greek bridges – the greater part is still standing – it spanned the stream in the form of a triangular or trapezoidal arch (the top is missing)'. Akurgal (1978: 34) says, 'The handsome remains of a stone bridge uncovered at Pergamon date from Greek times.' One would expect that the Egyptians also might have built bridges, although their needs would have been less because of their dry terrain. What we can say is that the Romans were the world's first major bridge builders, and that much of their work remains. It is surprising, therefore, that there is no book in English devoted to them.

There are many historical accounts of bridge construction, such as the books by Beckett (1969), Hopkins (1970), Leonhardt (1982), Naruse et al. (1964), Robins (1948), Sadler (1986), Shirley Smith (1953), Steinman and Watson (1941) and Tyrrell (1911), and most of these start with Roman bridges. However, it is impossible to write a full account of bridge construction in one book, and it is not unfair to say that these works tend to be anecdotal and selective, rather than systematic. A careful reading of them will show that many major bridges have been ignored. There have also been at least two books with an emphasis on the social aspects of bridge construction – by W. J. and S. R. Watson (1937), and by Boyer on French bridges (1976). Frank Brangwyn was a noted artist, and his book with Sparrow (1915) includes many of his bridge paintings. *Songs of a Bridge Builder* (1959) may come as a surprise, for it was written by D. B. Steinman, an experienced bridge engineer as well as writer.

There is, however, a definitive work on Roman bridges in Italian, by Gazzola (1963b). It is thorough; each bridge is numbered, and the book is divided into sections, one for each bridge. It has useful drawings and references. There are, in addition, many classical works which include accounts of Roman bridges, such as those by Blake (1947, 1959 and, with Bishop, 1973), and the accounts of roads and bridges in the classical literature.

Although this book relies heavily on that by Gazzola, it

differs in many respects, for its objectives are not the same. It sets out first, as does Gazzola, to be thorough, and to make reference to all bridges for which information has been found. Its arrangement is, however, different, for it is largely geographical whereas Gazzola is at least partly chronological. This book also deals with many matters that are not discussed by Gazzola.

The best way to study Roman bridges is to see them, and the book generally describes how they may be found. The second way is to study them through illustrations, and so there are here many photographs and a few sketches.

It has always appeared to the author that the Romans must have had rules for arch design. An attempt has been made, therefore, in Chapter 8 to elucidate these rules by an empirical study of the bridges themselves. In Chapter 9, these rules have been subjected to modern analysis, to see whether they are correct.

The design of a bridge will depend on the materials that are available and the technology which may be used to form them into a bridge. The materials the Romans used were not only stone but also timber and concrete. Timber was, of course, used for falsework for masonry arches, but not only so, for the Romans built many of their bridges in timber. The bridge of Apollodorus (Section 6.3) was one of the greatest of their bridges, and had a timber superstructure. The availability to them of a naturally occurring pozzolan meant that they knew how to build in concrete, a knowledge which was lost with the fall of the Roman empire and not rediscovered until the eighteenth and nineteenth centuries.

Ideally, it is desirable to go with the Roman engineer in spirit to his site, see it as he saw it, select his spans, design the bridge and arrange the supply of materials and labour. To do so involves a number of matters which are not commonly addressed – surveying, metal-working, computation, falsework construction – and these and other matters are discussed here.

It is also desirable to place the bridges within their geographical and historical contexts. The production of maps of the major Roman roads (Chapter 2) has, of course, been made possible only by the works of others, but it has been a useful exercise, for it brings into sharper focus a realisation of the geographical difficulties which the Romans faced and conquered. The maps include many Roman place-names. These are keyed to modern names in the text, and also in the index. An account has been given also of the historical events that led to the construction of the roads.

Finally, there is a short account of Roman aqueducts. Many of them included bridges, and one of these, the Pont du Gard, is among the most famous. At the time of writing, no consolidated account of Roman aqueducts has been published, although it is believed that this deficiency may shortly be redressed by the forthcoming book by Hodge (1992).

The bridge dimensions used in this book have been taken from published accounts. If these accounts are compared, it will be found that they are often inconsistent. Accordingly, it cannot be certain that the dimensions quoted here are correct. A glossary of technical terms is given in Appendix 1.

A bridge today is valued chiefly for functional and aesthetic reasons. It also has a social significance, for it joins communities, is a focus for routes of travel, and has provided the initial impetus for the establishment of many cities and towns (see for example Boyer, 1976). In older times it had also a religious significance.

Appendix tables A.1–8 list the names of many Roman bridges. Only a few of these are old, many are connected with place-names, and only a few are repeated. Some of the repeated names are predictable, such as Roman bridge, stone bridge or old bridge. More surprising is Devil's bridge, which occurs seven times in these lists, and is the name most often repeated. It is also not an uncommon name for medieval bridges (O'Connor, 1975). In many of these cases there is a legend attached to the name, although this could be, of course, a later addition. Such legends typically took the form that the builder, or some other interested party, sold his soul to the Devil to secure his cooperation in the building of the bridge (see, for example, Steinman and Watson, 1941: 67).

People have commonly seen spirit life in inanimate and animate objects, particularly if the objects appeared to have a life of their own, like moving water or fire. Galloway (1914: 90), for example, in his book on the philosophy of religion, says that 'among the peoples of classical antiquity vestiges of an original worship of rivers and of the element of fire survived'. The columns of Trajan and of Marcus Aurelius, in depicting the Roman armies crossing the Danube by way of a floating bridge (Section 6.2), each show the operation watched by the god of the river or Father Danube as he has been called. Statius (*Silvae* IV.3.67–71), referring to the god of the River Volturnus near Capua, says, 'But Volturnus, his yellow head and wet hair all entangled with pliant weed, raises his face ...' To build a bridge, therefore, could be seen as a possible cause of offence to the river-god, and a matter to be approached with fear. To quote Dilke (1971: 33), 'Bridge-building was regarded as needing some sort of magic powers.' Caesar built his bridge across the Rhine (Section 6.3) in ten days and destroyed it after another eighteen, and it has been suggested that this rather wanton feat was a demonstration to the Germans that he was superior to their river-god. It could, of course, be interpreted alternatively as a demonstration of his overwhelming technical and military strength.

The Pons Sublicius in Rome had an acknowledged religious significance (Section 6.3). Allegedly built by Ancus Martius about 600 BC, it was placed in the care of the high priest, who received the title *pontifex*, commonly translated as 'bridge builder', from the Latin *pons* (bridge) and *facere* (to make or build). The related title, *pontifex maximus*, passed on to emperors and to the present popes, leading to the related words 'pontiff' and 'pontificate'. It has been suggested that the continuation of the title arose from an allegorical interpretation – the pontiff as one who builds a bridge from God to man (see 'pontiff' in the *New Catholic Encyclopaedia*, 1967). In this connection there is some value in observing that 'bridge' is not a word that occurs in the Bible, despite the opportunity it gives for allegory or parable (Watson and Watson, 1937). Jesus said, 'I am the way', and the term, 'way', continues to be used throughout the New Testament as a description of the Christian faith. Paul, for example (Acts 9:2), wished to persecute those who were 'of the way'. The concept of the road is, therefore, biblical, but not that of the bridge, and this is of some interest in the dating of bridges in biblical lands.

On a lighter note, Robert Burns, the Scottish poet, describes in 'Tam O'Shanter' the incident when the farmer, Tom, and his horse, Maggie, were chased by a mob of witches and bogles (or spirits) whom he disturbed on his way home from the inn one dark night. To quote Kinsley (1968, vol. II, poem 321), 'It is a well-known fact that witches, or any evil spirits, have no power to follow ... farther than the middle of the next running stream.'

> Now, do thy speedy utmost, Meg,
> And win the key-stane of the brig;
> There, at them thou thy tail may toss,
> A running stream they dare na cross.
> [Maggie leapt, just in time.]
> Ae spring brought off her master hale,
> But left behind her ain gray tail.
> The carlin claught her by the rump,
> And left poor Maggie scarce a stump.

The bridge was the fine old fifteenth-century Brig o'Doon, near Alloway, south of Ayr (Johnson and Scott-Giles, 1933: 438).

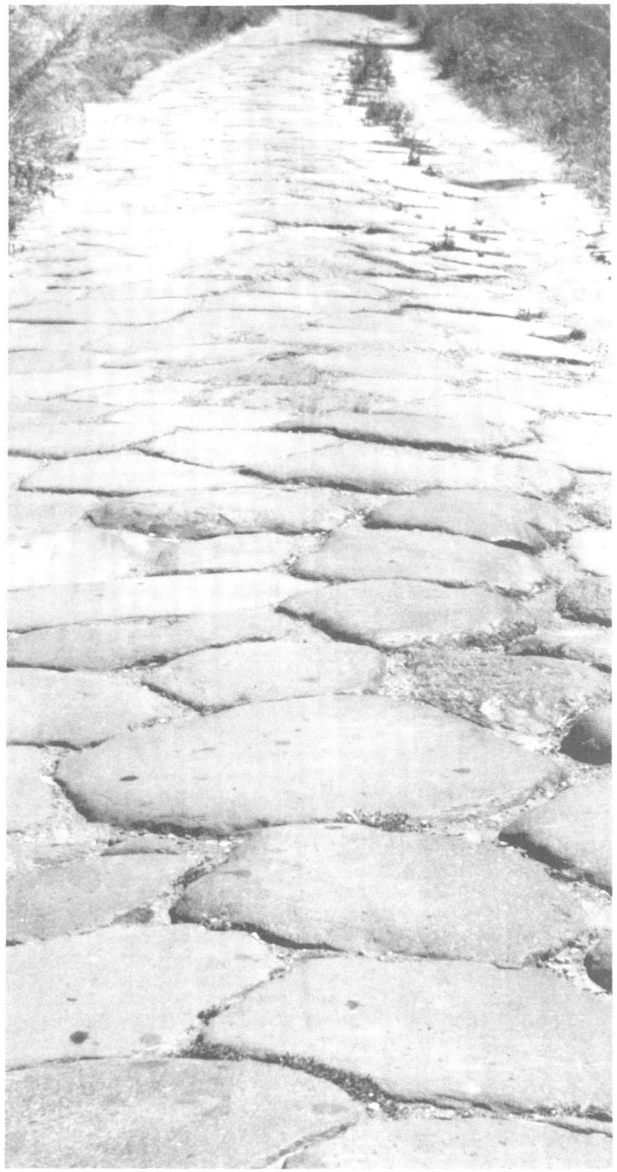

2

ROMAN ROADS AND THEIR HISTORY

2.1 Introduction

Bridges and roads go together, for it is the roads which require the bridges, and the bridges make possible the roads. Roman traffic bridges, as distinct from the aqueduct bridges described in Chapter 7, formed part of the Roman road system. A study of the roads is helpful here for a number of reasons.

1 It establishes the lines or routes along which many of the bridges are to be found.
2 It assists in dating the bridges, although it must be remembered that the bridges described here are often second-generation structures, replacing earlier bridges, ferries or fords.
3 It links the construction of the bridges with the construction of the roads. Their builders were, at least in some senses, the same and they shared a common technology.
4 It gives the bridges a better historical context, for it relates them to the growth and establishment of Rome and its empire.

The extraordinary extent of Roman roads is revealed in the maps in this chapter. These show in sequence the roads in Italy, from Italy west to Britain, down the Rhine and Danube, and through Greece and Turkey to Egypt and the east. They are based on the comprehensive accounts by Sitwell (1981), Chevallier (1976) and von Hagen (1967a, b) – see also Bergier (1626); Casson (1974); Forbes (1934); Gregory (1931); Hindley (1971); Finley (1973); Rostovtzeff (1967) – and the more particular descriptions of roads and road systems found in the literature. It does not take long to learn that these accounts differ in detail. Here, these differences have been resolved either by a judgement concerning the reliability of the source, a study of the most likely route based on topography, or, at times, by arbitrary choice.

No attempt has been made to go back to earlier sources, such as the Peutinger Table and the Antonine Itinerary, as this information has already been included in the reference material used here. Chevallier (1976), Dilke (1985) and others have given accounts of these documents. For those unfamiliar with the subject, the Peutinger Table is a kind of ancient road map, greatly distorted, with a length of 6.8 m and a height of only 0.34 m. It covers the entire empire, from the Rhine and Danube to North Africa, and from Britain and Spain to Persia. Although curiously diagrammatic, it contains a wealth of information, including an abundance of place-names and distances, so that it is possible to use it to reconstruct the road system on an undeformed scale. The present map, held in the Library of Vienna, is medieval, found at the end of the fifteenth century and given to Conrad Peutinger of Augsburg, after whom it is named. It is, however, clearly a copy of an ancient work and is surprisingly accurate. An excellent eight-page, photographic copy in colour of part of the Table – from the east coast of Britain to Rome – is given in the book by Sandström (1970), who attributes the original to Castorius, about AD 365. Dilke (1985: 113) also dates the map from the fourth century AD, but queries the ascription to Castorius. The Antonine Itinerary consists of lists of place-names and distances for various routes. It also dates from about AD 300. Chevallier (1976) gives an extensive bibliography for these and other sources. Some of these are also cited here (Cuntz, 1929; Miller, 1887, 1916; Levi and Trell, 1964).

The maps in this book show only the major roads. They show also the major streams, for it is suggested that the roads cannot be drawn sensibly without regard to topography. The best way of portraying the topography would be to use a contour map, with the roads superimposed. This has not been done here for two reasons: it would have required extraordinary effort; and within the space available the additional detail may well have been unclear. The major elements of the internal topography are the mountains and the streams. Sitwell (1981) has attempted to show both, but such a simplified drawing of the mountain chains can be misleading, for they are in fact of the most complex geometry. In any case, a map of the streams also indicates the mountains, for the mountains lie between the streams.

It requires only a glance at the maps to see that the roads are usually irregular in alignment. One of the most common misconceptions of the Roman roads is that they were generally straight. It is true that, on a smaller scale, the alignment of a road may often be formed as a series of straight segments. Margary (1973), for example, has described the evidence for this in Britain. These segmental alignments result, no doubt, from the use by Roman surveyors, or *agrimensores*, of the *groma* described in Section 4.7, which was capable of establishing and extending a straight alignment. It was, and still is, more convenient to specify a road that is straight between a sequence of designated points. But, on the scale used here, it is simply not true to say that the roads were generally straight, although there are exceptions, such as the Via Appia and Via Aemilia in Italy, and some of the British roads. The reality is that the Romans built straight roads only when they could and this was only in exceptional cases such as those mentioned above. The first section of the Via Appia from Rome to Terracina crossed land which was for the most part extraordinarily flat. The terrain of eastern Britain is generally flat, such as near Cambridge, as is also the valley of the Po, traversed by the Via Aemilia. The excessive emphasis on the straightness of Roman roads has arisen precisely because these exceptions are accessible and well known.

More generally, the alignment of the Roman roads was determined by the topography. Although the Roman engineer could properly accept steeper grades than those required for the roads of today, he was in the end bound by topographical constraints.

An engineer may often avoid steep grades by following either the ridges or the valleys, where the actual choice will depend on the nature of the terrain. If the higher land is flat, and the valleys steeply cut, then the high land will offer the better route. If the valleys are wide and flat, then a route along the valleys will be preferred, and this is often seen to be the case in the accompanying maps. A good illustration is to be found in the Via Flaminia north from Rome. A detailed topographical map (not, therefore, Map 1) shows a remarkable sequence of steep-sided valleys through the rough Apennine chain, north from Spoleto. The Via Flaminia followed this route, rising finally up the Topino valley to the Scheggia Pass – at 581 m the lowest and easiest of the more easterly Apennine passes – and then dropping with the Burano Torrent to the Metaurus River. Another example occurs in the Via Annia to the south, which rises from Salerno up the Val di Diano, crosses to join the valley of the Crati to Cosenza, passes over a sharp divide and then turns south west down the River Savuto to the coast.

In these examples the streams are small and rapidly flowing. In other cases they are large – such as the Rivers Rhône and Loire in France, the upper Ebro and Guadalquivir in Spain, the Danube and the Rhine, and the River Euphrates in Asia Minor. Often the association of the roads with river-borne traffic was important; the stream always provided water to drink; most commonly the road was effectively down in the valley with the stream; sometimes it was far up the adjacent highlands; but always it was the course of the river which defined the route of the road.

This common association of roads and streams is, of course, relevant to the present subject, for stream crossings required bridges. These were of two kinds: first, there were the minor crossings where the road crossed a multitude of minor tributaries running down from the sides of the valley into the main stream; second were the crossings of the main stream itself.

There are many examples where the Roman engineer sought to minimise the number of major crossings. If a road started on one side of a major stream, then it tended to stay on that side, and, as a result, the locations of the towns or settlements tended to lie on that one side. A striking example on Map 2 is the River Loire, where four of the five towns shown are on the right bank (that is, the right bank facing downstream). Another is the River Rhône south of Lyon (ancient Lugdunum). There is, however, an important exception. If a major route along a stream was crossed by another road at right angles, then a crossing had to be provided for this second road. If this was the case, then there was no reason why the first road also could not cross the stream and continue along the other bank. This can be seen at Lyon on the Rhône, where the road heading north crossed from the east (or left) bank of the Rhône to the west bank of the Saône, its tributary from the north. There were, of course, other cases where the terrain forced a crossing, or a crossing was made for other reasons.

There are some other points which should be made concerning the maps. They show the ancient names of towns, whereas in the text the modern names are generally preferred. The index shows the links between the two. The spellings of ancient names are generally those given by Sitwell (1981). The maps, or this discussion,

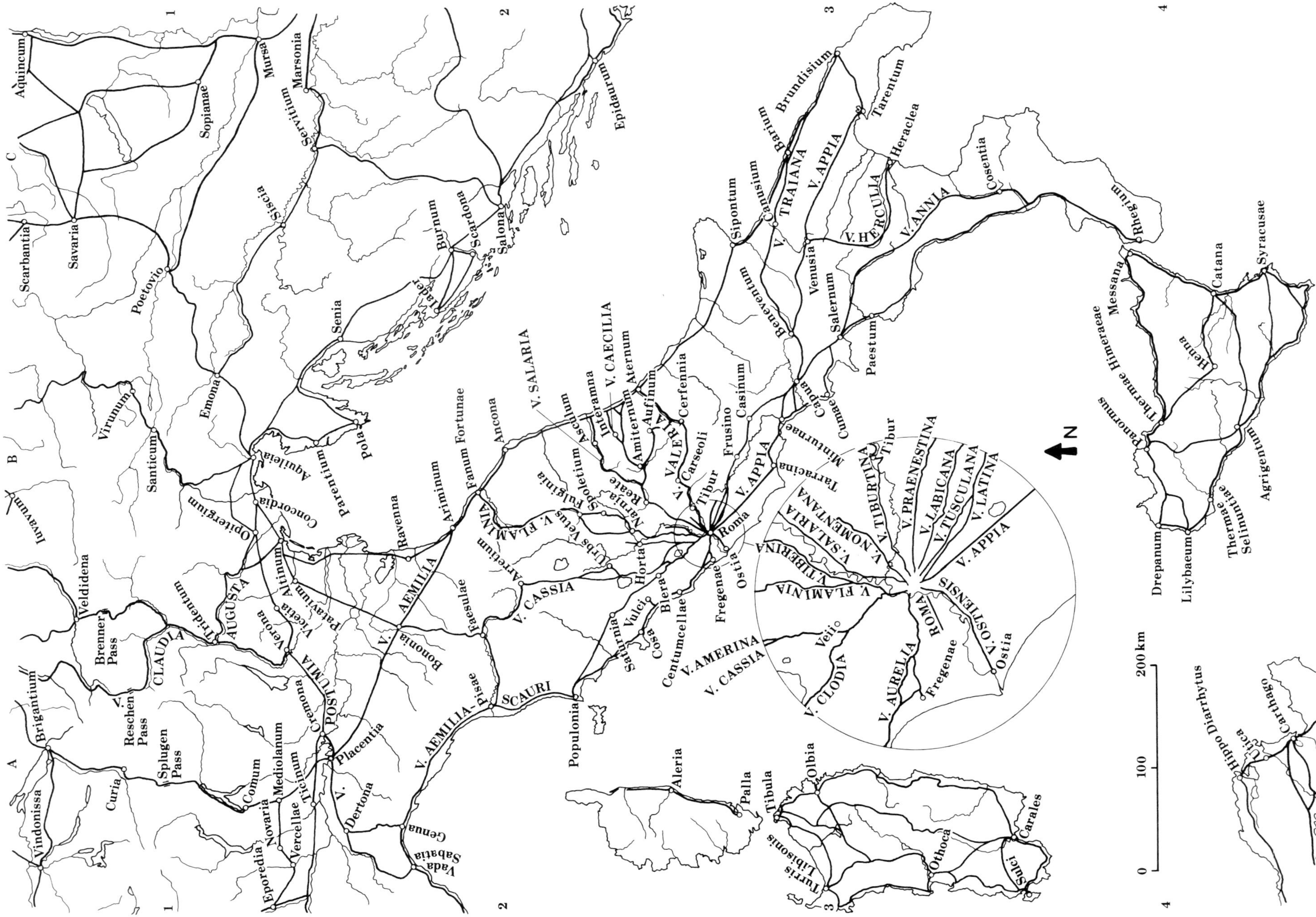

1 Major Roman roads of Italy

should not be taken to imply that there were no previous roads. There is no doubt that any settled community establishes favoured routes between centres. Trade routes became roads, like the great Royal Road (Charlesworth, 1970) across Asia Minor from east to west. However, the Romans took these routes and developed them as properly constructed roads. This development took some time. It would be incorrect to suggest that the roads shown on the maps existed as such from the earliest Roman times, and some, no doubt, fell into disrepair towards the end. To follow Sitwell (1981), the system shown here was generally in place by the middle of the second century AD, with the exception of the roads in Africa, which were not complete until early in the next century.

The following sections provide the background history to the roads and describe the roads themselves country by country, in geographic order, commencing with Italy itself. The historical account relies heavily on Cary and Scullard (1975). A summary of important dates is given in Table 1, which generally gives two dates for a country or region, these being the approximate dates when Roman settlement began and was then made complete. The order of the entries follows approximately the commencement dates, but is influenced also by other considerations so that the complete list follows the countries in the order of their effective settlement.

2.2 Italy

2.2.1 *Early history*

The traditional date for the founding of Rome is 753 BC. Standing on a fertile plain, it commanded the Tiber River, with access to the sea. Its powerful northern neighbour was Etruria, which stretched from the Tiber to the River Arno in the north, and from the Apennine mountains to the sea. Formed by a League of Twelve Cities, Etruria was composed of independent city-states, such as Veii and Caere near Rome, and Tarquinii and Vulci further north (see Map 1 for some of these towns).

Table 1. *Dates when Roman occupation began and was made complete*

Country or region	Dates	Incident	Section
Italy	753 BC	Rome founded	2.2
	AD 25	Defeat of the Salassi	
Spain	218 BC	Commencement of Second Punic War	2.4
	18 BC	Agrippa conquers the north and north west	
Macedonia	148 BC	Macedonia annexed	2.7
	129 BC	Dalmatian coast added	
North-west Africa	149 BC	Commencement of Third Punic War	2.5
	AD 42	Claudius annexes Mauretania	
Gaul	125 BC	Massilia requests Roman aid against the Ligurians	2.3
	49 BC	Massilia annexed by Rome	
Asia Minor	133 BC	Pergamum bequeathed to Rome	2.8
	AD 43	Lycia annexed	
Cyrenaica and Egypt	96 BC	Cyrenaica bequeathed to Rome	2.8
	30 BC	Egypt becomes a Roman Province	
Syria	62 BC	Becomes a Roman province	2.8
Lands south of Danube	15 BC	Raetia occupied	2.7
	AD 9	Noricum, Pannonia and Dalmatia occupied	
Thrace	9 BC	Reduced to status of vassal state	2.7
Palestine	64 BC	Pompey enters Jerusalem	2.8
	AD 70	Destruction of Jerusalem	
Britain	AD 43	Armies of Claudius land at Richborough	2.6
	AD 84	Agricola reaches Scotland	
Dacia	AD 85	Domitian crosses Danube	2.7
	AD 107	Dacia annexed	
Upper Mesopotamia	AD 166	Colony established at Carrhae	2.8

To the south was Latium, a land of farmers and herdsmen, who shared a common language with Rome. They occupied the fertile plain south of the Tiber to the Pontine (Pomptine) Marshes and the Lepini Mountains, the mountain chain which reached the coast at Terracina, just beyond the prominent Cape Circeo. They lived in villages which in time became cities and towns, and finally joined together to form the Latin League.

Rome's earliest rulers were kings. The kingdom lasted until 510 BC with seven kings, of whom the fourth, Ancus Martius (642–617 BC), is reputed to have built Rome's first bridge, the Pons Sublicius (Section 6.3) across the Tiber. The last three (616–510 BC) were Etruscans as their names indicate. The first and third were Lucius Tarquinius Priscus and Tarquinius Superbus. By the end of the kingdom, Roman territory had increased from perhaps 150 to 900 square km and Rome had formed alliances with many of the Latin towns. Under Etruscan rule, Rome had grown to be a city, with public buildings, streets and drains. The Cloaca Maxima, the great drain of Rome, was built in stone about the fourth century BC to replace an earlier open drain allegedly built about 530 BC by the second of the Tarquin kings. In 510 BC, the Romans rebelled against Tarquinius Superbus, the last of the Etruscan kings of Rome. He fled to Caere and then, with the help of Lars Porsenna of Clusium, returned to attack Rome. The legendary account says that, with the help of Horatius Cocles and his two friends (Section 6.3), the Romans successfully defended their city. Most probably Porsenna occupied Rome, but then in 506 BC, at the battle of Aricia (ancient spelling, modern Ariccia), the Etruscans' power in Latium was finally broken. This victory was won by the Latins, who formed a Latin League from which Rome was excluded, but in 493, after successive trials of strength between them, attacks from the neighbouring tribes caused the two Latin states to resolve their differences, and a combined Latin League was formed which, after some interruption and variation in the Latin war of 360, continued down through the centuries as a mainstay of Roman power.

The life of the Roman Republic extended from 510 to 27 BC. It is not the purpose of this chapter to follow its history in detail, but it is appropriate to describe the background to the construction of the first Roman road – the Via Appia in 314 BC – and the extension of the Roman road system through the whole of Italy.

In 479 BC the Romans attempted to take Fidenae, on the left bank of the Tiber 10 km from Rome, the only remaining Etruscan bridgehead east of the Tiber. They were unsuccessful, but in 426 BC they tried again and won. Prompted by this success, in 405 they attacked Veii itself, an Etruscan city equal in size to Rome, and finally entered it in 396. Cary and Scullard (1975: 72) regard this as 'the first definite step in Rome's career of world conquest'. By 264 BC they had control of the whole of the Italian peninsula. The central features of their advance were the Samnite Wars (I, 343–341 BC; II, 326–304 BC; III, 298–290 BC).

The Samnites formed part of the Sabellian tribes which came originally from the uplands south of Rome. In the late fifth and early fourth centuries BC they entered Campania, the coastal region extending from Latium to Paestum, and in 423 seized Capua. By 350 BC most of southern Italy was under their control. The nature of this control differed, however, from that of the Romans, which was characterised generally by a process of incorporation on reasonable terms, rather than by subjugation. In 343 BC the Capuans sought assistance from the Romans against the Samnites. Rome agreed, and for the first time was drawn into Campania. That First Samnite War was indecisive but led in 326 BC to a second war in which Roman armies suffered major defeats at the battles of the Caudine Forks (321 BC) and Lautulae (315 BC). Capua played a significant part in both – the Caudine Forks is the name of a mountain defile between Capua and Beneventum, entered by a combined Capuan and Roman force in an attack on the Samnites; the defeat at Lautulae, on the other hand, caused the confidence of the Capuans to be broken, and they changed sides. However, Rome and its Latin allies maintained their resolve; in 314 they re-entered Capua; and by 304 BC the Samnites had sued for peace.

Campania, in turn, was bounded to the south by Lucania, extending down to the Gulf of Taranto. Fearing the Samnites, they appealed to the Romans and induced them in 298 BC to enter Lucania and attack the Samnites again. Once more the Romans were in danger of defeat. A Samnite force travelled north past Carseoli (slightly north of Rome) to Camerinum (295 BC) – modern Camerino, north east of Foligno – being finally defeated in that same year at the battle of Sentinum (inland from Ancona), the largest battle fought until that date in Italy. Although the war continued for some years, the Samnites were now finally beaten, and requested peace in 290 BC.

It is instructive to observe the consequences of these wars. Roman dominion now extended south to the Gulf of Taranto, and north to Ancona. In 294 the Etruscan cities finally made peace with Rome. In an unpremeditated and largely involuntary way, Rome was now mistress of much of Italy.

But once more, Roman success led to further conflict. The establishment of a Roman settlement at Venosa caused friction with the Greek city at Taranto, and this was exacerbated in 282 BC when the Romans sent a naval force to Thurio (ancient Thurii) at the south-west corner of the Gulf of Taranto, again at the request of the citizens. The Tarantines engaged the services of King Pyrrhus of Epirus (western Greece), a mercenary soldier of great skill. Rome was again drawn into war – the Pyrrhic War from 280 to 275 BC. Pyrrhus achieved major successes – at Heraclea on the gulf south east of Taranto in 280 BC, and at Ascoli Piceno (ancient Asculum) in 279 – but in 276 he was confronted by superior Roman forces at Beneventum and retreated from Italy. Rome now held the entire Italian peninsula.

2.2.2 *Roads to the south*

The inset in Map 1 shows fourteen roads leaving Rome. The most famous of these was the Via Appia, which proceeded to the south east, to Terracina on the coast; to Minturno, Suessa and Capua; and then to Benevento, Venosa, Taranto and Brindisi (see also Guido, 1972; Rose, 1935).

The road was begun in 312 BC by the blind censor Appius Claudius Caecus to link Rome with Capua, and this date may readily be linked with the events described briefly in the preceding historical summary, for it was in 314 BC that the Romans re-entered Capua near the end of the Second Samnite War. Ancient Capua – the present Santa Maria Capua Vetere – was an influential city of the

Etruscans, who had established themselves there about 650 BC. In 524 BC they sought to occupy also the Greek settlement at Cumae, on the coast 20 km west of Naples. They were unsuccessful and by 420 BC both cities had fallen to the Samnites. Capua was a place of culture and trade, and the decision to build a road to it had two objectives. The Via Appia was primarily of military and strategic importance, but it quickly became a link for trade from Capua and the east. Capua itself dominated the major crossing of the River Volturno at Casilinum (modern Capua) and here a bridge was built that survived until recent times (Section 5.3).

The Via Appia was built as an all-weather, surveyed road, and as such it was the first of the Roman roads. While some roads followed earlier routes, here the straight alignment from Rome suggests that it was a new route, probably declared to be straight by an autocratic order from Appius (Ashby, 1927a: 176). Initially gravelled (Paget, 1973: 224), it was later paved, and Figure 1 shows part of the typically Roman paved surface at Suessa (modern Sessa Aurunca). According to Chevallier (1976: 87), the first paving of a Roman road did not occur until 174 BC, and on the Via Appia not until the time of Hadrian (AD 117–38); but others date it earlier. Ashby (1927a: 33), for example, states that in 292 BC it was paved in *silex* (Section 4.4) from the temple of Mars, just outside the city gates, to Bovillae, a distance of about 20 km. Even a gravelled road, however, would most likely have had a properly prepared subgrade of stones, sand or gravel. Chevallier speaks of a foundation course consisting of a rubble layer 'with stones set on edge to aid drainage', followed by a sand layer before the gravelled surfacing.

The road, of course, took years to build. Even the section to Terracina was not completed at first. Forum Appii, 70 km from Rome, was the place where Paul was met by his friends (Acts 28:15) as he came near the end of his journey from Jerusalem to Rome. It lay on the northern shores of the Pontine Marshes, which extended some 30 km to the south. Although the Via Appia continued through the marshes (Chevallier, 1976: 211 n. 19), it was regularly flooded. An adjoining canal was dug and the journey could be made either by road or on boats towed by mules; the poet Horace (*Satires* 1.5, see Chevallier, 1976: 24) described this passage in 38 BC. Drainage of the marshes was not achieved until the time of Trajan (AD 98–117).

Another significant engineering achievement was at Terracina. The original route crossed the mountains to the north of the town, but here Trajan also improved the road by taking it between the ridge and the sea in a rock cut shelf with a vertical face said to rise to 36 m (Blake and Bishop, 1973: 281; Paget, 1973: 230), as shown in Figure 2.

From Terracina the road turned inland to Fondi and Itri, near the Pass of Lautulae where the Romans were beaten in 315 BC. Rejoining the coast at Formi, the Via Appia continued on to Minturno, where it crossed the major Garigliano River (ancient Liris) by a timber bridge

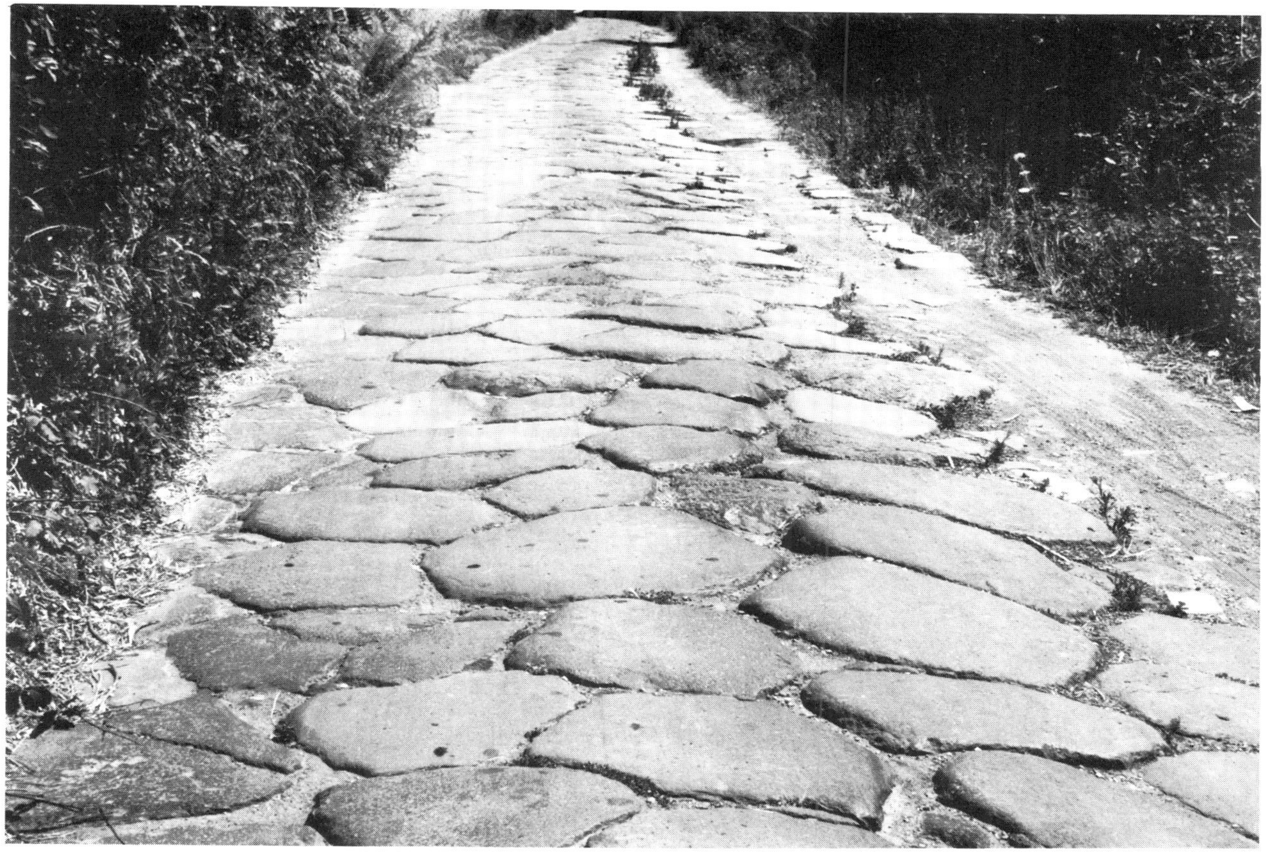

Fig. 1 Typical paved surface of Via Appia at Sessa Aurunca.

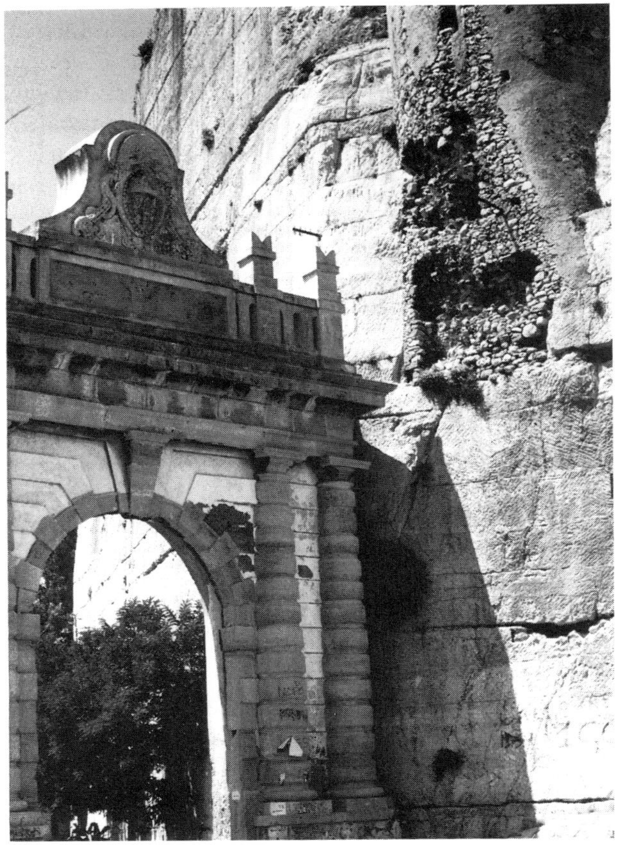

Fig. 2 Roman cutting at Terracina.

(Section 6.3). From here there were two routes to Capua – an older inland route through Suessa, as shown in Map 1, and a later coastal route through Mondragone (Sinuessa) built about 162 BC (Wiseman, 1970: 131).

The Romans established a colony at Venosa in 291 BC and another at Brindisi in 244 BC, following the successful conclusion of the Pyrrhic War (275 BC). The Via Appia was extended from Capua – possibly to Venosa – by Lepidus when he was consul (285 BC), and later to Taranto and Brindisi (Wiseman, 1970: 131–2; Vinson, 1972). Various dates have been suggested for the completion of the road – 264 BC by Ashby (1927a: 33) and Chevallier (1976: 133); soon after 285 BC (Wiseman, 1970: 131); or soon after 248 BC (Paget, 1973: 224). Wiseman's comment is made in the context of a dispute with Radke who places the completion in the second half of the second century BC. Overall a date of 264 BC seems the most reasonable choice.

The alternative northern road from Benevento through Canosa di Puglia (ancient Canusium) may possibly be identified with the Via Minucia mentioned by Cicero and Horace and may date from about 225 BC (Wiseman, 1970: 131, 140). If so, then it was rebuilt by Trajan (AD 98–117) and known generally as the Via Traiana (Ashby and Gardner, 1916). Its construction was commemorated by him by the magnificent triumphal arch known as Trajan's arch built in AD 114 in the city of Benevento (Fig. 3).

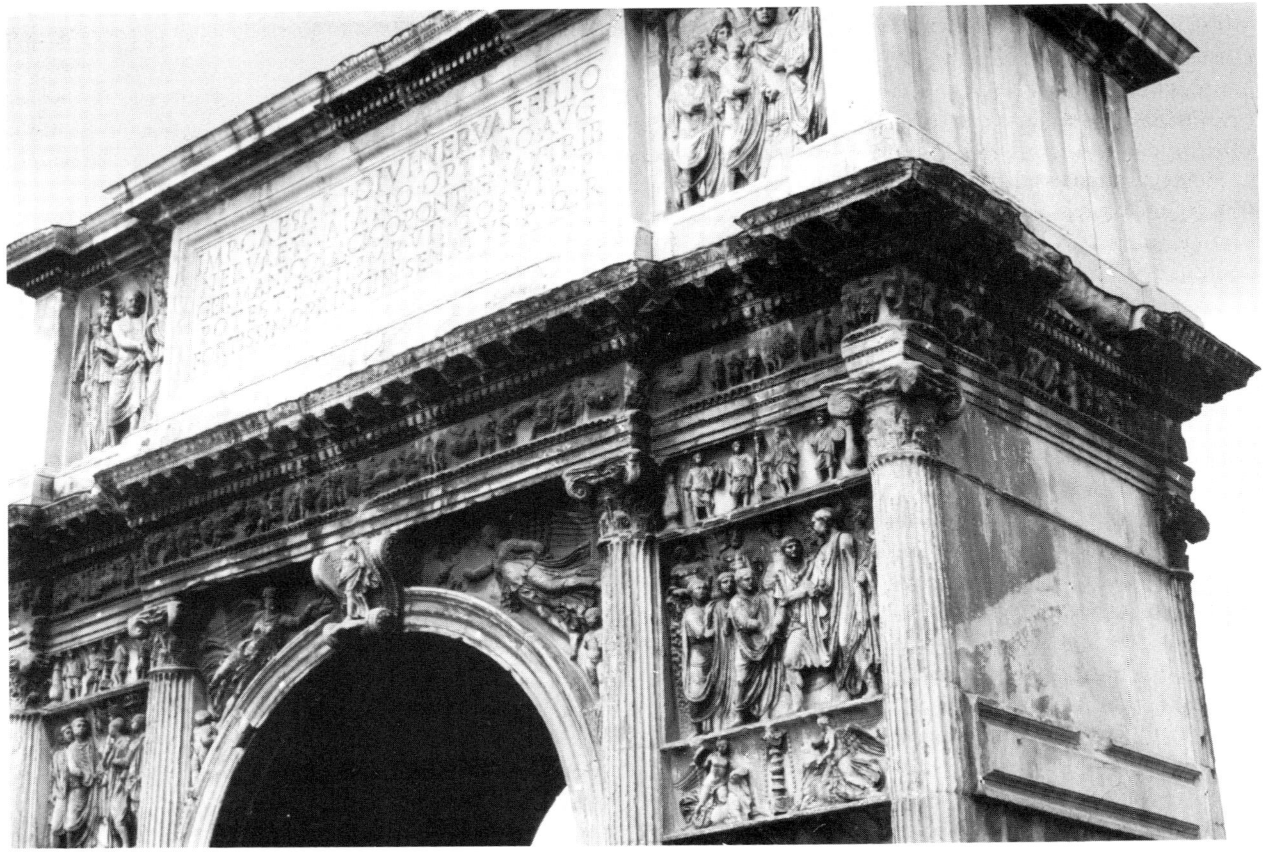

Fig. 3 Trajan's arch at Beneventum.

Map 1 shows three other roads in the south. The Via Annia, which joins Capua in the north to Reggio on the Straits of Messina in the south, runs inland for much of its route. Wiseman (1964: 33) identifies this as one of two Viae Anniae, the other being in the north, from Bologna (ancient Bononia) through Padua (Patavium) to Aquileia (Wiseman, 1969: 86). He dates the construction of the southern road as 131–128 BC, and suggests that it was begun by T. Annius Rufus as praetor in Sicily and completed by him as consul.

The parallel coastal road was begun by Domitian (AD 81–96), from Mondragone (Sinuessa) on the later Via Appia to Cumae and Pozzuoli. This road was called the Via Domitiana (see Rogers, 1984, for a discussion of its cost), and it is the description of the road by the poet Statius (*Silvae* IV.3.40–55) that is one of the most important accounts of Roman road construction (Chevallier, 1976: 83; von Hagen, 1967b: 39, 40). The translation given here is from Coleman (1988: 15, see also 102–5, 114–17).

> Here the first task is to start with the furrows, cut back the edges and hollow out the earth far down with deep excavation; next, to refill the scooped-out trenches with other material and prepare a bed for the topmost camber, so that the earth shall not wobble nor the spiteful ground provide a treacherous bed for the weight of slabs; then to bind the road with blocks rammed in on both sides and numerous pegs. How many hands labour together! Some cut down groves and strip mountains, some smooth stakes and beams with iron; others bind together the slabs and interweave the work with baked dust and dirty *tufa*; some dry up thirsty pools by hand and divert lesser streams far away.

Pozzuoli (ancient Puteoli) was the original source of a fine volcanic ash or *tufa*, which was mined here for use in Roman concrete (Section 4.5). The road was later extended to Naples (Neapolis), Herculaneum and Pompeii, then to Salernum and down the coast through Pesto. The atlas by van der Heyden and Scullard (1963, map 53) gives the name Via Domitiana to the whole of this road.

The final road shown here in southern Italy is the Via Herculia (Buck, 1971). Of much later date, it appears to have been begun during the reigns of Diocletian and Maximian (AD 284–305). Map 1 shows it beginning at Venosa, on the Via Appia. It probably extended further north, initially along the Via Appia, and then across to the Via Traiana, at Equum Tuticum, 30 km from Benevento.

Map 1 shows also the main Roman roads in Sicily (Verbrugghe, 1976), Sardinia (Meloni, 1975; Bouchier, 1917) and Corsica.

2.2.3 *Around Rome*

The following paragraphs refer often to the work of Thomas Ashby, whose studies of the Roman Campagna, the region near Rome, and its roads form the classical reference in English. His work was first published in 1902 and 1906 and then reissued in book form in 1927 (1927a). Both the journal articles and the book have detailed maps, but these are at a larger scale in the journal. The references here are to the book.

Although the Via Appia is the oldest Roman road, it was not the first route from Rome. The first appears to have been an old salt road, later called the Via Salaria, from the mountains down the left bank of the Tiber through Fidenae, either to the Pons Sublicius or an earlier crossing at this location, across to the west bank and on to salt marshes north of the river mouth (Ashby, 1927a: 30; Chevallier, 1976: 67; Martinori, 1931). Other primitive routes led to Tivoli or Tibur (the Via Tiburtina); to Mentana or Nomentum (the Via Nomentana, Martinori, 1930a); to Gabii (later the Via Praenestina through Gallicano to Palestrina or Praeneste, see also Kahane and Ward Perkins, 1972; Kahane, 1973); and to Labici and Tuscolana or Tusculum (the Viae Labicana and Tusculana) (Ashby, 1927a: 31–2; Paget, 1973). No doubt there were routes to the north as well. These early routes were short, and the names given to the subsequent roads were town names rather than the names of individuals.

The history of the Via Salaria itself is unclear. At some stage it was paved, possibly to a point near Passo Corese (Roman Cures), where the route shown in Map 1 leaves the Tiber (Wiseman, 1970: 134–6). The name, Via Salaria, was later given to the northern route to the Adriatic through Ascoli Piceno (Asculum). The southern route, the Via Caecilia through Amiternum and Teramo (Interamna) to the coast, is much older, being attributed by Wiseman to L. Caecilius Metellus, consul in 283 BC. This is the true continuation of the Salaria. In the years 17–16 BC Augustus built the road through Asculum; part of his road was new, part was a reconstruction with paving of the old Via Caecilia. Augustus described his road as extending to the Tiber.

Gabii, Praeneste, Labici and Tusculum were Latin towns to the east of Rome, joined to it by the treaties of 493 and 360 BC. All – but more particularly Labici and Tusculum – lay on a natural route north of the Lepini Mountains to the upper valley of the Garigliano (Liris) River which runs east and then south to join the sea at Minturno. This route was an obvious and easier alternative to that followed by the Via Appia. The early routes which became the Viae Labicana and Tusculana were extended to form a military road to Frosinone (Frusino), Cassino and Capua, joining the Via Appia at Teano near Suessa. This route became the Via Latina. It is not clear when this happened, but it is known that in 127 BC it was rebuilt by the consul L. Cornelius Cinna (Wiseman, 1970: 39) as the Via Latina Nova. Ashby states, 'it seems to me clear that the Via Latina must have existed before the Appia' (1927a: 34) and again, 'The Via Latina is, like the Via Appia, obviously a military highway ruled straight on the map, and therefore of artificial origin. Its establishment must be connected with the fighting with the Aequi for the pass of Algidus in the fifth and fourth centuries BC' (p. 31). The Aequi were a mountain tribe who threatened Latium. They were defeated finally in the battle of the Algidus Pass, near Tusculum, in 431 BC. The Via Latina Vetus was, then, of great antiquity.

The first dated road after the Via Appia appears to have been the Via Valeria (Ashby, 1927a: 34; Wiseman, 1970: 139; Gardner, 1913, 1920; van Essen, 1957), also from Rome to the Adriatic. It was first built in 307 by M. Valerius Maximus, censor, through Carsoli to Alba Fucens (near Avezzano), which was formed as a Roman colony in 303 BC near the end of the Second Samnite War. It was later extended to Cerfennia, and then to Pescara (ancient Aternum) by the emperor Claudius (AD 41–54).

The Via Clodia is the oldest Roman road to the north of the Tiber. It preceded the Viae Flaminia and Cassia, shown in Map 1 as occupying part of its course. Passing south of the Bracciano lake, it continued to Bieda (ancient Blera) and Saturnia (30 km inland from Orbetello). Closer to Rome it passed a little south of Veii, the Etruscan city defeated and destroyed by the Romans in 396 BC (Ward Perkins, 1957, 1961; Hemphill, 1955; Kahane et al., 1968). Etruria finally made peace with Rome in 294 BC, and there seems little doubt that the Via Clodia was built from Rome to control this area. Wiseman dates it about 287–285 BC (1970: 137). It probably followed the line of earlier Etruscan roads.

The Via Amerina is another old road, of about 240 BC (Ward Perkins, 1957; Paget, 1973). It is shown in Map 1 as leaving the Cassia north of Veii. It went due north to Nepi (ancient Nepete), passed some 5 km west of Civita Castellana to Orte (Horta) on the Tiber, which it crossed, and then on to Amelia (ancient Ameria). Frederiksen and Ward Perkins (1957) have given a detailed description of the lower part of the road and the towns it served.

The third major road of this system to the north was the Via Cassia, to Arezzo and Florence (ancient Faesulae). Much later than the Clodia and Amerina, the section to Arezzo is believed to have been built by C. Cassius Longinus, consul in 154 BC. Again it can be assumed that there was an earlier Etruscan road to Arezzo, for in 187 BC that city was joined through Florence to Bologna (Bononia) by a road built by C. Flaminius. This initial route served the north east, but in 150 (or 123) BC a further road was built down the Arno River to Pisa to provide the first link by Roman road to that town (see also Ashby, 1927a: 35; Harris, 1965: 114; Martinori, 1930b).

The coastal road to the north west was the Via Aurelia, eventually extended as the Viae Aemilia-Scauri and Julia Augusta to Arles on the Rhône and across to Spain – one of the great Roman roads. Built first by the consul C. Claudius Cotta in 241 BC, it initially ran west from Rome to the coastal towns of Fregene and Altino (ancient Altinum) and then to Civitavecchia (Centumcellae) and Cosa (Fentress, 1984); a branch led to the old Etrurian city at Vulci. About 119 BC, this Via Aurelia Vetus was rebuilt by another L. Claudius Cotta and extended to Cosa (Wiseman, 1970: 123–4; Ashby, 1927a: 35) as the Via Aurelia Nova. This was further extended to Populonia about 125 BC (see also p. 14).

The port of Rome was originally at Ostia on the left bank of the mouth of the Tiber. It was joined to Rome by the Via Ostiensis, whose date is unknown; but the settlement at Ostia dates probably from the fourth century BC (Ashby, 1927a: 215). The emperor Claudius (AD 41–54) built a new harbour to the north of the mouth, which was further developed by Trajan (AD 98–117), who constructed a new basin between the harbour of Claudius and the sea. This new harbour became known as Portus and was connected to the Via Aurelia by the Via Portuensis (Ashby, 1927a: 219; Sitwell, 1981: 13).

2.2.4 *The north*

Before considering the Viae Tiberina and Flaminia, it is necessary to extend further the history of Rome. In the year 276 BC, Roman control extended from Etruria and the River Arno in the north to Brindisi in the south. Rome was not long left in peace. Just as the year 396 BC marked the beginning of her growth, so 264 signalled that of her overseas conquests and her spread to Europe and north Italy.

Her neighbours to the east and south were Greece and Carthage. The Carthaginians, Punics or Phoenicians were a seafaring and trading people, originally from Tyre on the Palestinian coast, but had settled (from perhaps 814 BC) at Carthage (on the outskirts of modern Tunis) with control of North Africa, Corsica and Sardinia, the south-east coast of Spain, and western Sicily. The Sicilian city of Messana (modern Messina) dominated the strategic Straits of Messina. Attacked by Hiero, king of Greek Syracuse, it called on the Phoenicians for help, but then in turn appealed to Rome for deliverance from the Phoenicians. Thus started the First Punic War (264–241 BC). Rome moved into Messana, achieved a treaty with the Greeks and defeated the Phoenicians at sea, first in 260 BC at Mylae off Sicily, and then in 256 BC in the great naval battle of Ecnomos. The Romans landed in Africa briefly in that year, but the chief action of the war was in and about Sicily, and in 241 BC, after a bitter conflict, this was ceded by Phoenicia to Rome. Shortly after, both Sardinia and Corsica also became Roman, but in circumstances that led to long-standing resentment by the Carthaginians.

To the north, the valley of the Po had for years been occupied by the Gauls, and was known as Cisalpine Gaul. The Gauls entered Italy in the early fourth century BC and in 391 BC defeated the Romans at the battle of the Allia and captured Rome. Fortunately for the Romans, they accepted a ransom and withdrew north of the Apennines and by 331 BC were at peace. They were, however, drawn into the Third Samnite War and fought with the Sabellians at the battles of Camerinum and Sentinum (295 BC). Now, at the end of the First Punic War, they rose again, first at Rimini (Ariminum) in 236 and then in 225 BC they entered Etruria and advanced as far south as Telamon near Cosa. Once more the Romans were fortunate, for they could mobilise their entire strength against them, and not only forced them back, but occupied and colonised the valley of the Po. By 220 BC there were Roman colonies at Piacenza (Placentia) and Cremona,

Roman naval stations at Spezia (Luna) and Genoa, and Roman control extended well into the foothills of the Alps.

It is this background which lies behind the construction of the Via Flaminia by Gaius Flaminius, censor in 220 BC. The history and route of this fine road have been presented in detail by Ashby and Fell (1921), and its bridges described by Ballance (1951 – see also Martinori, 1929). Near Rome it was was preceded by the old route of the Via Tiberina which followed the right bank up to a crossing at Otricoli (ancient Ocriculum). Further north, Narni (Narnia) had been established as a Latin colony in 299 BC. Flaminius, who in 223 BC had led the first Roman army across the Po, but who in 217 BC died at the hands of Hannibal in the Second Punic War, decided in 220 BC to build the road which bears his name.

Leaving Rome by the Pons Mulvius (Section 5.2), which, however, did not exist in its present form at that time, the road went due north, across the Valchetta or Cremera from Veii, over the rivers Treia and Tiber – no bridges remain over either of these streams – to Otricoli on the left bank. It climbed the ridge between the Tiber and the Nera, its tributary from the north east, and dropped to the magnificent Augustan bridge at Narni. From here there were two routes: the older continued due north to Massa Martana and then to Forum Flaminii, just north of Foligno; the later followed the Nera for some distance, and then went north through Spoleto to Foligno, dropping as it did so down the valley of the ancient Clitumnus, a northerly-flowing stream which joined the Topino River west of Foligno, and then ran further west to the Tiber itself.

Foligno lies on the Topino, which turns here from a southerly to a westerly course. The road rises up this valley and then continues to reach the Scheggia Pass, its highest point, 130 Roman miles from Rome. It passes Cantiano and now drops into the Foci di Cagli, the narrow gorge of the upper Burano River. At 155 Roman miles from Rome, this narrows even further to form the Furlo Pass, a narrow cleft 1.5 km long between mountains about 1000 m high on either side. The road is cut into the cliff wall 30 m above the left bank of the northerly flowing stream and at one point passes through a tunnel around a sharp bluff.

It appears that there are here three parallel tunnels. Ashby and Fell (1921: 186, pl. 18) have a photograph of the north portal of Vespasian's tunnel, built in AD 76 or 77, with a length of 38 m, a height and width of 5 m, and with an inscription by Vespasian above. They describe also an older and much shorter tunnel at the south end of Vespasian's, 8 m long, 3.30 m wide and 4.45 m high. Figure 4 shows the south end of one of these tunnels, which, from the proportions, appears to be the earlier. It resembles that shown by von Hagen (1967b: 50), which he describes as the first tunnel, 'probably built by Gaius Flaminius in 220 BC'. These were not the only Roman tunnels, for there were two at Cumae (von Hagen, 1967b: 38; Blake, 1947: 329) and another on the Reschen-Scheideck Pass (von Hagen, 1967b: 119).

The road now drops rapidly to the Metaurus River and follows it, some distance from its left bank, to reach the coast above Fano. It then follows the coast to Rimini (ancient Ariminum).

Ballance (1951: 115–17) has dated the existing Roman bridges along the Via Flaminia. He quotes Livy (41.27) who speaks of the censors improving the road and building some bridges in 174 BC. In 109 BC Aemilius Scaurus built the existing Pons Mulvius and other bridges; about 65 BC the Pons Minucius was built, which probably carried the road over the Tiber south of Otricoli; and then in 27 BC Augustus carried out major repairs. There is a fine triumphal arch in Rimini to commemorate his work. Ballance dates none of the existing bridges earlier than 109 BC. Chevallier (1976: 135) claims that the road reached Rimini in 220 BC, but this is surely unlikely, even though the route from Rome to Rimini would have been used before this date.

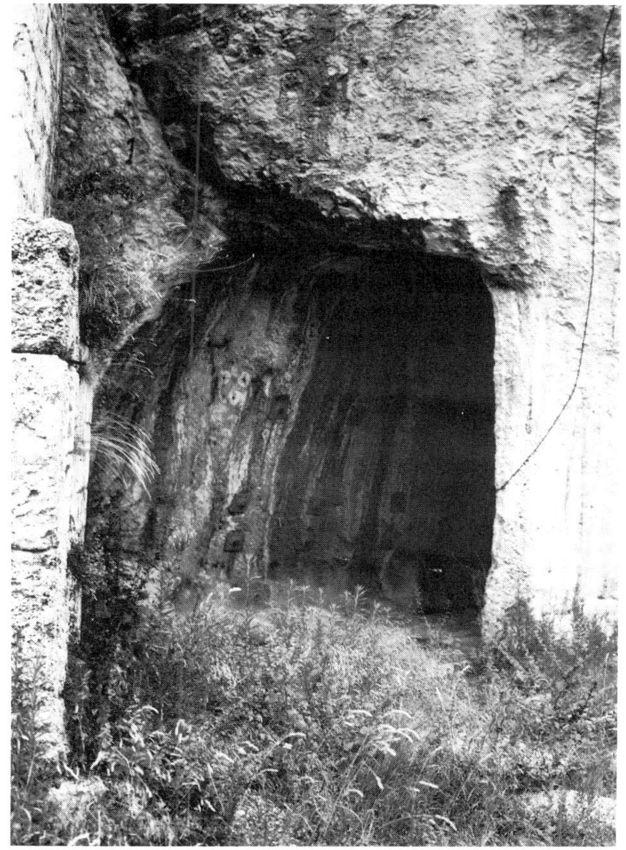

Fig. 4 Entrance to Roman tunnel in the Furlo Pass.

The Carthaginians had exercised influence in the south of Spain at least as early as the sixth century BC (McEvedy, 1967: 48). Following their defeat in the First Punic War (241 BC), they extended this influence inland and north to the Ebro under the leadership of Hamilcar Barca's son, Hannibal, and son-in-law, Hasdrubal, who established New Carthage at Cartagena in 221 BC. In 226 BC they had reached an agreement with the Romans that they would move no further north, with at least the implication that the Romans would leave the south to them. Saguntum, an independent city well south of the Ebro, quarrelled with Carthage in 221 BC and called on the Romans to help. The Carthaginians attacked the city and in 218 BC formally declared war on Rome: the beginning of the Second Punic War which was to last until 201 BC. Hannibal, now leader of the Carthaginian armies, surprised the Romans by moving his army overland through southern France, across the Alps in that year – probably at Mont Cenis or Mont Genèvre – and invaded Italy. There followed that extraordinary period when he was able to defeat any army the Romans could field against him, but was unable to enter Rome, or to establish permanent control over any substantial region. He marched through Italy to the south, occupied Capua in 216 BC and entered Taranto in 212 BC, but was defeated at the battle of the Metaurus River south of Rimini. Although Hannibal continued to fight – he occupied Bruttium in the toe of Italy for a further four years – the war in Italy was lost.

Meanwhile in Spain, the Romans under the Scipios had continued the fight. The elder Scipios were defeated by Hasdrubal at separate battles between the Guadalquivir (Baetis) River and Cartagena, but in 210 BC Rome sent a new army to Spain, under one of the sons, P. Cornelius Scipio, later called Africanus. He attacked Cartagena in 209 BC and conquered the Carthaginians in Spain by 206 BC. This Scipio landed in Africa near Utica in 204 BC, but when the Carthaginians were on the point of defeat, they recalled Hannibal with his remaining army from Italy in 203 BC. A final confrontation took place at Zama Regia, south of Musti, in 202 BC. Scipio won, and a treaty was imposed by Rome on Carthage in 201 BC.

When the Carthaginians passed through northern Italy in 218 and 207 BC, they received little help from the Cisalpine Gauls, but in 200 BC some of these Gaulish tribes attacked the Roman colonies at Piacenza and Cremona. In 197 BC, two Roman armies advanced against them. One crossed the Apennines from the coast at Genoa. The other advanced north of the Po and achieved a victory over them near Como. The uprising was finally quelled in 191 BC, and from that time the Po valley became truly Roman. New colonies were established at Bologna (189 BC); Modena and Parma (183 BC); and to the east at Aquileia in 181 BC.

Map 1 shows two major land routes from Rome to the valley of the Po: one was along the Via Flaminia to Rimini and then north west along the Via Aemilia; the other was from Florence on the Via Cassia across the mountains to Bologna. Both the Via Aemilia and the mountain road to Bologna were consular roads built in 187 BC – the first by M. Aemilius Lepidus, the second by C. Flaminius, son of Gaius Flaminius, the builder of the Via Flaminia in 220 BC (Wiseman, 1970). Three further roads completed this central road system in the north. Aquileia was joined to Bologna by the northern Via Annia – for there was a second in the south to Rhegium – passing through Concordia and Padua (Patavia). This route is that suggested by Wiseman (1964: 9), who attributes the road to Annius Luscus, consul in 153 BC. There was an east–west connection from Genoa through Piacenza (Placentia) and Cremona, to Verona, Oderzo (ancient Opitergium) and Aquileia. This, the Via Postumia, is attributed to Sp. Postumius Albinus, consul in 148 BC. It preceded the Via Aemilia-Scauri, not built until 107 BC, which joined Populonia to Genoa, and continued a little further west to Vado (Vada Sabatia). At the time of the Via Postumia, therefore, Genoa was reached by sea. Finally, there was the coast road around the Adriatic from Rimini through Ravenna, joining the Via Annia near Altino (Altinum), north east of Padua, on its way to Aquileia; it was possibly called the Via Popillia and built by P. Popillius Laenas, consul in 132 BC (Wiseman, 1964, 1969, 1970).

The roads into the upper Po valley and to the mountains were relatively late. They continued eventually to passes through the Alps, which surrounded northern Italy from Ventimiglia (ancient Albium Intimilium) on the Ligurian arm of the Mediterranean Sea at the west, to Austria in the north east and the Julian Alps branching down the Dalmatian coast of Yugoslavia to the east. In his book *Roman Alpine Passes* (1935), Hyde states that, of twenty-three major passes now in use, seventeen were known by the Romans. The accompanying maps show roads through nine of these, counting the coastal route taken by the Via Julia Augusta as one.

The Via Julia Augusta was not completed until 12 BC (Hyde, 1935: 45; Hall, 1898: 170) by the emperor Augustus, linking the older Via Aemilia-Scauri (107 BC) and Via Domitia (118 BC), from Vado in the east to Arles on the River Rhône in the west. The coastline both here and to the east of Genoa is extraordinarily rough, with mountains plunging steeply into the sea. Hyde describes the Riviera as then 'a wild, desolate and dangerous place', threatened by repeated attacks from the Ligurian Alpine people. Between La Spezia and Genoa the road rose to 680 m, and then again, west of La Turbie (near Monte Carlo), to 540 m.

An equally important road ran west from Cremona to Turin (ancient Augusta Taurinorum – see Map 2) and on to the pass of Mont Genèvre (1860 m) in the Cottian Alps. Known to the Romans prior to 100 BC, it was used by Pompey in 77 BC, and then Caesar in 58 BC. A road was built here by M. Julius Cottius, a local client king, for Augustus about 13 BC (Hyde, 1935: 51).

The Roman names given to the two St Bernard passes were Alpis Graia for the Little or more southerly pass (2190 m) to Vienne and Lyon (Lugdunum) and Alpis Poenina for the Great St Bernard (2470 m), leading to Lake Geneva (Lemanus Lacus), north-east France and Britain. Both were approached by the valley of the Dora-Baltea River, through Ivrea (Eporedia) and Aosta (Augusta Praetoria). The Salassi, who inhabited the valley, permitted the construction of a road through the Little St Bernard, possibly in 122 BC by Gaius Gracchus, but continued to attack those who used it. In 100 BC the Senate established a fort at Ivrea, and finally in 25 BC A. Terentius Varro, acting for Augustus, defeated the Salassi and sold them into slavery. To quote Strabo (4.6.7),

The Salassi used to wreak havoc with their brigandry on travellers going through their land in order to cross the mountains ... These folk even went so far one day as to steal money belonging to Caesar Augustus and roll boulders down onto columns of soldiers, with the excuse that they were constructing a road or building bridges over the torrents. However, to end the matter, Augustus subjugated them completely and had them all sold off as warbooty. He then sent in 3000 Romans who founded the town of Augusta ... and today there is peace in all the surrounding district as far as the top of the Alpine passes.

Aosta was built in 25 BC and the road to Ivrea greatly improved. Figure 5 shows an extraordinary archway cut into the solid rock on this road a few kilometres west of Pont St Martin (Brown, 1961: fig. 59). When Aosta was built, the Little St Bernard was the major route, and it was not until the times of Drusus, about 15 BC, and Claudius, AD 47, that the road over the Great St Bernard was constructed (Hyde, 1935: 59–68; von Hagen, 1967b: 103–4 gives a conflicting account).

Ivrea and the Aosta valley could be approached either through Milan (Mediolanum) and Vercelli or from the road to Turin. Alternatively, from Milan one could go north by Como through the Splugen Pass (2120 m) to Chur, Bregenz (Brigantium) and the head-waters of the Rhine and Danube. The road has not been dated (Hyde, 1935: 95).

Further east were the Reschen or Reschen-Scheideck (1460 m) and Brenner (1370 m) passes, approached through Trento (Tridentum) on the Adige River. One approach was from Verona, the other from Altino or Oderzo (Opitergium), up the valley of the Piave to Feltre and then across the divide to Trento. The road from Altino, the Via Claudia Augusta, was the original route. It was started by Nero Claudius Drusus, one of two step-sons of the emperor Augustus (the other being Tiberius Claudius Nero) and completed in AD 47 by the emperor Claudius, son of Drusus. One of the great Roman roads, it continued north west at the junction north of Trento, up the Adige valley, across the Reschen Pass, on through Landeck and the further Fern Pass (1209 m), and then down to distant Augsburg (Augusta Vindelicorum) (see Maps 1 and 8). It continued in use until the end of the second century AD when Septimius Severus and his son Caracalla, over the period AD 195–215, constructed the shorter road from the Via Claudia Augusta over the Brenner Pass to Veldidena near Innsbruck, and Regensburg (Castra Regina) on the Danube (Hyde, 1935: 140).

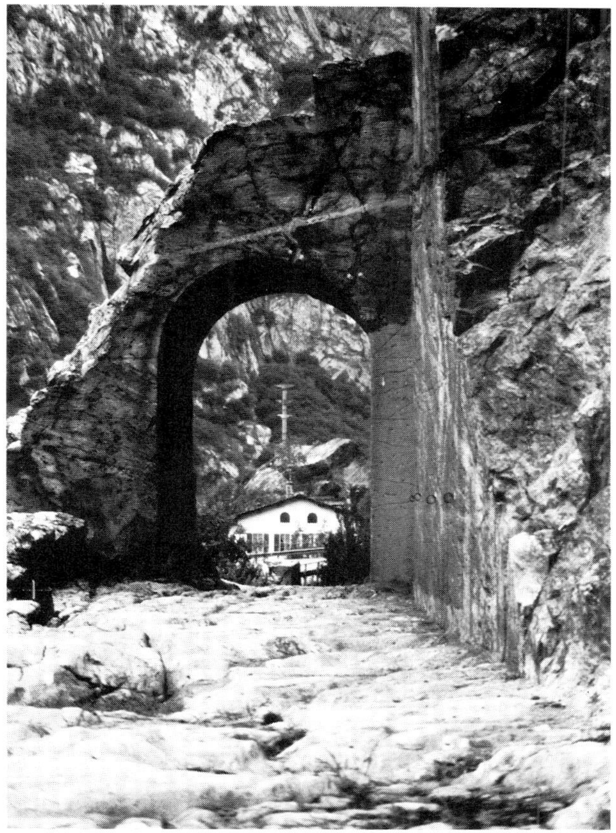

Fig. 5 Arch cut in the solid rock, in the Aosta valley west of Pont St Martin.

A second Via Julia Augusta was built by Augustus (27 BC–AD 14) north from Aquileia to Aguntum on the River Drave, upstream of Villach (Santicum). It was possible to travel from here upstream to join the road to the Brenner Pass (not shown on Map 1). Alternatively, turning east, a road crossed the Pontebba Pass, at 804 m the lowest of the Alpine passes, to Villach (Santicum) and Virunum (near Maria-Saal), the most important Roman town in the province of Noricum (Hyde, 1935: 162). As shown, a road led from here to Lorch or Lauriacum on the Danube (Maps 1 and 8), but there were also other roads (Sitwell, 1981: 94–5).

From Aquileia there was another Augustan road east over the Birnbaumerwald Pass (883 m) to Ljubljana (Emona), with important connections across to the Danube or down the River Save. Finally, from Aquileia there was the coastal route, down the Dalmatian coast to Durrës (Dyrrhachium) and the old Egnatian Way (148 BC).

It was mentioned earlier that there is uncertainty concerning some of the routes. One such case is in the triangle of roads west of Cremona, and the neighbouring roads to the west. Chevallier (1976: 135) shows the Via Aemilia terminating where it meets the Via Postumia at Placentia, with a further road along the north bank from Cremona through Ticinum to Augusta Taurinorum, and no direct connection from this road to Eporedia. The normal Roman principle was to minimise the number of river crossings and this principle would suggest that Chevallier may be broadly correct (although he is surely in error in showing the road from Placentia to Cremona crossing a loop in the Po, causing a total of three crossings on this short road). However, Brookes (1974) has stated that there are no traces of bridges crossing the Po, either here or elsewhere. The most likely conclusion is that these crossings were by boat. If this were so, then it may not have been unreasonable to have two crossings in close proximity, as shown in Map 1.

2.3 Gaul

Marseille (ancient Massilia – see Map 2) in southern France was settled by the Greeks about 600 BC, and was long an ally of Rome (Brogan, 1953: 26). Although Gaul was crossed by both Roman and Carthaginian armies in the Second Punic War, it was not occupied by Rome in 201 BC, but left to the Massilians, with whom Rome had a significant trade. Even by 125 BC, when the Romans held at least 80 per cent of Spain, they had not occupied Gaul; but in that year Massilia called for aid against the Ligurians, who were attacking from the Alps. A Roman army crossed from Italy by the pass of Mont Genèvre, defeated the Ligurians and in 123 BC established a settlement at Aix-en-Provence (Aquae Sextiae) to defend Massilia. This brought them into conflict with two of the Celtic tribes of Transalpine Gaul, the Allobroges who lived east of the Rhône and north of the Isère, and the Arverni, from the Auvergne, further to the west. The Romans defeated them in turn and, in 118 BC, established the province of Gallia Narbonensis, extending around the Mediterranean coast to Spain, inland to Toulouse (Tolosa), up the Rhône to Lake Geneva and east to the Alps. Massilia continued as an independent allied state.

The Roman proconsul in Gaul was Gnaeus Domitius Ahenobarbus. The chief town in the new province was Narbonne (Narbo Martius), and in 118 BC Domitius built a road around the coast from Arles (Arelate) through Narbonne to Spain; he gave his name to the new road – the Via Domitia.

It did not take long for the peace to be disturbed, for about 113 BC the new province was invaded by two wandering German tribes, the Cimbri and Teutones. They inflicted on the Romans one of the greatest of their defeats, in 105 BC at the battle of Arausio (modern Orange), and at one time threatened the whole of Italy's northern frontier. In the west they were defeated by Marius in 102 BC at Aix and in north Italy near Vercelli; but about 58 BC the Romans received word of a further imminent Germanic invasion. The Helvetii, then occupying Switzerland, were being forced to move by pressure from the north, and decided to migrate west into Gaul. They were forced back by Roman armies under Julius Caesar, but it was this incident that prompted his Gallic Wars (58–51 BC) and the total Roman occupation of Gaul.

Caesar, who, with Pompey and Crassus, formed the so-called First Triumvirate in 59 BC, and who was appointed consul in that same year, had been granted by the Senate the governorship of both Cisalpine and Transalpine Gaul. The movement of the Helvetii disturbed the Gallic tribes. Another German group, the Suebi under Ariovistus, had moved east across the Rhône. The Gauls prevailed on Caesar to oppose them and when he defeated them, also in 58 BC, the effects of this defeat spread north. By the end of 57 BC it appeared that Caesar held the whole of Gaul to the Rhine and the Atlantic coast, but further unrest was to follow. In 56 BC the Veneti of the Atlantic and others of that coast rebelled and were subdued only when Caesar built a fleet on the Loire strong enough to defeat the Venetic fleet off Quiberon. In 55 BC two Germanic tribes crossed the Rhine into northern Belgium. Caesar suppressed them brutally and in the same year made a brief excursion by bridge across the Rhine into Germany (see Section 6.3). In that year and again in 54 BC he entered Britain. But the real test came the following year when Vercingetorix, king of the Arverni, united the tribes of central Gaul under him, and opposed Caesar with significant success. He was defeated finally in 51 BC.

Victory in Gaul was followed rapidly by the Civil War, with Caesar making his decision in 49 BC to cross the Rubicon (see Section 5.5) and march on Rome to occupy the whole of Italy in only two months; and Pompey defeated at Pharsalus and dead in Egypt the following year. Caesar instituted a spate of vigorous reforms, but was assassinated in 44 BC after he declared himself permanent dictator. A Second Triumvirate was formed, leading to the effective end of the Republic in 43 BC. Antony and Octavian, Caesar's great nephew and adoptive heir, continued as confederates but rivals from 44 to 30 BC, and finally Octavian, now called Augustus, became emperor in 27 BC.

Massilia had sided with Pompey but in 49 BC was forced to surrender to Caesar. From that date it was part of the Roman empire. Outer Gaul, the name given to that part of Gaul outside Gallia Narbonensis, was divided by Caesar into the three Gauls – Gallia Belgica in the north, Gallia Lugdunensis occupying the greater part of central France, and Aquitania on the Atlantic to the south west. The boundaries set by Caesar were modified partially by Augustus and were centred effectively on Lyon (ancient Lugdunum), established in 44 BC by the Senate as a colony for veterans. In 40 BC the province came under the control of Octavian, who appointed governors, the first of whom was his friend and able supporter, M. Vipsanius Agrippa, governor from 39 to 37, and again from 19 to 17 BC. To the east were three small Alpine provinces extending from the sea to Mont Blanc, of which one, the Cottian Alps, continued as a semi-independent state to the time of Nero (AD 54–68).

Gaul was a land of major waterways, shielded from the Romans by the Alps and its hostile tribes. The first approach to it was from the sea, to Marseilles (Massilia), and to Arles (Arelate) on the mouth of the Rhône, and its first road was the Via Domitia from Arles through Narbonne to Spain, built in 118 BC (see Map 2). Branches led from the Via Domitia to Toulouse, extended eventually to the Atlantic at Bordeaux (Burdigala), and to Rodez (Segodunum) – the first of these at least must have been early. The original Gallia Narbonensis of 118 BC stretched from Spain east to Aix-en-Provence (Aquae Sextiae, 123 BC), up the Rhône to Vienne and north east to Geneva (Rivet, 1988). The Rhône had been used for Greek trade at least as early as the fourth century BC (McEvedy, 1967: 54) and there is little doubt that Roman vessels also plied the river in Republican times. The road up the east bank is believed to be Republican (Brogan, 1953: 30). The Alpine

2 Major Roman roads of Gaul (see also Maps 1, 3 and 6)

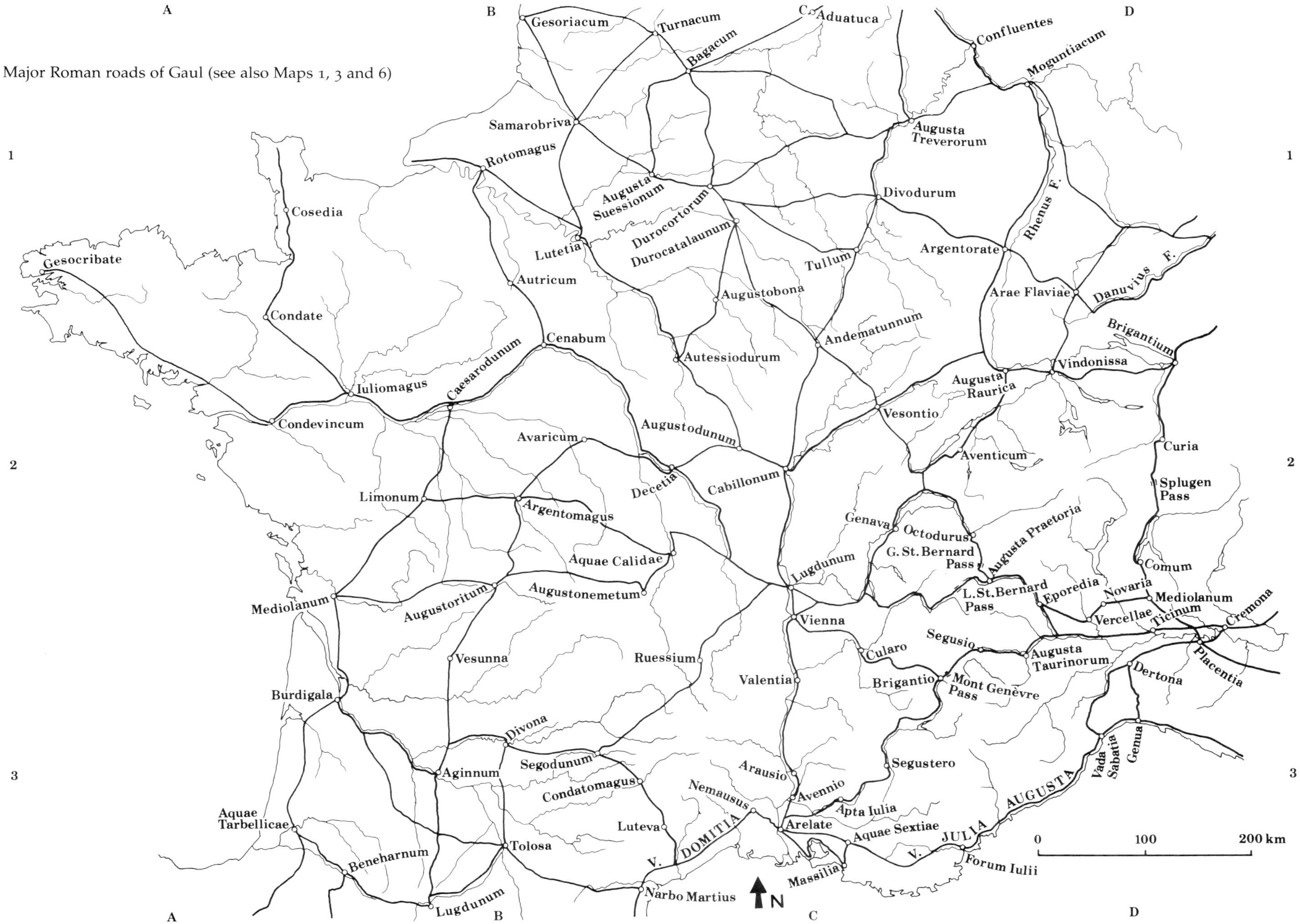

routes were in use early, but the first road, if it deserves the name, appears to have been built by Gaius Gracchus in 122 BC (Hyde, 1935: 60) across the Little St Bernard Pass, through Aosta to Vienne.

Much of the road system shown in Map 2 would date, then, from Augustus (27 BC–AD 14) or later. This is certainly true for most of the Alpine crossings (Section 2.2). The Via Julia Augusta was built by Augustus in 12 BC; the road through the Mont Genèvre Pass about 13 BC; Aosta (Augusta Praetoria) was established in 25 BC; and a 'steep and narrow' road (Hyde, 1935: 67, quoting Strabo 4.6.7; 4.6.11) across the Great St Bernard about 15 BC. Of these, it was the Mont Genèvre route which was the most travelled.

The framework of the main Roman roads in Gaul was established by Agrippa in 27 BC. As described by Strabo (4.6.11; Chevallier, 1976: 162), it was centred on Lyon (Lugdunum) and had four major roads – from Lyon down the Rhône to Arles; from Lyon west to Saintes (Mediolanum Santonum); from Lyon north up the Saône to Chalon (Cabillonum), across to Toul and Metz (Divodurum) on the Moselle and then down that valley to the Rhine; and diagonally to the north west, from Chalon-sur-Saône through Lutetia (now Paris) to Amiens (Samarobriva) and Boulogne (Gesoriacum). Strabo mentions also the road from Vienne and Lyon to the north east, joining the road from the Great St Bernard, and passing on to Augst (Augusta Raurica) near Basel, also on the Rhine.

The remaining chief roads of Gaul are shown in Maps 1, 2 and 6, and of these three deserve special mention: the road from Lyon to the Rhine and down the Rhine itself, from Chur (Curia) and Bregenz (Brigantium) all the way to the sea; the system centred on Bavay (Bagacum) in the north; and the road from Decize (Decetia) down the Loire, with connections to the peninsulas of Brittany and Cotentin. The Alpine passes were described previously in Section 2.2.4.

2.4 Spain

Section 2.2 has described the circumstances of the Second Punic War – that the Phoenicians had entered Spain as early as the sixth century BC, that Cartagena or New Carthage had been established by them in 221 BC, and that three years later Carthage declared war on Rome and Hannibal marched his army around the Spanish coast to Gaul and Italy. It was in 208 BC that Hasdrubal evaded Scipio Africanus near Castulo and marched to aid his brother in Italy. By this action Spain was effectively lost to Carthage, but it was not until 18 BC, almost 200 years later, that the whole of Spain was truly Roman.

The early conquest of Spain was sealed in 206 BC by the battle of Ilipa, near Seville, and the surrender of Cádiz (ancient Gades). Roman Spain was divided in 197 BC into two provinces. Hither Spain, or Hispania Citerior, stretched round the eastern coast from the Pyrenees to a little south of Cartagena and at the Ebro inland to Zaragoza (Caesaraugusta – Wiseman, 1956: 17). Further Spain, Hispania Ulterior, corresponded roughly to modern Andalusia, from the southern coast beyond Cartagena, to the River Guadalquivir or Baetis and its right bank. Italica near Seville (Hispalis) was founded by Scipio in 206 BC as the first Roman town in Spain. Roman Spain in 197 BC included, then, only a small part of the peninsula, with a long and ill-defined frontier facing the independent peoples to the north and west (see Map 3).

In 197 BC these peoples rose in revolt, beginning in the extreme south, then in the north near the Pyrenees, and involving finally the Celtiberians of the centre. M. Porcius Cato was sent with a new army in 195 BC. He suppressed the revolt in the north and marched to the south across the Ebro, along the line of the Jalón (Salo) River towards the Tagus. In central Spain, the Celtiberians were joined by the Lusitanians from the far west, and fighting continued until 179 BC. The final settlement was achieved by T. Sempronius Gracchus, who the preceding year had established a Roman settlement at Graccuris on the upper Ebro.

From 179 to 154 BC there was peace, but then Roman oppression caused further uprisings and war until 133 BC. Its end was prompted by the fall of Numantia, a fortified city south of Graccuris which had been besieged for nine years before its surrender, and was destroyed by Scipio Aemilianus, the same man who in 146 BC had destroyed Carthage. Roman Spain extended now to the upper Ebro, along the line of the Tagus, and to Lisbon (Olisipo) on the Atlantic. Its power reached further, across virtually the whole peninsula, except for the northern coast, north of the valley of the Douro. The date of this settlement in 133 BC should be compared with the dates of the founding of Aix-en-Provence in 123 BC and the establishment of Gallia Narbonensis in 118 BC.

It was in the early part of the first century BC that demands for the franchise to be extended to all Italians led to the first of the civil wars. Marius, who was consul for the first time in 107 BC and had fought in Spain, was by 90 BC out of favour with the Senate. In 86 BC he took Rome by force but died the same year. L. Cornelius Sulla Felix, the eastern general, returned to Italy in 83 BC and by 82 BC had defeated Marius' armies. One of these, in Spain, was led by Q. Sertorius. In 81 BC he was forced to flee to North Africa but in 80 BC he returned, and from then until 73 or 72 BC conducted a remarkable campaign against the Roman armies of Pompey and Metellus. He was finally murdered by one of his own followers.

Julius Caesar had also served in Spain, in 61 and 60 BC, and after the death of Pompey had defeated an army led by his sons at Munda, south of Córdoba, in 45 BC; but it was not until the time of Augustus, from 26 to 18 BC, that Roman armies under Agrippa finally conquered the north and north west. A military station was then established at León (Legio VII Gemina, or Castra Legionum) and colonies of veterans were placed at Mérida and Zaragoza. The peninsula was now divided into three provinces: Baetica to the south, corresponding closely to the old Hispania

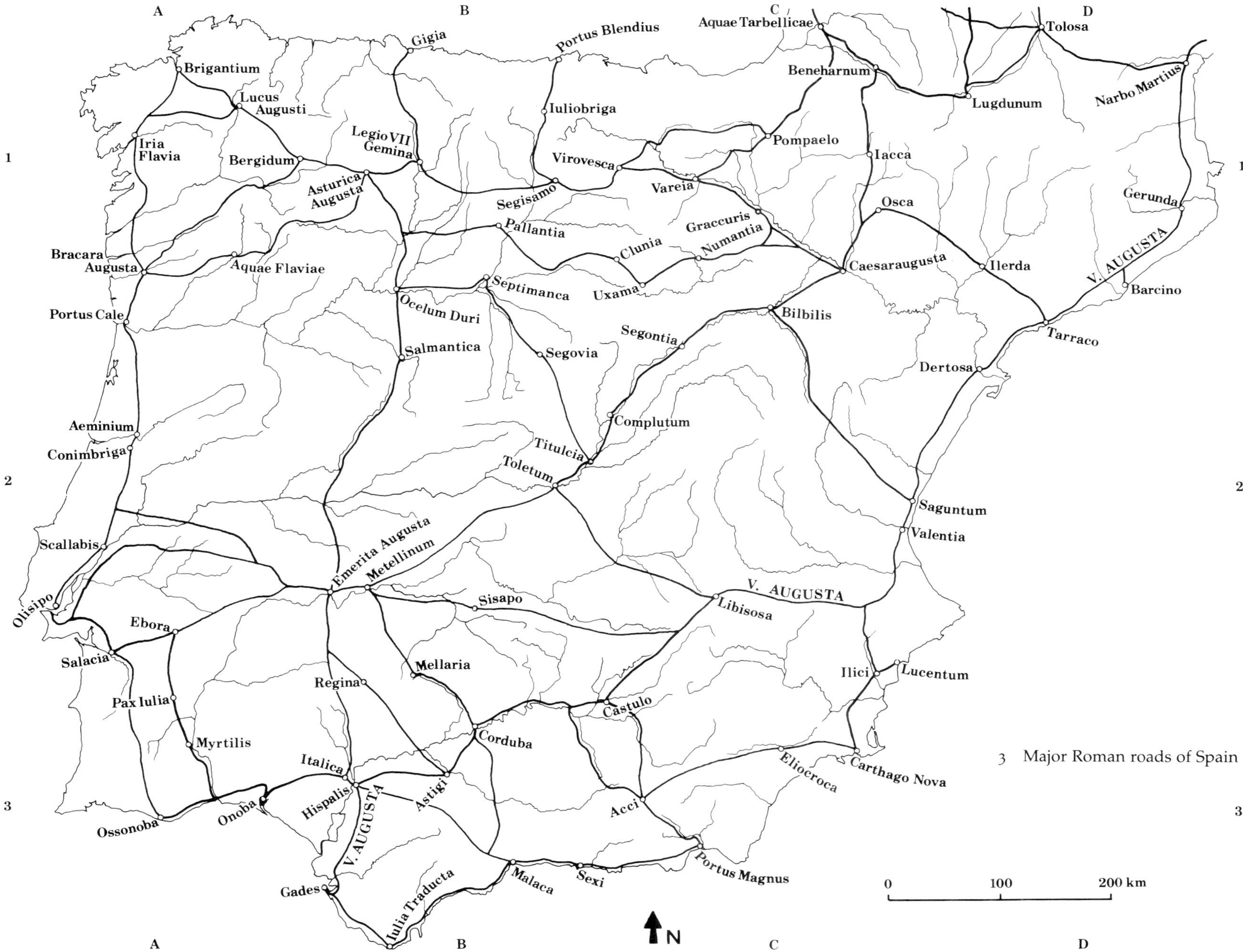

3 Major Roman roads of Spain and Portugal

Ulterior; Lusitania, including modern Portugal to the Douro, but extending to the north east almost as far as Toledo; and Tarraconensis, covering the rest.

The more important Roman roads in Spain are shown in Map 3, extending also into south-west Gaul. This is based chiefly on the map given by Casado (n.d.) but there are other maps in Sitwell (1981: 54, 56), Chevallier (1976), von Hagen (1967b: 114) and van der Heyden and Scullard (1963: 165) – see also Bates (1900), Bouchier (1914), Thouvenot (1940). The road system was determined not only on strategic grounds but also for the access provided to the extensive mineral deposits. In the south there was silver near Cartagena, at Castulo near Linares and at Almadén (ancient Sisapo). There was also gold in the Sierra Morena and copper at Río Tinto north west of Seville. In the north west was gold, with silver at Huesca (Roman Osca). Mercury and iron were also found (e.g. Wiseman, 1956: 17, 53, 58; Forbes, 1965, vol. 7: 150–3, 157). Wiseman quotes annual productions at the end of the first century AD of 8 or 9 tonnes each of silver and gold.

In view of this history, it is not surprising that the first of the Roman roads was from Narbonne around the coast to Cartagena, with a branch across to ancient Libisosa west of Albacete and then down the River Guadalquivir to Cádiz (Gades). Map 3 shows only one road from Castulo (near modern Linares) to Córdoba and Seville (Hispalis) but it is clear there were more. Sitwell (191: 56) shows roads down both banks, and there are Roman bridges on either side (Section 5.7). This road dates from 120 BC (Chevallier, 1976: 156; Wiseman, 1956: 52). Originally called the Via Maxima, it was reconstructed by Augustus and renamed for him. The road from Tarragona through Huesca (Osca) to Zaragoza (Caesaraugusta) is also believed to be Republican (Chevallier, 1976: 156).

The basic framework of the Spanish road system seems to have been established by Augustus (27 BC–AD 14) and the succeeding emperors of the first century AD. Including the Via Augusta, there appear to have been four major Augustan roads. There were two in the north. One followed the upper Ebro from Zaragoza through Graccuris to Vareia near Logroño, turned west through Burgos to ancient Segisamo, and then possibly north to Reinosa (Iuliobriga). The second left the Ebro further south and ran west through Numantia and Palencia to Astorga and then by the long northern route through Lugo (Lucus Augusti) to Braga (Bracara Augusta) near the western seaboard (see also González de Riancho Mazo, 1988). The third, in the west, ran due north from Seville (Hispalis) and Italica to Mérida (Emerita Augusta) and Salamanca, joining the road to Astorga. The final element in this major network was the diagonal cross-country road from Zaragoza through Toledo to Mérida, built by Vespasian (AD 69–79) to follow the route first used in 195 BC by M. Porcius Cato.

It was within this framework that the other roads shown on Map 3 were constructed: from Italica and Mérida across to Lisbon (Olisipo) on the mouth of the Tagus; the shorter road from Astorga to Braga through Chaves (Aquae Flaviae); the further links to the coast; the additional roads through central Spain, such as that through Segovia; and the crossings of the Pyrenees. It was one of these that left Toulouse (Tolosa) to pass through St Bertrand-des-Comminges, the second Lugdunum of Gaul, built by Pompey in 72 BC as a settlement for the defeated troops of Sertorius. From this road one could cross to Jaca or Pamplona (Pompaelo).

2.5 North-west Africa

When Scipio defeated Hannibal in 202 BC at the battle of Zama Regia, in Tunisia south of Musti (see Map 4), it was with the help of Masinissa, a chieftain of eastern Numidia. As a result, Masinissa was installed as king of all Numidia, with his capital at Constantine (Cirta) to the west of an enfeebled Carthage. Over the next fifty years he sought to extend his kingdom to include the whole of modern Algeria, Tunisia and Tripolitania. This was partly at the expense of Carthage which, by the treaty of 201 BC, was deprived of the right of self-defence. In 150 BC Carthage finally fought back, but was attacked in turn by Rome. It was at the end of this, the Third Punic War (149–146 BC), that Carthage was laid waste and its people disappeared from history. As far as Masinissa was concerned, the result was counterproductive, for Rome now settled in Africa, with Africa Vetus (as it was later called) a province, and Utica its capital. It covered only the north east of modern Tunisia, from Tabarka east of Annaba (ancient Hippo Regius) in the north to the Gulf of Gabès south of Sfax (Taparura) (Sitwell, 1981: 149; Raven, 1969: 35).

Despite the subsequent wars (112–106 BC) between Rome and Jugurtha, the effective heir of Masinissa, the territorial position remained unchanged until 46 BC. Under the Triumvirate of Caesar, Pompey and Crassus, Africa was given to Pompey. When war broke out between Caesar and Pompey, Juba, king of Numidia, sided with Pompey, and when Pompey's army, after his death, was defeated by Caesar in 46 BC at the battle of Thapsus (south of Sousse), Juba was defeated as well. Much of Numidia was annexed by Rome to become Africa Nova, with Carthage reinstated in 29 BC as capital of the enlarged province (Fentress, 1979).

In 25 BC Augustus gave Mauretania to Juba II, the son of Juba, who had been taken to Rome by Caesar and educated as a Roman. He held the kingdom for Rome until his death in AD 23 and built the cities of Cherchell (Caesarea Iol) and Volubilis (Map 5). In AD 42 his kingdom was annexed by Claudius and divided into two provinces – Mauretania Caesariensis in the east with Cherchell as its capital, and Mauretania Tingitana based on ancient Tingis, modern Tangier. Although notionally contiguous, they were in fact divided by rugged mountains, and the western province was usually reached by sea and came to be administered from Spain (Sitwell, 1981: 153).

Tripolitania received its name from the three cities of

4 Major Roman roads of North-west Africa, from Caesarea (Cherchell) to Carthage

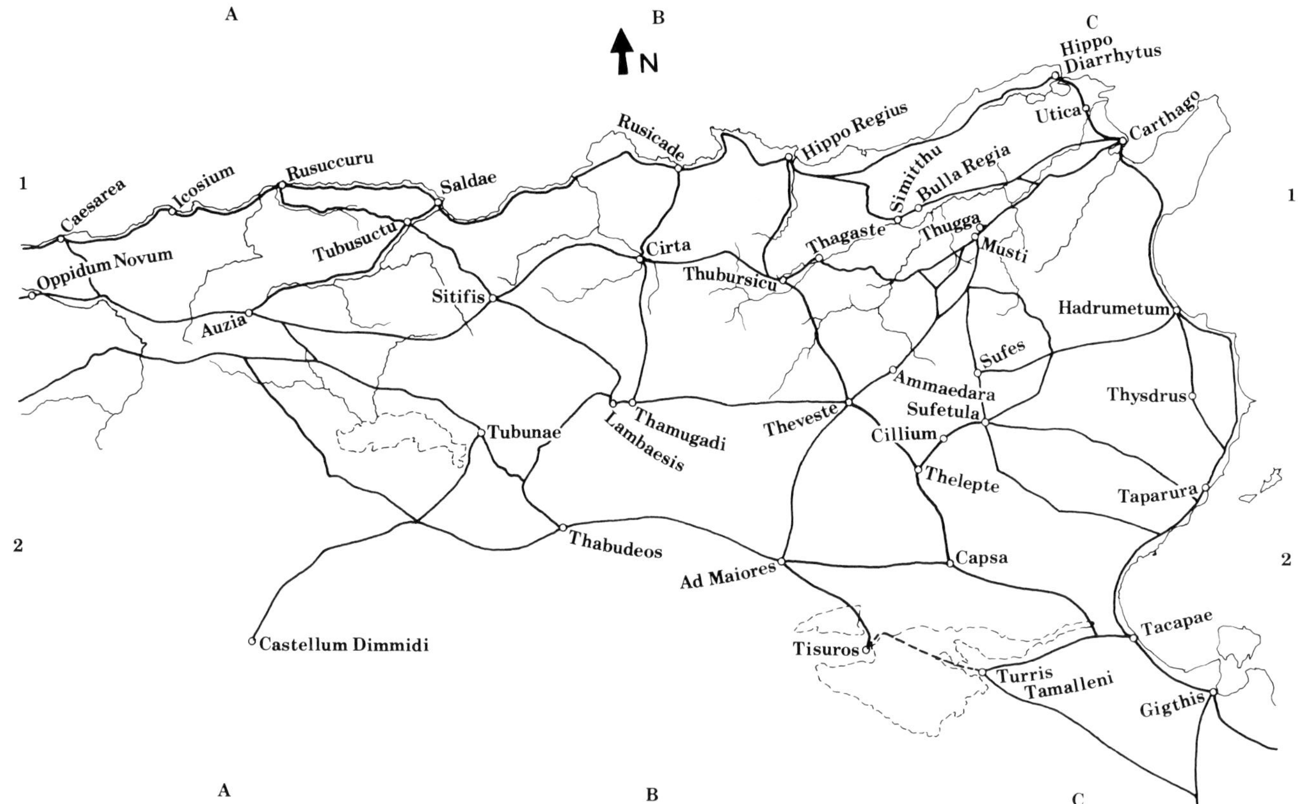

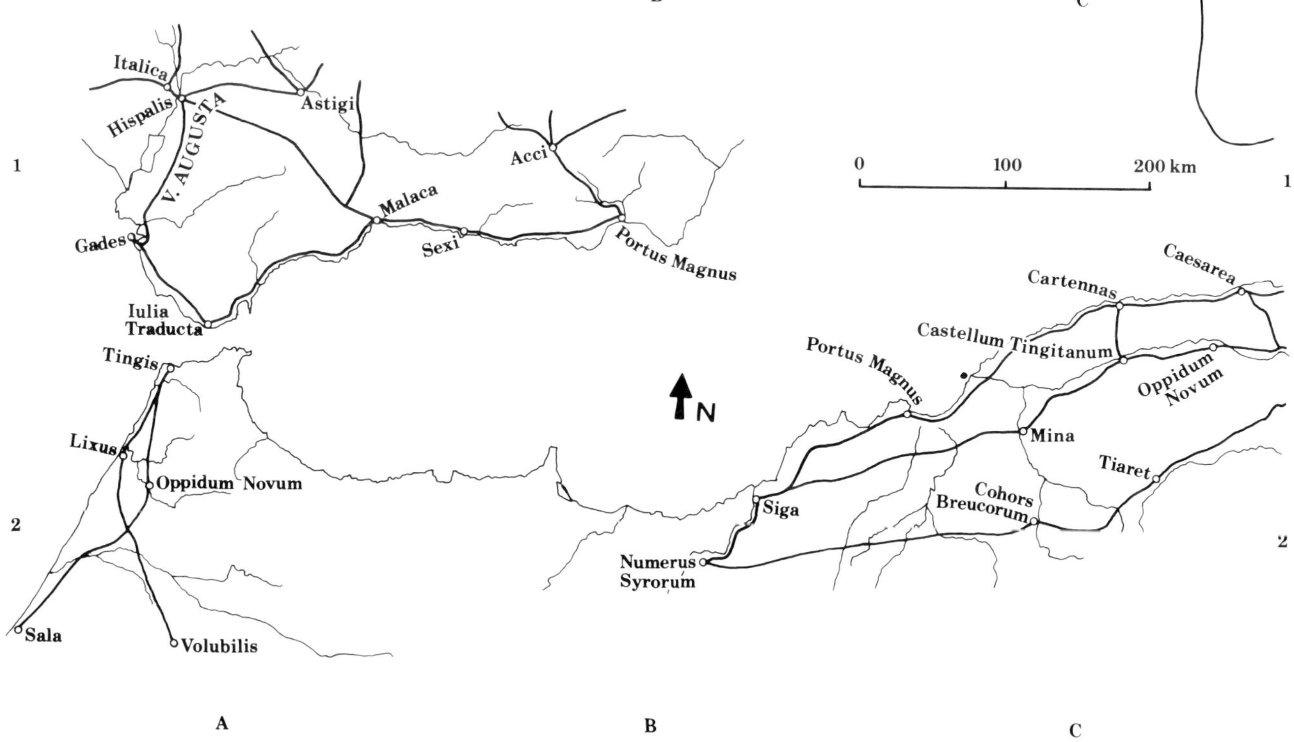

5 Major Roman roads of North-west Africa, from the Atlantic to Caesarea

Sabratha, Oea and Leptis Magna (Map 11). The coastal strip was included in the Africa Nova annexed by Caesar in 46 BC. Further east, Cyrenaica had been ceded to Rome in 96 BC. The southern boundaries of the province were pushed towards the Sahara by the early emperors, chiefly in the interests of defence. Timgad for example, Roman Thamugadi, was founded in AD 100 by Trajan (Chevallier, 1976: 150) on the northern flank of the Aurès Mountains, to support a defence line further south. Indeed, a study of a topographic map shows that the southern boundary was largely determined by the southern edge of the mountain range. The only real extension beyond these mountains was in the east where Septimius Severus (AD 193–211) formed the Limes Tripolitaniae swinging 150 km south of the Gulf of Gabès to defend the cities of Tripolitania. He also pushed south the borders of Mauretania Caesariensis to form the largest empire ever achieved by Rome. By AD 235 most of these extensions had been abandoned.

The major roads of north-west Africa are shown in Maps 4 and 5. Other maps may be found in Sitwell (1981: 54, 154), Chevallier (1976: 151), BNI (1945a: 130), Raven (1969, end map) and Webster (1979: 90–1). The last is a map of the Limes of Mauretania, Numidia and Tripolitania; it shows frontier roads and forts, and also interior roads.

The order of construction of the roads was determined largely by strategic considerations. The Third Augustan Legion (Legio III Augusta) was stationed first at Haïdra (Roman Ammaedara) about AD 6; it was moved to Tébessa (Theveste) about AD 80, and then finally to Lambèse (Lambaesis), probably by Trajan (AD 98–117). There were soon roads from these centres to Carthage, other ports and the frontiers. The topography also was important. Tunisia has only one major river, the River Mejerda (Roman Bagradas) running parallel to the northern coast to Utica. Its course is determined by the east–west mountain chain, and this is characteristic for most of the coast. The north coast has a reasonable rainfall, but this reduces rapidly to the south. Corn from the north was important to the Romans, as also were olives from regions such as near el Djem (Thysdrus), further south (Raven, 1969: 72).

No doubt the Carthaginians built roads (Chevallier, 1976: 150). Of the Roman roads shown on Map 4, some are believed to have been Republican. These were the road north from Carthage to Bizerte (Hippo Diarrhytus), from Carthage up the right bank of the Mejerda and then to Tébessa (Theveste), and from Carthage south east across Cape Bon. Chemtou (Simitthu) was an Augustan colony, and it is believed that the road from Carthage here and to Annaba (Hippo Regius) was in existence in Augustan times. In AD 14 the road from Haïdra (Ammaedara) and Tébessa (Theveste) south east to Gabès (Tacapae) through Gapsa (Capsa) was begun (Chevallier, 1976: 150).

Many of the roads shown on Map 4 were completed by the end of Hadrian's reign (AD 117–38). Some time after AD 80, the legion was moved to Tébessa and by that time there was a road from there to Annaba. Another road was built west to Timgad (Thamugadi) and Lambèse, near Batna, following a line north of the Aurès Mountains. Timgad was established in AD 100. Both it and Lambèse lay east of modern Batna, to the north and east of a rugged mountain chain. It was Trajan (AD 98–117) who established the more southerly limes east through Ad Maiores and west to Thouda (Thabudeos) near Biskra (ancient Vescera). There are today two major roads from Biskra to Batna, both running north east through rugged gorges. The more westerly of these follows the line of the Roman road shown in Map 5. This road is marked with an inscription of AD 145 (Raven, 1969: 52–3) and crosses the fine el Kantara bridge, built later in the time of Caracalla (AD 211–17) (see Section 5.8); it reached further north also to Constantine (Cirta). Hadrian also built roads in Africa. By AD 122 there was a through road to Mauretania, from Musti through Constantine to Sétif (Sitifis) and Auzia (Chevallier, 1976: 150). To the east, Gabès (Tacapae) had been reached soon after AD 14 (see above). In AD 98 Nerva started on the great road to Egypt. Much of this road, called the Via Nerva, or Nerva-Traiana, was built by Trajan; it reached Egypt some time after AD 113 (von Hagen, 1967b: 66–74).

The complete extent of the road system shown in Map 4 may be summarised as follows: (i) there are the four major roads radiating out from Carthage; (ii) there are the east–west roads of defence, with the earlier inner line through Capsa, Theveste and Lambaesis, and the later outer line formed by the Limes of Tripolitania, Numidia and Mauretania (Webster, 1979: 90–1); (iii) there are the north–south cross connections, and (iv) finally the roads for agriculture and commerce, linking the towns in the safer regions near the coast.

The major road to the west left Carthage up the right bank of the Mejerda River to Dougga (Thugga) and Musti (Figs. 6 and 7), and then ran on to Constantine (Cirta), Sétif (Sitifis), Sour-el-Ghozlane (Auzia) and Relizane (Mina). It stopped at Maghnia (Numerus Syrorum) west of Tlemcen, at the end of the Limes Mauretaniae. Some 400 km further on was Tangier (Tingis) on the Atlantic, beyond the Pillars of Hercules on either side of the Straits of Gibraltar at the extremity of the Mediterranean. From here roads stretched south to Volubilis and Chella (Sala) near Rabat. Probably there was a connecting land link between these two road systems, possibly from Tangier along the coast through Melilla (Rusaddir) to Siga near Tlemcen (von Hagen, 1967b: 114), but it is doubtful if this had the status of a road.

2.6 Britain

The excursions by Caesar into Britain were of short duration – in 55 BC he landed briefly in Kent, and again in 54 BC, when he advanced across the Thames to the region of St Albans. As a matter of deliberate policy he then left Britain alone, and this policy was followed by Augustus and Tiberius. It was not until AD 43 that an army sent by Claudius landed at Richborough (Roman Rutupiae), and by AD 49 had advanced to a line joining Exeter (Isca

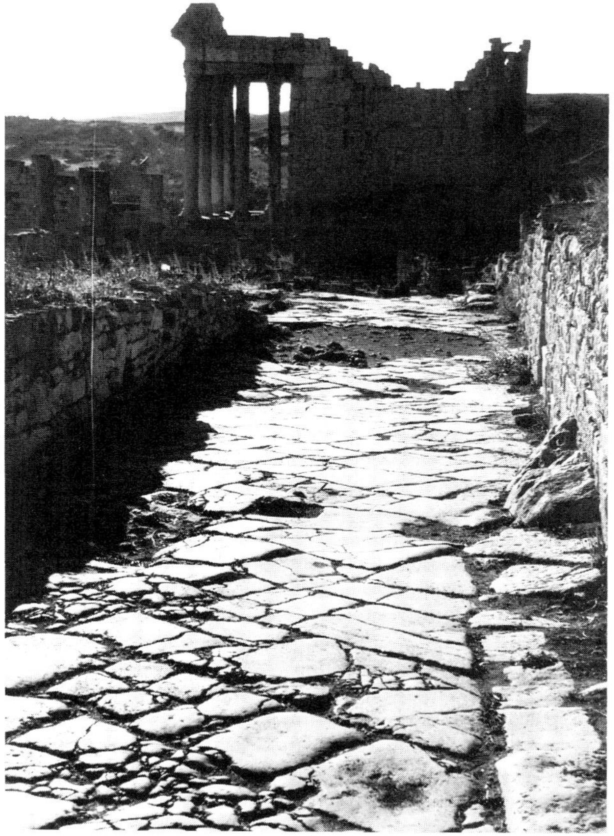

Fig. 6 Roman road and temple at Dougga in North Africa.

Dumnoniorum) to Lincoln (Lindum) (see Map 6). Despite the revolt of Boudicca (or Boadicea) in East Anglia in AD 61, this line was held for some twenty years. Under Vespasian (AD 69–79) there was a new advance, first to York (Eburacum) and Caerleon (Isca) and then, with Sextus Iulius Frontinus as governor, over the whole of Wales. It was this Frontinus who became later the

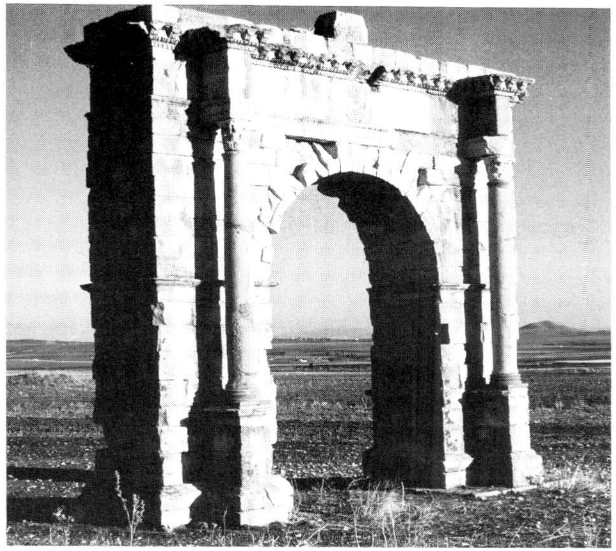

Fig. 7 The fine Roman arch at Musti stands alone beside the modern road.

chronicler of the aqueducts of Rome (Section 7.2). In AD 77 or 78, Frontinus was replaced as governor by Gnaeus Iulius Agricola, who by AD 84 had not only taken Roman armies north of Inverness, building a legionary fortress at Inchtuthil, the most northerly in the Roman empire (Pitts and St Joseph, 1985), but had also sent a fleet to circumnavigate the island (Burn, 1953; Birley, 1981).

This advance was shortlived, for disturbances elsewhere in the empire caused the Romans to withdraw, first to the line across lowland Scotland from the River Clyde to the Forth, and then about AD 105 to the Stanegate, the road built by Agricola from the Solway to the River Tyne. This line was later fortified by Hadrian between AD 122 and 126 (or later, see Webster, 1979: 73) to become what is now known as Hadrian's Wall (Sitwell, 1981: 40–1). In AD 139, the emperor Antoninus Pius (AD 138–61) moved north again and the Clyde–Forth line was fortified with a turf wall now called the Antonine Wall. For a while both walls were defended but then in AD 184 there was a final retreat to Hadrian's Wall. (Histories of Roman Britain may be found in Haverfield, 1912, 1924; Richmond, 1963; Salway, 1981 and Frere, 1987.)

The best maps of Roman roads in Britain are those by Margary (1973) who has described all these roads in detail (see also Viatores, 1964). Map 6 shows only those which are the more important. It shows also the roads of northern Gaul, including that down the left bank of the Rhine from Mainz (Moguntiacum) to the sea. This river may be followed upstream to its source in Map 2.

These two figures show that the chief overland access to Britain was along one of Agrippa's roads of 27 BC (Section 2.3), from Lyon (Lugdunum) through what was then the small town of Lutetia on the site of modern Paris to Boulogne (Gesoriacum). Alternatively, armies could be moved rapidly from Britain to the frontier along the Rhine, from Boulogne through Bavay (Bagacum) to either Cologne (Colonia Agrippinensis) or Mainz.

Margary (1973, vol. 2: 224) suggests that construction of the roads followed rapidly after conquest, with the first roads built from the Kentish ports to London, followed by roads from London to the south coast, and throughout the south-east corner. He believes that the Fosse Way (or Foss in Margary's spelling), along the line from Exeter (Isca Dumnoniorum) to Lincoln (Lindum), would have been in place at the latest by AD 47, with the connecting roads from the south built at about the same time. This line was held for some years, and was a boundary between the relatively flat land to the east and the mountains to the west and north. By AD 78 the frontier, with its supporting roads, had been moved north to the line from Chester (Deva) to York (Eburacum), and under Agricola (AD 77 or 78–84) the remainder of the British road system had been established, including the two great roads to the north.

Curiously, Margary gives the name Via to only one

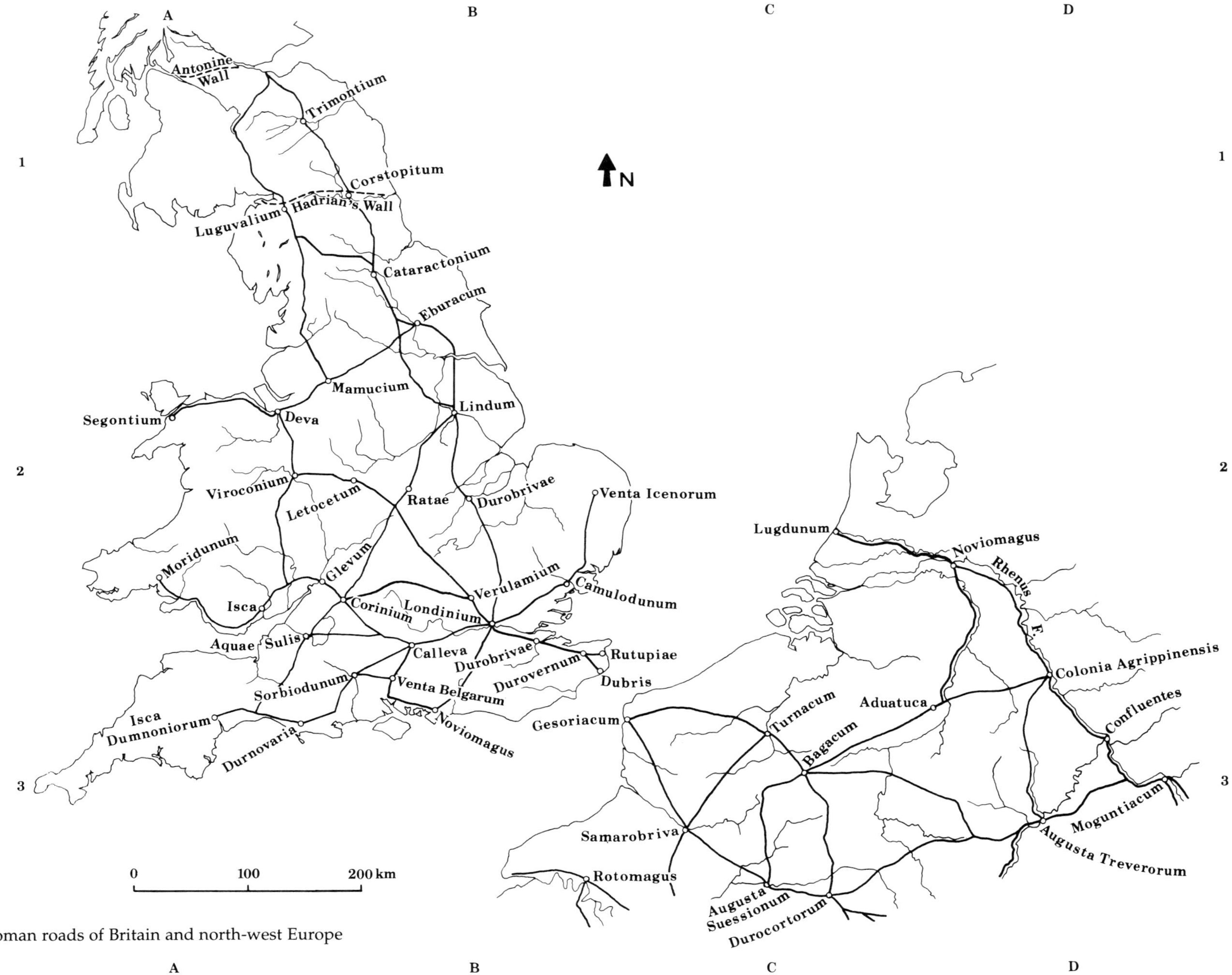

6 Major Roman roads of Britain and north-west Europe

British road, the Via Devana, a relatively minor road from Colchester (Camulodunum), which headed north west through Cambridge to Huntingdon. This name derives only from the eighteenth century and arises from a belief that its destination was Chester, Roman Deva (Margary, 1973, vol. 1: 183). He mentions also a Via Julia at Bath, but this name also is spurious and of modern origin. The term 'street' came from a Saxon word, *streat*, which was the name given to a road with a paved or metalled surface. The only roads of this kind were Roman, so the name Street implies a Roman road. The term 'stane' comes from 'stone' and has a similar derivation (Margary, 1973, vol. 1: 20).

In the time of Edward the Confessor, laws were made to protect travellers on the four main roads of England; these were listed as Watling, Foss and Ermine Streets and the Icknield Way (although there is some uncertainty concerning the last). Watling Street ran from Canterbury (Durovernum Cantiacorum) through London to Wroxeter (Viroconium), near Shrewsbury. At its eastern end, it was approached by roads from Dover (Dubris) and Richborough (Rutupiae), the Roman Channel ports. Fosse Way left the south coast in the west near Exeter and travelled diagonally right across the country, through Bath (Aquae Sulis), Cirencester (Corinium) and Leicester (Ratae Coritanorum) to Lincoln (Lindum). Ermine Street ran north from London to Lincoln and then followed the more easterly route across the Humber to York (Eburacum). The final road in the list is believed by Margary (1973, vol. 1: 22, 231) to be the Icknield Way, which he describes as 'one of the best-known and most important prehistoric trackways in Britain'. It left the Thames near Goring and followed an irregular route north to a point near Hunstanton on the Wash, passing to the east of Cambridge. Only parts of the route were Romanised, and so it is not shown in Map 6.

Three other major roads are shown leaving London. Stane Street went south to Chichester (Noviomagus Regnorum, or Regnum). The main road to the south west went initially to Silchester (Calleva Atrebatum), then down the Port Way to Old Sarum (Sorbiodunum) near Salisbury, and on to Dorchester (Durnovaria) and Exeter. From Silchester, one branch led to Bath (Aquae Sulis) and another, more importantly, through Cirencester (Corinium) and Gloucester (Glevum) to Carmarthen (Moridunum) in South Wales. Finally, there was the road from London east and then north to East Anglia, terminating at Venta Icenorum, 5 km south of Norwich.

St Albans (Verulamium) on Watling Street was an important centre in both Roman and pre-Roman times. An alternative route from London to the west was along Akeman Street, from St Albans to Cirencester (Corinium).

In AD 78, the Roman frontier followed roughly the line of a road from Caernarvon (Segontium) through Chester (Deva) and Manchester (Mamucium) to York (Eburacum). The central portion of this road, from Chester to Manchester, formed part of the western road to the north, through Carlisle (Luguvalium) and the valley of the upper Clyde to Cramond, a Roman fort and port on the Firth of Forth. The alternative route to the north was along Dere Street, from York through Corbridge (Corstopitum) and the fort at Newstead (Trimontium, see Section 4.2) on the River Tweed, just east of Melrose.

These northern roads were interconnected at a number of points; two of these are shown in Map 6. Just north of Catterick (Cataractonium) is the road junction called Scotch Corner. In Roman times, as today, there was at this point a choice of roads, either due north along Dere Street west of the modern road, crossing the Tees at Piercebridge, or north west, across Stainmore Pass to Penrith and Carlisle. Further north, the Stanegate, between Corbridge and Carlisle, formed part of Agricola's road system. It was along this line that Hadrian's Wall was built between AD 122 and 126. One of the finest of all Roman monuments, the wall is considerably longer than the road (Margary, 1973; Webster, 1979: 69; Sitwell, 1981: 40–1; Bruce, 1978; Breeze and Dobson, 1987). Its eastern end is now at Wallsend on the Tyne, but possibly it was originally planned to finish 5 km west, at the Pons Aelius, below the present Swing Bridge in Newcastle (see Section 6.4). At its western end it reaches Bowness, about 20 km beyond Carlisle, the total length being 118 km. The wall was not built for static defence but was designed so that troops could move rapidly along it – the Stanegate would have continued in use. Typically it lay about 2 km south of the wall.

Map 6 shows the Roman road system terminating at the Firth of Forth, with the Antonine Wall further to the west. Margary (1973, vol. 2), from whom most of this discussion has been taken, describes a further road extending north of the Antonine Wall, through Stirling and Perth to Kirriemuir, 24 km north of Dundee. This was not a track but a made road – to quote from Margary's concluding paragraph, 'the most northerly Roman road so far known anywhere in the Empire'. In AD 83 a Roman fort had been begun at Inchtuthil, near Kinclaven 16 km due north of Perth, but was abandoned while still incomplete, four years later (Pitts and St Joseph, 1985).

2.7 Greece and the northern frontiers

It was in the year 276 BC that the Romans came to control the whole of peninsular Italy (Section 2.2.1). By contrast, the Greeks came to peninsular Greece as early as 1600 BC (McEvedy, 1967: 30) and in the period 800 to 500 BC they had colonised the Aegean seaboard of Asia Minor and much of the Black Sea coast. In 338 BC, when Rome was recovering from the effects of the First Samnite War, Philip II of Macedon took control of Greece, and by 323 BC his son, Alexander the Great, had conquered Persia, and advanced through Asia Minor, Palestine and Egypt, and as far as India. His sudden death in that year led to the breakup of his empire. By the time of the Second Punic War (218–201 BC), most of the Greek cities were united in a Hellenic Confederacy under King Philip V of Macedon. Greece then consisted essentially of the Greek peninsula and Macedonia, to the north of Thessaly; it adjoined

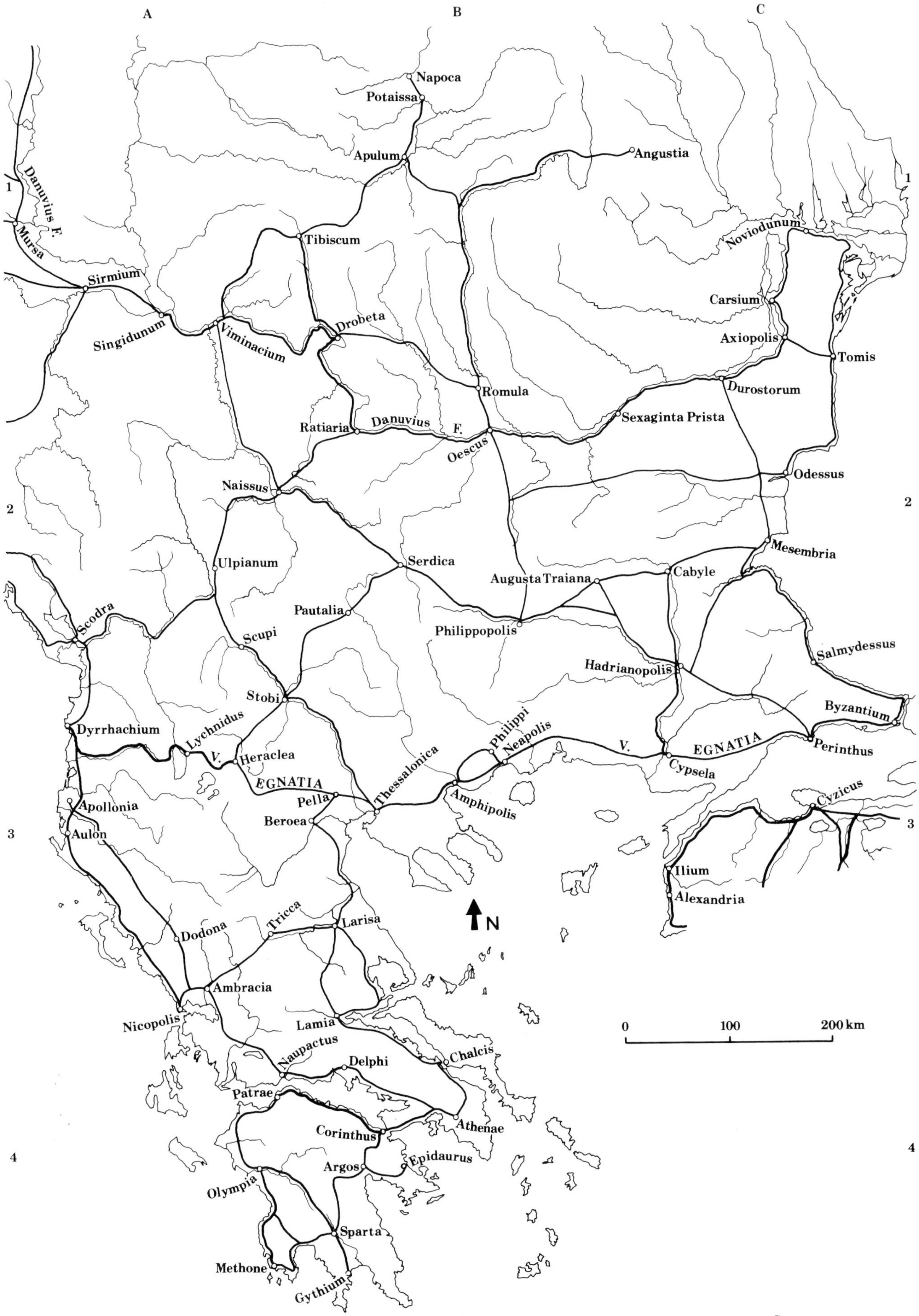

7 Major Roman roads of Greece, Macedonia and the lower Danube

Epirus, south of modern Albania and Thrace, between the Aegean and Black Seas. To the north west was Dalmatia (earlier called Illyria) along the Adriatic. In 229 BC Rome moved against the pirates of this coast, and established a protectorate over the Greek cities opposite Brindisi (see Map 7).

With Rome defeated at the battle of Cannae in 216 BC, and Hannibal apparently free to roam with his army through Italy, Philip entered into a treaty with him (215 BC). As it happened, this action had no influence on the outcome of that conflict, but it led to a series of Macedonian Wars between Rome and Greece – in 214–205, 200–196, 172–167 and 148 BC – and an intervening war from 192 to 189 BC on Greek soil, to defend the Greek homeland against the Seleucid king Antiochus III. In 194 and 188 BC Rome withdrew her troops, because of her admiration for Greek culture and from a genuine desire to maintain the liberties of the Greek cities. From the withdrawal of 188 BC to the Third War of 172 BC was a period of seventeen years, and during this period mistrust arose between the Romans and Macedonians, leading eventually to war. Following 167 BC, Rome again set Greece free, but under its own terms, which proved unacceptable. Finally, in 148 BC Macedonia was annexed. In 169 BC Epirus had been taken, and by 129 BC the Dalmatian coast had fallen under Roman control. Thrace was an independent but allied country.

By the time of the death of Caesar in 44 BC, the Roman frontier included the whole of Gaul, and extended in the north west to the Atlantic and the Rhine. It followed the Rhine down all the way to modern Basel, and then east to Bregenz (Brigantium Raetiae) on Lake Constance. To the north and east of Italy, however, the bounds of the empire were much more restricted. From Bregenz the frontier turned again sharply to the west, to include the land of the Helvetii between Lake Geneva and the Rhine, and then south from Geneva to the Mediterranean. The Alps were hostile territory until Augustan times. Roman lands in northern Italy stopped short of the Alps in both the west and the north, and then from Aquileia followed the Dalmatian coast down to Greece and Macedonia.

The independent northern lands were, from west to east, Raetia, from the upper Danube to the Alps, and from the Rhine east to the River Inn (Veldidena to Castra Batava on Map 8); Noricum, from the Danube to the Alps, with its eastern boundary running south roughly from Vienna (Vindobona) to the corner of the Adriatic; Pannonia, occupying the great corner of the Danube from Budapest (Aquincum) to Dalmatia; Moesia, south of the lower Danube, from Belgrade (Singidunum) all the way to the Black Sea, with southern borders to Macedonia and Thrace; and Dacia, north of the lower Danube (Webster, 1979: 53, 60, 62). By the end of Augustus' reign most of these lands were Roman.

In 15 BC, Tiberius and Drusus, stepsons of Augustus, occupied Raetia. Augustus himself (then Octavian) had earlier subdued the north Adriatic coast to Salona near Split in 35 BC, and moved inland to Siscia on the River Save downstream from Zagreb in the same year. In the period 12 BC to AD 9, Tiberius moved this border to the Danube, and occupied Noricum, Pannonia and Dalmatia. Moesia had been occupied in 28 BC, and Thrace reduced to the status of a vassal state in 9 BC. By AD 9, then, the Roman empire had advanced to the Danube. There were two further extensions. Vespasian moved the north-western frontier across the Rhine from Augusta Raurica (near Basel) in AD 73–74, to annex the Black Forest area of southern Germany. Domitian (AD 83 and 89) advanced further up this right bank of the Rhine to the Main. It was Domitian also who first crossed the Danube into Dacia in AD 85, but this country did not become a Roman province until AD 107, after Trajan's Dacian Wars of AD 101–2 and 105–6.

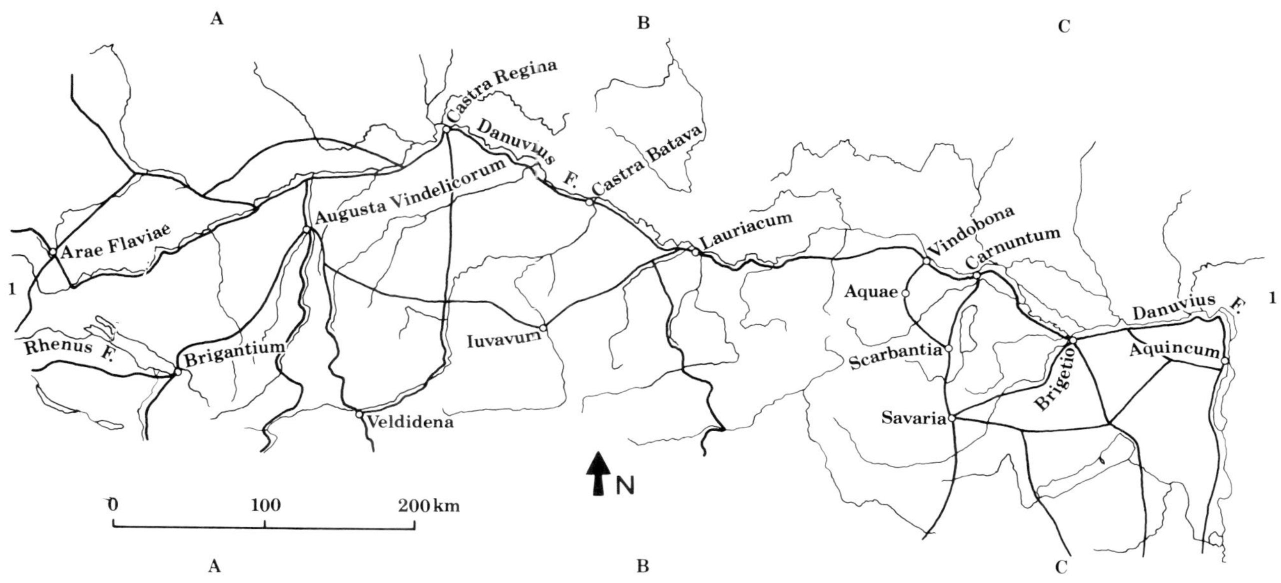

8 Major Roman roads of the upper Danube

The roads of Greece and Macedonia are shown on Map 7 from the Adriatic to Byzantium or Constantinople, and north through Thrace and Moesia to the Danube and across to Dacia. Dalmatia is shown with Italy on Map 1. At the northern frontier, the Rhine may be followed upstream from the North Sea to its source through Maps 6 and 2, and the Danube downstream in Maps 2, 7 and 8 (Sitwell, 1981: 92, 94–5, 125, 138–9; Chevallier, 1976: 140, 174; Webster, 1979: 48, 53, 60). The dates of these roads follow closely after the historical events summarised above: to Macedonia in 148 BC; to the left bank of the Rhine in 44 BC; and then east down the Danube under Augustus (22 BC–AD 14). Roman influence had spread across the Adriatic by 229 BC, but most of the roads of Dalmatia probably date from Augustan times.

The Via Egnatia, one of the great roads of antiquity, followed an ancient route used by Roman armies in 199 BC in the Second Macedonian War. The road is generally dated from 148 BC, although Hammond, who has studied its western part in detail, puts it later, about 120 to 100 BC (1974: 192). It ran from the Adriatic coast opposite Brindisi up the Shkumbi River, over a pass to Lake Ochrid at 695 m, and down to Thessalonica on the Aegean Sea. There is some diversity of opinion as to its western terminus – Apollonia or Durrës (Roman Dyrrhachium). Hammond believes that Durrës is the correct choice, with the Via Egnatia following the right bank of the Shkumbi in its lower reaches. He shows also roads south to Apollonia, on the right bank of the old course of the Aous (Vijose) River. The Via Egnatia was extended east to Byzantium, possibly in the time of Augustus. The apostle Paul travelled this route from Philippi to Beroea about the year AD 50 (Acts 16.12–17.10).

On that same journey, Paul continued south to Athens, but this was by ship (Acts 17.14). Ten years later he returned by land from Greece through Macedonia to Philippi (Acts 20.2–6). Presumably he travelled the eastern road up the peninsula through Lamia and Larisa. He would then have joined the Via Egnatia at Pella, capital of Macedonia under Alexander the Great. This eastern road, with the Via Egnatia, formed two sides of the triangle of roads which made up the main framework of the Greek road system. The third followed the western coast down from Durrës through Nicopolis (near Prebeza) and Delphi to Athens. It extended also north up the Dalmatian coast to Aquileia, passing through the substantial Roman town at Salona and the minor neighbouring village at Split where Diocletian built his palace following his retirement in AD 305 (Sitwell, 1981: 122). There were also roads south from Athens through Corinth, and a multitude of minor roads, only some of which are shown in Map 7.

The roads across the Alps, north from Italy, have been described in Section 2.2.4. Three of these crossed the Splugen, Reschen and Brenner Passes to distant Augsburg (Augusta Vindelicorum) in Raetia. Initially a strategic centre near the northern frontier, Augsburg became an important city for trade (Charlesworth, 1970: 174). From Augsburg it was an easy matter to reach the Danube – either to travel downstream through Regensburg (Castra Regina), which also could be reached directly from Veldidena near Innsbruck; or upstream to Rottweil (Arae Flaviae), and the lands of the Rhine frontier.

The next road to the north was from Aquileia through Villach (Santicum) and Virunum to Lorch (Lauricum), but, to quote Charlesworth (1970: 174), 'the way was difficult and the passes dangerous'.

The preferred route was from Aquileia across the Julian Alps to Ljubljana (Emona) near the River Save, settled by Octavian in 34 BC. There were two routes from this important centre. One led north east through Ptuj (Poetovio) to the Danube at Vienna (Vindobona), Petronell (Carnuntum), Szöny (Brigetio) and Budapest (Aquincum). The other followed the navigable Save River to Siscia and Belgrade (Singidunum). From here there were three major routes: into Dacia; down the Danube to Turnu-Severin (Drobeta), the site of the great bridge of Apollodorus, built for Trajan about AD 104–5 (see Section 6.2); and the equally important road through Niş (Naissus) and Sofia (Serdica) to Byzantium and the east (Charlesworth, 1970: 120–1).

The second of these followed the right bank of the Danube down to where, near Orsova, upstream from Drobeta, it penetrates the arc of mountains which surround its lower plain. It turns suddenly north through the rocky gorge of the Kazan Defile, and then east down the Iron Gates of Orsova, a set of rapids almost 3 km long. This obstacle divided the Danube into two and required the Romans to keep two river fleets, one based on Belgrade, and the other near the river mouth (Sitwell, 1981: 140; Chevallier, 1976: 174; Rossi, 1971: 25, 32, 49). About AD 101, Trajan built a canal here to bypass the worst of the rapids (Sasel, 1973), and upstream of Orsova a road which was partly cut into the southern cliff face, partly cantilevered out in timber extending over the water (see photographs in Rossi, 1971: 24–5; see also Rossi, 1968). He claims that this form of construction extended for 12 miles (19 km), one of the greatest achievements of Roman road building. It appears that it was used also as a towpath, to tow vessels upstream against the rapid current. Unfortunately it is now flooded by a dam. The bridge of Apollodorus was downstream from these rapids. As shown in Map 7, the road continued down the Danube to the mouth, with a branch to Constanţa (Tomis) on the Black Sea.

Charlesworth (1970: 119–20) has drawn attention to the importance of the third of these roads, for it connected the armies on the Danube with those in the far east. Nero (AD 54–68) either laid out or improved this road, while Hadrian (AD 117–38) built the connecting link from Byzantium to Tomis, up the shore of the Black Sea (Charlesworth, 1970: 120). Other interconnecting roads can be seen on the maps.

2.8 The east and Egypt

It was mentioned in Section 2.7 that between 192 and 189 BC, Rome fought a war with the Seleucid king Antiochus III on Greek soil. When Alexander died in 323 BC his European and Asian empire was divided between his generals: Lysimachus in Thrace, Antigonus in Asia Minor and Palestine, Seleucus in Mesopotamia and the east, and Ptolemy in Egypt. By 197 BC only two of these kingdoms remained, Egypt and Cyrenaica were still Ptolemaic, while King Antiochus III the Great had extended his Seleucid kingdom south into Palestine and west through Asia Minor to the Aegean. His advance into Greece after the settlement between Greece and Rome, following the Second Macedonian War, caused Rome to retaliate and force him to withdraw. Pergamum (see Map 9) was an independent city-state which had seceded from the Seleucid kingdom about 260 BC. Feeling threatened, Eumenes II of Pergamum in 196 BC had called on Rome for help. In 190 BC Roman troops for the first time set foot in Asia and, with the aid of Eumenes, defeated Antiochus at Magnesia near Izmir (ancient Smyrna). Although the Romans then withdrew, this was the beginning of the long series of events which made Asia Minor part of the empire by AD 43.

The first part of Asia to become Roman was, in fact, the kingdom of Pergamum, for in 133 BC it was bequeathed by Attalus III to Rome. Three years later it was constituted as the province of Asia, extending over the whole Aegean coast and its islands, west of a line drawn roughly north–south through Patara. It included the ancient lands of Mysia in the north, centred on Cyzicus; Lydia in the centre, with its towns of Pergamum, Ephesus and Smyrna; and Caria in the south.

The spread of Roman control east along both the north and south coasts of Asia Minor was not the result of planning, but of Roman response to aggression. In the south, following the decline of Rhodes as a sea power under Roman influence, from 165 BC, the pirates of the eastern Mediterranean became a major threat to Roman trade. In 102 BC Rome had established patrol stations in Pamphylia (whose coast ran from Antalya (Attalia) around the gulf past Side) and western Cilicia (further to the east), and these were re-established about 76 BC. Pompey received his commission to settle the affairs of the east in 67 BC, and his first action was systematically to destroy these pirates. Crete became a province in the same year.

In the north, the province of Asia was bordered by Bithynia, which extended from the shores of the Sea of Marmara (the Propontis) to about Ereğli (Heraclea) on the Black Sea. Bithynia, in turn, joined Pontus, which ran east along the remainder of that coast. From 120 to 63 BC Pontus was ruled by King Mithridates VI, of part-Persian, part-Greek descent. In 104 BC he embarked on a programme of expansion in Asia Minor and by 88 BC, the year when the First Mithridatic War (88–85 BC) began, he had occupied not only the whole of Asia, but also much of Greece. Defeated then by Sulla, he made peace with Rome, and retired to Pontus. In 74 BC, Bithynia also was bequeathed to Rome; Mithridates saw this as a threat and again embarked on war. He was opposed this time by Lucullus and then Pompey, who advanced through Pontus and into Armenia in the east. The settlement this time joined western Pontus to Bithynia as a new province, called Bithynia et Pontus, in 62 BC Eastern Pontus was given to the independent state of Galatia, which now occupied much of central Asia Minor, with Cappadocia to the south and east, another vassal state. Cilicia, expanded westwards along the Mediterranean coast to include Pamphylia, was now a Roman province (as was Syria). In 62 BC, then, Roman control of Asia Minor extended effectively to the Euphrates River, with the provinces of Asia, Pontus et Bithynia, Cilicia around the coast, Crete offshore, and Galatia and Cappadocia as allied states in the interior. To the east lay Armenia in the north, and Parthia, which extended far into Iran. Lycia, a small state in the south west between Pamphylia and Roman Asia, was still independent. Cyprus was annexed and added to Cilicia in 58 BC.

Development of Roman authority after the settlement of 62 BC was essentially by a process of evolution and peaceful adjustment. Galatia was made a province by Augustus in 25 BC; Cappadocia by Tiberius in AD 17; and finally Lycia was joined with Pamphylia to form a separate province by Claudius in AD 43.

The roads of North Africa from Tripolitania west were described in Section 2.5. Those further east are included here – they are shown in Map 11.

Tripolitania, the land of the three cities – Sabratha, Oea and Leptis Magna – was part of the Africa Nova formed by Caesar in 46 BC. Further east, Cyrenaica and Egypt made up the empire left by Alexander to the Ptolemies. As early as 273 BC, when Roman control had first reached out to cover the Italian peninsula, Ptolemy II made a pact of friendship with Rome and this was renewed in 173 BC. Four years later Rome stopped Antiochus IV Epiphanes when he sought to advance into Egypt. At that time, two brothers were claimants to the Egyptian throne. Their dispute was settled by Rome, when in 163 BC it awarded Egypt to Ptolemy VI and Cyrenaica to his younger brother Ptolemy VII. In 154 BC this younger brother published a testament, leaving his kingdom to Rome if he died without issue and in 96 BC, following the death of his son, this bequest came into effect – by 74 BC Cyrenaica was a Roman province.

During this period, Egypt was clearly subservient to Rome. After his defeat by Caesar at the battle of Pharsalus in Greece in 48 BC, Pompey fled to Egypt. From then until 30 BC Egypt was caught up in the wars of the Roman succession. Caesar beat Pompey, and Cleopatra became queen of Egypt. By 33 BC, during the Second Triumvirate, her influence over Antony became so great that he re-established the kingdom of Egypt under the rule of Cleopatra and her son, but all came to an end when Octavian (later Augustus) finally defeated Antony. In 30 BC Egypt became a Roman province (Baines and Malek, 1980).

9 Major Roman roads of Asia Minor

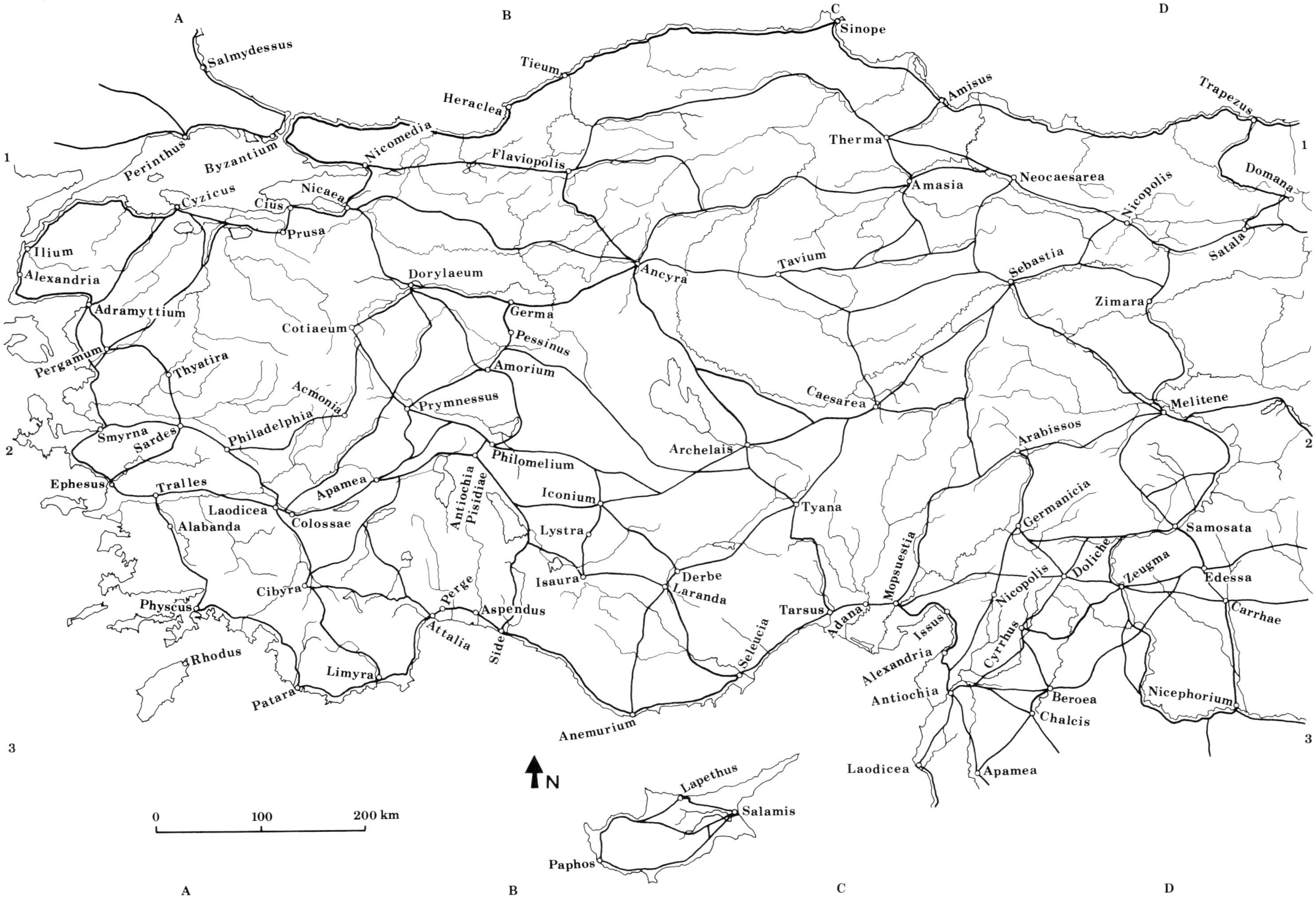

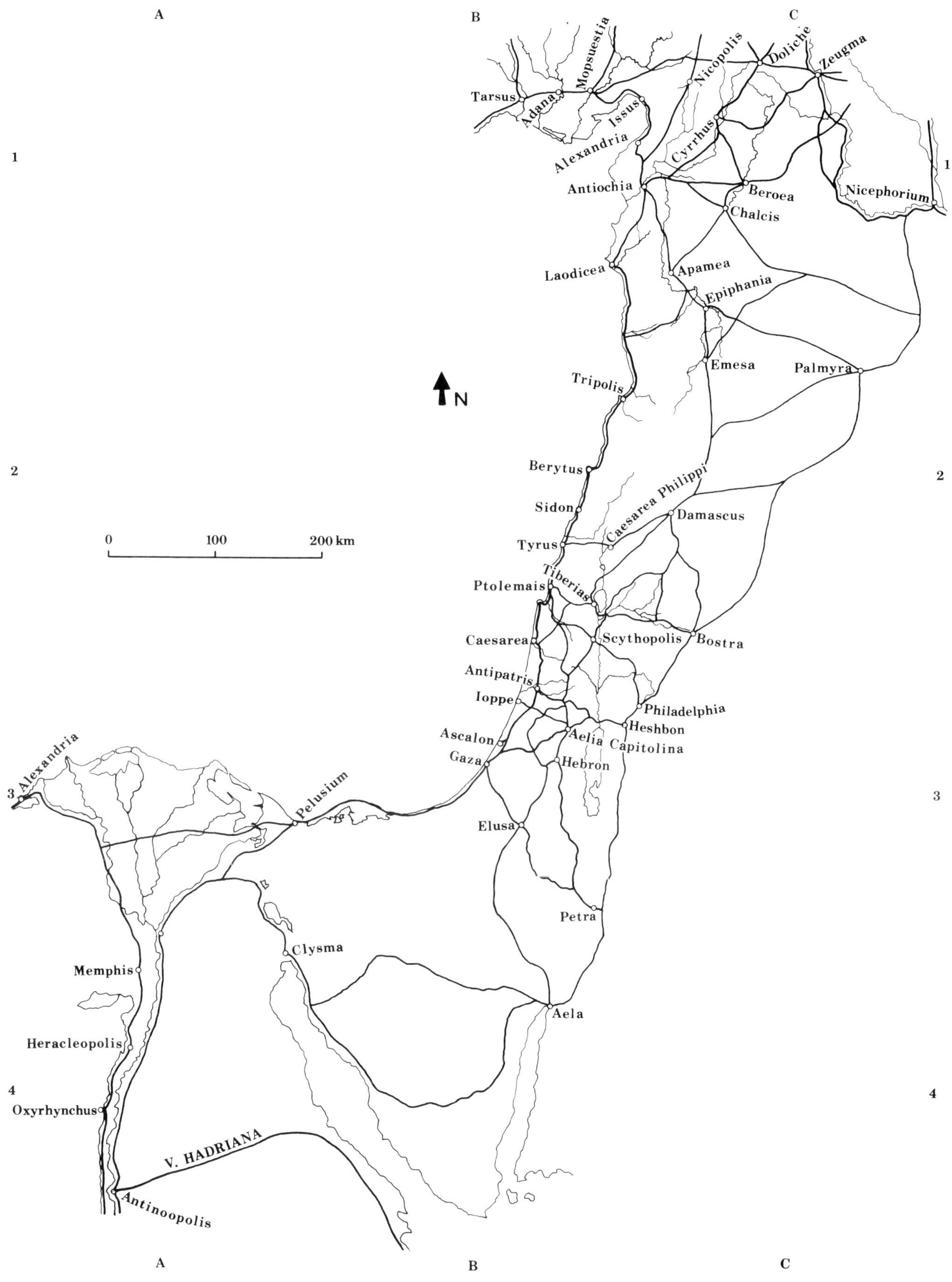

10 Major Roman roads of Palestine and Egypt

Rome's first involvement in the affairs of Palestine (Map 10) came in 161 BC when it entered into a treaty with Judas Maccabaeus, the leader of an uprising against Antiochus IV Epiphanes, the hated Seleucid king. From 150 to 129 BC Palestine was independent. Her neighbour, Syria, remained part of the Seleucid kingdom but was taken briefly by Armenia in 83 BC. Pompey's defeat of Mithridates in 66 BC gave him Syria, which in 62 BC became a Roman province (Bouchier, 1916). Although Pompey had entered Jerusalem in 64 BC, Palestine retained its independence in his settlement. In 37 BC, Antony's armies freed Palestine from a brief period of Parthian rule and Herod became king. After Herod's death in 4 BC, Augustus allowed his three sons to divide the kingdom: Archelaus ruled Judaea and Samaria; Herod Antipas was in Galilee and Peraea; and Philip received the lands east and north of the Sea of Galilee (Avi-Yonah, 1977: 103). The independence of Judaea and Samaria was soon shattered, for in AD 6, at the request of the Jews, Augustus made it into a province. Agrippa I, grandson of Herod, inherited the kingdom of Herod Antipas in AD 39, and in the time of Claudius was granted title to much of Herod's kingdom, but on his death in AD 44 the whole of Palestine reverted to Roman rule (Avi-Yonah, 1977: 106). There was, however, continued unrest, which reached a climax with the revolt of the Jews, the destruction of Jerusalem by Titus as general for Vespasian in AD 70, and that of Masada three years later. From that time on there was no separate Jewish state.

By the middle of the first century AD, therefore, the Roman empire had spread around the shores of the eastern Mediterranean to the Euphrates with the acquisition of Cappadocia in AD 17. It included Syria in 62 BC, Egypt in 30 BC, and Palestine in one form or another by AD 44. There were to be only three further significant advances.

In AD 106 Trajan annexed Arabia Petraea, formed of the lands east of the Jordan, south to Petra, and including the peninsula of Sinai. He advanced also across the Euphrates and by AD 115 had annexed Armenia, stretching to the Caspian Sea, the whole of Mesopotamia, and part of the left bank of the Tigris to form the province of Assyria. These annexations were shortlived, for Hadrian (AD 117–38) decided immediately to withdraw. His boundary ran from the south-eastern corner of the Black Sea to the Euphrates, down the Euphrates to Syria, and then along a southerly line to Aqaba (Aela) on the northern tip of the Gulf of Aqaba (Webster, 1979: 86). In AD 166 Verus annexed the land in the elbow of the Euphrates opposite Syria and established a colony at Carrhae, the Harran of Abraham (Genesis 11.31), today one of the oldest continuously inhabited places on earth.

The charts of major roads shown in Maps 9 to 11 derive from a variety of sources. In addition to the general references used throughout this study, there has been substantial use made of the classical map of Asia Minor by Calder and Bean (1958), the major book on Roman rule in Asia Minor by Magie (1950), the older work by Ramsay (1890) and Avi-Yonah's study of the historical geography of Palestine (1977) – see also Ballance (1958), French (1974), Jones (1971), Levick (1967), Robert (1962) and Starr (1962).

The most obvious routes across Asia Minor are those shown in Map 11: from Istanbul (Byzantium) along the Black Sea coast; from Byzantium to Ankara (Ancyra) and then to the east; or from Ankara to Antakya (Antioch or Antiochia in Syria). All three were of the greatest importance in Roman times. In the fourth century AD the third became a route for pilgrims from Europe to Jerusalem, and was called the Pilgrim's Road (Magie, 1950: 454, 491). It reached the Mediterranean by way of a pass through the high Taurus Mountains, the famous Cilician Gates. These gates had been used by Cyrus of Persia when he swept in from the east to conquer Asia Minor in 546 BC, and then by Alexander in 333 BC as he advanced to the east. Indeed, it was at Issus at the head of the Mediterranean, south east of the Cilician Gates, that Alexander defeated the main Persian armies (Magie, 1950: 276). It needs only a glance at Map 11 to see the importance of this passage to Antakya (Bouchier, 1921), and then on, either to the east and Mesopotamia, or south to Palestine, Egypt and the trading vessels of the Red Sea (Charlesworth, 1970).

However, it would be a mistake to regard these as the important routes of earliest times, for the gateway to Europe was then not by way of Byzantium, but through the Aegean ports. The most famous of early roads was the so-called Royal Road, which ran from the Aegean through Asia Minor to Susa, north of Charax on the Persian Gulf. Described by Herodotus (5.52–4), it was used by the Persians but predated them. Unfortunately its route has not been established with any certainty (Magie, 1950: 39, 786–9; Ramsay, 1890: 27–35; Anderson, 1897: 41–4; Charlesworth, 1970; Winfield, 1977). At its western end, it passed through Sardis (Akurgal, 1978), with branches from Izmir (Smyrna) and Selçuk (Ephesus). It then proceeded north east, possibly through the vicinity of Ankara (Ancyra) to Hattusa, modern Boğazköy (also referred to as ancient Pteria), some 20 km north east of ancient Tavium. Hattusa was capital of the Hittite state (second millennium BC), and this may have determined the western line of the road. From here it may have gone through Malatya (Melitene) across the Euphrates to the Tigris, but others have suggested more southerly routes. It does not appear to have formed the basis of a Roman road (see also Munro, 1901).

A later road which became more important for trade is called by Magie the Southern Highway, and this was also pre-Roman. Its route can be traced from Map 9 – from Ephesus through Aydin (Tralles), Laodicea near modern Denizli, Dinar (Apamea), Yalvaç (Antiochia Pisidiae), Aksaray (Archelais), and then down through Kemer Hisar (Tyana) to the Cilician Gates (Magie, 1950: 40, 789–93; Ramsay, 1890: 29). Clearly it became a Roman road.

The western sections of these roads started in Pergamum, occupied by Rome in 133 BC. Within that kingdom there were also two important north–south routes. The

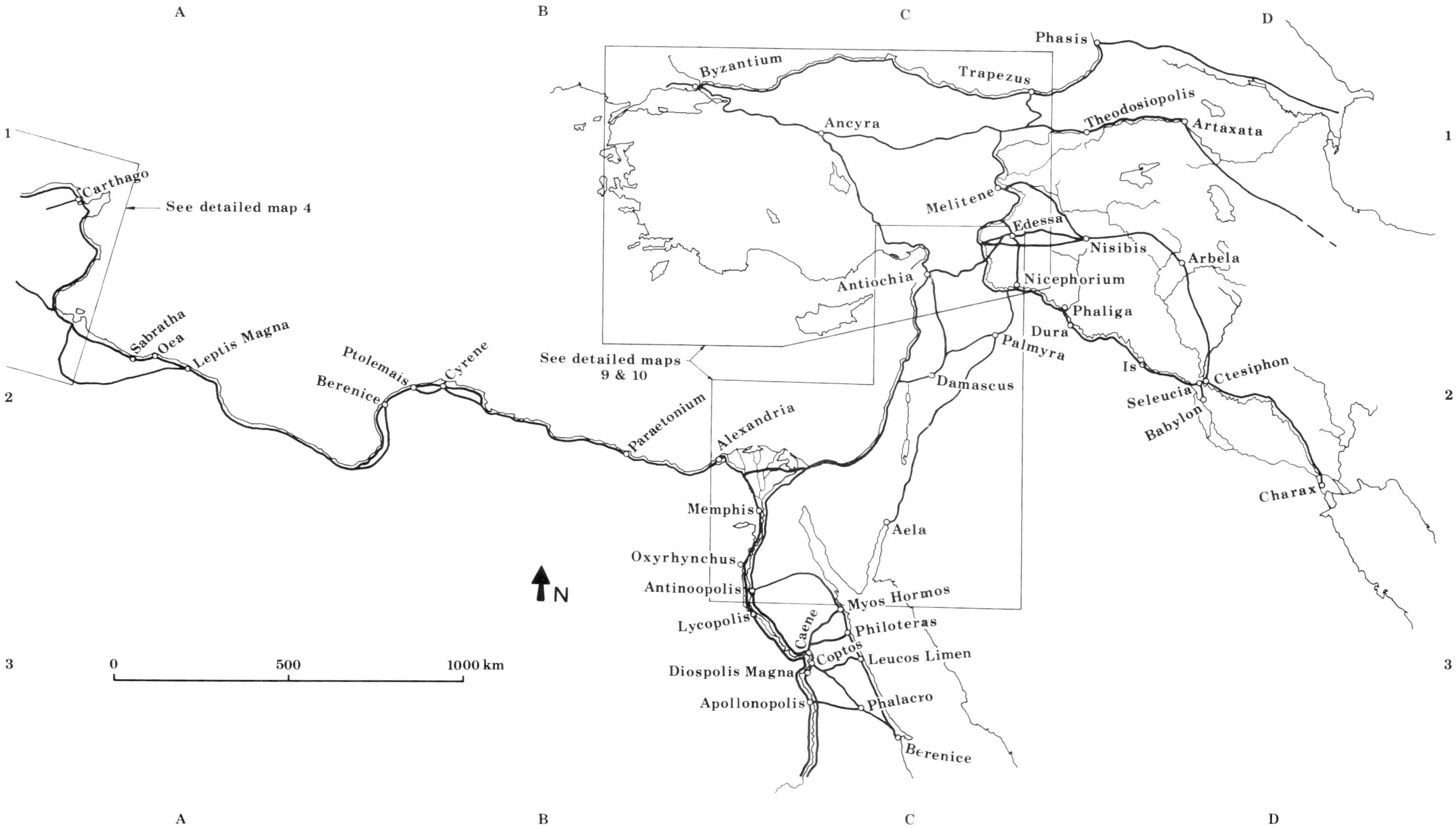

11 An overall view of the Roman roads to Egypt and the east

first of them ran from the Hellespont around the coast to Adramyttium near modern Burhaniye, and then continued as the coastal road to Ephesus. Here it joined the Southern Highway, from which it then branched south to Physcus, near Marmaris. The second, further inland, went from Cyzicus through Thyatira to Laodicea. Parts of these roads were repaired by Manius Aquilius, consul in 129 BC, whose milestones form the earliest official record of Roman rule in Asia Minor (Magie, 1950: 41, 157). Magie also gives other examples of the way in which Roman occupancy of Asia Minor was rapidly followed by the construction of roads.

Map 9 shows these roads. It shows also how the final system of Roman roads moved its emphasis to Istanbul (Byzantium), Izmit (Nicomedia) and Iznik (Nicaea), for from these centres there were roads to the east, the south east and the south. No doubt they combined the purposes of defence, communication and trade, but this map shows also another system of roads whose importance was chiefly strategic. From Adana or Antakya (Antiochia in Syria), it was possible to follow a series of routes to the adjacent reach of the Euphrates, with the important crossings (no doubt by ferry) at Ar-Raqqah (Nicephorium), Zeugma near Birecik and Samsat (Samosata). These were joined by a road along the west bank of the Euphrates and this formed the Limes of Hadrian through Malatya (Melitene) and Zimara, and then on through Sadag (Satala) to Trabzon (Trapezus) on the Black Sea (Webster, 1979: 86; Mitford, 1974; Yorke, 1896). Most of these towns were legionary fortresses.

It was Vespasian who had first fortified the line of the Euphrates. Following the fall of Jerusalem in AD 70, he transferred legionary troops to both Satala and Melitene and a road was built from Syria around the river to Samosata a few years later (Magie, 1950: 1430). In the north, A. Caesennius Gallus built an extensive road system in Pontus, Cappadocia and Armenia Minor (inland from Trapezus) while he was governor of Cappadocia from AD 80 to 82 and this road construction continued for a further twenty years; Mitford (1974: 173) refers to a milestone, possibly of AD 92–4, on the frontier road itself west of Satala. Trajan marched from Samosata to Satala in AD 114, possibly along the frontier road (Mitford, 1974: 171; Magie, 1950: 1464).

The roads further east form a simple system clearly shown on Map 11. Similarly in North Africa, there was the coastal road from Tunisia through Cyrenaica to Egypt, a road system in the delta of the Nile, and roads upstream on either bank. Trade by ship up the Red Sea was important (Charlesworth, 1970), from India and further east. Although these ships could reach to the heads of both of the northern gulfs and did carry out an important trade with Aqaba (Aela), at some times of the year this northerly passage was difficult. It was easier to shorten the voyage and land on the western shore, with the goods trans-shipped by road to the Nile and then carried upstream by boat, and thence by sea to Rome. Maps 10 and 11 show the ports which were established, and the system of roads (Murray, 1925).

The roads in Syria and Palestine (see Map 10) have been described by Chevallier (1976: 141–5) and Avi-Yonah (1977: 181–7) – see also Gardiner (1920), Guest (1899) and Speidel (1987). Antioch (Antiochia) formed part of the earlier Seleucid kingdom and could be reached from the west by pre-Roman roads. The coastal road appears to be the earliest of the roads to the south. The section to Acre (ancient Ptolemais) was constructed in the early days of Nero (AD 54–68) and was probably extended to Egypt by AD 70. When Jerusalem was destroyed in that year it was replaced by Aelia Capitolina and this was linked to the Judaean provincial capital at Qesari (Caesarea) by a branch road, which left the coast road at Qalat Roshha Ayin (Antipatris). At about the same time (prior to Trajan; that is, before AD 98) part of the road east of Tell Bet She'an (Scythopolis) was built. Avi-Yonah argues that there must have been an earlier connection from this road to Damascus and, presumably, from there to the north.

Inland, the road through Palmyra and Bostra formed part of the frontier. Between AD 111 and 114 this was joined to Aqaba (Aela) on the Red Sea by the construction of the great Via Nova by Trajan. There were then, prior to Hadrian, two Roman road systems, one along the coast, the other well inland. By the end of his reign in AD 138, most of the other roads shown on Map 10 were complete.

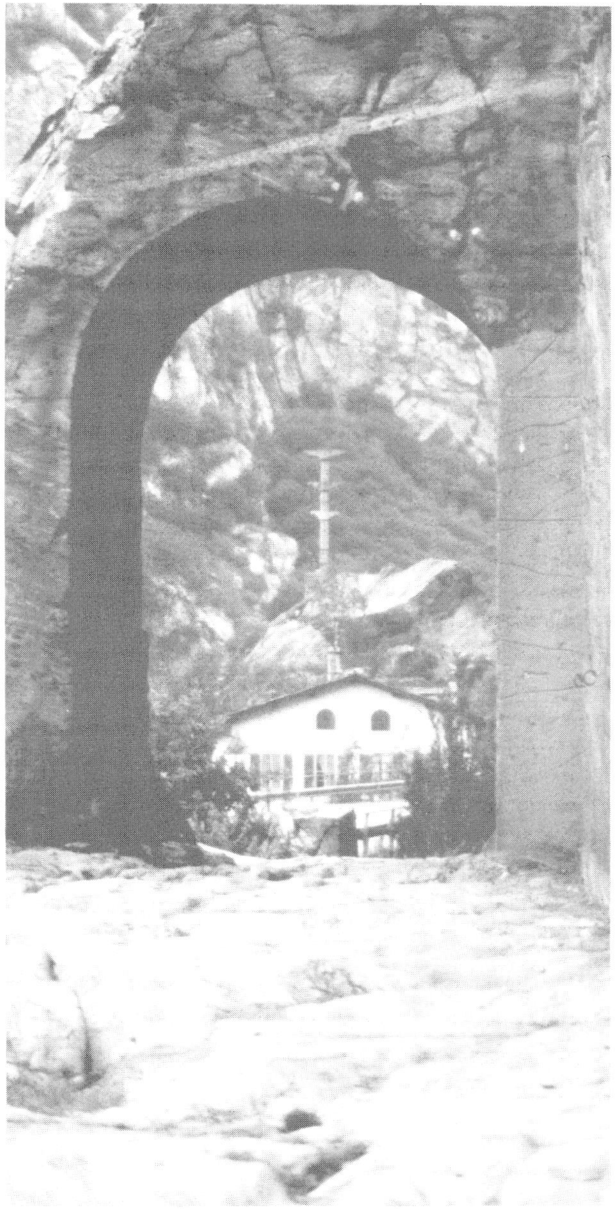

3

BUILDERS OF THE ROADS AND BRIDGES

3.1 The higher authorities

The high officials of both Republican and Imperial Rome included the consuls, praetors, aediles, quaestors and censors (all Roman terms for magistrates).

Following the close of the monarchy, five centuries before Christ, the *imperium* or royal authority passed to a collegiate magistracy of two consuls who served together for one year, each with equal authority, each able to veto the decisions of the other. Although their power was absolute, at least in secular matters, it was moderated in effect by the existence of the Senate, a council of magistrates and former magistrates holding office for life, which came to exercise both a consultative and a supervisory role. It alone had the power to allow a consul, or other high official, to extend his term and to act *pro consule* for a further period in exceptional circumstances. There was no restriction, however, on a consul seeking appointment for additional terms that were not consecutive, and it became the rule under the Republic that these should be separated by at least ten years. The consulship continued in Imperial times as the highest public appointment, even though it had lost ultimate power to the emperor. Augustus himself held the office until 23 BC, but then relinquished it.

In 366 BC, the consuls were given an assistant, the praetor, whose role was to free the consuls from civil administration. There were also four aediles. The original two were plebeian officers active in municipal administration and responsible for certain commercial and trading matters. In 367 BC, two patrician or 'curule' aediles were added and the four had a general responsibility for supervising the streets and markets of Rome, and later for providing entertainment. With time, the distinction between the orders became blurred and, to quote Cary and Scullard (1975: 659), the term 'aedile' came to mean 'commissioner of highways, sewers, public works, etc.'.

The quaestors were essentially financial officers. As early as the time of the kings, there was an *aerarium*, a strong-room or treasury in Rome which stored money in the form of weights of copper or bronze bars, and this was in the care of *quaestores*. Their number grew – in 447 BC there were two, in 409 BC four, and as Rome's empire was established, the first officer under each provincial governor was a quaestor, appointed to receive revenue and act as paymaster. By 81 BC, the number of quaestors had risen to twenty. It is appropriate here to observe that silver and bronze coins were introduced by Greeks into Sicily and Italy about 550 BC, and by 300 BC Italian cities had set up mints to manufacture coinage (Cary and Scullard, 1975: 106–7). Workmen could, therefore, be paid in money, and contracts could be let for civil and other works (Loane, 1979: 79).

Censors – two in number – whose primary role was to maintain a census or roll of citizens, were first appointed in 443 BC. Their authority to rank citizens by class, and to remove them from the roll for unworthy conduct, gave them great power, and as a result, the office was reserved for men of the highest distinction, typically for those who had previously been consul. They also assumed responsibility for the collection of indirect revenue and let contracts for public works.

Wiseman (1970) has listed the names and ranks of the builders of twenty Republican roads: twelve were consuls, six censors and two praetors. He lists also eleven known Republican programmes for road repairs; three were by curule aediles, two by censors. Chapter 2 has identified the builders of many Roman roads, in the Empire and in the Republic, and includes a majority of those named by Wiseman. They may be described as censors, consuls, praetors, emperors, governors and proconsuls (that is, governors with consular rank) and, in a few cases, persons with special imperial authority such as Drusus, stepson of Augustus.

When it is said, for example, that Appius Claudius Caecus, censor in 312 BC, built a road – in this case the Via Appia – what does it mean? It means simply that he decided and directed that the road should be built. It does not imply that he had the technical competence to build the road; nor indeed that he did not, for Appius Claudius was a remarkable man. To quote Cary and Scullard (1975: 79), he 'was a man of immense distinction: he was twice consul, and once dictator [a consular appointment in times of national peril], as well as censor, cultured, literary and expert in legal matters'. He revised the Roman constitution and 'built' not only the Via Appia but also the Aqua Appia, the first aqueduct to supply water to Rome.

Construction of the roads and bridges required three elements: the higher authorities to make the initial decisions; the technical experts to put these into practice; and the labourers to do the actual work. The identification of the second and third of these will be addressed later in this chapter, but it is important here not to dismiss too quickly the possibility of technical competence among those who made the first decision – for two reasons.

The Roman system did not permit an individual to reach the highest ranks without prior training and experience. As early as the fourth century BC, the positions of public office had been arranged into an orderly sequence called the *cursus honorum*. The normal process of advancement began with a period of military service, typically about the ages of 17 to 25; then a quaestorship, or more probably a series of appointments at this level; then aedile, praetor and possibly consul; followed in some cases by censor or dictator. By 180 BC, minimum ages were set for the higher magistracies – 36 for a curule aedile, 39 for a praetor and 42 for a consul. Before these offices were held, there was normally an appointment as quaestor, possibly at a minimum age of 25; it was necessary to be praetor before being considered for a consulship. The actual nature of these appointments would vary, but it is difficult to believe that in no case was there opportunity for technical experience, which could be supplemented by personal study and reading.

The second reason for a belief that there could be technical competence among the Roman higher authorities is that in at least two cases there is evidence that this was so. The first case is that of Marcus Vipsanius Agrippa, friend, colleague and helper of Augustus. In his biography, Wright (1937) quotes Dio Cassius, 'Of all the men of his time Agrippa manifestly was the best,' and adds, 'In his character the chief qualities are those which the Romans called *industria*, *perseverantia*, *fortitudo*, and *prudentia*; a love of work, persistent energy, stubborn endurance, and political foresight.' Born about 64 or 63 BC, he died in 12 BC, before Augustus (63 BC–AD 14). In 23 BC, Augustus, when he was seriously ill, handed his signet ring to Agrippa as a sign that he should succeed him (Wright, 1937: 158), and there is little doubt that in other circumstances this might have happened. Indeed, Augustus arranged later for Agrippa to marry his daughter Julia and intended that the succession would pass to a son from this marriage. In 18 BC Augustus recognised Agrippa as co-regent. Although this cannot be interpreted as implying that they were equal, it shows that Agrippa was the highest of all high officials, subject only to the emperor.

As a young man, in 46 BC he was taken (with Augustus, then Octavius, and Quintus Salvidienus Rufus) by Caesar into Spain and took part in the battle of Munda. Following Caesar's murder in 44 BC, he accompanied Octavius, then known as Caesar's heir, from Apollonia to Italy and aided him in the initial battles with Antony and his brother Lucius. From 39 to 37 BC he was governor of Gaul; in 37 BC he was consul; and from 37 to 36 BC he was builder of the Roman fleet which he led to victory over Sextus and the pirate fleets at Naulochus near Messina: 'the largest and most decisive of ancient naval encounters in western waters', to quote Cary and Scullard (1975: 293). He was with Augustus during his campaigns in Illyria and Pannonia from 36 to 34 BC, and then in 33 BC was appointed aedile at Rome, the longest, and in some ways the greatest, of his appointments. In 31 BC, Agrippa returned to sea and led the Roman fleet in the decisive defeat of Antony at Actium. In 28 BC he was co-consul with Augustus; from 23 to 13 BC he was in charge of the east; from 20 to 16 BC he spent much of his time in Gaul and Spain; and when he died as a comparatively young man, 51 years of age, in 12 BC, it was at the end of a series of travels which he had undertaken for Augustus from 16 BC through much of the eastern countries (Wright, 1937).

There is no doubt then that Agrippa was a successful administrator and leader, but on what grounds may it be suggested that he had some degree of technical competence as well? After Pompey's action in destroying the pirates of the eastern Mediterranean in 67 BC (Section 2.8), Rome had allowed her naval power to decline. Sextus, son of Pompey and ally of Antony, had organised the pirates of the Mediterranean into a disciplined fleet and attacked the corn supplies of Rome. Agrippa's task in 37 BC was effectively to build a fleet from nothing. Not only did he have to build the ships, but he had to locate a spot where they might be built. He chose the Lacus

Avernus, a deep lake inland from the shallower Lacus Lucrinus which joined the sea near Cumae and Puteoli. He dredged the shallow lake, joined it to the Lacus Avernus, and constructed a new harbour. To join Avernus to the town of Cumae he built a road, and this road passed through a tunnel, 1 km long, built by Agrippa. Further, in the battle of Naulochus which followed, Agrippa won partly because of two technical innovations – a collapsible tower on board ship, from which the men could fire down on their opponents, and a new grappling iron, which could be shot by catapult to engage the enemy's ship (Wright, 1937).

In 33 BC, when he was made aedile in Rome, one of Agrippa's first tasks was to clear the *cloaca maxima*, the great drain. It is said that he himself took a boat, and made an underground voyage of inspection. He reconstructed three of the aqueducts of Rome – the Appia, the Anio Vetus and the Marcia – and built two new ones, the Julia and the Virgo (see Section 7.2). He also built the Thermae of Agrippa, the first bath house of the Romans, and many of the great buildings of Rome, including the original Pantheon, which was reconstructed by Hadrian (Wright, 1937; Robertson, 1943: 258). He also 'built' a bridge, or there was a bridge built for him and named after him – the Pons Agrippae – across the Tiber in Rome, whose foundations lie in the stream 160 m north west of the Ponte Sisto (Nash, 1968: 184; Section 5.2).

While he was consul with Augustus in 28 BC, a census of the whole Roman people was taken. The following year, Agrippa rearranged the boundaries of Gaul and established the basic framework of its roads (Section 2.3), radiating from Lyon. During his second governorship, which began in 20 BC, he made Nîmes much more important than it had been. Amongst the buildings he constructed was the Maison Carrée, which is still standing. Nearby is the Pont du Gard, the aqueduct bridge which is one of the greatest of Roman building achievements, generally attributed to Agrippa (Emerson and Gromort, 1925: 27) although it is believed by some to be later (Rivet, 1988: 116 – see also Section 7.3). At about the same time he was responsible for a reorganisation of the administration in Spain, and as in Gaul, played a major part in the establishment of its road system (Wright, 1937).

Finally, his most surprising achievement was the publication of a treatise, now lost, the *Commentarii geographici*, described by Cary and Scullard (1975: 397) as 'probably the best treatise on a scientific subject by a Latin author ... a systematic exposition of the geography of the Roman Empire, in explanation of a large map set up by him in Rome' (see also Dilke, 1985: 39–54). To quote Wright (1937: 165), 'Agrippa wanted by actual measurements to discover the size of each country, the number of its inhabitants, the extent of its material resources', and for this purpose he sent out four Greek surveyors, who returned these particulars to Agrippa, to be classified by his assistants. Wright suggests that Agrippa's map formed the basis of the original Peutinger Table and the Antonine Itinerary mentioned in Section 2.1. The significance of this achievement in the present context is that it shows something of Agrippa's mind, and suggests that he was far more than an administrator, that he was, in fact, deeply interested in the detailed aspects of the work which he controlled.

Sextus Julius Frontinus is the second example of a high official who was also technically competent. Indeed, he is known more for his technical work than for his administration, for it was Frontinus who wrote *De aquis urbis Romae*, two books on the water supply of the city of Rome. In 1897 Clemens Herschel discovered a manuscript of this book dating from about the twelfth to the fourteenth century. He published a translation, and with this included a biography of its author (Herschel, 1913: 4–12, 75–101; Eck, 1985: 141–2; see also text by Kunderewicz, 1973).

Frontinus is believed to have written seven works – a treatise on surveying; 'The Art of War'; 'Strategemata', on military science; essays on farming; a treatise on boundaries and roads; a work on the Roman colonies; and the two books of the *De aquis*, on Roman water supply and the aqueducts of Rome. The first and last are clearly technical; the first follows an earlier work on surveying by Hero of Alexandria (Drachmann, 1963; Dilke, 1971; see Section 4.7); both were probably included in 'an official library' of Roman handbooks, of which most have since been lost.

Frontinus was born about AD 35, possibly in Sicily. His knowledge of Hero's work led Herschel to speculate that he might have been educated in Alexandria. In AD 70 he was appointed *praetor urbanus* for the city of Rome and was probably consul in AD 73. About this time he wrote his treatise on surveying. From AD 74 to 78 (Cary and Scullard, 1975: 420) he was governor of Britain (Section 2.6, see also Tacitus). He then returned to Rome, but from AD 78 to 97 his career is largely obscure, except that it appears he was a successful general, and was in Germany with Domitian in AD 83 and proconsul of Asia in AD 84. His best known appointment was by Nerva in AD 97, when he was made *curator aquarum* or commissioner for the water supply of Rome, a position established by Agrippa (see Rodgers, 1982). He retained this appointment while he was consul for the second and third times in AD 98 and AD 100, and died about AD 103 or 104, at an age of approximately 69 years. His book, *De aquis*, was written about AD 97.

Frontinus describes his attitude to his work in the following words, taken from Herschel's translation (Herschel, 1913: 27) of the *De aquis*: 'I therefore consider it to be the first and most important thing to be done, as has always been one of my fundamental principles in other affairs, to learn thoroughly what it is that I have undertaken ... It is for this reason that I have set down in this commentary all that I could gather as bearing on the subject matter.' He was a man, then, of intelligence, with a disciplined and enquiring mind.

The younger Pliny (C. Plinius Caecilius Secundus) also showed an interest in technical work. In AD 104, at the age of about 42, he became *curator alvei Tiberis et riparum et cloacarum urbis*; that is, commissioner for the bed and banks of the Tiber, and for the city drains. To quote from the introduction by Radice to the *Letters of the Younger Pliny*, he 'was responsible for keeping the river banks in repair to prevent flooding and for maintaining Rome's sewers. Pliny must have enjoyed this; practical problems of water supply and drainage had always interested him and he probably knew the works on civil engineering of his friend Frontinus' (1963: 15). He had previously been a lawyer from AD 80, senator and quaestor in 88 or 89, tribune of the people in 91, praetor in 93, treasury official from 94 to 96, and from 98 to 100, and then consul.

In AD 110–11, Pliny was sent by Trajan as special commissioner to Pontus-Bithynia (Cary and Scullard, 1975: 430), and his correspondence with Trajan from then until his death in AD 113 has been recorded (Sherwin-White, 1966). They refer on a number of occasions to technical matters. For example, Book 10.37 of Pliny's letters refers to neglect in the construction of an aqueduct to Nicomedia.

> I have been myself to look at the spring which could supply pure water to be brought along an aqueduct, as originally intended, if the supply is not to be confined to the lower lying parts of the town. There are very few arches still standing, but others could be built out of the blocks of stone taken from the earlier construction, and I think some ought to be made of brick, which would be easier and cheaper. The first essential is for you to send out an engineer (*aquilex*) or an architect (*architectus*).

In Book 10.39 he makes a similar request concerning public works in Nicaea and Claudiopolis. Trajan replies, 'You cannot lack architects: every province has skilled men trained for this work. It is a mistake to think they can be sent out more quickly from Rome when they usually come to us from Greece' (Pliny 10.40).

Another reference to engineering work is in Letters 41 and 61, referring to the connection of a lake near Nicomedia with the sea, to provide access for shipping. The water in the lake was well above sea level, and Pliny (10.61) addresses a concern expressed by Trajan (10.42) that if they were connected the lake might drain away.

> If we decide to cut a longer canal, deepen it and bring it down to sea-level so that the water will flow direct into the sea, instead of via the river, the counter-pressure from the sea will check the outflow from the lake. Even if we had none of these advantages we could manage to regulate the flow of water by sluices. But these and other details can be much more accurately worked out in a survey by the engineer (*librator*).

Trajan replied, 'I am sure Calpurnius Macer will not fail to send you an engineer (*librator*)' (Pliny 10.62).

It would be difficult to infer that the younger Pliny was himself a technical man, but his interest is evident and the references to engineers and architects are of some value: they will be discussed further in Section 3.2.

Finally, it may be seen that even the emperor Vespasian did not see himself as being above the manual side of technical work (Suetonius 8.8.9). 'He began the restoration of the Capitol in person, was the first to lend a hand in clearing away the debris, and carried some of it off on his own head ... He also undertook new works.'.

3.2 Bridge builders

The suggestions made in the preceding section, that some at least of the Roman higher authorities were technically competent, cannot be taken to imply that they were in direct control of the building of roads, or more particularly of bridges – for one thing, they would not have had the time. To determine who provided the technical leadership in the design and construction of bridges is surprisingly difficult. In only a few cases is the name of a bridge builder known, and it is only rarely that his career may be determined. The aim of this section is not chiefly to list names – although it is appropriate that they should be acknowledged – but rather to elucidate what kind of man was responsible for Roman bridge construction, for the capacity and experience of those in technical leadership largely determines the nature of the design process that was undertaken.

The words used by Pliny in Section 3.1 for engineer or architect have the following meanings: *architectus*, a master-builder or architect (literally, a leader of the builders); *aquilex*, a conduit master, a man employed to find sources of water; and *librator*, a leveller or surveyor. The connection between engineering and the construction of aqueducts is obvious. Procopius uses also the Greek words *mechanikos* and *mechanopoios* (see, for example, 1.1.24, 2.3.2, 2.3.11). In his translation, Downey replaces these by the English word, 'master-builder', and discusses this choice in pages xiii to xv of his introduction. He adds,

> By long tradition each Roman and Byzantine Emperor was regarded as the direct source and origin of all the building operations executed during his reign ... the assistance which he received from his master builders is but another manifestation of God's watchfulness in providing the Emperor with the best means for the execution of his mission, which was, as Procopius says, 'to watch over the whole Roman Empire and, so far as was possible, to remake it'.

Procopius also uses the word *architecton*, which he applies (uniquely) to Apollodorus of Damascus (see p. 142), who built the great bridge over the Danube for Trajan (4.6.11–18). He names Anthemius of Tralles, Isidorus of Milesia, and Chryses of Alexandria as master-builders (1.1.24, 50, 70; 2.3.1–15), but none in connection with bridge building, even though the building of bridges is described in various places in his book (for example, 5.2.13; 5.3.5–6, 8–9; 5.4.3; 5.5.4–20). Indeed, in connection with the bridge of Apollodorus and the adjacent town called Pontes, he adds (4.6.16), 'for the Romans call a bridge (or *gephura*) *pontem* in the Latin tongue'.

It is not uncommon to find one or more inscriptions on

a bridge, but these more commonly record the name of the emperor or other dignitaries rather than the builder of the bridge. Some examples may be given. The Ponte d'Augusto at Rimini (Section 5.4, no. I117) lay at the beginning of the Via Aemilia and the end of the Via Flaminia. Although it is more truly part of the Aemilia than the Flaminia, it was begun in the time of Augustus, who built the Flaminia, and completed in AD 20 by Tiberius, his successor. Both names appear on the bridge. The fine but little-known bridge over the River Bibey in the north west of Spain (Section 5.7, no. SP36) has two original inscribed columns remounted at its eastern end, with transcriptions beneath. One is a milestone; the other refers to the reconstruction of the bridge and ends:

IMP.CESAR.DIVI
NERVAE.FILIVS
NERVA.TRAIANVS
AVG.OPTIMVS
GERMANICVS

The reference is, then, to Trajan, adopted son of Nerva, 'who reconstructed this bridge at his expense'. It was probably built about AD 98. Finally, the magnificent aqueduct at Segovia (Section 7.4, no. AQ60) has an enlarged panel half-way along with words now partly effaced by time, but believed to include (Casado, 1972: 71)

TI.CLAUDIVS.PONT.MAX. — OMNIVM FECIT.

It was built in the time of the emperor Tiberius Claudius, and the words of ascription appear to attribute the construction to him.

Table 2 lists a few cases where the name of the builder may be known, but even this short list probably includes some where the identification may be incorrect. It is not known (Section 3.1), for example, whether Agrippa actually built the bridge named after him.

The Ponte Lucano is another such case (Section 5.2, no. I17). Now spoilt, it was until recently a fine, picturesque bridge, standing beside the Tomb of the Plautii, a circular stone tower in the valley below Tivoli, Roman

Table 2. *Bridge builders*

Sequence	Date	Name of bridge	Name of builder	Country	Reference number*
1	65 BC	Pons Minucius	Minucius Thermus	Italy	I85
2	62 BC	Pons Fabricius	L. Fabricius	Italy	I3
3	62–27 BC	Pons Cestius	Cestius	Italy	I4
4	33–20 BC	Pons Agrippae	Marcus Vipsanius Agrippa†	Italy	
5	18 BC	Pont du Gard	Marcus Vipsanius Agrippa†	Gaul	AQ30
6	12 BC	Pont Flavien	C. Donnius Flavos	Gaul	G2
7	2 BC	Ponte Lucano	M. Plautius Silvanus†	Italy	I17
8	3 BC	Ponte del Pondel	C. Avillius C. F. & C. Aimus Patavinus	Italy	I159
9	AD 104	Puente de Alcántara	Caius Julius Lacer	Spain	SP21
10	AD 104/105	Bridge of Apollodorus	Apollodorus of Damascus	Romania	T13
11	*c.* AD 384	Taşköprü (Adana)	Auxentius	Turkey	E28

* See Chapters 5–7 † Uncertain

Tibur. Its name derives from the belief that it was built by M. Plautius Lucanus, one of the Plautii, in the late Republican period. Blake (1947: 213), however, points out that the name Ponte Lucano did not appear until the twelfth century, and prefers to ascribe the bridge to M. Plautius Silvanus, named in an inscription on the adjacent tower. He was consul in 2 BC, which raises again the question of whether he is named as the appropriate authority rather than as builder of the bridge.

The Ponte del Pondel or Il Pondel (Section 5.5, no. I159) lies in the valley of the River Cogne, west of Aosta. It is a remarkable bridge for a number of reasons, one of which is that it was built privately to provide access to a privately owned mine. It carries the word PRIVATVM, possibly in three places, and the inscription (Blake, 1947: 218; *CIL* v(2).6899)

IMP.CAESARE.AVGVSTO.XIII.COS.DESIG
C.AVILLIVS.C.F.C.AIMVS.PATAVINVS

Its builders, then, were C. Avillius C. F. and C. Aimus Patavinus, but nothing further is known of them.

The bridge listed as no. 10 in Table 2 is the famous bridge of Apollodorus (Section 6.3, no. T13), which was built by Apollodorus of Damascus for Trajan about AD 104–5 across the Danube, to aid his entry into Dacia. One of the greatest, if not the greatest of all Roman bridges, it is recorded as having twenty-one timber arch spans, each 32.3 m clear. Unfortunately, its superstructure was removed by Hadrian but parts of its substructure still exist. It was segmental in form. Bruce (1978) has speculated that the Pons Aelius in Newcastle upon Tyne was similar, although this is not known. The stone bridge at Alconétar in Spain (Section 5.7, no. SP20) also had segmental arches and was built about the same time. For these reasons Casado (n.d.: 242–3) has suggested that this was also built by Apollodorus.

Apollodorus was a Greek. His work was carried out chiefly under Trajan and included the Forum of Trajan, the Baths of Trajan and Trajan's column, all in Rome. Loane (1979: 84) describes him as one of the Greek *peregrini*; de Camp (1977: 220f.), Robertson (1943: 258) and others refer to him as an architect. Following Trajan's death in 117 BC, he continued to work with another architect, Decrianus, but was banished and then put to death by Trajan's successor, Hadrian (Dio Cassius 69.4; Lepper and Frere, 1988: 187–93). His worth can best be measured by the quality of his work, particularly the Danube bridge which shows him to be practical, innovative and with a sound knowledge of structural behaviour. As discussed in Section 6.2, the column of Trajan is more than an architectural feature, for it is carved to form a continuous spiral of 155 scenes. Richmond (1982: 2–5) suggests that these scenes, although influenced by Trajan's Dacian campaign, were chosen to show the army at work: 'they illustrate . . . the adage that the Roman army owed as much to the spade as to the sword'. If they were chosen by Apollodorus, then they also tell us something of his knowledge and experience.

Number 6 of the bridges listed in Table 2 is the Pont Flavien, on the Via Julia Augusta in south-east France (Section 5.6, no. G2). A small bridge, with a single span of 12.3 m, it is decorated at the ends by two fine triumphal arches, one of which carries the inscription (Emerson and Gromort, 1925: 30; Rivet, 1988: 205, 210; *CIL* XII. 647)

C.DONNIVS.C.F.FLAVOS.FLAMEN.ROMAE
ET.AVGVSTI.TESTAMENTO.FIEREI.IVSSIT.

It was, then, built according to the instructions of the will of Donnius Flavos, a priest of Rome and Augustus. Despite the comments in Chapter 1, the reference to a priest is surprising. The bridge itself is of minor significance compared with its triumphal arches, and it may be that the inscription refers as much to these as to the bridge itself.

Bridge no. 5 is the Pont du Gard, generally attributed to Agrippa, as recounted in Section 3.1. The ninth bridge listed is at Alcántara in Spain (Section 5.7, no. SP21), and was built about AD 104 by C. Julius Lacer for Trajan. The attribution is based on an inscription over the entrance to a small temple at the southern end of the bridge, quoted in Section 5.7 and given in full by Casado (n.d.) and the *Corpus Inscriptionum Latinarum* (*CIL* II. 761). Unfortunately, as in so many of these cases, nothing further is known of him.

An inscription held in the museum at Adana states that bridge no. 11 was built by Auxentius (Akurgal, 1978: 345). The bridge itself is mentioned by Procopius (*Buildings* 5.5.8–13), who speaks of it as 'an enormous and very notable bridge'. He attributes it to Justinian (AD 527–65), but this is probably incorrect. Akurgal says that it was 'probably built by Auxentius (*CIG* 4440); he was perhaps identical with the "*comes et mechanicus*" Auxentius, who lived and worked around AD 384 in Rome, as a constructor of bridges'.

The remaining three bridges, nos. 1, 2, and 3, have this in common, that they were built by men bearing the title *curator viarum*. The evidence for this follows.

The Pons Minucius (Section 5.4, no. I85) no longer exists. It lay on the Via Flaminia, which, having crossed the Tiber over the Pons Mulvius as it left Rome, crossed it again near Otricoli, where there is now no visible trace of a bridge. Ballance (1951: 91) quotes Nissen as suggesting that this crossing was by the Pons Minucius, 'presumably a bridge of some importance, perhaps built by the Minucius Thermus who is mentioned by Cicero in a letter of 65 BC as being at that time *curator* of the Flaminia' (see Cicero, *Ad Att.* 1.1.2). He goes on to say, 'This identification seems as likely as any other.' Wiseman (1970: 145, 149) is a little stronger. He includes the work of Minucius, ex-praetor, curator of the Via Flaminia in 65 BC, in his list of Republican repair programmes. He adds, 'Whether the Tiber bridge built by Thermus in 65 replaced an older one, or a ferry, is not certain' (1970: 149). Ashby and Fell (1921: 127), writing before Ballance, regard the site of the Pons Minucius as unknown but state that it was probably built by Q. Minucius Thermus. They quote Cicero, who wrote in 65 BC that this man was a strong candidate for the consulship because of his office as *curator* of the Via Flaminia. There is good reason, then, to believe that, although the precise location may be uncertain, there was a Pons Minucius across the Tiber on the Via Flaminia, and that this was built by Minucius, a *curator viarum*.

The Pons Fabricius is the only bridge of this group which may still be seen in its original condition. In Rome, it joins the left bank of the Tiber to the Isola Tiberina, the Island in the Tiber (Section 5.2, no. I3). Both faces of the left or north-eastern arch carry the inscription (*CIL* VI.1305)

L.FABRICIVS.C.F.CVR.VIAR
FACIVNDVM.COERAVIT

The adjacent small span over the central pier adds the words

IDEMQVE.PROBAVIT

It was therefore L. Fabricius, *curator viarum*, who had it built and approved it.

These inscriptions leave no doubt that Fabricius was in some sense responsible for the construction of the bridge, for his name would not otherwise have been there at all. They do, however, leave his exact role unclear. The bridge carries a further inscription naming the consuls of 21 BC, M. Lollius and Q. Lepidus, who apparently at that time repaired damage due to floods.

Bridge no. 3, the Pons Cestius (Section 5.2, no. I4), proceeds from the Isola Tiberina to the right bank of the Tiber. The present bridge is a reconstruction of the Pons Gratiani, built in AD 370 to replace an earlier Pons Cestius. To quote Nash (1968: 187), 'Its builder was presumably a certain Cestius, who was *curator viarum* between 62 and 27 BC.' Nothing further is known of him (see also Platner and Ashby, 1929: 399–400).

To these three may be added the reference by Wiseman (1970: 145) to C. Vibius Temuudinus, urban quaestor and *curator viarum* in the early first century BC, who carried out repairs of bridges and of the gravel surface of the Via Caecilia. He also mentions the bridges built 'in many places' by the censors of 174 BC.

Ertman (1976) has attempted to list all those who held the office of *curator viarum* and has given an authoritative account of their function. She lists six (p. 14) who held the office under the Republic, including Minucius and Fabricius but not Cestius, and, interestingly, C. Julius Caesar, but argues that these were special rather than regular appointments. Cary and Scullard (1975: 321) name five offices of *curator* established by Augustus about 20 BC as part of his new executive; all had responsibilities of an engineering kind. They were

(a) *curatores alvei Tiberis*, commissioners for the bed of the Tiber;
(b) *curatores riparum Tiberis*, commissioners for the bank of the Tiber;
(c) *curatores aquarum*, commissioners of aqueducts;
(d) *curatores operum publicorum*, commissioners of public works; and
(e) *curatores viarum*, highway commissioners.

The *curatores viarum*, who, to quote Ertman, existed only in Italy (with one possible exception), had the responsibility of maintaining, rather than of building roads; but this maintenance included the construction of bridges, for many of the bridges described in this book are clearly second-generation structures, built to replace either fords and ferries or earlier bridges.

There is no doubt that a *curator viarum* was a senior official and it may well be argued that the ascription of a bridge to a *curator viarum* carries no further implications than those of the previous section. Ertman (1976) quotes a statement by Frontinus (*De aquis* 2.64) as indicating that 'a *curator* did not necessarily have other than casual knowledge of the public works which he might be concerned to maintain'. She goes on, 'Frontinus, nonetheless, took his *curatio* very seriously' and it may be shown that the career structure for at least some of the *curatores viarum* was such as to allow them to have developed technical competence.

It is particularly helpful that Ertman has listed the *cursus honorum* for a great many of those who held this office: six Republican and 107 Imperial curators (thirteen of the latter held the office twice). Some statistics may be noted. For the *curatores* of Imperial times, their *cursus* included typically eight or nine appointments, of which *curator viarum* was fifth (these are average figures). The position was, then, a relatively high office. This can be shown in other ways. More than half of those who held it were also praetor. In every such case the position of *curator viarum* followed that of praetor, typically two appointments later. Many of the holders also became consul. Although a few became consul first, more normally the position of *curator viarum* was a precursor to this appointment, with consul coming shortly afterwards, typically as second or third subsequent appointment. Thirteen advanced further, to the distinguished appointments of proconsul of Asia or Africa, or *praefectus urbi* – prefect in charge of the city of Rome (Ertman, 1976: 314–15).

Even more important in the present context is the nature of the other appointments. Nine either had been or later became *curator rei publicae*; seven became *curator operum publicorum*; one became *curator alvei Tiberis et cloacarum sacrae urbis* (that is, of the bed of the Tiber and of the drains of Rome) and another *curator aquarum*. More directly again, two held concurrent offices – *curator viarum et pontium Umbriae et Piceni*, or curator for the roads and bridges of Umbria and Picenum (Ertman, 1976: 256, 258).

These appointments were chiefly civil (Ertman, 1976: 318–9). It is interesting to note, however, that two men from Ertman's list had also held the office of *praefectus fabrum* (Ertman, 1976: 209, 257), a high army appointment. The class of *fabri* in the Roman army will be discussed further in the next section, but to quote Smith (1901–14: 821) *fabri* were workers in wood, stone or metal and were the 'engineers in the army'. They fell under the command of an officer called the *praefectus fabrum*, who was either the officer of a legion, or 'more probably – the engineer-in-chief to the army as a whole'. There is every reason to believe that the *fabri* and their officers provided Rome with a major source of engineering expertise (MacMullen, 1963; Webster, 1979; Watson, 1969; et al.). The following is taken from MacMullen.

> But the legions provided also a school of engineering. The high position of this science is suggested by the use of the word, *philosophus*, as the equivalent of 'building expert' or 'technician' ... Architecture, philosophy, mathematics, physics and a number of allied disciplines, so far as the ancient world knew them, formed an almost indivisible association of great prestige; and its practitioners were both rare and valuable. Many were soldiers. (1963: 34–5)

In his description of Trajan's Canal at the Iron Gate, Sasel (1973) has provided an example of the work of a *praefectus fabrum*.

> All these constructions [the road in the cliff face (see Section 2.7), the canal, flood gates, watchtower and fort] were carried out by technical units from the Moesian garrisons ... The question of the identity of their commander can also be answered. He was C. Manlius Felix ... whose service as *praefectus fabrum* on the emperor's staff falls in the years 100 to 101 (AD) ... His next post was *praefectus classium Pannonicae et Germanicae*.

He goes on to say, 'Those who executed his plans on the individual work-sites were army units, directed and advised by specialist technicians and *libratores* ... For instance, the rock-cut tow-path at the old Tabula Traiana was executed by *lapidarii* of both Moesian legions.' (See also Jarrett and Dobson, 1966.)

Many of the Roman roads were built initially for military purposes (Chapter 2), and their construction would have involved the army (see also Section 6.2). The army was also involved in civil works. To quote Webster (1979: 272),

> But the army was also the main source of the technical skills needed for large-scale engineering projects. Engineers, architects and surveyors would doubtless have been made available for advice and technical assistance ... This would apply specially to those provinces, such as Britain, where these skills were non-existent. There is, however, little direct evidence of army participation in the early period, except on road and bridge building (see Tacitus, *Ann.* 1.20).

The army, then, provided a major part of the technical leadership in bridge construction (see next section). Nevertheless, the bridge builders listed in Table 2 appear to be predominantly civilians. This may be simply because they were the appropriate higher officials; it may be by chance, or alternatively by policy, for it may not have been considered appropriate to name soldiers on bridges.

3.3 The provision of skilled labour

The army provided three things – expertise, materials and labour. Watson (1969: 144) describes Roman armies as 'torchbearers of civilization', 'in possession of skills and technical expertise which were uncommon in the surrounding population', able 'to carry out tasks which would be of immediate practical value to the local community as well as to the Empire as a whole'.

Many of these skills are described in Chapter 4. This refers also in passing (Section 4.2) to the use of army manpower to provide stores of valuable materials, in this case metal implements and fasteners (Curle, 1911). MacMullen (1963: 31–4; see also Watson, 1969: 144) speaks also of the use of the army in quarries to produce stone, either in the supervision of labour or to provide the labour itself. 'In many of the tasks ... soldiers appear as a source of mere muscle. Their labours were considered healthy from the point of view of discipline, desirable from the point of view of the treasury, and essential in problems requiring enormous numbers of men.'

Surprisingly, the army had no separate specialist engineering corps. Specialists were not separated out but the legion as a whole was responsible for all building and engineering work (Webster, 1979: 118; Watson, 1969: 75). Further, these specialists were not so much given a special rank but rather a special privilege – they were *immunes* from heavy fatigues (Tarruntenus Paternus, *Digest* 50.6.7, quoted in Watson, 1969: 76). They were, then, classed as *immunes* and this immunity had a special significance, for they were thereby exempt from the arbitrary control of the centurion. The *immunes* included a variety of specialists: the *haruspices* or priests, for example, were *immunes*, but some of those of engineering interest were *mensores* (surveyors), *architecti* (master-builders), *naupegi* (shipwrights), *fabri* (craftsmen in the *fabricae* or workshops), *aerari* (bronzesmiths), *carpentarii* (carpenters), *aquilices* (hydraulic engineers), *plumbarii* (leadsmiths), *ferrarii* (blacksmiths), *lapidarii* (stonemasons) and various *librarii* (book-keepers or clerks). The *optio fabricae* was the optio (the rank below a centurion) in charge of a workshop (Webster, 1979: 119; Watson, 1969: 76–86).

The term *principalis* was also used, either as an alternative to *immunis*, or as a level of promotion (Watson, 1969: 78). It is clear that this was followed by further promotion paths to centurion and beyond. Breeze (1974) has discussed this career structure. The highest engineering rank appears to have been *praefectus fabrum*, or engineer-in-chief, as mentioned above either within a legion or for the army as a whole (Smith, 1901–14: 821).

Clearly the army built bridges, especially in remote areas, strategic sites and the smaller towns; it would have been less active in the towns, particularly in Italy.

Slaves provided a second source of labour. To quote Finley (1973: 79) in his excellent chapter on 'Masters and slaves', classical Greece and Italy were slave societies. The number of these slaves was great. Finley says, 'Caesar alone is said to have been responsible for one million during his campaigns in Gaul between 58 and 51 BC, a not wildly incredible figure.' Again, estimates of the numbers of slaves in classical Athens vary widely, but lie in the range 20,000 to 400,000. 'The impossibly low estimate of 20,000 slaves ... gives a ratio of slaves to citizen households of not much below one to one.' Finally (p. 191, n. 20), 'The slave population of Italy may have been twice that of adult male citizens at the death of Caesar.' Loane (1979: 11) estimates the numbers of slaves in Rome in the first and second centuries AD as about one million (see also Hopkins, 1978).

It would be incorrect to regard slaves as cheap, harshly oppressed, unskilled labour. First, their labour was not free in economic terms, for they had to be housed and fed. Secondly, they were not necessarily oppressed. Both Finley (1973: 64) and Barrow (1968: 100) refer, for example, to the *peculium*, property in one form or another assigned by the owner to a slave for his use and management. To quote Finley again, 'those who had a *peculium* were working independently, not only for their owners but also for themselves. And if the business were on any scale above the minimal, their *peculium* was likely to include other slaves along with cash, shops, equipment and stock-in-trade.' He argues that the Roman slaves were neither truly slaves, in the sense that they were simply property, nor truly free. A slave could be highly trained, for not only might he have been already highly trained when taken captive by the Roman armies at war, but there is evidence that slaves often took pride in their work and profited from it. Indeed, they could manage the work, for 'management throughout the classical period, Greek as well as Roman, urban as well as rural, was the preserve of slaves and freedmen, at least in all larger establishments' (Finley, 1973: 75–6, 82).

Slaves were employed in public works. Agrippa employed 240 slaves to guard and maintain the aqueducts (Loane, 1979: 80), and this number rose to a permanent staff of 700 slaves, including slave architects (Finley, 1973: 75).

However, this example is of permanent employment. For specific tasks, and this would include bridge building, freedmen were also used. Finley (1973: 192, n. 33) quotes Frank as saying,

> That free builders continued to make their livelihood in the capital is proved by the unusually large and active *collegium fabrum tignuariorum* [a carpenters' association] ... From a study of the membership lists it seems likely that these 1000–1500 free or freed *fabri* were successful carpenters who controlled the services of numerous slaves ... It is probable, then, that in public works large numbers of free labourers were employed.

Loane also makes this point (1979: 81), but in both cases the reference is to the city of Rome. Although the use of free labour in rural areas and the provinces may not have been as great as in the city, it did exist there. Chevallier (1976: 84) speaking generally of road construction, says, 'To carry out the work there must have been a work force of some size, provided by the army and helped, of course, by conscripted civilians, as Cicero's *Pro Fonteio* tells us (7.17). All were obliged to share in the work ... Nobody was exempt.' Indeed the tax system, as applied to many of the population, required the provision of services by labour as well as in cash or in kind (Cary and Scullard, 1975: 501, 531), as in the *corvée* or statute labour of the later feudal law. This labour could be used for public works, the *operae publicae*, including the upkeep of roads and bridges and, no doubt, for their construction in some cases.

3.4 Some closing observations

The previous sections provide only a little evidence of those who were in direct charge of the construction of bridges – those who provided technical leadership. By inference they could be private contractors, as possibly Avillius and Patavinus were at Pondel, skilled slaves, *peregrini* like Apollodorus, possibly high Roman officials, like the *curatores viarum* in some cases, and undoubtedly army engineers in the more remote areas, as one would guess Lacer of Alcántara to be. They do, however, provide clear evidence that there was ample opportunity for the transmission of information: through the interest of some high officials, through the promotion structures in the civil corps and the army, and through the opportunities made available to slaves of ability. It is suggested that this would have led both to the presence of a skilled technical leadership and to the growth of technology.

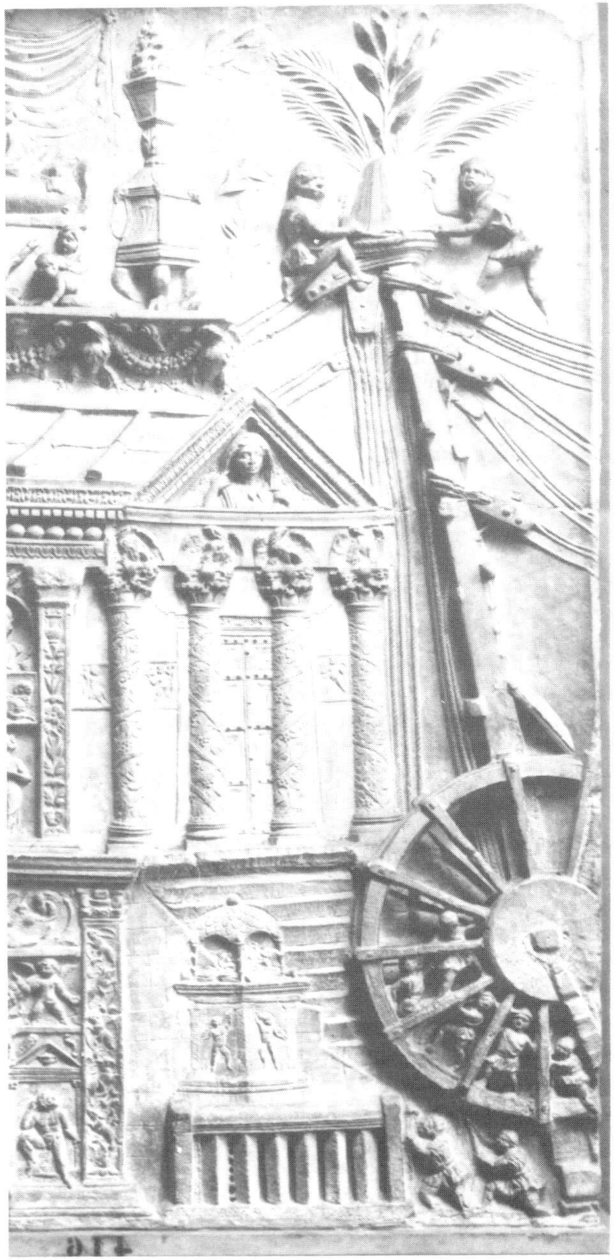

4

Roman technology

4.1 Introduction

A designer must work within the constraints imposed by his materials, and his ability to shape, transport and handle them.

The form of a bridge is largely influenced by material characteristics. Timber, for example, consists of longitudinal, hollow fibres, parallel to the original length of the tree trunk and with weak linkages in the radial and circumferential directions. It possesses both tensile and compressive strength in this dominant longitudinal direction. Its transverse and shear strengths are small and it can be readily split longitudinally but not transversely. Its major dimension, limited by the height of the tree, is longitudinal. Its transverse dimensions are relatively small and it cannot easily be bent. It is best used, therefore, in the simple girder bridge, with horizontal straight members spanning between abutments and piers and supporting transverse planking which serves as the deck. Stone is essentially different. Its major strength is compressive. It is heavy, and this, together with the difficulties of its procurement, means that it is usually available in comparatively small pieces. Although the direction of the bedding plane is important with regard to durability, it does not dominate the material to the same degree as the longitudinal direction dominates the piece of timber. The compressive strength of stone is typically great in all three directions and in none of these three directions can the length of the stone be easily extended. Stone is best used, therefore, in small pieces, and in a manner which capitalises on its heavy weight, in either the characteristic Roman stone arch or the vertical masonry wall. Without support it can scarcely be used to form a bridge deck, but forms a suitable wearing surface when placed on earth or some other supporting material.

It is not only in this fundamental sense that the designer is controlled by his material: the entire feasibility of his work is dependent on all aspects of material procurement. If the material is timber, then the designer is constrained by its availability, type and size, and must have at his disposal methods for cutting, shaping, transportation and fixing. The same types of constraints apply also with stone or other materials.

A study of Roman bridges is incomplete if it considers them only in their finished form. Ideally it should retrace the steps of the designer through all stages of his work from site selection to the finished bridge. The evidence available today is such that, in the absence of a written text, it is hardly likely that this re-enactment can be carried out with precision for any particular bridge. At the least, however, it is necessary to identify for all stages of the work feasible procedures that are either known to have been available at that time, or at least by reasonable conjecture may be assumed to have been available. A most salutary exercise is to imagine oneself at a virgin site – perhaps a rocky gorge, or again perhaps at a wide sandy river bed interrupted by a stream meandering at its low level flow – and to enquire what must be done before the bridge can be built. The first question is surely to ask: What is to hand? With what materials, tools and equipment can we build this bridge?

The problem of bridge construction is closely linked, therefore, with the question of available technology. The purpose of this chapter is to review this technology. First of all, the question of tools and fastenings is considered and it is suggested that the use of iron was crucial to Roman technology. Attention is then given to materials used for bridge building – stone and related materials, timber, metals and concrete – and the methods used to shape, handle and join them. Lifting equipment in the form of cranes, lifting tackle and jacks is studied, together with associated special purpose machinery. Finally, attention is given to methods of surveying and dimensional control, and to computation. The matters of centering and bridge erection are reserved for a later chapter.

The ten books by Marcus Vitruvius Pollio form a Roman manual for some of the matters considered here. The quotations are from the translation by Granger (1931–4) in the Loeb Classical Library. The translation by Morgan (1960), and the books by Gwilt (1860), Fleury (1990) and Plommer (1973) are included in the references.

4.2 Tools and fasteners

The nature of the Roman tools can be determined both from carved representations of artisans at work, and from samples which have been preserved to this day. They include the hammer, axe, adze and pick; the knife, scythe, spokeshave and plane; and the chisel, drill and file (see, for example, fig. 206 of Singer et al., 1954–84, vol. 2). The best proof of their efficiency is the quality of the work they produced, particularly in such fields as furniture making and fine metalwork (Singer et al., 1954–84, vol. 2). The same basic skills, modified by the pressures of economy and scale, were applied also in other fields such as building and bridge construction. In this latter field, the availability of tools in hardened iron was essential, for picks and chisels in copper and bronze were inadequate for the working of stone.

It is appropriate in the context of bridge construction to see what the Romans could and could not do. Clearly, they could cut down a tree, shape it into a member and then saw, notch or drill it. In the case of stone, they could cut, shape and handle it, using techniques that were essentially similar to those used in the smaller quarries today – as will be described in Section 4.4 of this chapter.

It has been suggested (see, for example, Singer et al., 1954–84, vol. 2: 604, 636, 638) that the Romans had a poor regard for manual and practical skills and that their techniques were taken from others. Both observations appear to be of doubtful validity, at least in the field of civil engineering, for their vigorous construction programme is, in itself, strong evidence of the esteem in which it was held, and it is hard to believe that those responsible for these major works did not themselves add to the body of knowledge they received from others. What is certain is that the Romans had no shortage of ideas, principles and skills.

Their limitations lay in two major areas: they were effectively without mechanical power, and they were restricted in their ability to work hard metals. Both need some qualification. Forbes (in Singer et al., 1954–84, vol. 2: 598–9 and Forbes, 1965, vol. 7: 176) describes the use of the water mill by the Romans, for example, in the corn mills at Arles in France (see also Rivet, 1988: 196, pl. 29). He also quotes Ausonius, speaking in the later empire of the stone saws on a tributary of the Moselle near Trier, 'As he turns his mill-stones in furious revolutions and drives shrieking saws through smooth blocks of marble, hears from either bank a ceaseless din' (Ausonius, *Mosella* 362–4). These references to the use of waterpower show that the Romans had at least this skill, but it appears that they did not choose to use it to any great extent. To quote Forbes again (Singer et al., 1954–84, vol. 2: 601), 'During the early Empire unemployment was still a serious problem. Vespasian (AD 69–79) is said to have refused the building of a water driven hoist "lest the poor should have no work"' (see Suetonius, *Vespasian* 18). And again he says, 'Mass production of goods would in fact have meant over-production in the ancient world, which invested its capital in shares and land' (p. 604). In place of mechanical power, the Romans chose to use the large reservoirs of labour provided, at some considerable cost, by their army and slave system, supplemented to some extent by the use of animal traction.

Similarly, too, the observation concerning their limited ability to work hard metals needs some qualification. The Romans could work most materials, as their instruments, fine metalwork, and metal furniture fittings show (Singer et al., 1954–84, vol. 2), but not on a large scale. In the case of ironwork, they relied on forging, bending and filing rather than on cutting or machining, and in regard to bridgework, this deficiency is most apparent in the area of fasteners or connections.

The Romans did use metals for connections. In stonework, for example, it was not uncommon to cut a suitably shaped recess into the upper horizontal face of adjacent stones, insert a metal link or cramp, and bond it into the stonework by running molten lead into the recess. Vitruvius (2.8.4) refers to this practice using iron, but bronze was also used. Lugli (1957), Blake (1947, 1959) and Blake and Bishop (1973) have many examples.

Curle (1911) has described a large deposit of Roman implements and relics from Newstead in Scotland, on the River Tweed, north of Hadrian's Wall. They include Roman chisels and axes, picks of various kinds, carpenter's tools, a hand auger about 30 mm diameter and 380 mm long, an anvil, scythes, and a manufactured whetstone 300 mm long. They also include (pl. 57) a variety of holdfasts and nails. One, very like a modern railway dog spike, is about 15 mm thick and 100 mm long; another is about 12 mm by 200 mm long. There is also a large range of nails, typically square in cross-section, tapered, and with flat heads. The two largest are about

8 mm by 150 mm, and 10 mm by 90 mm. There is also the equivalent of what would be called today an eyebolt, about 75 mm long, with a shank about 10 mm diameter. The timber to which this eyebolt was fastened, has, of course, long disappeared. What is to be seen, therefore, is the metal piece as it would have appeared after it was installed. The point that is of interest is the mode of its attachment. It is not threaded, nor does it have a nut. Rather, on the remote side of the timber the shank is passed through a washer and then riveted over.

Further north at Inchtuthil (Section 2.6), a large hoard of iron nails was discovered in 1960 in a deep pit in the *fabrica* (manufacturing workshop) of the Roman fort (Pitts and St Joseph, 1985). Their total number exceeded 850,000, with lengths from 38 to 370 mm. Most had lengths in the range 38 to 100 mm but there were 26,000, or about 3 per cent, longer than 216 mm. The very long nails, 248 to 370 mm long, were square in cross-section, with sides in the range 11 to 17 mm and round heads from 32 to 54 mm in diameter. Despite their large size, they had the form of tapered nails.

These fasteners lead to a series of conclusions:

(1) the Romans could drill holes of large diameter through thick timber;
(2) they had an abundance of nails;
(3) there is no evidence here of screws, threaded bolts or nuts; and
(4) they knew how to pass a round bar through a washer and rivet over the end.

It would also appear that they could, if they wished, simply bend over, or clinch, the end of a round bar, as it protruded through the timber; but in this case, without a washer, it would be expected that the timber would crush, and the bar would bend either to an angle, or, if it were longer, to a large radius.

There is other evidence concerning the Roman use of nails. Brookes (1974), in describing underwater surveys where a Via Appia bridge crossed the Garigliano River at Minturnae, mentions that nails were recovered there by the kilogram, mostly used, bent, clinched, twisted or broken. About 60 per cent were of bronze; the rest, apart from a few made of lead, were of iron. Hero, as quoted by Drachman (1963: 102), warns against the use of nails or plugs to join the tops of sheer legs where heavy loads are to be lifted. Vitruvius refers to nails (1.5.4; 7.1.2; 7.3.1).

It should be remembered also that the Romans, although initially not a seafaring nation, came to amass wide experience in shipbuilding. Lethbridge (in Singer et al., 1954–84, vol. 2) describes the use of wooden pegs – treenails or trenails – for the fastening together of wooden frames and the attachment of the planking. White (1984) provides more detail of the attachment of the planks, which were fitted edge to edge, not overlapping, with plate-like wooden dowels or keys between adjacent edges, and fixed to the frames by wooden treenails pierced by bronze spikes. Certainly this procedure would have been effective. The central spike would cause the timber peg to expand radially and fit tightly within its hole. Even without the spikes the treenail could, if well seasoned, be tightly driven through the plank and into the frame, and could also, if necessary, be glued in place by the use of material such as pitch. A large wooden peg would have a considerable capacity both in tension and in shear, even compared with a metal bolt, for in the case of a metal bolt in timber, its capacity is determined largely by bearing against the timber, that is, by the timber strength. However, the point remains that spikes were used, and that the Romans were able to produce them in bulk. They used nails also for the attachment of sheathing. The two large wooden ships preserved beneath the water of Lake Nemi (Singer et al., 1954–84, vol. 2: 572) had a thick padding of cloth fixed outside the planking; outside the cloth was a sheathing of lead fastened to the planking with bronze studs. The purpose was, of course, to guard against attack from marine borers.

What the Romans appear to have lacked is the equivalent of the modern bolt and nut, or, more particularly, the nut.

The concept of the screw was known to them. Archimedes (Drachman, 1963; Dijksterhuis, 1987), who died in 212 BC, is credited with the invention of the endless screw. Vitruvius (6.6.3) refers to the screw press. In his *Mechanica* of about AD 62, Hero discusses the theory of the five powers – the windlass, lever, pulley, wedge and screw – and gives 'examples of their practical use' (Drachman, 1963, Book 2: 50f.). He describes how to mark out a screw by wrapping a wedge-shaped sheet around the cylinder from which the screw is to be cut, and states correctly that 'we must take the screw to be just a twisted wedge . . . worked not by a blow but by turning'.

However, there is a distinction to be drawn between the invention of the screw and the invention of the nut. Hero (Drachman, 1963: 74) speaks of placing 'the piece of wood that is called the *tulos*' in the 'first furrow that is in the screw': 'this is what lifts the weight, by the turning of the screw it will lift the piece of wood'. His description is, in effect, of a nut with a smooth hole into which protrudes a suitably shaped peg that engages the thread of the screw. Theoretically, of course, this is entirely sound, and in principle enables the conversion of a rotary motion (of the screw) into a linear motion (of the nut) at a considerable mechanical advantage. It is known that this system was actually used in instruments and machines. However, the use of a simple peg would greatly limit the magnitude of the axial load, for the advantage of the nut is that it spreads the load over a considerable length of the screw.

The Romans, at least by Hero's time, were also aware of the principle of the female thread or nut. Drachman (1963: 135–40; see also White, 1984) has constructed an instrument for the cutting of a female thread, based on a description given by Hero (Book 3). A central cylindrical shaft carries both the cutter and a portion of the male thread. The male thread engages two pegs like the *tulos*, so that, as the shaft is rotated, it advances at a rate determined by the pitch of the male screw. At the other end, the shaft is pierced by a central longitudinal hole, into which is fitted a wedge, whose sloping face engages a transverse pin or cutter, proceeding through a radial hole in the shaft into the material in which the female thread is to be cut. Initially the wedge is adjusted so that the protrusion of the cutter is small. The shaft is rotated, and in its first pass the cutting of the female thread is begun. The wedge is then adjusted, the cutter is advanced and the thread is deepened. The process is repeated until the female thread is completed.

Despite Hero's claim that he offered not only the theory of the five powers, but also examples of their practical use, it must be remembered that his examples were simply examples in principle. They were not final designs, and it does not follow that his machines were actually built or used. This may be so in the case of the screw-cutting machine.

To quote Gille (Singer et al., 1954–84, vol. 2), 'apart from the origin and history of screws is the problem of how the early examples were made'. He refers to the drawing of a press found at Pompeii used for the pressing of cloth, and described by Patterson in chapter 6 of the same work. The press had two columns and two crossheads, the upper fixed, the lower moving. Two large-diameter screws extended downwards from the upper crosshead, through which they passed, and could be used to press the lower crosshead down against a bottom platform or base plate. Patterson describes the screws: 'a second-century illustration shows that pegs in the cross-piece engaged with the threads cut in the uprights, so avoiding the need for a female screw or nut'. It appears likely that the columns, crossheads and screws were made of wood.

Male threads were simpler to cut than the female threads. Gille suggests that, after a wedge-shaped sheet had been wrapped around the cylinder, this sheet was used 'as a guide to the file, chisel, or gouge cutting the spiral groove'. There is no reason to doubt that these techniques were used to cut male threads in both wood and metal. The female thread poses a much more difficult problem. Three options were available. Hero's screw cutter could also be used with soft materials such as timber. For screws of large diameter the female section could be made in two parts, split along the axis of the screw so as to give access for shaping the female screw by hand with a chisel or gouge. The third possibility for a metal nut was to form the thread as a spiral of wire and solder this within the hole – for the use of tin–lead solders was not uncommon (see Singer et al., 1954–84, vol. 2: 47). All three presuppose that the hole itself could be cut, but in hard metals such as iron this itself would pose problems.

It seems reasonable to draw the following conclusions: the principles of the screw and the nut were known to the Romans; they had the ability to use both in special circumstances, such as in presses and instruments; they could cut male threads in quantity, if they so wished, in both timber and metal; they would have found it difficult to cut female threads in metal, and for this reason the modern bolt and nut was not part of their available technology.

Some further things need to be said. The lathe had been known since about the middle of the second millennium BC (Singer et al., 1954–84, vol. 2: 232), but there is no evidence either of its being used to turn metals or of the invention of the tool-feed mechanism necessary to cut a thread. The use of metal fittings to join and decorate timber furniture was common (see for example fig. 209 of Singer et al., 1954–84, vol. 2), and included hinges, nails and socketed junctions. Vitruvius appears to refer to the use of bolts. For example, in 5.1.8, he describes the use of a compound beam made of three large pieces of timber 'bolted together', and in 10.2 to the use of bolts in the construction of cranes. It would appear, however, that there is a problem in translation. His reference could be to heavy rods of metal, clinched over or riveted through washers at their ends. He does refer to iron nails (1.5.4; 7.1.2; 7.3.1) (see p. 45f.).

4.3 Lifting apparatus and special equipment

The five powers of Hero (*c.* AD 62) were the windlass, lever, pulley, wedge and screw. To these should be added the roller, wheel and axle, and the gear or toothed wheel. The oldest are the wedge, the roller and the wheel and axle. The last is believed to date possibly from about 3000 BC (Singer et al., 1954–84, vol. 1: 206) and the other two may be expected to be still older. By 2000 BC, the Egyptians used levers, sledges, rockers and rollers in quarrying (Lloyd in Singer et al., vol. 1). They also used the Spanish windlass, which applied tension by the twisting of a multiple set of ropes, but not the capstan, conventional windlass or pulley.

The conventional windlass consists of a drum on a horizontal axle which is anchored against displacement. A rope from the drum is tensioned by rotating the drum with handspikes or levers. With the pulley, it was known to Aristotle (384–322 BC) or his school (Drachman, 1963: 15; Singer et al., 1954–84, vol. 2: 658; Hodges, 1970). The screw is commonly attributed to Archimedes (287–212 BC) but may have been earlier, invented by Archytas of Tarentum (Gille in Singer et al., 1954–84, vol. 2: 631). The most primitive gear or cog is the toothed wheel, used for example in a traditional Egyptian system for lifting water by transforming rotation about a vertical axis to rotation about a horizontal axis. It has also been attributed to Archimedes (Forbes in Singer et al., 1954–84, vol. 2: 676) and certainly it appeared at about his time. More complex gear systems are discussed by Hero (Drachman, 1963).

The theory of these devices is discussed by Aristotle

and Hero. In Problem 11 of his Mechanical Problems (Drachman, 1963) Aristotle describes the roller – 'on the rollers there is no friction at all, but on the carts there is the axle, where there is friction . . . The burden upon the rollers is moved on two points, the ground supporting from below and the burden lying above, for the circle turns at both these points and is pushed forward the way it travels.'

The principle of the lever was well understood. In the case of the windlass, Aristotle (Problem 13) knew that it was better to have a thin windlass and longer hand-spikes. The action of pulleys and gears was correctly understood by reference to the lever. Hero set himself the task of devising an apparatus which could lift a weight of 1000 talents by a force (power) of 5 talents and did in fact propose four. The first was a sequential system of gear wheels with, at each stage, a large gear wheel driving one that was small. The second used multiple pulleys and the third multiple levers. The fourth combined four of his powers, with a screw driving a winch which pulled a rope through multiple pulley blocks to operate a final lever. Now it should not be understood that these were practical devices. The provision of large forces requires the availability of large reactions, and Hero did not often indicate how these reactions were to be supplied. Furthermore, it would appear that neither the gear nor the screw was used in the cranes and hoists of his day (Drachman, 1963: 91). However, he had two outstanding achievements to his credit.

The first was that he did indicate systems that were theoretically sound – at least, if friction were neglected. The second was that he appeared to understand the principle of mechanical advantage. At one point he said, 'The proportion of the moving power to the power moved is inverse' (Drachman, 1963: 86). The true principle is that, in a frictionless system, the product of the force by the distance moved is constant; that is, the magnitude of the force is inversely proportional to the distance moved. It does not require much revision to translate Hero's principle into these terms.

In Roman bridge construction, the most pressing problem was to move heavy weights. A stone block, for example, would have to be moved on the quarry floor, lifted, carried to the site, and then placed in position. Excluding for the moment the transportation problem, which could be solved either by cart or by boat, the most relevant lifting devices were the roller, wedge, lever, windlass and pulley. Two of these required the use of rope. Gilbert (in Singer et al., 1954–84, vol. 1) has described ancient methods for the manufacture of rope. In Egypt the earliest known specimens, made of reed, date from about 4000 BC. Later ropes were made from the fibre of the date palm (most commonly) and also from flax, grass, halfa, papyrus and camel hair.

The manufacture of rope requires the spinning of fibre into yarn, then yarn into strands and strands into rope. Egyptian representations of *c.* 2400 BC and *c.* 1500 BC show these processes. The first, on a tomb at Thebes, is entitled 'Twisting the ropes for boat-building'. A later painting, of *c.* 1500 BC, shows two men spinning yarn, with one (at the left) feeding fibre into a whirling tool which twists the fibre into yarn held in tension by the man at the right. In the subsequent stages, the direction of twist is successively reversed, so that the fibres and strands lock themselves into the rope. Another Dynasty XVIII tomb shows the final formation of the rope.

It is possible to form some idea of the size and strength of these early Egyptian ropes. In an underground lime-stone quarry at Tura, ropes 60 mm in diameter have been found with three strands, each strand of forty yarns, each yarn of about seven fibres, made from papyrus. These larger ropes were used for the haulage of heavy stones or stone statues by gangs of slaves. One representation (Singer et al., 1954–84, vol. 1, fig. 283) shows the trans-portation of a colossus on rollers. Another shows a stone statue estimated to weigh 60 tons, being hauled by 172 men on four long ropes.

The force a man can pull depends on his weight and the coefficient of friction between his feet and the ground. For a man with a mass of 90 kg, and a coefficient of friction of 0.5, the maximum force is 441 N (approxi-mately 100 lbf – see Appendix 1).

A gang of 172 men could, as an upper estimate, achieve a pull of about 76 kN, or about 19 kN (4300 lbf) per rope. This total force of 76 kN would succeed in moving the 60 ton load, provided that the coefficient of friction between the stone and the ground did not exceed about 0.13. It would be difficult in fact for 172 men to achieve an even pull, and the friction coefficient between their feet and the ground might be less than 0.5. Even so, a reasonably conservative estimate would give a force per rope of 9.5 kN (2100 lbf), and a total force of 38 kN, able to overcome a coefficient of friction at the load of 0.06. Given that such a load would be on rollers, on a specially prepared track, and that assistance would be available from men with levers behind the load, this appears to be feasible.

It appears, therefore, that the Egyptians could make a rope able to pull 9.5 kN or 2100 lbf. There is no reason to believe that the Romans could not achieve the same or even better. For comparison, the minimum breaking loads specified by the British Standard BS2052: 1990 (British Standards Institution, 1989) for a modern hemp rope vary from 16.7 kN for a 16 mm diameter rope, to 36.8 kN for 24 mm diameter, and 63.7 kN for 32 mm diameter. The best modern, natural-fibre ropes have capacities about 20 per cent higher than these for Grade 1 manila ropes. All of these have three strands – as did the Egypt-ian ropes. In modern practice, for lifting tackle a factor of safety of six would be applied. A safe working load of 9.5 kN (the lower estimate for an Egyptian rope) would then require a rope diameter of about 30 mm for a hemp rope, or 28 mm for the best manila rope. These figures are reasonably consistent. It is possible, therefore, that the capacities of ancient ropes may not have been greatly different from those of modern ropes.

The cases quoted here do not exhaust the capacity of

the ancients to undertake heavy lifts. An excellent example is given by Parsons (1968: 155–67) in his chapter entitled 'Moving the Vatican obelisk' (see also Lewis, 1984 5). This obelisk now stands as a central feature of the piazza which serves as the main entrance to St Peter's in Rome. In 1586, it was moved a distance of 240 m to its present location from an earlier site in the Circus of Nero at the foot of Vatican Hill. The operation was planned and supervised by the Italian architect Domenico Fontana, and took a period of five months, from 30 April, when the first lift was undertaken, to 10 September, when it was finally dropped into position. Fontana, followed by Parsons, has described the operation in detail. He records the obelisk as having a main shaft cut from a single stone, 24 m high, with a square section tapering from 2.79 m square at the base, to 1.8 m square at the top. It is surmounted by a pyramidal top 1.3 m high. His estimate of its mass is equivalent to about 310 tonnes in modern units.

The early history of the obelisk is that it was cut during the reign of Noncoreo, king of Egypt, early in the tenth century BC, and about AD 37–41 was transported from the city of Heliopolis to Rome, by order of Caligula, and set up below Vatican Hill. The task of moving the monument in 1586 was a major achievement, but it is not to be compared with that of the Egyptians who quarried it, or that of the Romans, who 'carried it down the Nile, thence across the open Mediterranean and by land from an Italian seaport to Rome' (Parsons, 1968: 155; see also Suetonius, *Vespasian* 18).

The best evidence for Roman cranes comes from Vitruvius, Hero (Drachman, 1963: 100) and two Roman reliefs, and is summarised by Drachman in an excellent article in Singer et al. (1954–84, vol. 2). Vitruvius (10.2.1–7; see also Drachman, 1963, fig. 55a) describes a set of sheerlegs, with two heavy timber poles, attached together at the top and held in a semi-vertical position by rope stays running forward and backward from the top of the poles. A pulley block is also hung from the top, and a windlass is attached between the poles near their base. This windlass could be operated by men pulling on rotating arms. Alternatively, a treadmill could be used.

Figure 8 shows part of a relief held in the Vatican, dated about AD 100. It shows the mast (either a single pole, or a pair of poles in side view), with its stays and lifting blocks. The treadmill has a diameter of the order of 6 m, and is being powered by five men. Both stays and hoist ropes include multiple blocks. The number of parts is unclear, but in the main hoist there are certainly at least five ropes, and probably more.

It would be difficult for five men to work effectively in a 6 m diameter treadmill unless two could work side by side. In this case, five men could exert a torque of about 7 kNm. The diameter of the capstan would depend on the size and length of rope. Suppose a diameter of 1 m. Then the force in the final stage of the hoist rope would be 14 kN. The force in a rope passing over a pulley is influenced by friction in the pulley. Assuming an increase per pulley of 5 per cent due to friction, then a five part lift provides a total lifting force of about 4.3 times the final rope force; that is, approximately 61 kN in this case, sufficient to lift 6.2 tonnes. Alternatively, three men in a 6 m treadmill could exert a torque of about 4 kNm, apply a rope force of about 8 kN, and lift a load of 3.6 tonnes. For comparison, it was estimated earlier in this section that the Egyptians may have had rope strengths in the range 9.5–19 kN. There is a reasonable consistency between these figures.

Alternatively, the final hoist rope could have been taken directly down to a group of men. These men together could possibly apply and control a force of about one and a half times their individual weights; that is, about 1.3 kN. The lift would then become about 0.6 tonnes, or possibly up to 0.8 tonnes.

As with many Roman representations, the scale of Figure 8 varies, with the men at the top shown about twice the size of the men in the treadmill. However, it seems reasonable to scale the length of the crane jib, from

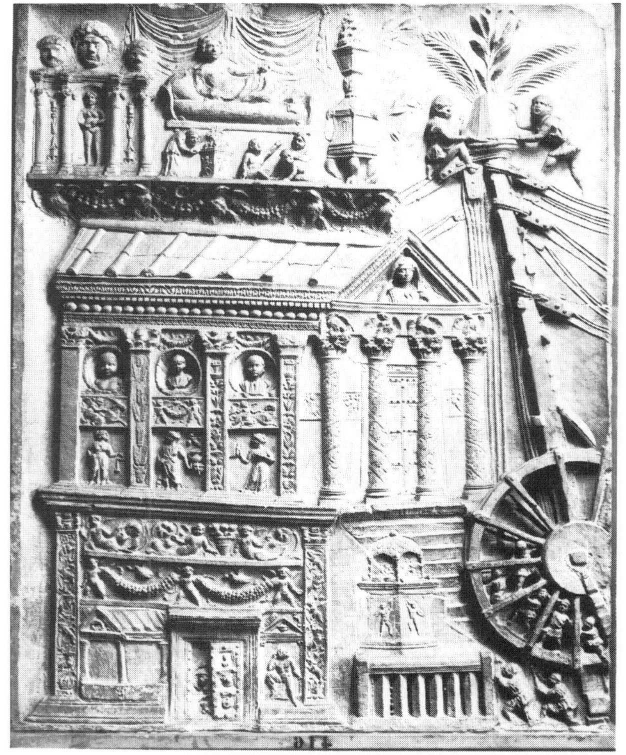

Fig. 8 Roman crane, *c.* AD 100, from relief held in the Lateran Museum.

the attachment of the lifting block to the pivot point below the treadmill, by reference to the treadmill itself, and then the treadmill diameter from the men within it. On this basis the jib length would be about 16 m. If the crane were operated with the jib at 35° to the horizontal, then it could reach to a radius of 13 m. At lower angles there could be interference, either between the treadmill and the ground, or between the treadmill and the ropes used to control the jib. If the centre of the jib were set 1.5 m back from the edge of a supporting platform, then the clear reach would become about 11.5 m. The Romans may have achieved longer reaches than this in bridge construction, possibly as much as 15 to 18 m clear – see Section 8.6. It would not have been difficult to construct longer jibs than that shown in the illustration.

The density of a good building stone may be of the order of 2700 kg/m^3. Then the cranes described above could lift stones with the following volumes: 2.3 m^3 for five men in a treadmill; 1.3 m^3 for three men in a treadmill; or possibly 0.3 m^3 for three men at the end of a six part rope. If the stones were cubes (and it is unlikely that this would be the case) then the corresponding block dimensions would be 1.3, 1.1 and 0.7 m. Suppose instead that the stones were square, with a thickness equal to one half of the sides of the square, then the square (or large) dimension would be 1.6, 1.4 and 0.8 m. These capacities would be entirely suitable for bridge construction in stone.

There are, of course, many other possibilities. The final rope could be taken around a pulley at ground level and then drawn by a gang of men. Assuming 220 N per man (the lower of the two earlier estimates) and allowing again for friction, a gang of sixty-seven men would be required for the largest of the above lifts. A second Roman relief (Singer et al., 1954–84, vol. 2: 637) shows a treadmill set back in this way from the crane, and operated by two men. There is no reason why the treadmill or a windlass could not be set even further back, if this were convenient.

The systems of sheerlegs described above have the disadvantage that the load can be moved in only one horizontal direction, using the stays to adjust the angle of the sheerlegs. However, small adjustments of position in the transverse direction could readily have been achieved by pulling the load sideways. Before a lift, the location and direction of the sheerlegs could be adjusted. Final adjustment of the position of the load could also be made by sliding or rolling it into position. These techniques would give sufficient flexibility in practice.

Vitruvius (10.2.8), however, also describes an alternative. A single post could be used, with stay ropes in all four directions. The hoist, he suggests, could be made by using three sets of blocks, each operated by a gang of men. He warns of dangers with the method and suggests its use only by 'skilled workmen'. The handling of heavy lifts requires not only the provision of an adequate prime mover or crane, but also the availability of suitable lifting tackle at the load. In many cases the use of ropes, either tied or with metal fastenings, would suffice, but special equipment was required for stone (see Section 4.4).

Heavy loads could be carried either by sea or by land. Forbes (in Singer et al., 1954–84, vol. 2: 514) has given an excellent summary of Roman wheeled vehicles. A heavy ox-wagon, or *clabularia*, was drawn by eight to ten oxen (or horses) and carried about 0.5 t. The *raeda* was a four-wheeled cart drawn by eight to ten mules, and carried up to 0.33 t. A lighter four-wheeled cart was the *carrus*, carrying 0.2 t. Another, the *vereda*, drawn by four mules, carried about 0.1 t and the two-wheeled *birota*, with three mules, carried about 0.07 t and one or two passengers. Assuming the load was a single stone, the largest of these capacities corresponds to a stone with a volume of only 0.19 m^3, that is, a cube of side 0.57 m, or a square thin stone 0.72 × 0.72 × 0.36 m.

The vehicles listed by Forbes were used for long distance transport. It would be expected that heavier loads could be carried by wheeled vehicles over shorter distances, or alternatively drawn on sleds or rollers. Vitruvius (10.2.11–13) describes the transport from a quarry of a stone of size 12 × 8 × 6 Roman feet – approximately 3.5 × 2.4 × 1.8 m, totalling 14.6 m^3 with an estimated weight of about 40 tonnes. The ends of the stone were built into huge timber wheels 15 feet (4.4 m) in diameter. Timber slats were used to join the wheels around their circumference; ropes were passed around these slats and used to draw the whole by oxen. Vitruvius criticised the method, for he said it was impossible to keep the machine straight along the road. He preferred the system used earlier for heavy stones, with the stone mounted as the axle of two 12 feet (3.5 m) diameter wheels mounted in a heavy wooden frame drawn by oxen. He goes on to say, 'This expedient would not have been possible unless, to begin with, the distance had been short. It is not more than eight miles from the quarries to the temple, and there are no hills but an unbroken plain.'

Forbes also speaks of the pulling capacity of horses and mules. He suggests that, owing to inadequate harness, an animal could pull a maximum load of 62 kg – 'only a quarter of the modern figure'. He is referring presumably to the load on a wheeled vehicle, not to the horizontal force exerted by the animal. His figures may be compared with some records from early Australia (Ruhen, 1980: 194). A team of fifteen horses pulled for 7 miles a load of 32.5 tonnes on a four-wheeled wagon of mass 4 tonnes. The total load of 36.5 tonnes corresponded to 2.4 tonnes, or 2400 kg, per animal. Forbes' 'modern figure' of 248 kg may be his estimate of a reasonable working load under normal conditions. Ruhen's figure, of 2400 kg, is probably an extreme value. However, this extreme value may be the figure that is more relevant in bridge construction. Even if this were discounted to a quarter for inadequate harness, it follows that in Roman times an animal might pull more than 62 kg, possibly up to 600 kg. This is close to the estimate by Burford (1960), of 1100 lbm (500 kg).

As far as sea transport was concerned, Roman ships could carry loads much heavier than those mentioned here. The maximum single load was determined not by the ship, but by the means necessary to get the load onto the ship, that is by carts and cranes.

The Romans not only had efficient cranes but also other special-purpose equipment including the piledriver. The underwater investigation at Minturnae (Brookes, 1974), for example, revealed the heads of timber piles. Vitruvius refers to piles at Ravenna (2.9.11), and then in Book 3 says, 'If a solid foundation is not found, and the site is loose earth right down or marshy, then it is to be excavated and cleared and remade with piles of alder or of olive or charred oak, and the piles are to be driven close together by machinery, and the intervals between are to be filled with charcoal' (3.4.2).

He has another interesting passage in Book 5:

> The masonry which is in the sea must be constructed in this way. Earth [by which he means pozzolan – see Section 4.5] is to be brought from the district which runs from Cumae to the promontory of Minerva [Sorrento] and mixed in the mortar, two parts to one of lime. Then in the place marked out, cofferdams, formed of oak piles and tied together with chains, are to be let down into the water and firmly fixed. Next the lower part between them under the water is to be levelled and cleared with a platform of small beams laid across and the work is to be carried up with stone and mortar as above described until the place for the structure between the dams is filled. (5.12.2–3)

He continues (5.12.5–6), 'Where however the earth in question is not found', double cofferdams were to be used. 'Clay in hampers made of rushes' was to be pressed down in the intervening space. 'When it is well pressed down and as closely as possible, the place marked out by the enclosure is to be emptied with water screws [that is, the Archimedean screw – Singer et al. (1954–84), vol. 2, fig. 617] and waterwheels with drums and so dried. Here the foundations are to be dug.' He goes on again to speak of the use of charred piles of alder.

The use of these construction techniques is confirmed by the bridge at Trier (Chevallier, 1976; Cüppers, 1969). Remains found beneath the Moselle show that the earlier bridge had seven piers supported on 0.5 m diameter oak piles, with their tips protected by iron shoes. The piles were driven to form a pentagonal base (rectangular with a pointed upstream face) 10.2 m × 19.5 m, with flat planking laid over a clay floor rammed down with broken stones into the gaps between the piles. Ashlar stone blocks were then laid on this base.

Piers for a later bridge used the earlier piers as a construction platform. Caissons were constructed around these piers, made up of double walls of timber sheet piling, with the intervening space filled with well placed clay. Similar remains were found at Newcastle, England, of the original Pons Aelius, when the old swing bridge was built in 1872 across the River Tyne (Dymond, 1961; Bruce, 1978).

In another place Vitruvius (10.2.3) refers to sloping piles. The ability to drive sloping piles is of particular interest in bridge construction, where they may be required to support lateral loads. In Book 4 of his *Gallic Wars*, describing his timber bridge across the Rhine, Caesar speaks of 'piles 1 ½ feet thick, slightly pointed at the lower ends', driven obliquely into the river bed.

4.4 Stone, brick and tile

Building stone was quarried as early as 2800 BC (Singer et al., 1954–84, vol. 1: 569). The Egyptians successfully quarried not only soft rocks like limestone, but also harder materials like sandstone, serpentine, basalt and granite (Forbes, 1965, vol. 7: 163, 168–71), using only wood, stone and bronze tools. An unused but completely quarried obelisk at Aswan, 42 m long, and other similar stones, have provided evidence of their methods. The upper horizontal face was initially formed by pounding with balls of dolerite. The same technique was then used to cut trenches 1 m wide and 2.75 m deep around the obelisk, forming it to its correct shape on all but the bottom face. It was then split off the parent rock, using pounders to cut horizontal holes for wedges around the base. The wedges were either of wood between plates of metal, expanded by wetting, or metal between flat metal plates. The Egyptians were also able to saw limestones using copper blades fed with sand or set with emery teeth.

This technology was passed on to the Greeks and then to the Romans. Porter (1907), Blake (1947, 1959), Blake and Bishop (1973), Tenney (1924), and others (Hill, 1984; Landels, 1978) have described the early Roman building stones.

Tufa, or tuff, is a compressed volcanic ash, common in Italy and present in three forms – stony (*tufa litoide*), employed as a building stone; granular (*tufa granulare*), too soft for a building stone, but forming the chambers of the catacombs; and sandy (*pozzolana*), used in an hydraulic cement. Blake has subdivided the building *tufas*: *capellaccio*, a widespread, very soft rock, was taken from the first of the layers of ash that fell near Rome; Fidenae *tufa* came from the second layer; and Grotta Oscura, Monte Verde and Anio *tufa* from the third. The fourth discharge of ash gave rise to a stone consisting of a relatively hard, dark grey *tufa* containing small fragments of dark lava, white limestone and occasionally other materials (Blake, 1947: 34). The resulting speckled appearance gave rise to the name *peperino*, used instead of the original name, Alban stone. A similar but coarser stone is Gabine stone. Generally these stones came into use in the order in which they have been cited above. The better materials for building purposes appear to have been Monte Verde and Anio *tufa*, and *peperino*.

Vitruvius (2.7) lists some of these stones. He also refers to travertine (*lapis travertinus* or Tiburtine stone), quarried near Tivoli or Tibur, on the banks of the River Aniene. This high-quality building stone, consisting of calcium carbonate deposited by hot springs, is essentially a form of limestone. It is not only strong and durable but has a good appearance, creamy white in colour, and a fine texture (Blake, 1947: 44). Whereas the *tufas* were commonly faced with other materials, such as travertine itself, travertine could be used as both structure and facing. Vitruvius refers to its strength but states also that it is easily injured by fire – scarcely a significant disadvantage in bridge construction. One of its earliest known usages is in a bridge, the Pons Mulvius, in 109 BC.

Other limestones were also used. Marble (*marmor*) was not used in early Roman building construction. A 'mania for rare and costly building materials brought from a distance of thousands of miles began about 144 BC' (Porter, 1907: 4; Blake, 1947: 51) and flourished particularly about the time of Augustus (27 BC–AD 14) who could say, 'I found it [Rome] of brick and left it of marble.' Italian marble, from the now famous Carrara quarries (ancient Luna), was first used in a Roman building of known date in 48 BC (Blake, 1947: 53). It differs from other limestones only in its crystalline structure, which allows it to be polished but makes it difficult to work. It was hardly used for bridge construction, except possibly occasionally as a decorative facing.

The term *lapis silex*, used by the Romans for any hard stone, was applied particularly to selce, cut from a lava stream which reached close to Rome near the Via Appia. It was used for paving, as also was basalt or millstone (Forbes, 1965, vol. 7: 175). Granite was sometimes used, more particularly in the period following Augustus (Blake, 1947: 60). Although quarried in Italy, it is found more commonly elsewhere, in Spain, Germany and North Africa. The use of sandstone in bridge construction is rare.

The Romans did not hesitate to import stone and to carry it long distances. Marble in particular, chosen for its colour and appearance, was brought to Rome by sea in large quantities from Greece, Asia Minor, Egypt and North Africa (Blake, 1947: 53).

The importance of these various stones can be gauged from a study of 106 bridges whose material has been identified in the literature. Of 91 Italian bridges, 32 used travertine (in 7 cases in association with *tufa* or limestone); 29 used limestone (including 2 with either *tufa* or a conglomerate); and 23 were of *tufa* (including Gabine stone and *peperino*). The remaining 7 used conglomerate (3), gneiss (2), *puddinga* (1) and trachyte (1). That is, of these 91 bridges, 84 (92 per cent) were constructed of travertine, limestone or *tufa*. Another 2 bridges are cited as having lava, selce or conglomerate paving.

Identification from the literature of the stones used in colonial bridges is more difficult. Of 13 bridges, 7 were of granite (6 in Spain, 1 in Tunisia); 4 of limestone (3 in France, 1 in Syria); and 2 of basalt (1 in each of France and Syria).

It is clear, therefore, from these statistics, that bridge construction tended to use the softer rocks.

Modern research indicates that the durability of building stones is not necessarily related to their hardness or strength, and that the higher strengths are not necessary for either masonry walls or bridge construction. To quote Clifton Taylor and Treson (1983: 63), 'a widespread fallacy ... is that the harder stones are necessarily the most durable'.

The British Building Research Establishment has published a series of papers on the durability of building stones: in particular, British and French limestones and British sandstones (Schaffer, 1972; Honeybourne, 1982; Leary, 1983, 1986). With regard to limestone, Leary has said (1983: 6), 'Strength tests are rarely necessary. Durability is not determined by chemical composition; what is more important is physical structure. A stone with a high proportion of very small pores is less durable than a stone which has larger pores.' The best single test is the crystallisation test in which crystals are formed in the pores by cycles of immersion in a 14 per cent solution of sodium sulphate, followed by drying in a humid oven. Crystal formation causes the fretting off of stone, with weight loss which is greatest in the less durable stones. The test is a comparative one in which one stone is compared with another whose durability is known by experience.

Honeybourne (1982) has applied these methods to an evaluation of over 100 French limestones; he quotes the results of previous French tests for ultimate compression strength and superficial hardness, and also French recommendations for use. These recommendations are in reasonable agreement with performance as indicated by the crystallisation test. However, the data show that durability is not closely related to strength and hardness.

Strengths quoted by Honeybourne vary widely, from 10 to 201 MPa (see Appendix 1). Stones recommended for the most severe locations (exterior paving, river piers, base courses) have strengths varying from 19 to 197 MPa; the second range of use (in balconies), 23 to 58 MPa; the third (cornices, sills and string courses), 18 to 137 MPa; and the fourth (general cladding), 11 to 201 MPa. These also vary widely, indicating that the recommended usages are not determined by strength.

On the other hand, strength is reasonably well related to superficial hardness, as measured by the scratch width caused by a sharp, conically pointed tool under a 3 kg load (the Martens Sclerometer). For these French limestones, scratch width varies from 0.6–0.7 mm for the strongest stones (200 MPa) to 1.4–2.2 mm for the weakest (10 MPa). Weathering causes a significant increase in hardness in the outer layers – the *calcin*, to use the French name. For a soft stone, the scratch width reduces from about 2.0 to 1.1 mm, suggesting an increase in surface strength from 10 to about 50 MPa (found by plotting Honeybourne's results). This confirms what is common knowledge in the stone industry, that seasoning can significantly change the properties of freshly quarried rock – either by exposure to the atmosphere or by draining out some of the moisture that is present in the native rock. For example, limestone from an underground quarry at Corsham has a natural moisture content of 11 per cent (O'Neill, 1965: 77) and cannot be brought to the surface during the winter months. It is brought out between May and October, and then left to season for about six weeks (see also Price, 1975).

Sandstone is far less common in Roman bridge construction. Leary (1986) points out that the quartz that forms the grains is very durable and the durability of the stone is determined by the cement that binds the grains together. Calcareous or dolomitic sandstones may perform badly; the majority of sandstones are siliceous and are generally very durable.

These observations have an important bearing on the quarrying and shaping of stone for bridge construction: a majority of Roman bridges were made from the softer rocks; the durability of a stone may be high even if its strength is relatively low; a stone is often softer and easier to work when it is first removed from the quarry. Vitruvius makes similar observations:

When we have to build, let the stone be got out two years before, not in winter but in summer, and let it lie and stay in exposed places. Those stones, however, which in the two years suffer damage by weathering, are to be thrown into the foundations. Those which are not faulty are tested by Nature, and can endure when used in building above ground. (2.7.5)

An important parameter in the quarry is the depth of bed available. It is only rarely that unbroken solid rock will persist through depths of more than a few metres. Although deep quarries exist, the more normal situation is where the rock is separated by horizontal planes of weakness at intervals of about 1.5 to 2 m. The most important tool for quarrying this rock is the plug and feathers. Figure 9 shows a set from the limestone quarry at Ketton (Leary, 1983), a quarry in Leicestershire, England, that has been used for centuries, possibly since Roman times, and is now operated by the firm of Rattee and Kett. Figure 10 shows the use of two sets of plugs and feathers to move a stone that has been already separated from the parent rock. Previous rock faces can be clearly seen with the holes used to split off the rock. In this case the holes are deep and closely spaced, and have been drilled with a modern pneumatic tool. In Roman times the holes would have been made with a hand drill and hammer. The depth of bed can also be seen. Quarrying without a horizontal bed would require splitting with wedges along both horizontal and vertical planes, and might require the cutting of distinct vertical slots or trenches to separate the stone from the parent rock.

A row of plug and feathers, progressively struck by hammer, can achieve an immense splitting force. The principle of the wedge struck by impact is at the crux of the quarrying operation and has been known from earliest times. Hero said,

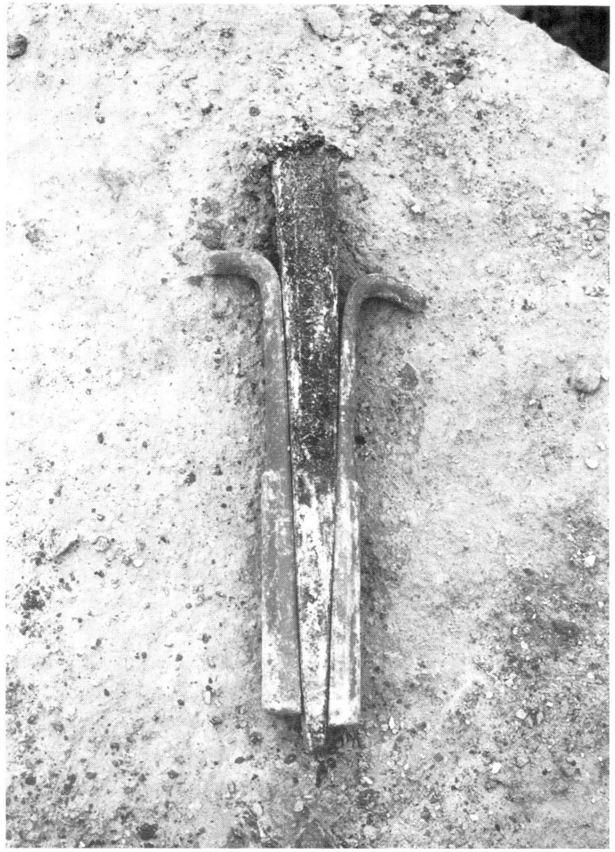

Fig. 9 Plug and feathers at Ketton limestone quarry, England.

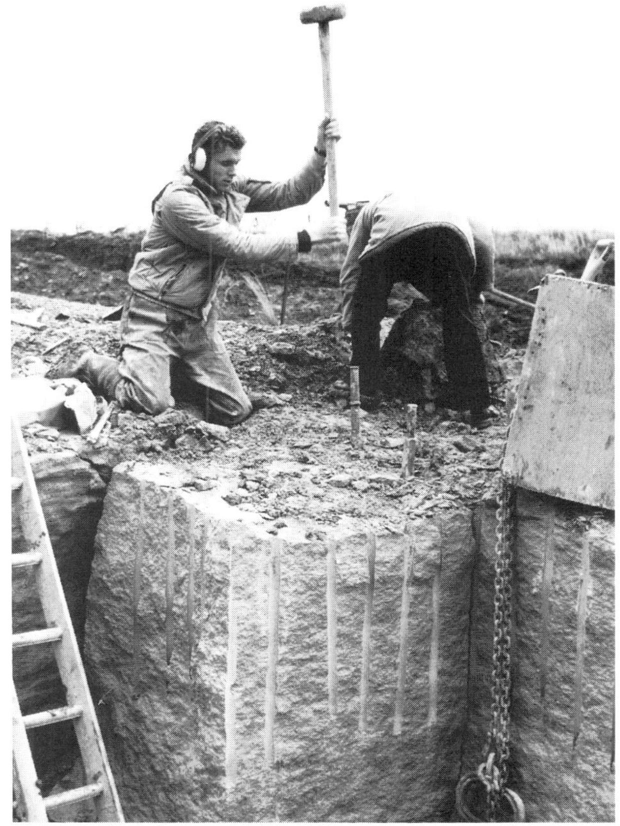

Fig. 10 Plug and feathers in use at Ketton.

Its greatest use is when we want to split the bottom of the stone which we want to cut and we have cut its sides free from the mountain from which we cut it ... Its effect is through blows that strike it, however these blows are, and its effect does not end when the blow stops, and this is evident to us that, without its being struck often, there will be noise and splitting of that which its power is bursting apart. And the smaller the angle of the wedge, the easier is its working. (Drachman, 1963: 55)

The sawing of stone has been known since Egyptian times. It may be surprising to those unfamiliar with build-

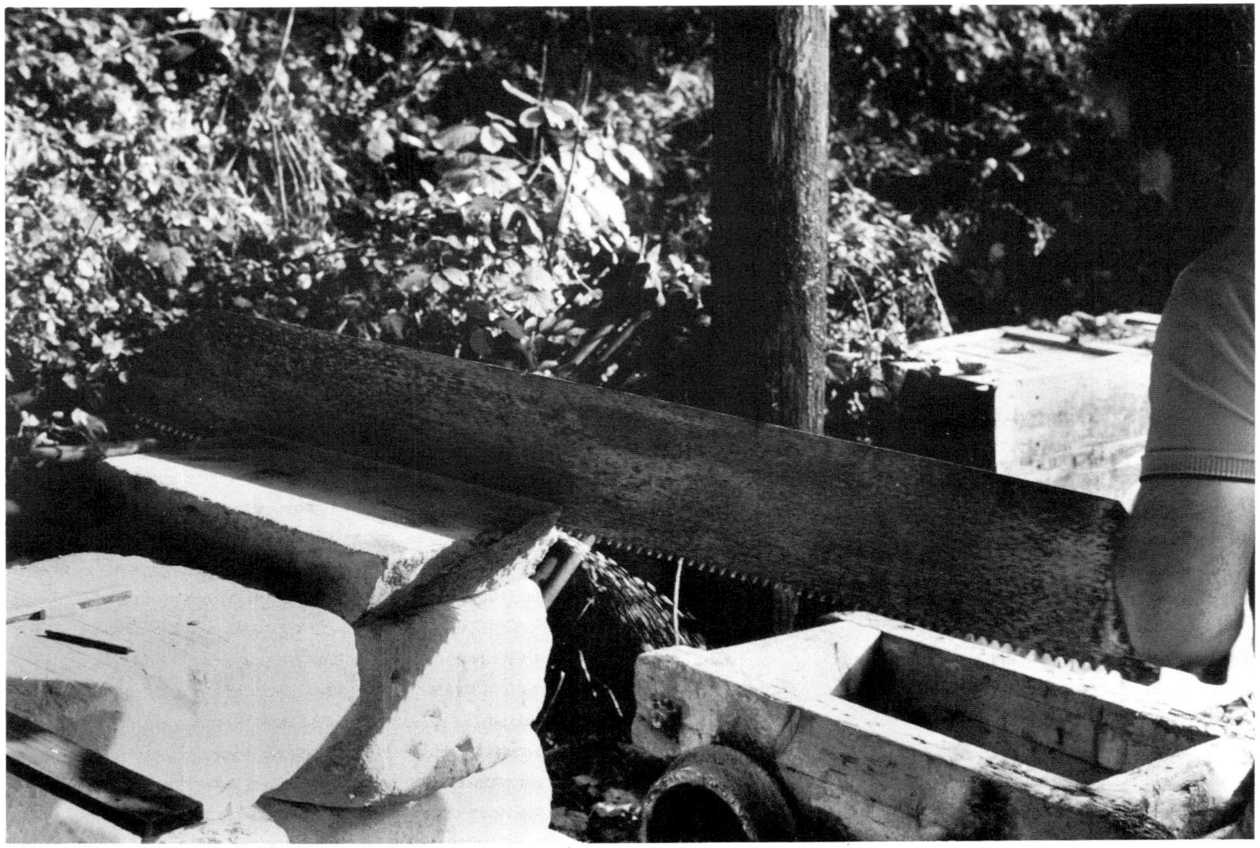

Fig. 11 The use of a frig-bob saw by hand to cut limestone at Corsham (Bath Stone Quarry Museum).

ing stone that durable stones such as limestone may be worked and sawn as readily as timber – indeed possibly with less difficulty because of the absence of grain. Figure 11 shows a traditional one-man frig-bob saw at the Corsham Bath Stone Quarry Museum (see also Purchase, 1904). It has a length of about 2 m, with teeth at a pitch of about 10 mm. Warland (1953) has a sketch showing an inclined wooden platform mounted above the slot cut by a saw of this kind. Sand placed on the platform is washed into the cut by a flow of water. A similar system, no doubt, could have been used by the Egyptians with their saws of soft metal. In modern practice power saws are used at all stages of the quarrying operation. Heavy chainsaws are used to cut the stones from the parent rock, and these stones are cut into slabs either by machine mounted chainsaws or multiple circular saws. The frig-bob saw was used underground in the Corsham Bath Stone Quarries. It may be expected that the Romans similarly used stone saws for a variety of purposes (Forbes, 1965, vol. 7: 176).

The harder rocks such as granite would have required different techniques. Bowles (1939) describes quarrying techniques used in the United States of America earlier this century. He points out (p. 108) that the quarrying of granite is generally assisted by two natural features: the presence of horizontal sheeting planes that separate the granite mass into sheets or layers; and the occurrence of two perpendicular planes of weakness in the granite. The direction of easiest splitting is called the rift, and the other less strongly marked fracture system is called the grain.

Fracture is a complex phenomenon which divides into two parts – fracture initiation and fracture propagation. Both are aided if the material is of low ductility. Tensile stresses applied to the material cause very high tensile stresses at the tips of naturally occurring notches or cracks (or their equivalent). When these are sufficiently high, fracture initiates and then propagates at a rapid rate. There is no necessary correlation between fracture toughness, as it is called, and compressive strength. Granite, for example, is strong in compression and resistant to abrasion, but is easy to split – it is said to be notch-sensitive, and has a low fracture toughness. Such a material may be expected also to be easy to shatter – that is, fragments will split off near the point of impact of a sharp blow, particularly if it is at a point or corner of the rock. Bowles (1939) describes how sheeting planes or 'bottom joints' greatly assist quarrying. Vertical fractures can then be achieved by wedging with the plug and feathers, preferably in the rift or grain directions. Holes 19 mm diameter, 100–125 mm deep, spaced 150–450 mm apart are sufficient to achieve major fractures. It is fortunate that these directions are generally at right angles.

Current techniques for the lifting of stones also date from Roman times. Figure 12 shows a three-legged Lewis

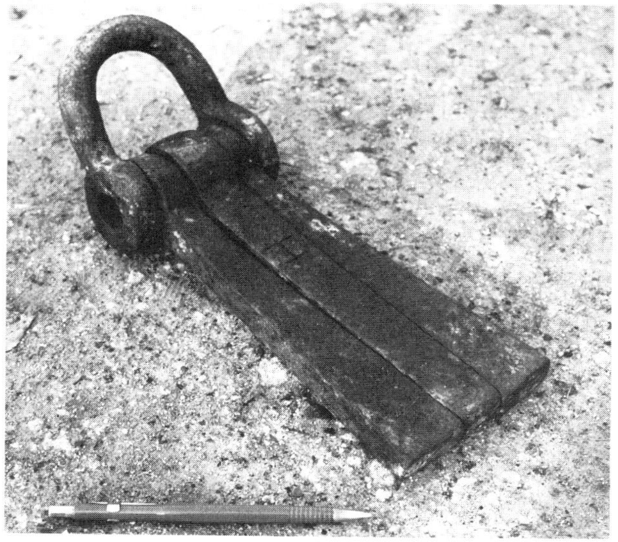

Fig. 12 Three-legged Lewis, similar to Roman Lewis (Bath Stone Quarry Museum).

at the Bath Stone Quarry Museum, Pickwick Quarry, Corsham. The outer pieces are wedge-shaped, being wider at the bottom. They are inserted in an undercut hole in the top surface of the block and are then separated, with the central plate pushed into place, and held in position by the top bolt. The presence of similar rectangular undercut holes in the top face of Roman stones is commonplace; they can be seen, for example, in stones at the Roman bath house at Bath itself. Other techniques were also used, either with fittings in the upper face, or tongs or dogs that entered into rectangular holes cut into the sides of the stone near their upper face. Figure 13 shows a set of chain dogs at Ketton. The current British code of practice for stone masonry, BS5390:1976 (British Standards Institution) gives recommended proportions for five lifting systems (Fig. 18) – the three-legged Lewis, the two-pin Lewis, the chain-linked Lewis and the split-pin Lewis, all of which lift from the upper face, and the

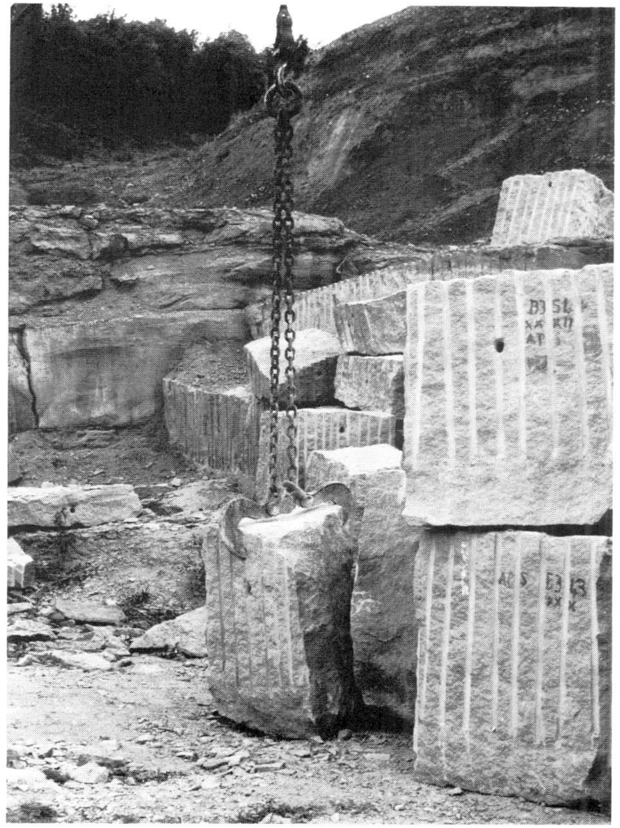

Fig. 13 Chain dogs in use to lift stone at Ketton.

chain dogs. All had their counterparts in Roman quarrying. Iron pincers were also used (Vitruvius 10.2.2). Some of these systems are illustrated by Lugli (1957, vol. 1: 227–35), together with techniques used to lever stones into place. Holes in the upper face of one stone could be used to engage the point of a lever to shift the stone above. The vertical face of a stone could be left proud to provide space for a similar hole used to lever and align the superimposed block; the surplus rock would then be removed in place.

Vitruvius (2.8.4) mentions the use of 'iron cramps run in with lead' to join two stones together. Both Blake (1947) and Lugli (1957, vol. 1: 236–9) have many examples. Figure 14 shows the downstream wing wall of the bridge abutment opposite Chesters on Hadrian's Wall, England. The groove for a long cramp can be seen along the upper face, finishing with a T-shaped keyed end. Lewis holes are also visible, although these are partly obscured by leaves.

The other area of Roman stone technology that needs to be addressed concerns the shape of the stones, and the way they were joined to form a wall or vault. A variety of terminology has been used to describe early stone walling. Lugli (1957), for example, has defined nine categories, from *opus silicum* to *opus cementizia*. Not all of these forms are relevant to Roman bridge construction. Blake (1947: 2), following van Deman, argues against the introduction of terms consisting of the Latin word *opus*, modified by an adjective, particularly when they were not used in antiquity.

The earliest walls (Blake, 1947: 71) consisted of large stones rolled together with their interstices filled with smaller stones. This cyclopean masonry, as it is called, was gradually refined by the smoothing of the surface and the trimming of the adjoining faces, and evolved into so-called polygonal masonry. Walls of this construction generally had their faces smoothed, but their joints were neither vertical nor horizontal. Rather, the geometry can best be described as a set of randomly located points joined by straight junctions. The resulting stones were polygonal, although this does not imply that their sides were equal in length. The walls of Alatri (Fig. 15) are a good but relatively late example (Blake, 1947: 97–8).

Squared stone masonry, often called *opus quadratum*, has horizontal and vertical joints and rectangular stones, with the vertical joints usually staggered to provide interlocking or bond between the stones. A less regular form is described by Blake as *opus quasi-quadratum*.

The interior of a squared stone wall could be filled with

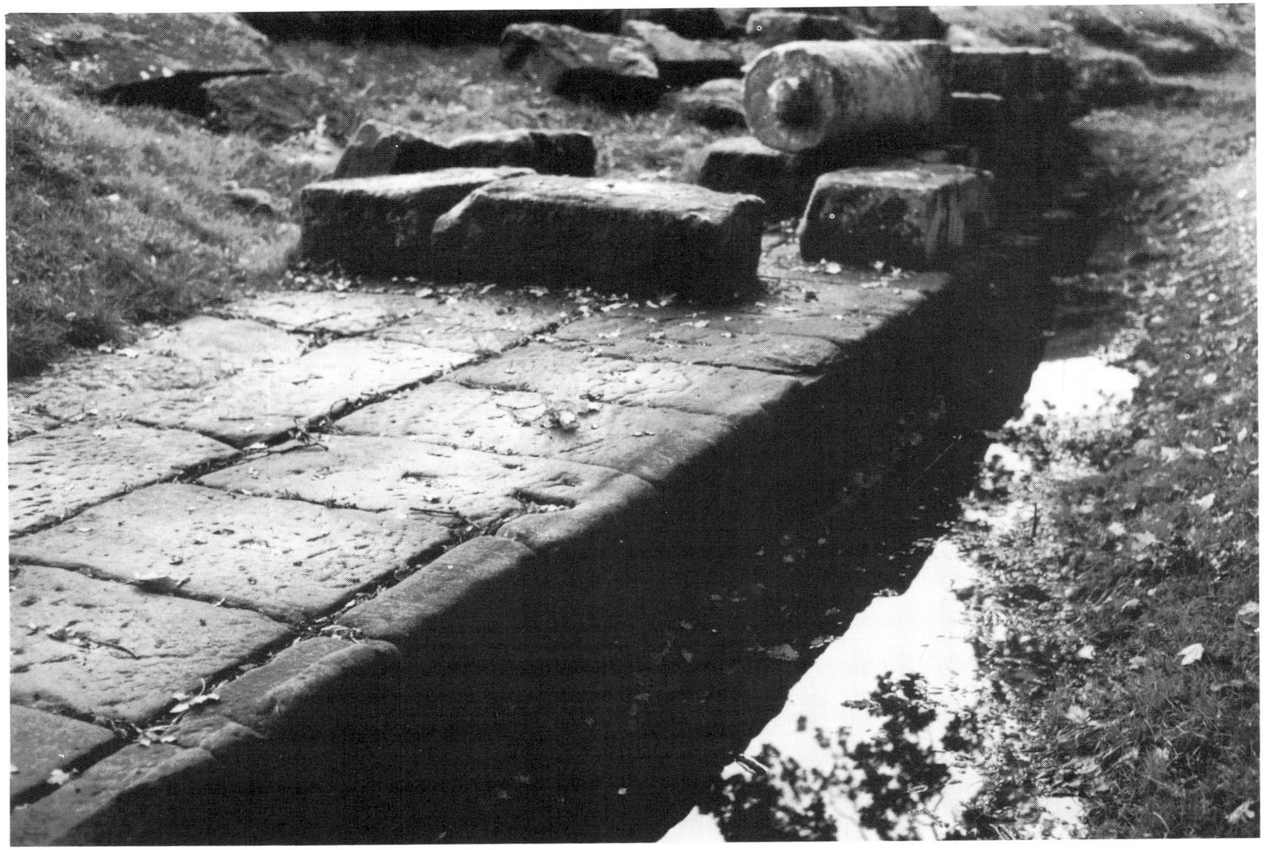

Fig. 14 Abutment of Roman bridge at Chesters on Hadrian's Wall (T30, Section 6.4). The long groove in the upper face in the foreground, with a T-shaped end in the distance, is for a metal cramp.

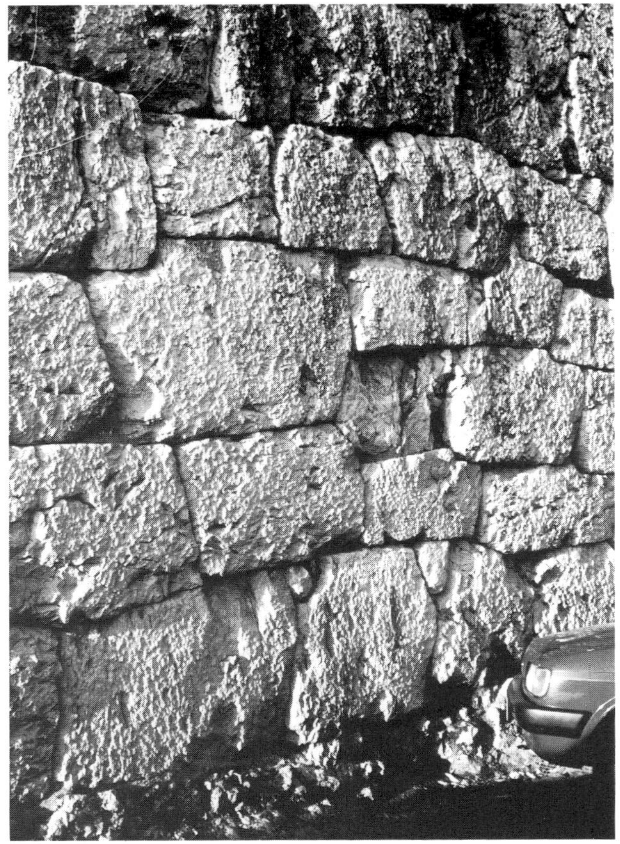

Fig. 15 Cyclopean masonry wall at Alatri (for scale, see the front of a car at bottom right).

more squared stones. Alternatively a wall faced with stone may be filled with concrete (see Section 4.5). In this case, the exterior stonework could become no more than a facing, subsidiary to the internal concrete core. This facing could be composed then of smaller stones – in *opus incertum* there is no discernible pattern in the jointing of these facing stones, whereas in *opus reticulatum* stones with small square faces were aligned with their diagonals horizontal – the joints formed a continuous pattern or net of mutually perpendicular lines, inclined at 45° to both the vertical and the horizontal.

Clay was used to form bricks and tiles, dried either in the sun or in kilns. The term *opus testaceum* has been used to describe concrete walls faced with bricks or tiles. The dimensions of Roman bricks varied widely, but the three common sizes were *bipedales*, about 600 mm or 2 Roman feet square; *sesquipedales*, 450 mm or 1.5 feet square; and *besales*, 200 mm square. The thicknesses also varied widely, from about 4 to 7 mm. Roofing tiles, made of kiln fired clay, were also converted for use in wall construction, by either chipping or sawing off the flange required in the roof but not in the wall. The resulting tile bricks were somewhat thinner, about 4 to 5 mm. One form of facing was achieved by cutting tiles or bricks to form isosceles triangles, with the bases showing at the faces of the wall, and the intervening space filled with concrete, the wall being about 600 mm thick. Bonding courses of *bipedales* were laid at intervals to join the two faces.

Generally these types of construction came into use in the order quoted. Vitruvius (2.8.1–4) mentions four of them – the use of squared stone, *opus incertum*, *opus reticulatum* and concrete faced with brick or tile. Bridges required the construction of piers, arch vaults and spandrel walls (see Section 8.1). Only some of these forms of construction were used. Squared stone facing with a core of either squared stone or concrete was by far the most common. Some later bridges had arch vaults faced with *bipedales*. Section 8.5 discusses the coursing of stones in Roman arches. In aqueduct construction, *opus incertum* and *opus reticulatum* were also common.

4.5 Concrete

Lime was manufactured by the Romans by burning limestone in kilns and then slaking in water. The first process reduced calcium carbonate to calcium oxide or quicklime; the addition of water converted this to calcium hydroxide, or slaked lime. The process has been described by Vitruvius (2.5.1–3; 7.2), who advised the selection of white stone, and knew of the importance of thorough slaking before use. Lime has the capacity of hardening on exposure to the air; the calcium hydroxide combines with the carbon dioxide of the air to form again the calcium carbonate of which it was originally composed.

The use of lime with sand and water to form a hardening mortar was known to the Greeks (Blake, 1947: 311), who passed this knowledge on to the Romans. They, in turn, were fortunate to find, presumably by chance, a means of converting this to a hydraulic cement, that is, a cement capable of hardening under water. Vitruvius (2.4.1–3) has a section on sands, which he divides into sea sand, river sand and pit sand, or sand dug from pits. Concerning this pit sand he goes on to say (2.6.1), 'There is also a kind of powder which, by nature, produces wonderful results. It is found in the neighbourhood of Baiae and in the lands of the municipalities round Mount Vesuvius. This being mixed with lime and rubble, not only furnishes strength to other buildings, but also, when piers are built in the sea, they set under water.' The material to which he refers is not truly a sand, rather it is a fine volcanic ash related to the harder *tufa* used by the Romans as a building stone (Section 4.4). 'It is found in all the volcanic regions of Italy, but takes its name (*pozzolana*) from Pozzuoli, ancient Puteoli, near Naples, where it was first mined' (Blake, 1947: 41). Vitruvius (5.12.2) also describes it as coming from Cumae and the promontory of Minerva – for underwater work he says, 'Earth is to be brought from the district which runs from Cumae to the promontory of Minerva, and mixed in the mortar, two parts to one of lime.'

The name 'pozzolan' has been transferred in modern practice to any similar material which hardens by combination with lime and water. It is rare to use a pozzolan alone. When it hardens with water, Portland cement releases lime. This may combine with certain aggregates in a way that is deleterious to the hardened concrete. If a pozzolan is used as an admixture with Portland cement, it takes up this surplus lime, and contributes also to the concrete strength.

Hopkins (1970), quoting from Lea and Desch, has compared the chemical composition of normal Portland cement with those formed by the combination of *pozzolana* from Baiae and Rome with lime and has shown that these compositions are similar. Vitruvius (2.6) speaks of four types of *pozzolana* – black, white, deep red and carbuncular. The original *pozzolana* from the Vesuvius area was coloured brownish to grey (Blake, 1947: 42). It was originally imported for use in Rome, possibly as early as about 200 BC, but it was then discovered, about the time of Julius Caesar (*c.* 60 BC) that the local red *pozzolana* gave better results. Blake (1947: 313) distinguishes the three phases: the introduction of lime mortar from Greece; the discovery of the hydraulic property of *pozzolana* from Puteoli with lime; and the observation of the greater strength achieved with the red *pozzolana* found near Rome.

In modern concrete, cement, aggregate and water are mixed together, and then placed in the forms in a semi-fluid condition. Roman practice was different, with the mortar (*pozzolana*, lime and water) handled separately from the larger aggregate, whose pieces were called *caementa*. Initially at least the pozzolan was not sieved (Blake, 1947: 314) and had a range of particle sizes, from fine to coarse sand and beyond. It was not washed, and could include soil or dust. The recommendations of Vitruvius (2.6) were that it be used in the proportions 1:2 of lime for underwater work and 1:3 for normal work.

According to Blake (1947: 328, 346) the mortar was commonly poured over the aggregate. However, she refers also to *caementa* laid in mortar (1947: 329, 333, etc.), and again to the *caementa* and mortar thrown in and 'rammed down into a compact mass' (1947: 351). The pouring of mortar over multiple layers of aggregate would inevitably have led to poor quality concrete, and it is difficult to believe that this would not have been realised by the Romans. More likely the mortar and aggregate were brought up together in lifts of about one course, at least for more important work. The *caementa* could be quite large, as much as 7 kg in weight; more normally they were smaller, of the order of 60–100 mm across (Blake, 1947: 350). They could be of stone, such as *tufa*, selce or travertine, but broken tiles were also often used.

The concrete was usually placed with permanent forms or facing of squared stonework, *opus incertum* or *reticulatum*, bricks or tiles. *Opus reticulatum* usually consisted of tapered stones embedded in the concrete. The easiest method of construction would be to take up the facing with the core – indeed it is difficult to see any reasonable alternative. *Opus incertum* also would hardly withstand the pressure from high lifts. With brick facing, bond was often achieved by using longer bricks or tiles extending into the core, and in this case also it is reasonable to expect that the core would be brought up before these bricks were laid. Only with a facing of large squared stonework is it likely that the facing would have been brought up well in advance of the core.

Concrete walls and foundations were also constructed between timber forms. Most commonly these consisted of vertical posts and horizontal boards, possibly with the posts inside the boards (Blake, 1947: 337). These faces could be left exposed, but were often faced subsequently with stone bonded to the concrete with metal ties.

In bridge construction, exposed concrete was rarely, if ever, used: the most normal form of construction was a concrete core within squared stone walls and with a squared stone vault. Alternatively, the vault could be constructed of brick or tile. In these circumstances neither the strength nor the durability of the concrete would be of great significance. Stresses in the concrete would be small, and the avoidance of any exposed faces would automatically lead to high durability.

Malinowski (1982) has discussed the durability of ancient mortars and concretes. He states: 'Despite centuries of use and exposure to adverse environmental conditions, these mortars and concretes have often survived better than the natural stone or burnt brick found in the same structures and sometimes even better than modern concretes exposed to similar conditions.' No doubt the explanation of these high durabilities lies in the fact that durability is often more a function of physical structure and pore size than of strength. He adds some further reasons. In the case of renders, he suggests that the practice of using multiple layers of high-quality polished mortar as described by Vitruvius (7.3.5–9) controlled shrinkage and gave reduced permeability. In concrete, the nature of the aggregate and the pozzolan itself was such as to lead to air voids, similar to those deliberately introduced into modern concrete by the use of air entraining agents. The presence of these voids reduces internal pressures produced by freezing and thawing, and by crystal formation due to absorbed salts. Other recent studies may be found in Malinowski (1979), Rayment and Pettifer (1987) and Davidovitis (1987).

4.6 Timber

The term 'timber' tends to be used by the engineer for sawn timber or for structural members, as distinct from 'wood' for the same material in the tree. Here the terms will be used interchangeably. It will be argued later (Sections 6.2, 6.4) that the use of timber by the Romans for bridge construction was more widespread than is commonly believed. It was used temporarily for centering, scaffolding, cranes and other erection equipment, but also for final structures.

Vitruvius (2.9) lists twenty timbers, of which, to judge from his comments, seventeen had some use in building construction – fir, oak, winter oak, white and black poplar, willow, lime, withy (*Vitex*), alder, elm, ash, hornbeam, cypress, pine, cedar, juniper and larch. He also lists Turkey oak, cork and beech, but dismisses them with the comment that they will soon rot. The 1971 British Standard Code of Practice for the Structural Use of Timber, CP112 Part 2 (British Standards Institution), gives basic allowable stresses for six of these timbers, when used green or in the fully seasoned or dry condition.

Of these, the strongest is ash (*Fraxinus excelsior*), with basic allowable stresses of 17.2 MPa green (1 MPa = 145 psi), 22.8 MPa dry, and a density of 720 kg/m^3. It commonly grows to 25 m high, exceptionally to 40 m, and can have a long, straight trunk. Vitruvius says that when it is green it bends easily, but on being well dried becomes much harder. Baterden (1908) describes it as 'too flexible for building work', but it is hard to substantiate this, as its modulus of elasticity – 6600–10,000 MPa green, 7200–11,400 MPa dry – is comparable with that of other timbers of similar strength.

Beech (*Fagus sylvatica*) has physical properties equal to those of ash. A very large, spreading tree, it often grows to heights of 30–40 m. Although rejected by Vitruvius, it appears to be a useful timber.

Oak (*Quercus robur*) has basic allowable stresses of 15.9 MPa green, 20.7 MPa dry, and a density of 720 kg/m^3. A very large tree, with a height of up to 45 m, its trunk is irregular in shape. Although this restricts the availability of large structural timbers, Baterden speaks of a piece 9.5 m long, with cross-sectional dimensions 560 mm × 460 mm. This irregularity may also cause the timber to warp when green, as remarked by Vitruvius, but if properly seasoned it forms an excellent building timber noted for its long life.

Larch (*Larix decidua*), with basic allowable stresses of 13.8 MPa green and 17.2 MPa dry, and a density of 560 kg/m^3, is a tall tree (to 40 m) with a straight trunk. Vitruvius speaks well of its durability and describes an incident when Caesar, besieging a fortress at Larigna, was unable to set fire to the timber; he enquired and named the timber 'larch' after the name of the place. He attributes its durability to its resin content, and this is confirmed by Baterden.

Pine (*Pinus sylvestris*) has basic allowable stresses of 11.0 MPa green, 15.2 MPa dry, a density of 540 kg/m^3, and again is a tall, straight tree, reaching 35–40 m in height.

Fir, as listed by Vitruvius, may be European spruce (*Picea abies*), 8.3 and 11.0 MPa, 380 kg/m^3, and another tall, straight tree. Of all trees, it is possibly the lightest for its strength.

The inclusion of timbers in the British Code of Practice is influenced by their availability today (most of these timbers, for example, are omitted from the current BS 5268, Part 2, British Standards Institution). No doubt others of Vitruvius' timbers, such as elm, were also excellent building timbers. However the list given above does give some indication of the strengths of suitable timbers. The term 'basic allowable stress' implies that these are the allowable stresses for defect-free samples.

For timbers used in bridge construction, more realistic allowable stresses would be about 0.65 times these values.

Consider a bridge with two girders, each 450 mm high by 300 mm wide, carrying a deck 3 m wide, with planking of the same timber as the girders, 100 mm thick, carrying a superimposed load of 250 kg/m^2. Then, for simply supported girders, the safe span will vary with the timber strength. For the six timbers listed above, the safe spans, calculated on the basis of strength and not deflection, are as follows:

ash, beech	12.6 m
oak	12.1 m
larch	11.8 m
pine	10.6 m
spruce	9.6 m

To assess the significance of the superimposed load, suppose that the central space, 3 m wide × 1 m of the deck, is occupied by a heavy load, and the remainder is occupied by men, each of mass 90 kg, with one for every 2 square metres of the bridge. Then the 12.6 m span bridge has a safe central load of 4.0 t while it carries also seventeen men. At the other limit, the 9.6 m span bridge could carry a central load of 3.1 t and thirteen men.

Although purely conjectural, these figures are sufficient to show that the timber available to the Romans was strong enough for bridge construction. In terms of durability, the first four species would probably perform better than the last two, and there are others in Vitruvius' list that would probably perform equally as well. A span of about 11 m would appear to be about what was actually used.

4.7 Surveying

Dilke (1971) has described the work of the Roman land surveyors or *agrimensores*, who were required to divide the land into squares or *centuriae*, each with sides of 2400 Roman feet, for subdivision into 100 plots of the early agricultural smallholdings. More simply, the surveyor might also be called *mensor* or *metator*, or possibly *librator* (leveller) (Hamey and Hamey, 1981: 14; de Camp, 1977: 165) from the work of levelling, so important for the construction of the aqueducts.

Roman surveying instruments have been described by Dilke (1971: 51, 66–81 etc.), White (1984: 170–2), Hill (1984: 117–23), Parsons (1968: 97–104) and others.

Dilke quotes Hero, who says that ropes or cords, twice tautened and then smeared with wax, were used for measuring long distances. Alternatively, for shorter distances, a wooden measuring rod 10 Roman feet long (1 Roman foot = 0.963 English feet = 0.294 m) might be used, with the ends subdivided into inches.

The *groma* was used to extend a line in the same direction, or to mark off a line at right angles. A restored *groma* has been reconstructed from metal parts found in the workshop of a surveyor at Pompeii and is held in the Museo Nationale at Naples. It has a vertical shaft which could be either pushed into the ground or mounted vertically on a base. At its top it supported a pivot, offset from the centre of the shaft by a distance of 235 mm, possibly intended as 1 Roman foot. This in turn supported a timber cross, mounted horizontally, with plumb bobs hanging from each of its four ends. The surveyor could use the two plumb bobs from one arm of the cross to sight along the line from the pivot to a preceding point, and then the same two plumb bobs to sight forward along the same line. Alternatively, he could use the other two plumb bobs to establish a line at right angles.

The Romans knew how to construct a 90° square. Vitruvius (9.Preface.6) quotes Pythagoras and says, 'If three straight rods be taken, of which one is 3 feet long, the second 4 feet, the third 5 feet, and let these rods, being jointed together, touch one another at their extremities in the form of a triangle, they will make a perfect set-square.'

Levelling was carried out by the combination of vertical staffs and level. Dilke (1971; see also White, 1984) has reconstructed a staff described by Hero, with a graduated timber rod of rectangular section and a sighting disc controlled by a cord passing over a pulley at the top of the staff, and fixed by a pin from the other end of the cord into the back of the staff.

The simplest levelling device was the *chorobates* described by Vitruvius (8.5). A deep horizontal timber member or plank 20 feet long (6.1 m) had vertical legs attached at each end. Lines were drawn on the legs accurately perpendicular to the upper face of the plank. Plumb bobs were hung in front of these lines from pins driven in near their tops. The slope of the *chorobates* was adjusted until the cords of the plumb bobs were aligned with the lines; sightings were then taken along the upper face. Vitruvius continues:

> But if the wind disturbs and, owing to their movements, the lines cannot give a certain indication, a channel is to be put on the top side of the plank, five feet long, an inch wide and an inch and a half deep [1.5 m by about 20 × 30 mm]. Let water be poured in. If the water evenly touches the lips of the channel, we shall know that the levelling is successful.

He was aware that the earth was spherical, but argued that the ends of his device would still form a horizontal line.

Hero (Dilke, 1971: 76) described a much more elaborate device, the *dioptra*, which he claimed not to have invented, but to have perfected. A circular table was mounted first on a horizontal axis, by attachment to a toothed vertical circular plate, which rotated about the axis and was controlled by a hand-operated screw which engaged the teeth on its perimeter. This whole assembly was then mounted on a vertical shaft about which it could be rotated by another hand operated screw. The whole rested on the ground (Parsons, 1968: 103–4).

A device giving a line of sight could be mounted across a diameter of the table. Alternatively, a levelling system, the *libra aquaria*, could be used. This had a horizontal bronze tube mounted within a member that rested on the table. Each end of the tube fed up into a short length of vertical glass tube. The whole was filled with water. Horizontal plates at each end were aligned with the water level by rotating the table about its horizontal axis and used to provide a horizontal line of sight.

It is not clear that Hero's devices were used in practice. Vitruvius (8.5) says, 'But the more accurate method is by the *chorobates* because the *dioptrae* and the water levels mislead.' In speaking of aqueducts (8.6.1) he says, 'the bed of the current must be levelled with a fall of not less than 6 inches in 100 feet'; that is, 1 in 200. This suggests that the accuracy of levelling was somewhat better than this. A conservative estimate might be about 1 in 500. Forbes has suggested 1 in 2000, but White (1984: 172) doubts if this was possible. A slope of 1 in 500 of the *chorobates* would cause a difference of level between the ends of the water channel, 1.5 m long, of 3 mm. This would be perceptible. Although there would be other sources of error, an accuracy of the order of 1 in 1000 should be achievable.

Another instrument described by Vitruvius (10.9.1–4) was the hodometer used for measuring distances, or in his words for 'measuring a journey'. It was a four-wheeled carriage with wheels 4 feet in diameter. In the carriage was a drum, geared to a wheel in such a way that 400 revolutions of the wheel caused one revolution of the drum. As it completed a revolution, a stone in the drum came opposite an aperture and fell into a bronze vessel. The resulting noise indicated the passage of a Roman mile, and the final number of stones in the vessel indicated the number of miles. The geometry was based on the principles that the circumference of the wheel was 12.5 feet, and 400 revolutions corresponded to 5000 feet or 1000 paces, equal to one mile. Assuming that one Roman foot was 0.963 English feet or 0.294 m (as quoted earlier from Dilke, 1971), one Roman mile would equal 0.912 English miles or 1.47 km. Chevallier (1976: 39) gives the slightly higher values of 0.915 English miles or 1.48 km. The corresponding equivalents of one Roman foot would be 0.972 English feet or 0.296 m.

Roman bridges were typically straight in plan. The deck could be straight and horizontal in elevation; it could be inclined; or it could rise to the centre. In any case it would be necessary to control levels from side to side of the crossing. The direction of the bridge with respect to some absolute reference axis, such as north and south, was hardly likely to be important; neither was its absolute location. The necessary surveying skills were therefore to establish a line that was straight in plan and parallel to the axis of the bridge; to mark off directions perpendicular to this line; to measure distances along and perpendicular to this line; to establish verticals; and to fix relative levels over the area occupied by the bridge. The equipment necessary to carry out these tasks could consist of cords for use as straight lines and for measuring distances, the *groma*, the plumb bob and the *chorobates*.

Some Roman bridges were skew; that is, the channel was not perpendicular to the length of the bridge. The direction of the channel could have been determined by measurement from lines perpendicular to the axis. Alternatively, the more elaborate *dioptra* could be used, although there is no direct evidence for this.

In modern bridge construction, long members made in a factory or at a remote site need to be fitted to foundations constructed in the field, and a high level of accuracy needs to be maintained. This was not so in Roman times; in a stone arch bridge, for example, the relatively small stones could be shaped in the field, and errors could be accommodated. The same was the case with the typical timber girder. Under these circumstances, an accuracy of 1 in 500 in either slope or distance would not normally cause difficulty. Perhaps the only exception would be with a long low bridge over a stream that was liable to flood. The bridge at Mérida, to quote an extreme example, is about 720 m long between abutments. An error in level of 1 in 500 would correspond to a difference in height between abutments of 1.44 m. Such an error would be significant in this case; one would have hoped for an accuracy of 1:2000, as suggested by Forbes, or 0.36 m. However, this is an exceptional bridge. A study of existing levels and dimensions, including those at Mérida, suggests that the Romans were not greatly concerned by quite substantial errors. The figure of 1 in 500 for bridge construction is probably a fair indication of what they had in mind.

4.8 Computation and drawing

The construction of a modern bridge is preceded by design calculations and the preparation of technical drawings. Design calculations are concerned with such matters as the geometry and strength of the bridge. Chapter 8 enquires whether the Romans had mathematical rules for design, but it is useless to ask this question unless it can be shown first that they had the necessary basic mathematical skills, that is, the ability to add and subtract, multiply and divide.

Modern technical drawings serve a number of functions – they aid the engineer in the design process, they provide the means whereby the engineer transmits his intentions to those who build the bridge, and they serve as a record. It is appropriate then to ask whether the Romans also produced design drawings.

The history of mathematics has been discussed by Smith (1919, 1925), Knott (1886) and others (Bunt et al., 1976). The familiar Arabic or Hindu-Arabic number system of the present day has existed from about the first century BC. It is a decimal system, with ten distinct characters. Counting proceeds in the order 0, 1, 2, 3, 4, 5, 6, 7, 8, 9 (i.e. ten characters), and then continues by placing 1 in a second column, and returning the first column to 0. The number 71, for example, means 7 times 10 plus 1.

The Roman system used letters as numbers – I for 1, V for 5, X for 10, L for 50, C for 100, D for 500 and M for 1000. It could be extended higher, to as large a number as required, using a C reversed or a preceding C as multipliers (to multiply by 10 or 20), by extending a line above the number (to multiply by 1000), or enclosing a number with vertical lines on either side (to multiply by 100,000). It could also express some fractions (1/2, 1/3, 1/4, 1/2 + 1/3, 1/2 × 1/3 etc.). Although more cumbersome than the Arabic system – compare, for example, 731 with DCCXXXI – it was also a decimal system, but with some important differences: it did not have nine distinct characters between 1 and 9; it had no symbol for 0; but it did have a character, X, for 10. Nevertheless, counting started again after ten – XI, XII, etc. – and the multipliers mentioned above were all decimal.

The Arabic decimal number system is so familiar that it is easy to forget that alternatives are possible. The binary system used in digital electronic computer systems has only two characters, 0 and 1, and counts in the order 0, 1, 10, 11, 100, 101, 110, etc. At each stage, when a column reaches the limiting character (in this case, 1), counting proceeds by adding 1 in a new column, and returning the original column to 0. To give another example, counting on the fingers and thumbs may be improved by using a system based on five (the quinary system): count 1 to 5 on the right hand; add 1 on the left hand, returning the right hand to zero; then start again on the right hand. The quinary number 24 for example – 2 on the left, 4 on the right – means 2 times 5, plus 4, or 14 in Arabic. In this way, one can finger-count to 30 instead of 10.

It is clear that, for arithmetic computations, the Romans used an abacus, of which three forms are known (Smith, 1925: 157–8). The modern abacus has a frame enclosing beads on wires; an earlier form had counters placed either on lines or in grooves on a table or board; the oldest or dust abacus had marks and lines in dust, either on the floor or on a table. Both the dust and counter abacus predated Roman times. The British Museum has a Roman abacus with pebbles on a grooved bronze plate (Fig. 16). The pebbles were typically of limestone or calx. To calculate means, literally, to pebble.

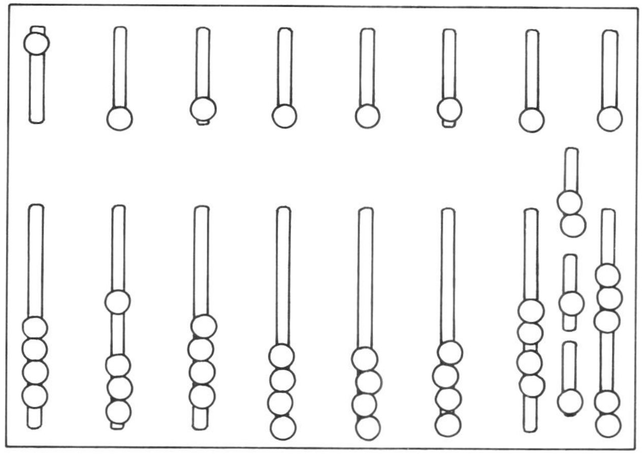

Fig. 16 Roman bronze abacus (based on Smith, 1925: 167).

The abacus is still in common use in many countries. In China it is called the *suan-pan*, in Japan the *soroban*. Both of these, and also the Romans use the base five, the quinary system, but with variations. Initial counting in the Japanese system uses five counters with one equal to 1, five equal to 5. The Roman and Chinese systems have only four pebbles – when counting has reached 4, the next number is 10; in effect the five characters are 0 to 4 but with the zero not expressed (zero means no pebble, for which no special symbol is required). This is illustrated below:

Decimal counting	1 2 3 4 5 6 7 8 9 10 11 12 . . .
Quinary counting	
Japanese	1 2 3 4 5 11 12 13 14 15 21 22 . . .
Roman	1 2 3 4 10 11 12 13 14 20 21 22 . . .,

where it should be remembered that, on the abacus, a number such as 4 is not a special character, as it is in modern arithmetic, but a number of pebbles or beads. Figure 17 shows the Japanese *soroban* with the number 371 in columns 2, 3 and 4, and 284 in columns 7, 8 and 9. Appendix 2 describes its use.

The Roman abacus (Fig. 16) is different in some respects. First, it has fewer columns – seven, if the two columns at the right are ignored. If this is the standard Roman instrument, then it is suited only to the simpler calculations. Second, as noted previously, the counter arrangement is different, although it may be operated to achieve the same effect. Finally, the purpose of the two right-hand columns is not known – although they may possibly be associated with a desire to store the remainder following division. The concept of the decimal fraction was not known until about 1600 (Bunt et al., 1976: 229).

The presence of the abacus in Roman times makes it

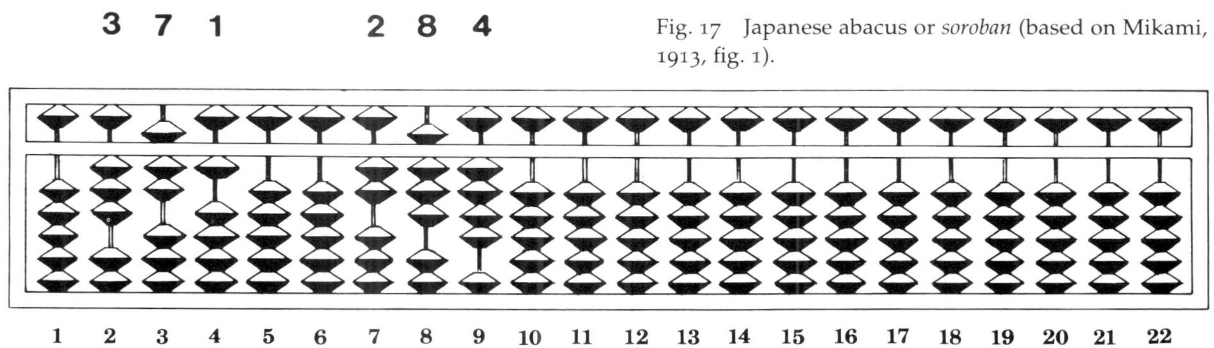

Fig. 17 Japanese abacus or *soroban* (based on Mikami, 1913, fig. 1).

certain that the Romans could add, subtract and multiply. The existence of Latin words for fractions suggests that they could divide by simple numbers, and with the abacus they had a device that could perform more complex divisions also. Although the concept of the decimal point and the decimal fraction was much later (see, for example, the article on 'Fractions' in the *Encyclopaedia Britannica*), anyone with experience in the division of decimal numbers would rapidly learn that it is easier to divide by ten than by any other number. For example, neglecting remainders,

Arabic 731 ÷ 10 = 73
Roman DCCXXXI ÷ 10 = LXXIII

Concerning the use of design drawings, Coulton has an excellent discussion of the whole problem of design in his book *Greek Architects at Work* (1977, ch. 3). He says, 'The idea of an architectural ground plan had certainly been developed in both Egypt and Mesopotamia, for examples have survived ... Two statues of Gudea of Lagash (*c.* 2200 BC) show him seated with a drawing table on his knees, equipped with stylus and ruler, and on one of these tables a plan is engraved' (Coulton, 1977, pl. 5). He refers also to a papyrus plan of the tomb of Ramasses IV (12C BC); such plans were also produced on tablets of soft stone. Haselberger (1985: 122) suggests that, before construction, Greek architects produced scale drawings, probably on 'papyrus, parchment, whitewashed wood tablets or even flat stones', and gives a specific example where a stone block was used. He states also that lines on the walls and floor of the temple of Apollo at Didyma in Greece were working drawings.

As far as the Romans were concerned, it is clear that they produced design drawings for some buildings: to quote Coulton again (1977: 68), 'Architectural drawings and drawing equipment have survived from the Roman period', and he has an excellent photograph (pl. 6) of a building plan on stone (see also Dilke, 1985: 102–11). Procopius (1.1.24), in connection with the church of Santa Sophia in Istanbul, mentions the master builder, Anthemius, 'preparing in advance designs of the future construction'.

Vitruvius (1.2.2) speaks of those things on which architecture depends. He refers to arrangement, which consists of *ichnographia*, *orthographia* and *scaenographia*, generally translated as plan, elevation and perspective. Although it would be possible to read all terms as applying to the building as a whole, either in finished form or as a concept, the paragraph suggests that they refer to drawings. Gwilt's rather free translation (1860) reads, 'The first is the representation on a plane of the ground-plan of the work, drawn by rule and compasses. The second is the elevation of the front, slightly shadowed, and shewing the forms of the intended building. The last exhibits the front and a receding side properly shadowed, the lines being drawn to their proper vanishing points.'

In Morgan's later translation of Vitruvius, the first sentence becomes, 'A ground plan is made by the proper successive use of compasses and rule ... An elevation is a picture of the front ... properly drawn in the proportions of the completed work.'

There is little doubt, then, that the Romans could and did produce design drawings, at least for some buildings. They may also have followed the Greek example of producing detailed specifications – *syngraphai* (Coulton, 1977: 54).

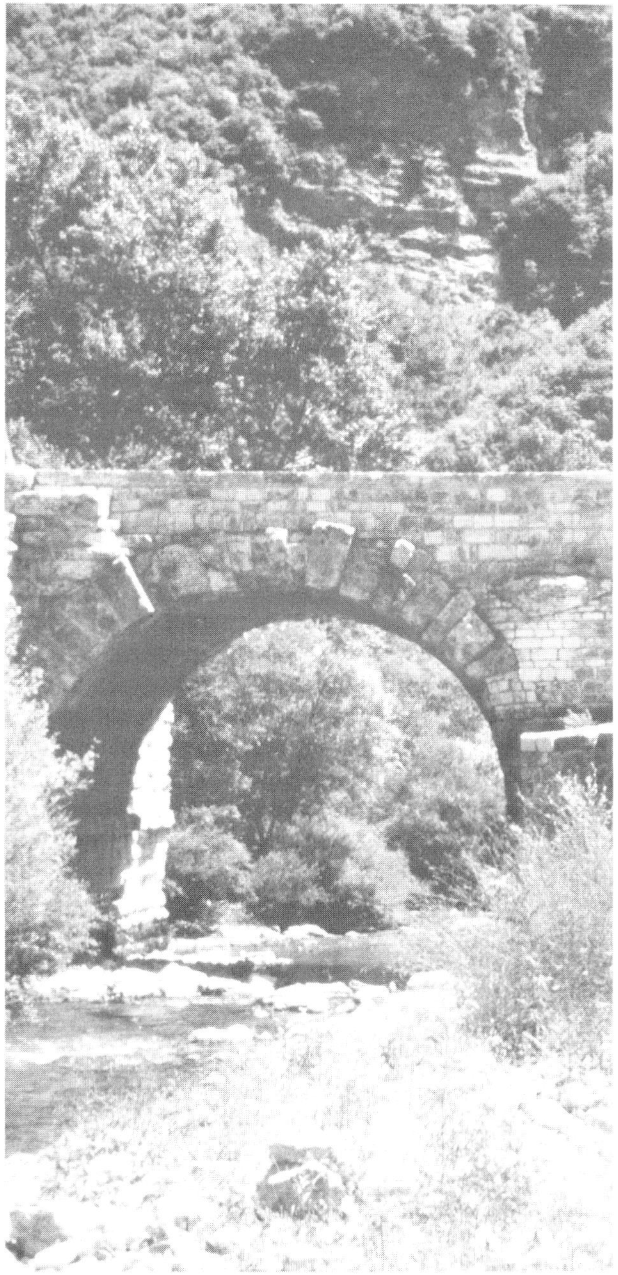

5

MASONRY BRIDGES

5.1 Introduction

This chapter has three objectives: to list existing bridges; to identify where they may be found; and to describe the more important ones.

Timber bridges are dealt with separately in Chapter 6. Most (if not all) British bridges had a timber superstructure and for this reason all British bridges will be considered there. Stone arch bridges will be discussed further in Chapters 8 and 9 in a study of their proportions, to clarify principles used in their design; and to see whether any evidence exists as to the method of their erection. Some arch terminology is defined there, in Figure 132 and pages 163–4; see also Appendix 1 (Glossary of technical terms).

The order of treatment is geographical, beginning with Rome and moving out. In many cases a group of bridges forms a sequence along a particular road, in which case that sequence will be followed, outwards from Rome. The section headings indicate the countries in which the bridges are to be found, with one exception. Italy is divided into four sections: Rome and its vicinity, southern Italy (including the islands), central and northern Italy.

All bridges (excepting some which no longer exist) are given an identifying number, I1, etc. – I for Italy; G for Gaul; SP, Spain and Portugal; A, Africa; and E, the east – and are listed in Appendix tables A.1–8. Some of the less important bridges are not described in the text. Others are included in the text, but in smaller type. The original source material used in this book is that provided by Gazzola (1963b), supplemented by a study of other sources. Gazzola is, however, remarkably complete and accurate, and cites many references.

5.2 Rome and its vicinity

The area considered in this section is bounded by lines from Civitavecchia to Carsoli, and Carsoli to Latina. It includes Rome itself, the Via Aurelia to Civitavecchia, Veio, Tivoli, Palestrina, Velletri, Cori and Ostia, and corresponds generally to Rome and the Roman Campagna, with a few excursions beyond. A list and summary of the bridges may be found in Appendix table A.1.

The ancient bridges of Rome across the River Tiber began about 5 km north, with the Pons Mulvius on the Via Flaminia, and then, in order downstream, Pons Aelius, Neronianus, Agrippae, Aurelius, Fabricius and Cestius, Aemilius, Sublicius and Probi; ten in all (van der Heyden and Scullard, 1963). There are now only substantial remains of five – Mulvius, Aelius, Fabricius, Cestius and Aemilius.

Pons Neronianus (*c.* AD 37–68) crossed the river on the bend immediately downstream of the modern Ponte Vittorio Emanuele (Nash, 1968: 193; Platner and Ashby, 1929: 401), on the line of the Via dei Coronari, and was probably destroyed before the time of Constantine (AD 306–37); the remains of its stone piers can still be seen at low water. Pons Agrippae (*c.* 33–12 BC) has been located 160 m north west of the Ponte Sisto, again with foundations remaining in the river bed (Section 3.2; Nash, 1968: 184; Platner and Ashby, 1929: 398). It survived until at least AD 147, when it was restored by Antoninus Pius. The Ponte Sisto was built in 1473–5 on the foundations of the Pons Aurelius attributed to Marcus Aurelius (AD 161–80). It was restored in AD 366 by Valens and Valentianus (Platner and Ashby, 1929: 398–9; Nash, 1968: 185). The ancient Pons Sublicius (about 600 BC) was of timber and is discussed in Section 6.3. It was located downstream of the Isola Tiberina and the Pons Aemilius but upstream of the modern bridge with the same name (Chevallier, 1976: 95). The furthest downstream of these bridges was the Pons Probi or Pons Theodosii or Pons Marmoreus Theodosii (Platner and Ashby, 1929: 401; Nash, 1968: 196), between the old and the new Pons Sublicius. It was rebuilt about AD 381 to 387, but destroyed by the beginning of the eleventh century. The bases of its piers were visible in the river until 1877.

The name of the **Pons Mulvius (I1)**, although substantially the same throughout its existence, has changed to some extent, variants being the Ponte Milvio, the modern name, and Ponte Molle. As shown in Figure 18, looking north from the upstream bank, the bridge has six spans, with spans 2 and 3 faced with white travertine. Ballance (1951) has described the history of the bridge.

Construction of the Via Flaminia, which passed across the bridge, began in 220 BC (Sections 2.2.4; 3.2; Ballance, 1951: 80; Ashby and Fell, 1921: 127; Blake, 1947: 134). The bridge was referred to by name by Livy (27.51.2) in 207. In 109 BC, M. Aemilius Scaurus carried out either a thorough restoration or, more likely, a complete reconstruction, and substantial portions of his work still remain. Extensive repairs were carried out between 1451 and 1481 by the popes Nicholas V and Calixtus III, and again in 1805 by Pius VII. During the defence of Rome by Garibaldi in 1849, the northernmost span was blown up (Nash, 1968: 191), but was repaired later in the same year. There was a later overhaul in 1871. At present the bridge is intact, but used by pedestrian traffic only.

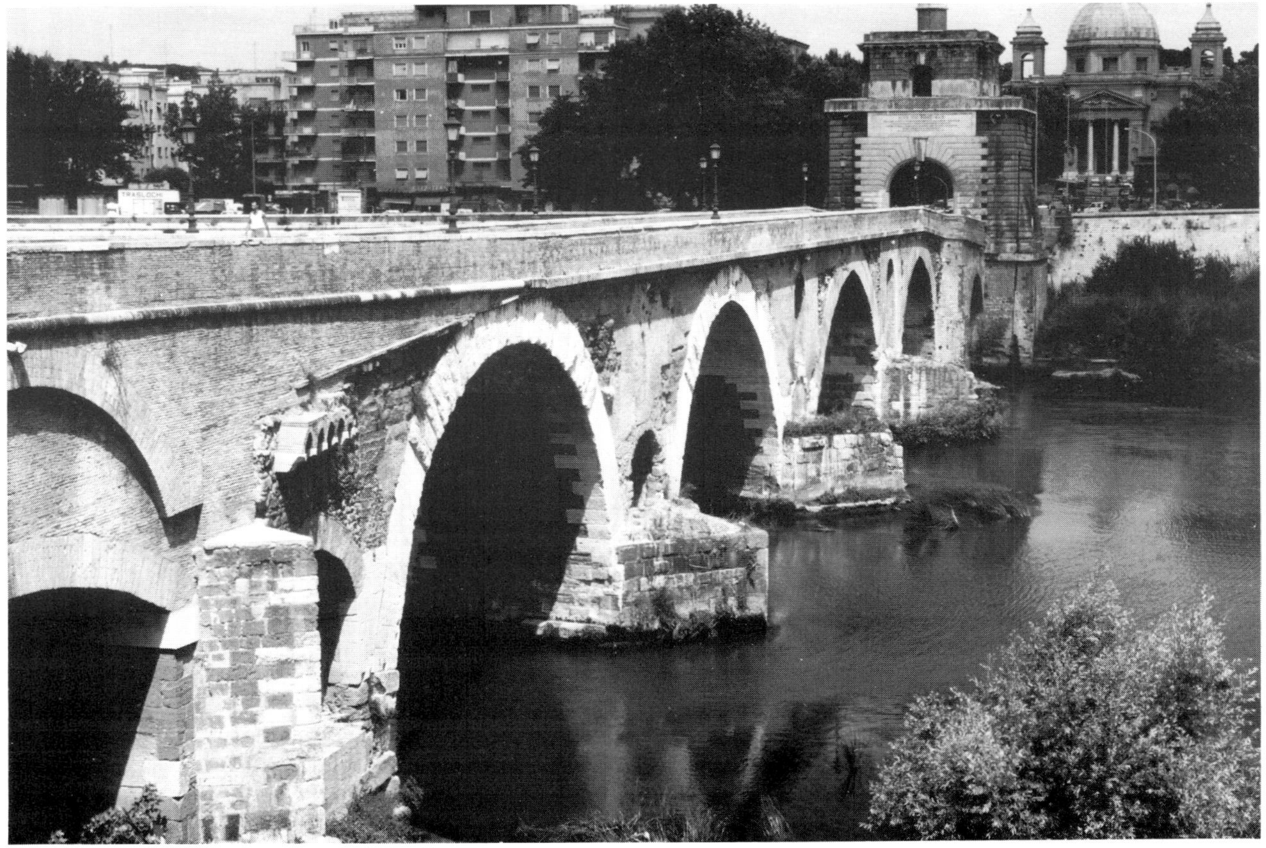

Fig. 18 The Ponte Milvio, over the River Tiber on the northern outskirts of Rome (I1).

Ballance numbers the six existing spans as 1–6 from the southern end, with the four central spans 2–5, each about 18 m clear, and the end spans about 9 m. The first span and southern abutment have been extensively reconstructed, and the original northern abutment is now buried. Gazzola (1963b, no. 29) shows a seventh span in the northern approach. There are also floodways in the four more southerly piers. They differ considerably in geometry: the first two are small in height and set low; the next has its crest high up, close under the roadway, but the bottom of the opening is also high; the fourth again has a high crest, but its base is somewhat lower, providing the tallest opening of the four. There is no flood opening between spans 5 and 6.

Span 1, which is of brick, is entirely new. Spans 2 and 3 (see Fig. 18) have travertine facing for the complete

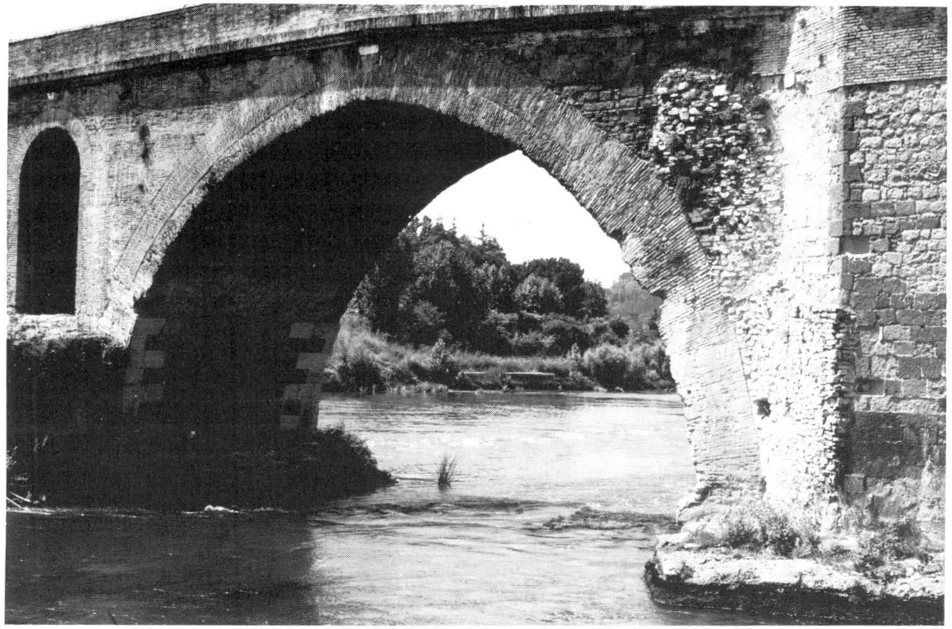

Fig. 19 Brick arch in span 5 of the Ponte Milvio (I1).

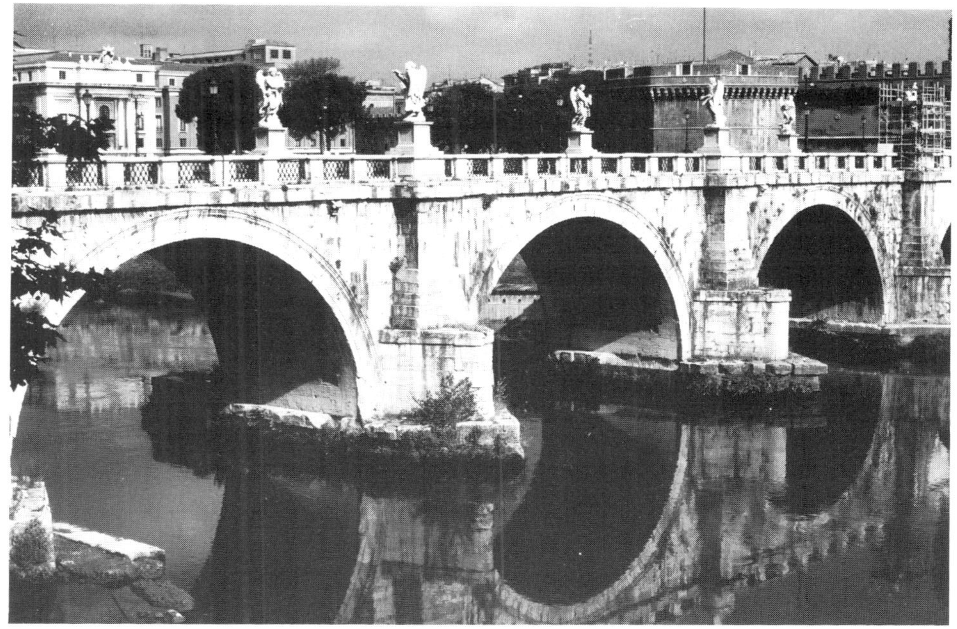

Fig. 20 The Pons Aelius or Ponte San Angelo in Rome, with sculptured figures added in the seventeenth century (I2).

arches, but in two different styles. The lower courses have alternate stretchers and headers; that is, each alternate voussoir is composed of two stones. The upper voussoirs are each formed of a single stone, with the outer face left rough. Travertine is used only for the facing and the first course of the vault; the remainder is Gabine stone. The fourth span has three styles of construction; the lower courses show the two stages of travertine facing found in spans 2 and 3; the upper courses are of brick. Span 5, apart from seven courses at its southern end, which are of the second period of spans 2–4, is entirely of brick (Fig. 19), as is span 6.

Ballance believes that the earliest work, corresponding to the first of the stages described above, is from 109 BC; the second travertine stage may belong to the first century AD; and the brickwork of spans 4 and 5 is from the time of Nicholas V.

The roadway was originally horizontal between the centres of spans 2 and 5, and then sloped down to each bank. This has since been modified at the southern end, where the roadway is now horizontal. Figure 18 shows part of a protruding cornice at this original deck level. The upper arch in span 1 presumably also forms part of this modification. The tower on the north abutment (Fig. 18) carries an inscription from 1849. There was an arch erected on the bridge in 27 BC to honour Augustus at the time of his restoration of the Via Flaminia, but this may have stood over the central pier.

The second of the existing ancient bridges, the **Pons Aelius** (Elio) or **Pons Hadriani** (**I2**), now known as the Ponte San Angelo (Fig. 20), joins the city of Rome to Hadrian's mausoleum (the Castel Sant'Angelo) and the Vatican. It was built by Hadrian and completed in AD 134 (Platner and Ashby, 1929: 396–7; Nash, 1968: 178) with seven spans – three central spans, each 18 m clear, flanked on each side by a span of 7.5 m, and then another of 3.5 m (Gazzola, 1963b, no. 179).

The structure stood in that form until 1892–4 when, in connection with the construction of new training walls for the River Tiber, the smaller end spans and the ancient inclined approaches were demolished and replaced by two 18 m spans to match the three ancient central spans. The inclined approaches were built and paved with stone and were quite substantial – 33 m long on the city or left bank, with an additional stone-arched floodway, and

22 m long on the right bank. Both Nash and Gazzola have photographs of the ramps taken before they were destroyed. The original roadway was 4.7 m wide, of polygonal lava-stone, between two 3.1 m wide sidewalks of rectangular travertine stones cramped together (Chevallier, 1976: 97). The bridge is still used as a footway.

It is decorated with ten sculptured figures. The two at the eastern end represent St Peter and St Paul, and were placed there in 1527 by Clement VII. The other eight represent angels carrying symbols of the Passion. They were commissioned by Alexander VII and placed in position between 1667 and 1669. One, carrying the inscription INRI, was carved by Bernini, the others by sculptors of his school. They replace earlier ancient statues, and an ancient balustrade (Nash, 1968: 178).

Blake and Bishop (1973: 53) have also described the bridge. 'Thus there were eight arches for a causeway about a hundred and thirty-five metres long. All this masonry was of concrete faced first with Gabine stone, which appears only in the foundations and in the intrados of the arches. The rest was covered with travertine fastened together with iron cramps leaded into place.' The Gabine stone in the intrados can be seen in Figure 20.

The Isola Tiberina, the island in the Tiber, was joined to the mainland by two ancient bridges, the Pons Fabricius (I3) to the east and the Pons Cestius (I4) to the west.

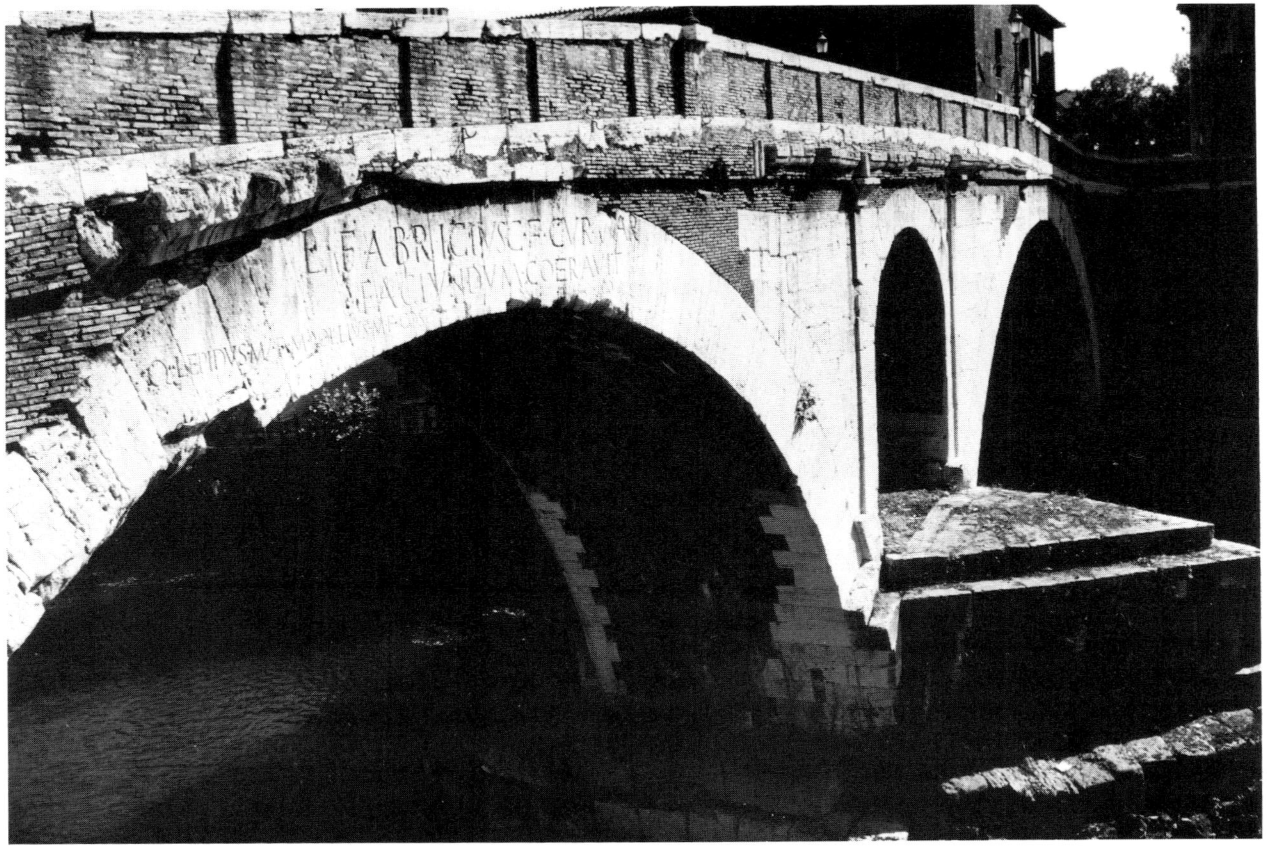

Fig. 21 The Pons Fabricius, 62 BC, crosses to the Isola Tiberina, or Island on the Tiber, in Rome (I3).

The **Pons Fabricius, I3** (Fig. 21), still retains much of its original form. Used now as a footbridge, it is referred to also by the name Ponte dei Quattro Capi, derived from the four-headed herms which served as piers for the original bronze balustrade, and were incorporated into the present balustrade built by Innocent XI in 1679 (Platner and Ashby, 1929: 400; Nash, 1968: 189). The name of the builder, L. Fabricius (see Fig. 21), and the date of construction, 62 BC, are carved into the bridge. It also bears an inscription of the consuls in 21 BC, M. Lollius and Q. Lepidus, referring either to a repair after floods in that year (Chevallier, 1976: 96), or to the final approval of the structure by the authorities (see Section 3.2). It has two main spans, recorded as 24.2 and 24.5 m, and a wide central pier, with a large floodway. It is built of *tufa* and *peperino*, faced with travertine and Gabine stone (Gazzola, 1963b, no. 40; Blake, 1947: 45).

The western bridge, the **Pons Cestius** (I4), is used to carry road traffic to the island. It receives its name from an earlier bridge believed to have been built by Cestius, *curator viarum* between 62 and 27 BC (Section 3.2; Platner and Ashby, 1929: 399–400; Nash, 1968: 187).

An early reference to this bridge records a restoration in AD 152, following damage by flood (Blake and Bishop, 1973: 70). It was replaced in AD 370 by the Pons Gratiani, which survived until 1885. It carried an inscription to Gratianus (AD 367–83) on the parapet. Nash has a photograph of the bridge as it appeared then, with a large central span and two much smaller side spans. Between 1885 and 1889 the channel to the west of the island was widened from 48 to 76 m. The bridge was taken down

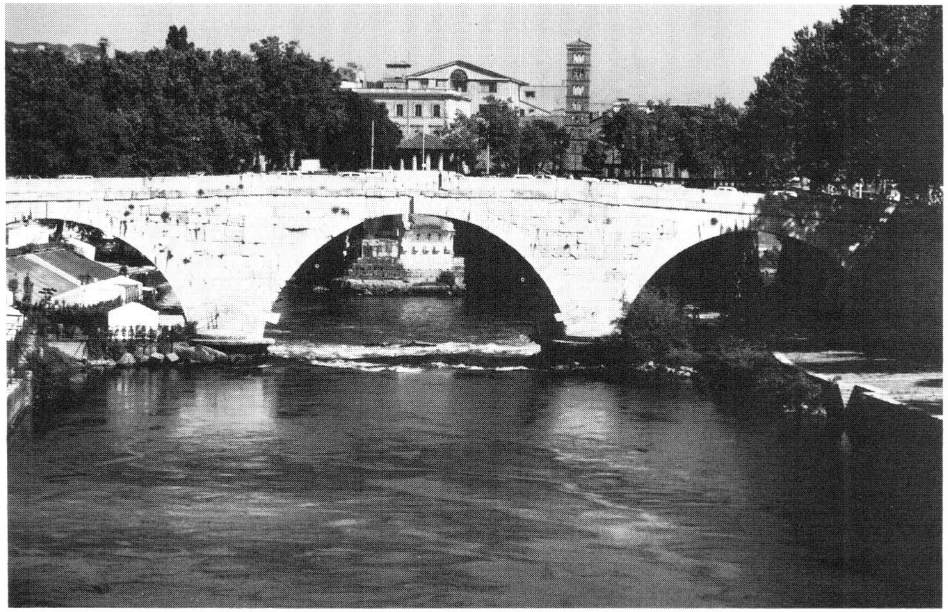

Fig. 22 The Pons Cestius, rebuilt in 1892, joins the Isola Tiberina (on the left) to the right bank of the Tiber (I4).

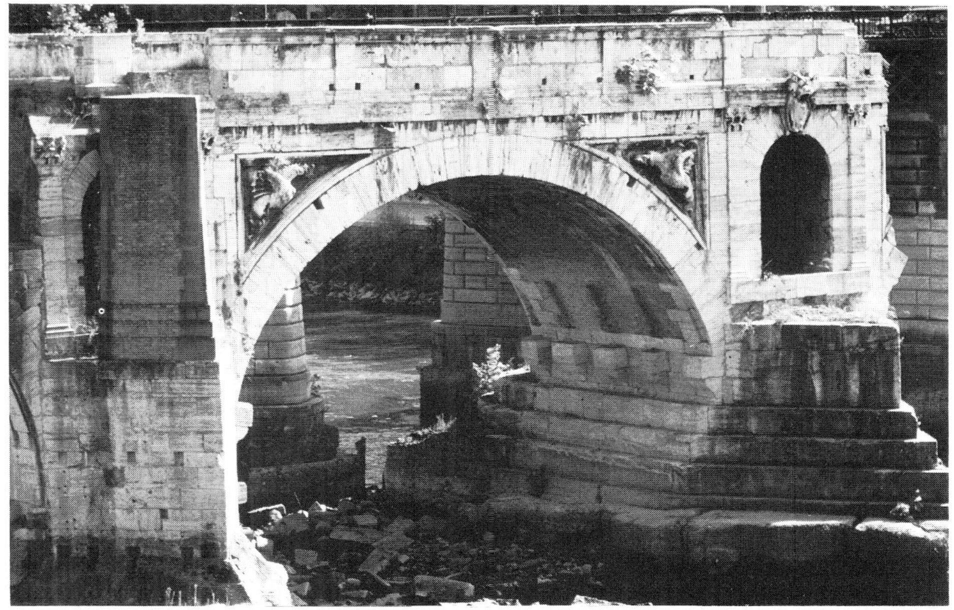

Fig. 23 The only remaining span of the Pons Aemilius in Rome, now known as the Ponte Rotto or broken bridge (I5).

and rebuilt in 1892. The new bridge (Fig. 22), has three spans, of which the central span was rebuilt to the original design and measurements, with 347 of the original 563 travertine blocks used again. This central span is 13.7 m clear (Gazzola, 1963b, no. 41), the same as that in the AD 370 bridge. The side spans are now of the same size, replacing the previous small 5.8 m spans. The original bridge was 48 m long and 8.2 m wide.

Only one span remains of the **Pons Aemilius, I5** (Fig. 23), downstream of the Isola Tiberina, and immediately upstream of the modern Ponte Palatino. It is referred to today as the Ponte Rotto, or broken bridge, and has also been called the Pons Lepidi. In 179 BC a bridge was built here by the censors, M. Fulvius Nobilior and M. Aemilius Lepidus, with stone piers and a timber deck (Platner and Ashby, 1929: 397–8; Nash, 1968: 182). This deck was replaced by stone arches by P. Scipio Africanus and L. Mummius in 142 BC, forming the first stone arch bridge across the Tiber in Rome (Chevallier, 1976: 96). It was reconstructed during the reign of Augustus some time after 12 BC (Blake, 1947: 178) and has since been restored on a number of occasions. Two of the arches were destroyed by flood in 1557 and rebuilt by Gregory XIII in 1575. Another flood in 1598 carried away the eastern half of the bridge. The three western arches were joined in 1853 by a suspension bridge to the eastern bank. This stayed in service until 1885, when it was removed, together with the two westernmost of the remaining three arches.

This account of the bridge is not universally accepted. Blake (1947: 178) says,

> Recent scholarship, however, seems to have proved that this arch formed part of a bridge with an orientation slightly different from that of an abutment belonging to the earlier structure. The abutment is probably all that is left of the bridges of 179 BC and 142 BC, and the existing arch and pier belong to an Augustan rebuilding, possibly referred to in an inscription. If so, it was made after 12 BC.

Gazzola (1963b, no. 28) has a drawing of the complete bridge with six arches. Hopkins (1970: 33) gives the main spans as 24.3 m; the two end spans were somewhat smaller, with the same springing level and lower crowns. The roadways sloped upward from the abutments to the first piers. The facing is of travertine, with arches of

Gabine stone and a *tufa* core (Blake, 1947: 178). As can be seen from Figure 23, which shows the upstream face, the bridge is highly decorated. The upper brick cutwater at the left is obviously new.

In chronological order, then, the oldest bridges of Rome were the timber Pons Sublicius of about 600 BC, and the original 179 BC Pons Aemilius with a timber deck. The first stone arch bridge was the reconstructed Pons Aemilius of 142 BC, followed by the 109 BC Pons Mulvius and the 62 BC Pons Fabricius, and shortly after by the original Pons Cestius (62–27 BC). It is remarkable that Rome was supplied with as many as ten major bridges, of which eight had been built before AD 150.

Moving from Rome to its environs, the Via Aurelia crossed either the Pons Fabricius and Pons Cestius, or alternatively the Pons Aemilius, and travelled west to the coast at Alsium. Following the flood of 21 BC, a viaduct was built to carry the road across the low land from the Tiber to the Janiculum Hill (**I6**). It has stone retaining walls of *tufa*, with a single arched opening, and a projecting course of *peperino* to form the sidewalk for the road above (Blake, 1947: 172). Ashby (1902: 227–8) refers to an ancient two-arched bridge of limestone that originally crossed the Fosso di Palidoro (**I7**) and now serves as the foundation of a large farmhouse, and to another ancient bridge two miles west. Gazzola lists four bridges at Santa Marinella (**I8–11**, see also Blake, 1947: 120). All are believed to be old, possibly from the second century BC.

The next road system (in a clockwise direction) to leave Rome was formed by the Viae Clodia, Cassia and Amerina (Map 1). The Via Clodia went north west to pass around the southern shores of the Lago di Bracciano. About 11 km from the Pons Mulvius it branched, with the Via Cassia and then the Via Amerina going off to the north. A short distance from the junction (2 km north east) was the ancient Etruscan city of Veii, 12 Roman miles from Rome, captured and sacked by the Romans in 396 BC (Paget, 1973: 141). The town was built on a *tufa* plateau, virtually surrounded by the gorges of two streams, the Valchetta Stream (ancient Cremera), which ran south west to join the Tiber north of Rome, and its tributary, the Fossa dei due Fossi. To the north of Veii, the Valchetta stream was diverted through a tunnel 70 m long, cut through a rocky bluff to eliminate a horseshoe bend and to provide a bridge of natural rock to the fields outside the town (Ward Perkins, 1961: 49, 50). Although not truly a bridge, the work is called the **Ponte Sodo** (**I12**). Ward Perkins dates it as earlier than the late fifth century BC, and it is important as one of the earliest known 'bridges'. It is located about 400 m west of the north-east gate.

The Valchetta is crossed by the Via Flaminia just south of the 12 km mark from Rome, 0.25 km north of the Castel Giubileo Station and here there was a small, 2-span bridge (**I13**, Ballance, 1951: 84, 117; Gazzola, 1963b, no. 88).

There was a series of three fine bridges across the Anio River north of Rome. The Via Salaria crossed the stream not far from its confluence with the Tiber. The **Ponte Salario** (**I14**) has been destroyed three times in the last three centuries, in 1798, 1849 and 1867. An engraving of the bridge (Gazzola, 1963b, no. 283) shows one major arch, with two minor arches in the abutments, and the steeply humpback road surmounted by a prominent castellated gateway. Tyrrell (1911: 36) quotes the main arch span as 26.7 m, with the two side arches as 8.5 m. According to Ashby (1927a: 65) 'little of the ancient structure now remains except the greater part of the small arches on each side'. It was built of *tufa* and travertine.

Two and a half kilometres east, the Via Nomentana was carried across the Anio by the **Ponte Nomentano** (**I15**), a picturesque bridge of three spans, with a large central span and two small side spans, and a medieval tower (Gazzola, 1963b, no. 282). Again, according to Ashby (1927a: 84), very little of the original bridge remains. A further 3.5 km east was the **Ponte Mammolo** (**I16**) across the Anio, carrying the Via Tiburtina. Ashby (1927a: 96) describes it as built of blocks of *tufa* and travertine: 'one of its ancient arches still remains in part, though damaged in 1849.' Gazzola (1963b, no. 59) quotes Nibby, who suggested that the name may have been derived from the name of the mother of Severus Alexander (AD 225–35), Julia Mammea, in which case the original bridge may date from about that time. It had three spans, a central 16 m span and two small side spans, and carried a tower at each end.

The Via Tiburtina crossed the Anio again 3.5 km west of Tivoli (5.5 km by road) by the **Ponte Lucano** (**I17**) (Gazzola, 1963b, no. 51). Ashby (1927a: 101) describes it as 'picturesque and fairly well preserved'. It stands beside the Tomb of the Plautii, a circular stone tower, and the common view of the bridge shows it with the tower behind. However, the site has degenerated and the bridge now stands in a polluted stream with industrial works nearby – (Fig. 24). The original solid stone

Fig. 24 The Ponte Lucano near Tivoli, now spoilt by surrounding development (I17).

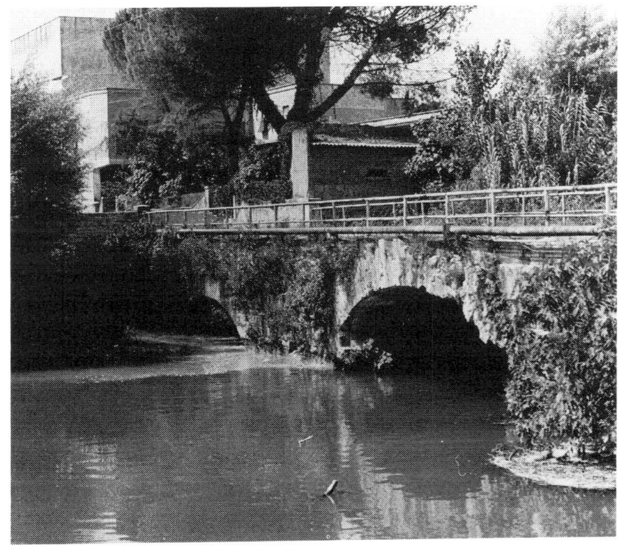

balustrades have been replaced by light metal railings, and the bridge carries a pipe and foot traffic. The bed and surface level of the stream have risen since Roman times and only the crowns of four arches can be seen; there are in fact five. The core of the bridge consists of lumps of *tufa* rather than squared blocks, and Blake (1947: 213) interprets this as 'a step toward the concrete core that became the rule later'. The facing is of travertine. The adjacent tower on the eastern bank carries an inscription to M. Plautius Silvanus, consul in 2 BC (see Section 3.2). Blake believes that the bridge was probably also built at that time (see also Paget 1973: 187).

Two and a half kilometres upstream, north of the Via Tiburtina and 1.2 km west of Tivoli, are the remains of another ancient bridge, the **Ponte dell'Acquoria (I18)**, believed to be contemporaneous with the Ponte Lucano. Only one of the original seven arches now remains (Ashby, 1927a: 115; Gazzola, 1963b, no. 52).

Van Essen (1957) has described the Via Valeria as it proceeds east from Tivoli to Collarmele (beyond Avezzano). Near Arsoli were two small bridges (Blake, 1947: 212), the **Ponte Scutonico (I19**; Gazzola, 1963b, no. 49) and the **Ponte San Giorgio (I20**; Gazzola, 1963b, no. 48). Blake believes that both may belong to the original construction of the Via Valeria in 154 BC, whereas van Essen places them later, in the reign of Nerva (AD 96–8), 'who carried out important restorations along the Valeria'.

About 3 km south of Arsoli, the Via Sublaquensis branched off to the south, following the right bank of the Anio River towards Subiaco. The village of Agosta is about 5 or 6 km along this road and west of the village is the **Ponte di Agosta (I21**; Gazzola, 1963b, no. 134) carrying a minor road across the river. As shown in Figure 25, this substantial bridge contains much modern work; the arch vaults, for example, are of modern concrete. The lower stonework and cutwaters appear old, and it seems that the bridge may be on Roman foundations.

The Via Praenestina travels about due east of Rome to ancient Praeneste, modern Palestrina. Twelve kilometres out are two bridges across the Torrente Marrana. The large **Ponte di Nona (I23**; Gazzola 1963b, no. 43) is still in use carrying modern traffic; built into its foundations (Fig. 26) is the small **Ponte di Nona (I22**; Gazzola, 1963b, no. 16), with a single 6 m span. Both bridges are very important. Blake (1947: 211) believes that the small bridge is old and that it was possibly one of the bridges 'mentioned by Livy as having been built in 173 BC'. She describes it as having alternate stretchers and headers in the voussoirs, but this is not apparent in Figure 26. Ashby (1927a: 132) describes the large Ponte di Nona (Fig. 27) as 'by far the finest road bridge in the neighbourhood of Rome'. It has seven main spans, each of 6 m, and another two small spans, one at each end, with a total length of 71.9 m and a height of 15.8 m. The arches are of Gabine stone. Lugli (1957: 309) dates it in the first century BC.

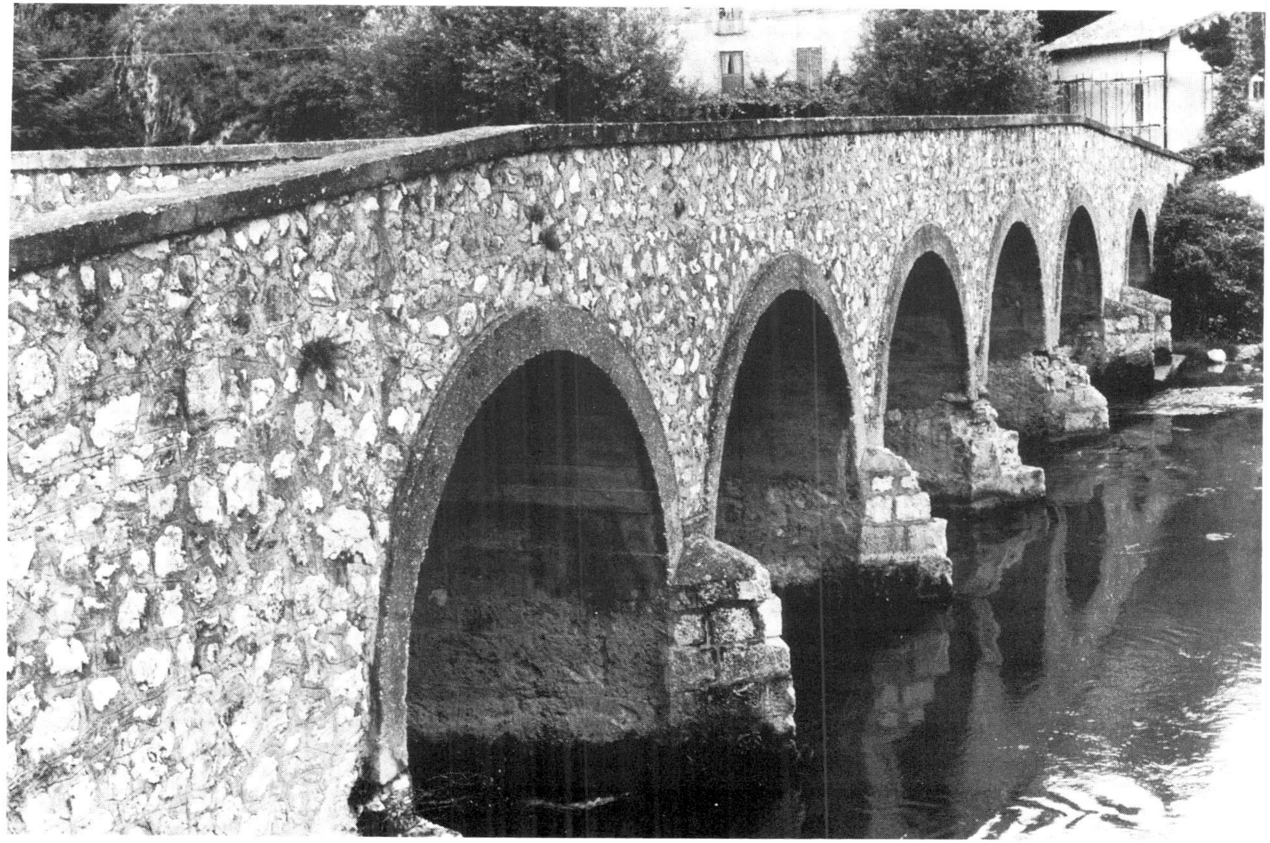

Fig. 25 The largely modern Ponte di Agosta (I21).

Further out, before the 15th milestone, 4 km east of the Lago di Gabii was the **Ponte di Terra (I24**; Gazzola, 1963b,

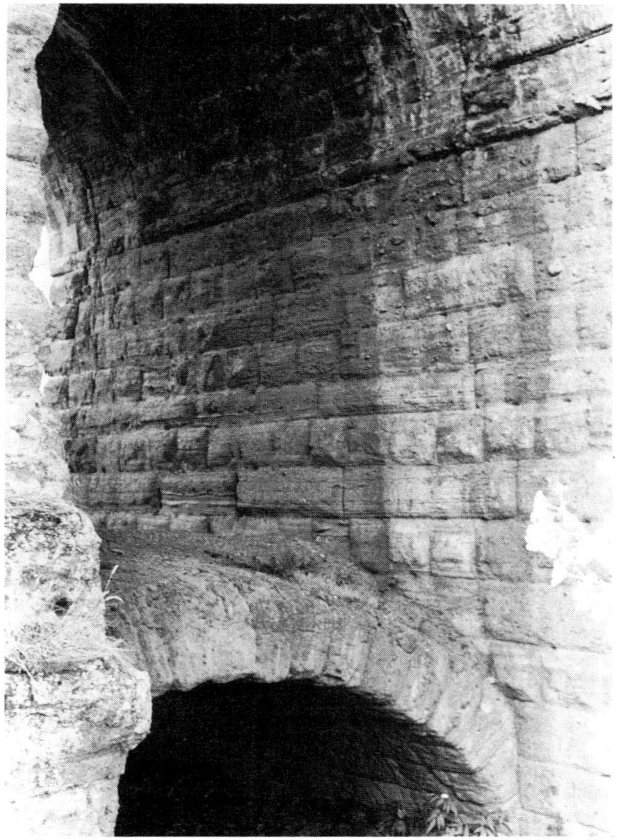

Fig. 26 The old and small Ponte di Nona, incorporated by the Romans into a larger bridge on the same site (I22).

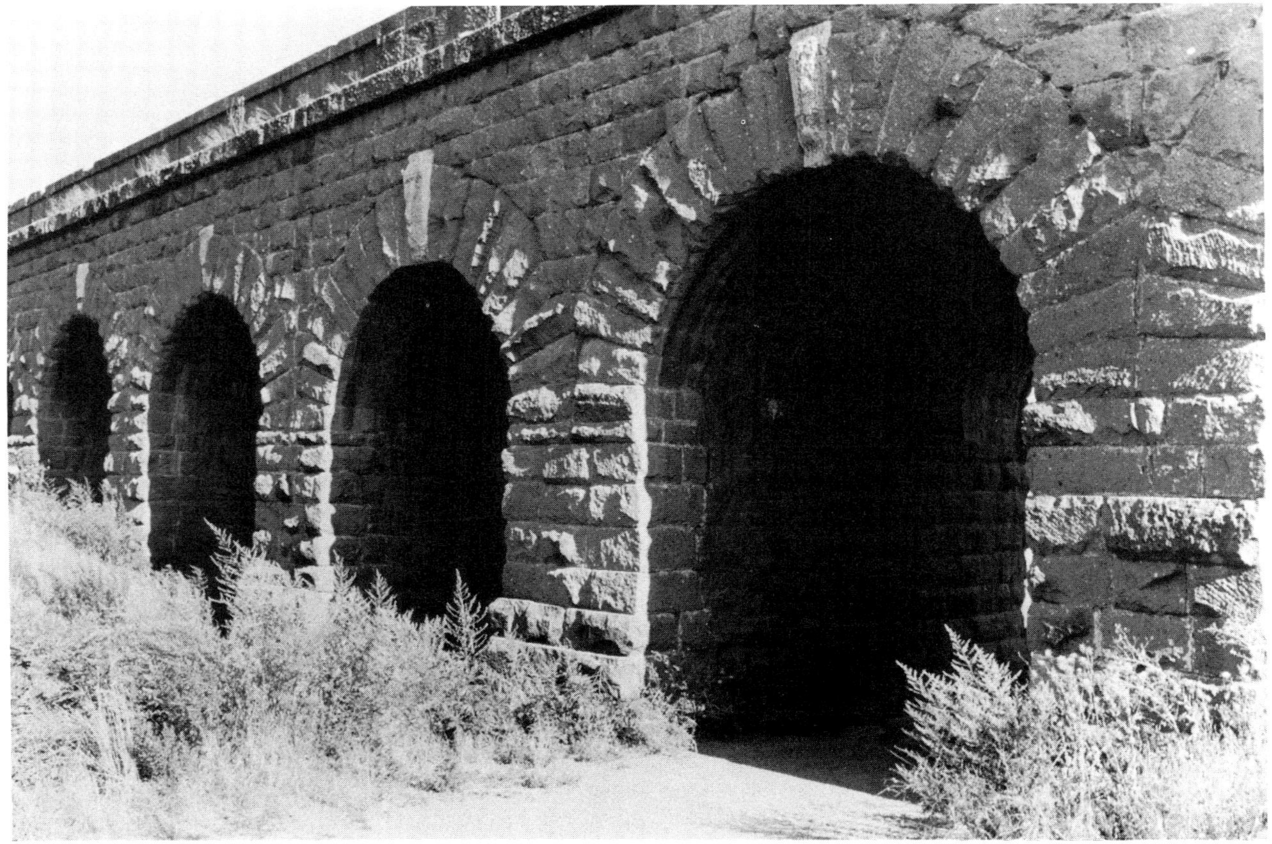

Fig. 27 The large Ponte di Nona, still in use carrying a major road near Rome (I23).

no. 20). As described by Ashby (1927a: 202), it had a span of 4.9 m, a height of 5.4 m, a width of 6.4 m and a total length of 16.2 m. The **Viaduct over the Fosso di Ponte Terra (I25**; Ashby, 1927a: 147, fig. 14) was the name given to another bridge, about 9 km to the north east, 1 km beyond San Vittorino. As illustrated by Ashby, it had the appearance of an aqueduct, with two superimposed tiers of arches, 17 m high, a single arch of 6.7 m in the lower tier, and four arches – only two intact – in the upper. Despite its appearance, and its relative proximity to the line of the aqueducts, it is believed to have been a road bridge – Ashby says 'it certainly carried a road and not an aqueduct'.

Further out again, at the 18th milestone of the Via Praenestina are the remains of the **Ponte Amato (I26**; Gazzola, 1963b, no. 17), east of the modern road across the Fossa Collafri, about 400 m south west of Gallicano. Ashby (1927a: 209) has a photograph of the bridge showing it in excellent condition – but this is presumably of the upstream face, as the side closer to the modern road has suffered damage. It is now also overgrown. The span is about 5 m, with a maximum height of 10 m, a width of 6 m, and substantial stone retaining walls.

The bridge is not in use, and some of the fill has been removed to expose the top of the arch vault. It is of Gabine stone, with divided voussoirs, like the small Ponte di Nona (Blake, 1947: 212). Blake believes it is a little later; Lugli (1957: 309) dates it about the first century BC (see also p. 152).

Finally, near Palestrina (0.7 km SW) is a structure called the **Ponte Sardone** (**I27**; Gazzola, 1963b, no. 18), described by Ashby (1927a: 138) as 'an ancient road embankment much repaired'.

South of the Via Praenestina is the Via Labicana, now known as the Via Casilina, with a small bridge (**I28**; Gazzola, 1963b, no. 19; Ashby, 1927a: 147) over the Fosso del Giardinetto near Torre Nova.

The Via Appia (Section 2.2.2) reaches the town of Ariccia (modern spelling), 27 km south east of Rome, by a stone-walled embankment 200 m long, the **Viadotta di Valle Ariccia**. A massive structure, but not really a bridge, it was probably built in the time of Augustus (31 BC – AD 14) with *peperino* walls (Blake, 1947: 212) pierced by two or three arched openings (**I29**, Gazzola, 1963b, no. 2; Lugli, 1957, pl. 47).

Concerning the Via Appia, Blake (1947: 211) says, 'It is impossible to say with confidence that between Rome and the vicinity of Terracina any bridge remains which was constructed for the Via Appia of 312 BC.' However, two old bridges are the **Ponte di Vigna Capoccio** (**I30**; Gazzola, 1963b, no. 33; Ashby, 1927a: 198; Blake 1947: 211); and the **Ponte Catena** (**I31**; Gazzola, 1963b, no. 99) outside the Porta Ninfina of Cori (ancient Cora), 10 km north east of the Via Appia town of Cisterna di Latina, with a single arch span over a very deep gorge (Blake, 1947: 211). Lugli (1957: 356) dates it between 100 and 70 BC; Paget (1973: 228) attributes it to Sulla, at about the same time.

The Via Ostiensis runs south of the left bank of the Tiber from Rome to Ostia, its port. About half-way there, 9 km from Ostia, it crosses the Fosso Malafede close to the Tiber, over an old two-span bridge (**I32**; Gazzola, 1963b, no. 3). About 4.5 km closer to Rome, near the Tor di Valle, is a small one-span bridge (**I33**; Gazzola, 1963b, no. 4) with similar features. Another similar bridge is the **Ponte di Mezzocamino** (**I34**; Gazzola, 1963b, no. 5), believed to be in the same area.

Travelling south from Ostia along the coast was the Via Severiana. After about 2 km, near the Castel Fusano, this crossed a stream which, according to Ashby (1927a: 217), 'may very likely be the Numicus'. Here there was, at least in Ashby's time, a fine three-span stone bridge (**I35**).

5.3 Southern Italy and the islands

For the purposes of this chapter, southern Italy is regarded as the area south of an east–west line through Rome, but excluding Rome and its vicinity (Section 5.2). Although this differs from the usual definition of southern Italy – the regions of Campania, Puglia, Basilicata and Calabria – the only effective differences are that a few bridges near Terracina and from there to Gaeta are included here rather than in central Italy, and this is more convenient.

One of the important routes east and then south east from Rome was the Via Latina (Map 1) which, having been joined near Valmontone by the Via Labicana, continued (as the modern Via Casilina) to Frosinone and Cassino. South west of Cassino, on the River Liris, was the Roman town of Interamna Lirenas (modern Pignataro). The connecting road crossed the River Spalla Bassa by the **Ponte del Diavolo**, a small travertine bridge with a single span of 5.5 m (**I36**; Gazzola, 1963b, no. 37). Only the arch rib remains.

To the north, towards Atina, a few kilometres from Cassino was the **Ponte di Cassino**, in the valley of the Rapido (**I37**; Gazzola, 1963b, no. 38). It was somewhat larger, with a single arch of clear span 12 m. Gazzola dates the bridge about the first or second century BC. Again only the arch rib remains. In Atina itself, 21 km north of Cassino, is the **Ponte di San Giuliano**, a substantial two-span bridge across the River Rapido (**I38**; Gazzola, 1963b, no. 87). The foundations and part of the retaining walls are believed to be Roman. Further west, near Arpino, the Ponte Marmone crossed the River Liris (Gazzola, 1963b, no. 50). It no longer exists (Blake, 1947: 212).

North east of Cassino at Isernia is the **Ponticello di Isernia** (**I39**; Gazzola, 1963b, no. 32) with the roadway rising from the low banks of the stream up and over the single span. Paget (1973: 201) gives the location as at the confluence of the Carpino and Sondo Rivers. The **Ponte del Diavolo** crosses the River Tiferno at Campobasso (**I40**; Gazzola, 1963b, no. 216). It is a single-span bridge of unusual appearance in a rugged site.

The great majority of bridges in southern Italy lie on the Via Appia and its tributary, the Via Traiana (Section 2.2.2). The Via Appia crosses the Pontine Marshes between Cisterna di Latina and Terracina along a 40 km straight road, La Fettuccia or ribbon. The original road, as built by Appius Claudius in 312 BC, required a journey by boat across the marshes, or a detour. Trajan drained the marshes in AD 98 and constructed this section of the road. The **Ponte di Tre Ponti** (**I41**, Gazzola, 1963b, no. 11) receives its name from the district, Tripontium, and is not unlike a small version of the Viadotto di Valle Ariccia (I29), with stone retaining walls of *tufa* pierced by two small arches.

Terracina had two ancient bridges, the **Ponte Alto** (**I42**; Gazzola, 1963b, no. 100) and the **Ponte Antico** (**I43**; Gazzola, 1963b, no. 101), both on the Via Appia. The first is of some interest. It is in flat country, with long retaining walls leading the road up and over the stream. The masonry is irregular, with sloping bed joints giving courses of variable height, built of stones whose height is often greater than their width. To quote Blake (1947: 211), 'It is connected with a substructure of quasi-quadrangular masonry, and its single arch is made up of rather roughly cut voussoirs, which have not been shaped to the curve of the arch.' She speaks of the bridge as one of a group of bridges near Terracina which 'certainly

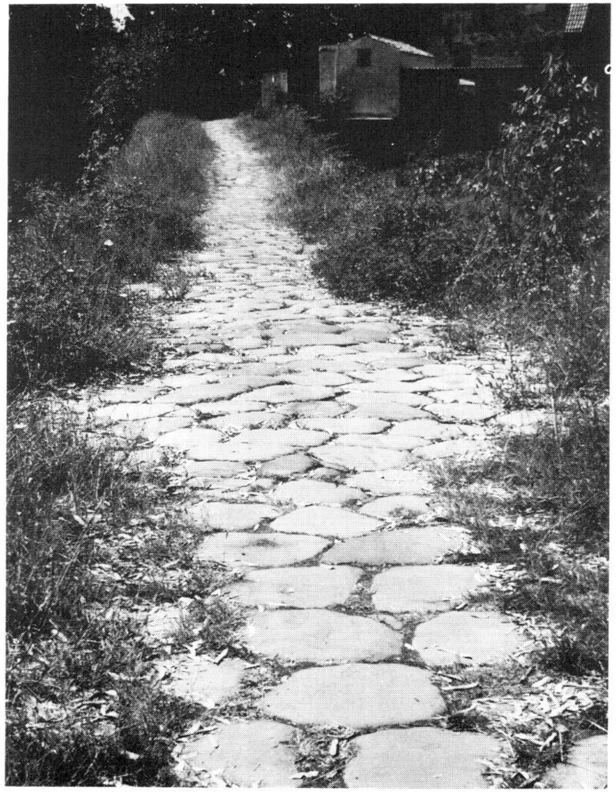

Fig. 28 Paving on the Pons Auruncus (I45).

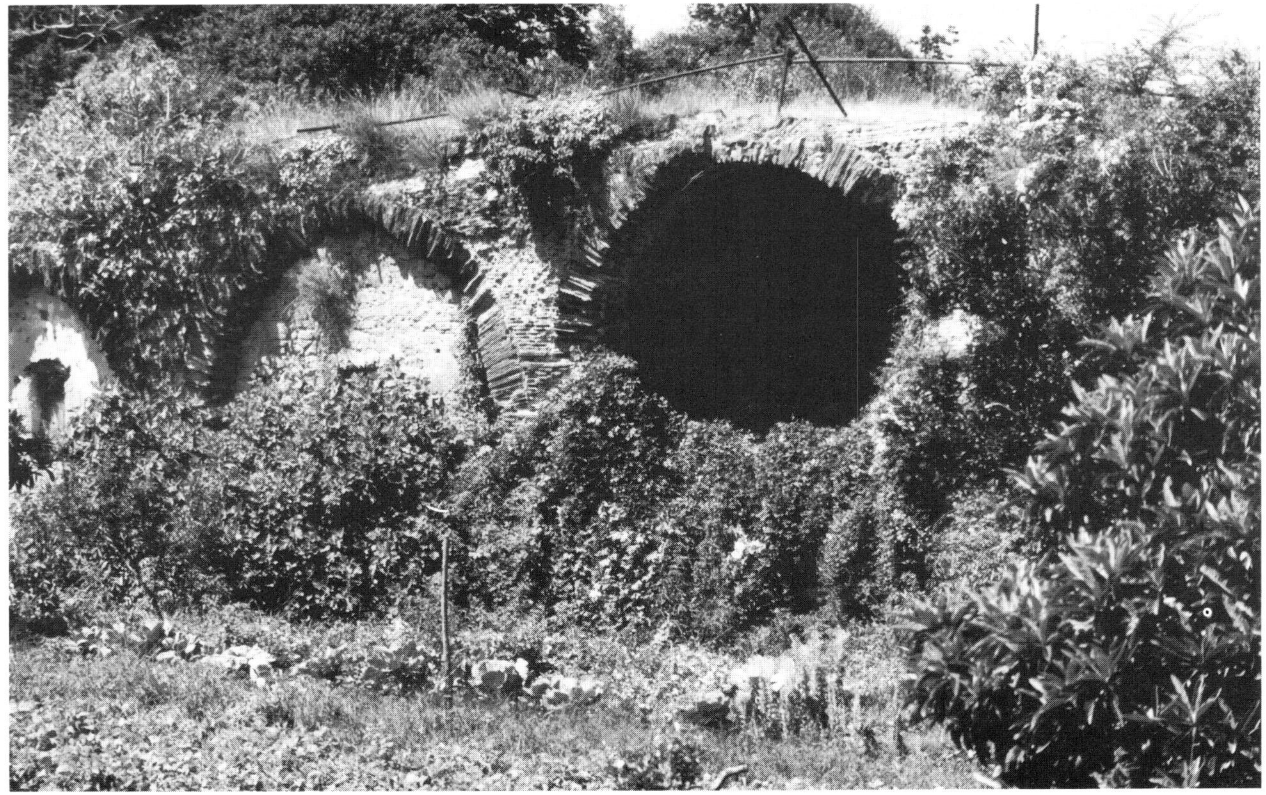

Fig. 29 The Pons Auruncus carried the Via Appia near Sessa Aurunca. Now partly overgrown, its arches provide storage space for an Italian farmer (I45).

represent an early, though probably not the earliest phase of the Via Appia'. Ballance also refers to the bridge (1951: 100) and cites a Trajanic inscription (AD 98–117) which refers to repairs on the bridge 'below the Punta di Leano'; he describes the bridge as 'remarkable for its irregular limestone masonry'. The other, the Ponte Antico, crossed the Fosso dell'Acqua Traversa near Terracina, with a single arch of local limestone and the same workmanship. It may no longer exist (Ballance, 1951: 100).

East of Terracina, the Via Appia turned inland to Fondi, returning to the sea at Formia. The coast road, the Via Flacca, passes through Sperlonga, where, near the Villa of Tiberius, is the **Ponte di Sperlonga** (**I44**; Gazzola, 1963b, no. 117). It also is unusual, with spandrel walls faced with *opus incertum*. Gazzola suggests it may date from the first century BC.

The Via Appia, after it rejoined the coast, crossed the Garigliano River at Minturnae by a timber bridge, described later in Section 6.3. Its original route (Wiseman, 1970: 130) then turned inland to Sessa Aurunca, Roman Suessa, and crossed the **Pons Auruncus** (**I45**; Paget, 1973: 236) about 1 km west of the town. The bridge can be found by following the right bank of the stream down from the minor road which leads from Route 7. The access is along the old Via Appia itself, and considerable lengths of the original paving can be seen, including that on the bridge – Figure 28 looks along the bridge, towards the eastern approach. The bridge is long, with twenty-one spans, some of which are quite large (Fig. 29). The road on the bridge is scarcely used, and for much of its length the spans cross farmland and are used by the local farm proprietor for storage. Some are walled up and

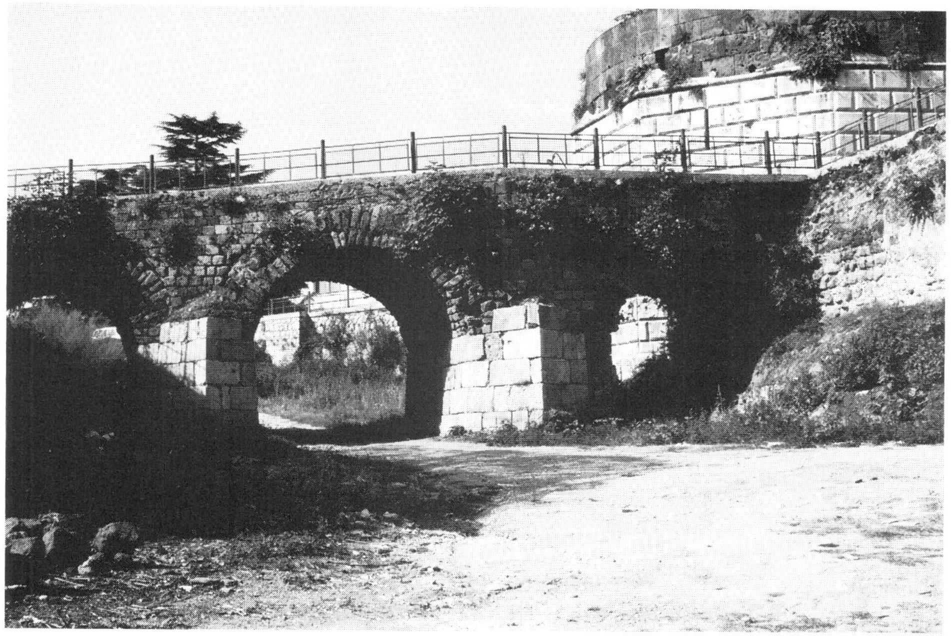

Fig. 30 This bridge, which appears to be Roman, lies near the northern approach to the large bridge at Capua, originally Roman but now modern.

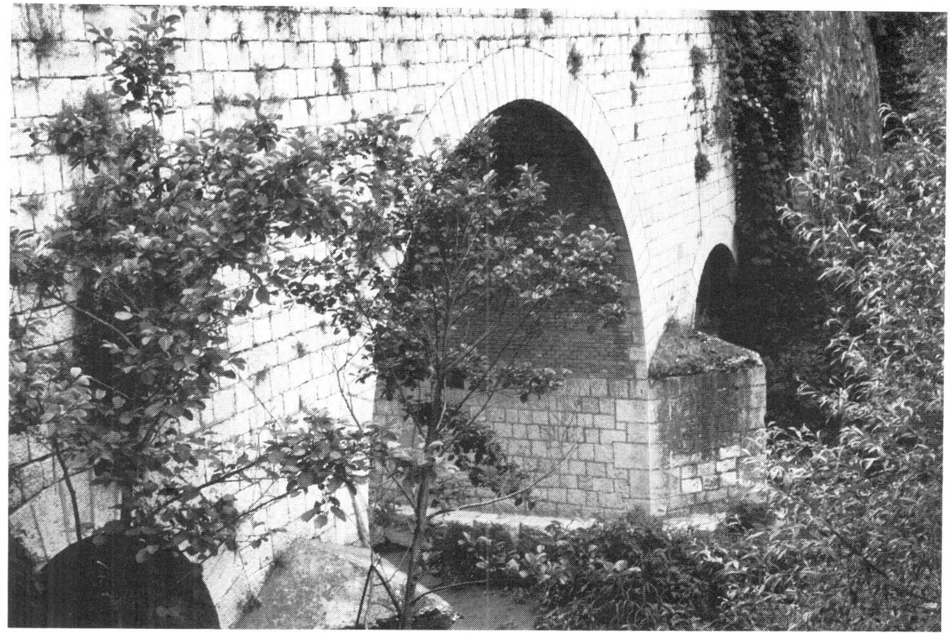

Fig. 31 The Ponte Apollosa near Beneventum was destroyed in the last war and later rebuilt (I47).

there are doorways cut through the piers (Cerruti, 1981: 732). The abutments are of stone; the piers and vaults are of brick *bipedales* with *opus incertum* in the spandrel walls.

The inland route of the Via Appia joined the Via Latina (Casilina) and then turned south east to join the coastal route, which went from Minturnae to Sinuessa along the coast and then turned inland, crossing the Rive Savo over the ancient Pons Campanus (Wiseman, 1970: 130). The junction was north west of ancient Capua (modern Sta Maria Capua Veteri), on the western bank of the Volturno River opposite Casilinum (modern Capua). Gazzola (1963b, no. 131) shows here a major bridge with spans of 10.2, 10.9, 10.3, 18.0 and 5.2 m, and piers 5.4 m wide, totalling 76.2 m between abutments. This has since been destroyed. The piers in the river beneath the present bridge are presumably those of the bridge described by Gazzola. However, at the northern end the bridge led to a fort or gate surrounded by a moat, and one of the bridges across this moat appears to be Roman; it is shown in Figure 30. It has three spans with piers and cutwaters of large squared stone; small, irregularly cut stones form the two concentric rings of the arches.

Between Montesarchio and Beneventum, the Via Appia crossed four bridges, all of which were damaged in the last war. The **Ponte Tufaro** or **Trifaio** (**I46**; Gazzola, 1963b, no. 102) had three spans, the side spans being somewhat smaller than the main span and set a little lower. It may be Augustan (Ballance, 1951: 100–1). It was destroyed and not rebuilt.

Ponte Apollosa (**I47**; Gazzola, 1963b, no. 103) has been rebuilt, as shown in Figure 31. The stone facing and the arches of brick are entirely new. However, the foundations, abutments and wing walls are original (Ballance, 1961: 99–100), as is the general arrangement of the arches and the spans: 3, 9 and 3 m. The original bridge was very fine and probably Augustan.

The two-span **Ponte Corvo** (**I48**, Gazzola, 1963b, no. 109) was destroyed and its brick arches still lie in the stream bed; its massive stone abutments are intact. **Ponte Leproso** or **Lebbroso** (**I49**; Gazzola, 1963b, no. 108) still appears to be intact, even though Gazzola describes it as having suffered considerable damage during the last war. It is located immediately south of Beneventum and crosses a channel of the Sabato River, whose flow has been interrupted, damming it up, raising its level, and

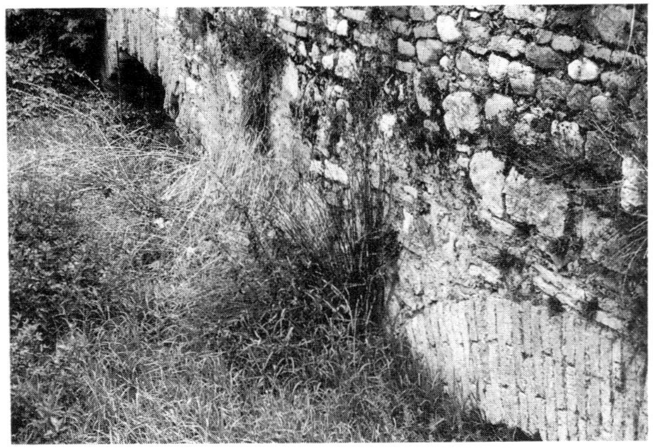

Fig. 32 The Ponte Leproso, immediately south of Beneventum, is now only partly visible (I49).

causing it to be highly polluted. The arches are formed of thin slabs of local limestone (Fig. 32). The bridge appears to have had four or five main arches with smaller flood arches through the piers. Gazzola dates it to the first century BC. Blake refers to three of these bridges: Tufaro, Apollosa and Corvo (Blake, 1947: 211; Blake and Bishop, 1973: 281). She also has an excellent photograph (Blake, 1947, pl.21) of another, the **Ponticello** at Beneventum (**I50**; Gazzola, 1963b, no. 107).

The Via Appia continues from Beneventum through Taranto to Brindisi; there are no further bridges recorded on this route. Ashby and Gardner (1916) have described the Via Traiana built by Trajan (AD 98–117) along the more northerly route from Beneventum to Brindisi through Canosa, and document a number of substantial bridges.

The first was the **Ponte Valentino** over the River Calore (**I51**, Gazzola, 1963b, no. 130) about 5 km east of Beneventum. There are today only a few remains, but these indicate that the bridge had three arch spans, of 15, 18 and 15 m. Ashby and Gardner give a length of 76 m between embankment walls. The structure had a concrete core, with the arches and part of the piers faced with *bipedales* (Blake and Bishop, 1973: 281). Seventeen kilometres from Beneventum are the remains of the **Ponte dei Ladroni** (**I52**). It is unusual in that it makes an abrupt turn in plan, through an angle of about 40°. A long causeway or embankment leads to an abutment and a span of 3 m to a triangular pier. The bridge turns and then spans possibly 10 m and 14 m to a substantial abutment; or alternatively, there may be three spans, of 10, 5 and 5 m.

The **Ponte delle Chianche** (**I53**) is a further 4 km from Beneventum, about 3.5 km north west of Montecalvo. The bridge had six spans, of graduated size, increasing to the centre and an overall length of about 120 m (Ashby and Gardner, 1916: 132; Blake and Bishop, 1973: 283). The base of the central pier was of squared limestone; the rest of the bridge was in brick-faced concrete, with two rings of *bipedales* to form each arch rib. It was substantially intact at the time of Ashby and Gardner (1916).

East of Troia and south to south east of Foggia, the road crossed the Rivers Cervaro and Carapelle by two very large bridges. Only ruins of each of them remain, and for this reason both are called **Ponte Rotto** (the broken bridge). In both cases the stream has moved since Roman times, leaving the bridge 1 km east of the Cervaro and 1 km west of the Carapelle. Ashby and Gardner have reconstructed the first (**I54**; Ashby and Gardner, 1916: 144). They believe it had seventeen arch spans and additional long stone faced approaches, particularly to the east. The overall length was about 300 m; the bridge proper was about 180 m long, with spans varying from about 4 to 15 m. The width of the bridge was 7.1 m. The Carapelle River bridge (**I55**) was equally as large, with a total length of nearly 450 m and an opening of ten arches totalling about 200 m overall. Both bridges were of concrete faced with brick *bipedales*.

Gazzola describes two bridges across the Ofanto River west of Canosa di Puglia. The **Ponte sul Ofanto** (**I56**; Gazzola, 1963b, no. 133), or **Ponte di Santa Venere**, with a narrow superstructure of seven spans, rising to mid-stream on massive piers with cutwaters at both ends, has only scanty remains of the original bridge (Ashby and Gardner, 1916: 153–4). **Ponte Pietra** (**I57**; Gazzola, 1963b, no. 132) was apparently on the Canosa side, and had two spans. It is not mentioned by Ashby and Gardner. These are the last bridges recorded on the road.

A major road left the Via Appia at Capua and, after passing through Nuceria near Naples, followed an inland route roughly parallel to the west coast to Rhegium, modern Reggio, on the Straits of Messina opposite Sicily. Commonly called the Via Popillia (Chevallier, 1976; etc.), it has been argued by Wiseman that its proper name was the Via Annia (Section 2.2.2; Wiseman, 1970: 130). It crossed the River Savuto not far from Scigliano south of Cosenza. The fine single-span bridge, the **Ponte sul Savuto** (**I58**, Gazzola, 1963b, no. 226), rises steeply from the low bank at one side to mid-span and then proceeds horizontally to the high bank at the other. The arch, with a span of 20.5 m and a width of only 3.58 m, has two rings of local limestone. The spandrel walls are of small, irregularly coursed stones, with a large buttress flanking the archway on the lower side.

The island of Sicily has only one recorded bridge – at the ancient town of Selinunte on the south-west coast, near Castelvetrano. The town is flanked by two streams, the Modioni or Selino to the west, and the Gorgo di Cotone to the east. Across the Selino is the **Ponticello sul F. Selino**, a roughly constructed stone bridge with a small span (**I59**; Gazzola, 1963b, no. 47).

In Sardinia, on the other hand, are three notable bridges. The bridge over the River Mannu (or Turritano) at Porto Torres (**I60**) is particularly important; Gazzola (1963b, no. 152) has given its proportions. It has an inclined roadway, rising at a constant slope over seven arch spans, of 5.4, 5.4, 5.0, 5.3, 5.2, 12.5 and 19.5 m. The pier widths are respectively 5.3, 5.3, 5.2, 3.3, 10.5 and 16.7 m, giving a total length between abutments of 104.4 m; the total length is 112.5 m and the width 6.6 m. The maximum height is 11.5 m. The deck has stone parapets; the piers have low set cutwaters on the upstream side only; and the whole is built of large squared stonework. It may be found to the west of the harbour on a road which joined the ancient towns of Torres and Nurra.

Near the centre of Sardinia, at Gavoi south west of Nuoro, is the **Ponte di Gavoi** (**I61**; Gazzola, 1963b, no. 154), a bridge which originally had five arches, of which four appear to be intact. At the south-west corner is the **Ponte di Sant'Antioco** (**I62**; Gazzolo, 1963b, no. 153). It has two widely spaced arches of moderate span in a stone bridge of considerable length. Gazzola thinks it may date from the time of Trajan (AD 98–117). On the west coast near Oristano are ruins of the **Ponte del Tirso** (**I63**; Gazzola, 1963b, no. 155) in the river of the same name.

5.4 Central Italy

Central Italy has as its northern boundary the natural barrier formed by the Tuscan Apennines which lie along the southern edge of the Po valley. Rimini, although more truly in northern Italy, lies at its southern border, and is the terminus of the Via Flaminia which spans central Italy from Rome; it will be treated here with central Italy. To the south, this section includes bridges as far as a line drawn east from Rome, but excluding those dealt with in Section 5.2, Rome and its vicinity. It is convenient to work clockwise around Rome (see Section 2.2.3; Map 1).

The Via Aurelia followed the coast north west from Rome. About 15 km beyond Civitavecchia it reached the ancient town of Graviscae, south west of Tarquinia, where the River Marta reached the sea. The road crossed the outlet of the Marta (**I64**; Gazzola, 1963b, no. 21) on a small drain-like structure called the **Cloaca di Porta San Clementino**. It has some resemblance to the Cloaca Maxima which enters the River Tiber in Rome by its left bank, south of the Pons Aelius (Blake, 1947: 159–60). This has attracted interest because of its possible early date, and has been dated by Frank (1924: 142) and Platner and Ashby (1929: 127) as about 100 BC or slightly before. Blake believes it is later and that its entrance to the Tiber was possibly built by Agrippa for Julius Caesar about 30 BC. Referring to the structure at Graviscae she says, 'Although [it] may possibly have been built by Etruscan engineers, it was almost certainly erected under Roman supervision . . . A misconception with regard to the age of the Cloaca Maxima in Rome has led most scholars to ascribe too early a date to it. It [i.e. at Graviscae] is a massive construction with *tufa* voussoirs more than 1.5m long laid without mortar . . .'

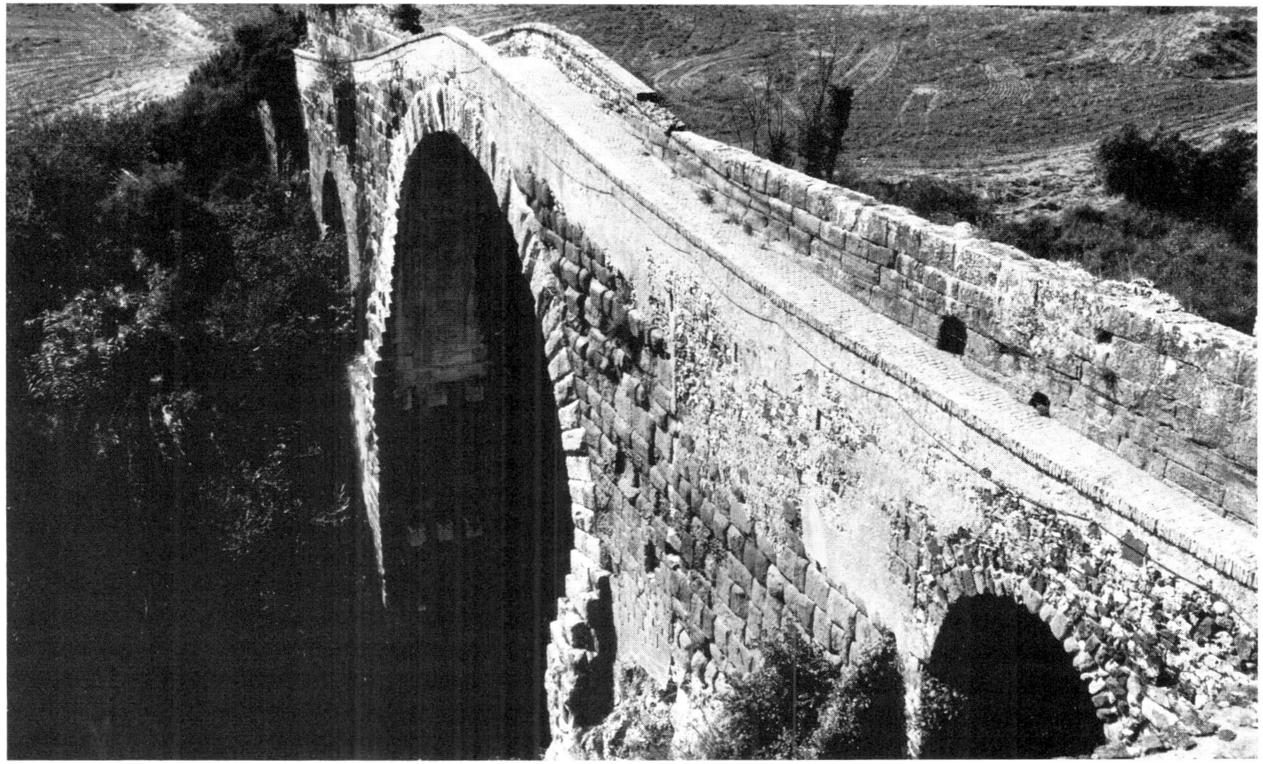

Fig. 33 The fine Ponte dell'Abadia adjoins the Etruscan museum at Vulci (I65).

Further north and inland from the Via Aurelia, about 8 km from the modern town of Montalto di Castro, was Etruscan Vulci. On the north side of the town, the River Fiora was crossed by the **Ponte dell'Abadia** (**I65**; Gazzola, 1963b, no. 30), possibly built on Etruscan foundations. Figure 33 shows the bridge as viewed from the adjacent Etruscan museum. It is high, slender and spectacular. Gazzola gives a dimension of about 35 m. If this were the span, then it was one of the largest ever built by the Romans, possibly a little less than Pont St Martin (I155) in northern Italy (35.6 m), and a little greater than that at

Kâhta, Turkey (E36, 34.2 m). It is rather more likely that Gazzola is referring to the height, which is striking. If this is so, then the span would be of the order of 25 m, and this figure will be taken here. The bridge also has two small side spans and the unusual feature, possibly unique in Roman bridges, that its width tapered with its height, the arch being considerably narrower than its foundations.

Blake (1974: 210) describes the bridge in the following terms:

This lofty bridge, with its impressive ramp somewhat restored, still serves its original purpose. The piers and the wall construction associated with the bridge show alternating rows of headers and stretchers, in which the headers are heavily rusticated. The masonry displays a clever use of the materials available: a rather soft reddish *tufa* for the core and for pillar-like projections of which the purpose is not clear, local grey *tufa* (nenfro) for the facing and the voussoirs of the west arch, and limestone for the voussoirs of the east arch and the great central span. Mortar was employed with the nenfro, but not with the red *tufa*, nor between the voussoirs. Dowels and cramps apparently were not used. These considerations suggest a date about 90 BC, which was the time when Vulci became a municipium ... The remains of a carved figure still appear on the keystone. This bridge is one of the outstanding Roman monuments in Etruria.

(See also Paget, 1973: 125.)

The Via Clodia, having left the Via Cassia near Veii, stayed to the south of the Lago di Bracciano, and then travelled direct to Bieda (ancient Blera) and Saturnia. West of the lake, a branch road led to Manziana, where it crossed yet another **Ponte del Diavolo** (**I66**; Gazzola, 1963b, no. 22). This limestone bridge was quite substantial, with a single 7 m span arch, buttressed retaining walls, and a total length in excess of 70 m. About 4 km south east of Bieda was another bridge, the **Ponte Piro** (**I67**, Gazzola, 1963b, no. 36), of which the piers remain but not the arch (Blake, 1947: 210).

The present road enters Bieda by a high-level concrete bridge over the deep gorge. Below and to the right the stream is crossed by the **Ponte del Diavolo** (**I68**; Gazzola, 1963b, no. 35). This had originally three spans of *peperino* with two smaller spans and a larger span over the stream. The spandrel walls and most of the concrete core have disappeared, but the main vault and that closer to Rome still remain (Fig. 34). It now carries a pedestrian track, which probably follows the route of the Via Clodia, and climbs to the left up the far side of the gorge below the town.

Fig. 34 The Ponte del Diavolo at Bieda (I68).

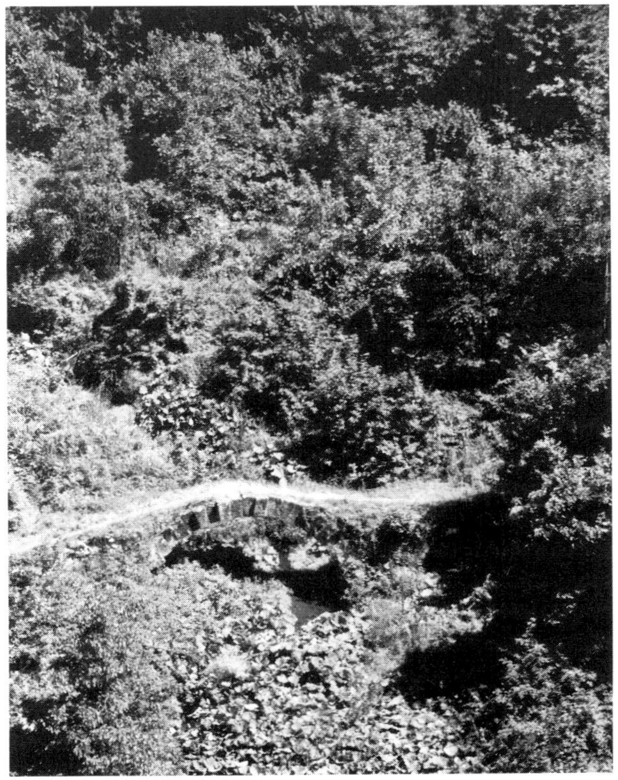

Some distance further on, perhaps a kilometre, the road crosses the single-span **Ponte della Rocca** (**I69**; Gazzola, 1963b, no. 34) (Fig. 35), dropping down through a narrow rock cutting onto the bridge, and continuing to fall to the other side. The bridge is of *tufa*; the balustrades have been reconstructed and stop short of the ends of the long retaining walls on the lower side. Blake (1947: 209) believes that this bridge belongs to the same period as the walls of the city. She refers to 'the systematization of the roads and bridges [in Etruria] undertaken by the Romans in 174 BC'; 'the Via Clodia ... was probably built in the early years of the second century before Christ'; and the bridge may have been built about that time. On the other hand the 'Ponte del Diavolo ... shows a different technique and is certainly somewhat later.'

The Via Cassia branches north from the Via Clodia near Veii, and then about 13 km further on veers towards the west, leaving the Via Amerina to continue to the north. A few kilometres from the junction, at Sette Vene, the Via Cassia crossed a small single-arched bridge (**I70**; Ashby, 1927a: 241). It continued through Viterbo, Arezzo and Florence to join the Via Aurelia, or its extension, near Pisa.

Viterbo has the remains of a single-span bridge, the **Ponte San Nicolao** (**I71**; Gazzola, 1963b, no. 156), restored by Vespasian (AD 69–79) and possibly dating from Claudius, AD 41–54 (Blake, 1947: 209). In neighbouring Bulicame, a kilometre to the north east, the Via Cassia crossed an old bridge variously named as the **Ponte San Lorenzo**, **Ponte d'Ercole** or **Ponte Camillario** (**I72**; Gazzola, 1963b, no. 39). It has a single small span. As described by Blake (1947: 209), it 'exhibits an arch of seven wedge-shaped blocks of travertine with the joints not radial, as though the voussoir principle was still in an experimental state. This bridge is certainly not later than the second century before Christ. Ponte San Nicolao in the same neighbourhood is definitely later.'

East of the Via Cassia, the remains of an ancient bridge, the **Ponte sul Tevere**, were found in 1936 in the Tiber at Sigliano in the province of Arezzo (**I73**; Gazzola, 1963b, no. 135).

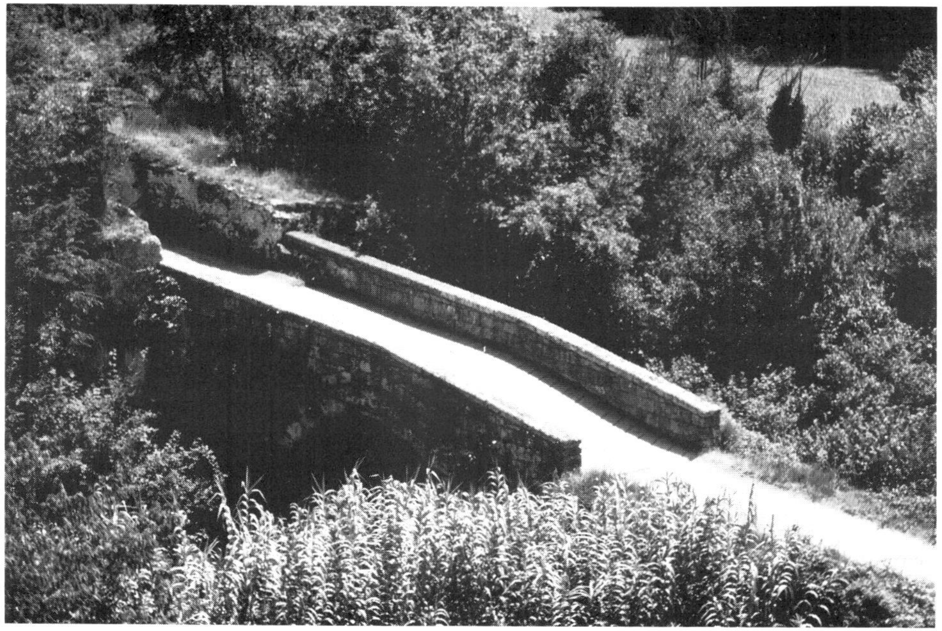

Fig. 35 Ponte della Rocca, on the outskirts of Bieda (I69).

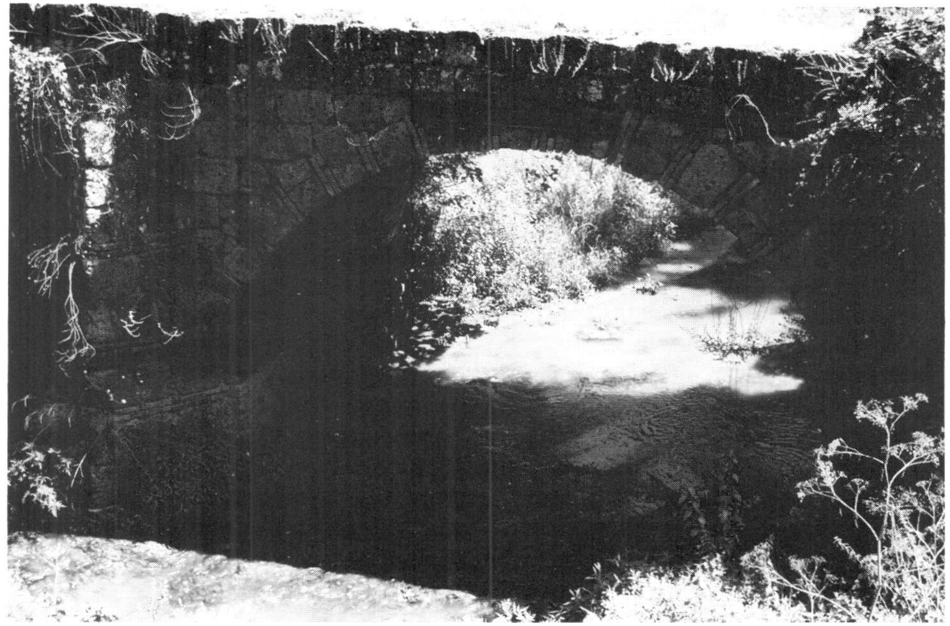

Fig. 36 Ponte Nepesino, near Nepi on the Via Amerina (I74).

Frederiksen and Ward Perkins have described structures on the Via Amerina and in its vicinity as it goes north into southern Etruria. The first is the **Ponte Nepesino** (1957: 79, 81–2; **I74**) over the Fosso Cerreto, 3 km south of Nepi, on the Via Amerina (Fig. 36). The present bridge, which is in good condition, has three larger spans, and a smaller span at the southern end. It has been reconstructed in recent times, but includes substantial Roman stonework. The geometry is probably Roman.

Gazzola (1963b, no. 25) has a photograph of a small bridge of *tufa*, the **Ponte di Nepi** (**I75**), presumably in Nepi itself, with a single small span. It is not mentioned by Frederiksen and Ward Perkins. They have, however, fully documented the fine small bridge over the Fosso Tre Ponti (**I76**), 4 km north of Nepi (1957: 97, 100; Gazzola, 1963b, no. 23). It has a single span of 4.55 m; a clear height of 9.1 m; and an overall height, including balustrades, of 11.2 m. The excellent stone retaining walls are buttressed beside the arch below the springing level, with additional full-height buttresses some 3 m back. It is similar to, but larger and in better condition than, the Ponte del Diavolo at Manziana, and is believed to date from the same time.

Gazzola cites another bridge (**I77**; 1963b, no. 24) in this area, the **Ponte sul Fosso della Mola**, near Castel S. Elia.

The next major road from Rome was the Via Flaminia (Section 2.2.4), built following the censorship of Gaius Flaminius in 220 BC. Having crossed the Tiber at the Pons Mulvius, the Via Flaminia runs north to Narni. It then divides, the original road continuing north through Massa Martana to Foligno, and a later diversion branching east and then north through Terni and Spoleto to the same town. The two branches combine and continue north to Cagli and the Furlo Pass; then north east to the coast at Fano; and along the coast to Rimini. The road itself has been described by Ashby and Fell (1921), and the bridges on the road in more detail by Ballance (1951). Blake (1947: 213–5) also discusses many of the bridges. As will be seen, it has more Roman bridges than any comparable length of road either in Italy or elsewhere in the Roman world.

After the bridge over the Fosso della Valchetta (I13, Section 5.2), the next on the Via Flaminia was the **Ponte Ritorto** (**I78**; Gazzola, 1963b, no. 31), 2 km east of the 50th kilometre stone of the modern road (Ballance, 1951: 85), over the Fosso di Chiavello near Civita Castellana. It has a single arch, of brown *tufa*, with a span of 5.9 m and a width of 5.8 m.

About 300 m further north are the remains of a small culvert (**I79**; Gazzola, 1963b, no. 6) of brown *tufa*, with a span of 3.1 m. Ballance (1951: 86) believes it to be quite old, earlier than 27 BC, possibly as early as the original construction of the road in 220 BC.

Having crossed the Tiber by the Pons Mulvius, the road continues up the river valley for some 14 km, beyond Prima Porta and the modern Grande Raccordo Anulare (GRA) or ring road. The river then swings a little to the east, while the road follows the high ground to the north. Beyond the Ponte Ritorto, the road approaches the next major obstacle, the valley of the River Treia. Ashby and Fell (1921, fig. 11) have a detailed sketch of the crossing. The road turns sharply left and drops about 80 m down the side of the valley to the stream. Part way down it crosses the Voltarella (**I80**, Gazzola, 1963b, no. 7), a stone-arched culvert with a span of 3.1 m and a width that was originally about 7.4 m. 'The material, as in the rest of the crossing, is a yellowish-brown *tufa*, presumably cut from the cliff on the far side of the valley, where there are extensive signs of ancient quarrying' (Ballance, 1951: 86).

The bridge across the Treia itself no longer remains. Ashby and Fell describe some remains on the left bank of an arched bridge, but conclude that they are not Roman. Gazzola (1963b, no. 8) reproduces a painting of an old three-span arch bridge (**I81**), but again this is hardly likely to be original. At the north bank, the road turns sharply from about north north west to north north east, and after some 500 m begins to climb out of the valley by an inclined causeway, retained on each side by ancient stone walls, called the **Muro del Peccato** (**I82**; Gazzola, 1963b, no. 9). Like the viaduct at Ariccia (I29, Section 5.2), it is pierced by an arched opening. The walls have a maximum height of 11 m and are built of massive *tufa* blocks, as big as 0.6 m high by 2 m long, with a thickness about 0.55–0.65 m. The width of 10.8 m is broken by a stone gutter, leaving a footway beside the parapet 1.25 m wide. The length was about 200 m. Ballance (1951: 89–90) discusses the dating of these works in some detail. He concludes, 'It is not impossible that the Muro del Peccato goes back to the original building of the Flaminia in 220 BC. On the whole, however, the viaduct and the two bridges [I79, I84] seem more likely to be at least a century later'; that is, possibly about 100 BC. Blake (1947: 109) seems to agree with this.

The Via Flaminia has now rejoined the Tiber, and continues north some 1 km west of its right bank. Very little now remains of the bridge over the Rio Fratta (**I83**; Gazzola, 1963b, no. 27; Ballance, 1951: 88) about 4 km north of the Treia River, but another kilometre further on is the **Ponte Picchiato** (**I84**; Gazzola, 1963b, no. 26) across the Rio Miccino north of Borghetto. Still in good condition (Ballance, 1951: 88), the bridge is slightly skew, by about 10°, and has a span perpendicular to the opening of 11.5 m, and a width of 7.8 m. The voussoirs of the arch have alternate stretchers and headers of *tufa*, with stones up to 2 m long. The joints are well fitted, with evidence of wooden swallow-tail cramps, but no mortar. As already mentioned, Ballance dates it with the Muro del Peccato, possibly about 100 BC.

The Tiber was crossed somewhere south of Otricoli over what must have been a substantial bridge, possibly the **Pons Minucius** (**I85**; Ballance, 1951: 91; Gazzola, 1963b, no. 44). 'When Augustus undertook the restoration of this important military highway in 27 BC, he found only the Pons Mulvius of 110–109 BC and the Pons Minucius of 64 BC in sufficiently good condition to be retained; the rest had to be rebuilt by his engineers' (Blake, 1947: 213). Although Ashby and Fell (1921: 162) describe the sighting of some travertine remains by Martinelli in 1673, later floods have completely altered the course of the river and these relics have now disappeared under the left bank; even the location is uncertain. There is some interest in the name of the bridge, which probably suggests that it was built by Q. Minucius Thermus, *curator viarum* for the Via Flaminia in 65 BC (see Section 3.2).

The road now leaves the Tiber and climbs to Otricoli and over the ridge between the Tiber and its tributary, the Nera. About 7 km beyond Otricoli it crosses the **Ponte Sanguinaro** (**I86**; Gazzola, 1963b, no. 45) over a bridge with a new arch on old abutments. The old road then went north and crossed the River Nera immediately north of the hilltop town of Narni.

The **Ponte d'Augusto** across the Nera River at Narni (**I87**; Gazzola, 1963b, no. 60) was one of the finest of all Roman bridges; the Frontispiece is an attempt to show it in its original form. At present only one span remains (see Figs. 37 and 38) but it appears that there were originally four. Perhaps the best description is that given by Ballance (1951: 91–7) based on the earlier account by Ashby and Fell (1921: 169–70; see also Blake, 1947: 213–14). It was originally about 160 m long at deck level, 8 m wide, with a maximum height above the water of 30 m. Most likely the deck sloped down from south to north and the spans (in the same order) were 19.6, 32.1, 18.0 and 16.0 m. Ballance numbers the spans in this way 1 to 4, and the piers from 1 to 3; that is, pier 1 was between spans 1 and 2. Figure 37, which is taken from the south bank, shows span 1, pier 1, the collapsed remains of pier 2, pier 3 (standing), and the north abutment. A railway has been carved along the far bank (the right bank) and has been cut through a new opening in the north abutment. The stonework beyond and to the right of pier 3 appears to be the remains of span 4. Pier 2 fell in 1885. Photographs exist which show it still standing (e.g. Blake, 1947, pl. 24). It is noticeable that it has fallen upstream, presumably because the pier was undermined by scour beneath its upstream face – none of the piers have cutwaters.

Eroli, as quoted by Ashby and Fell, believed that the bridge originally had only three spans, of 19.6, 32.1 and 43.0 m; that it collapsed about AD 800, was restored and fell again in AD 1053; and that after AD 800 the third span was converted to two spans of 17 and 18 m, with a 10 m wide pier between. These figures are inconsistent, for the

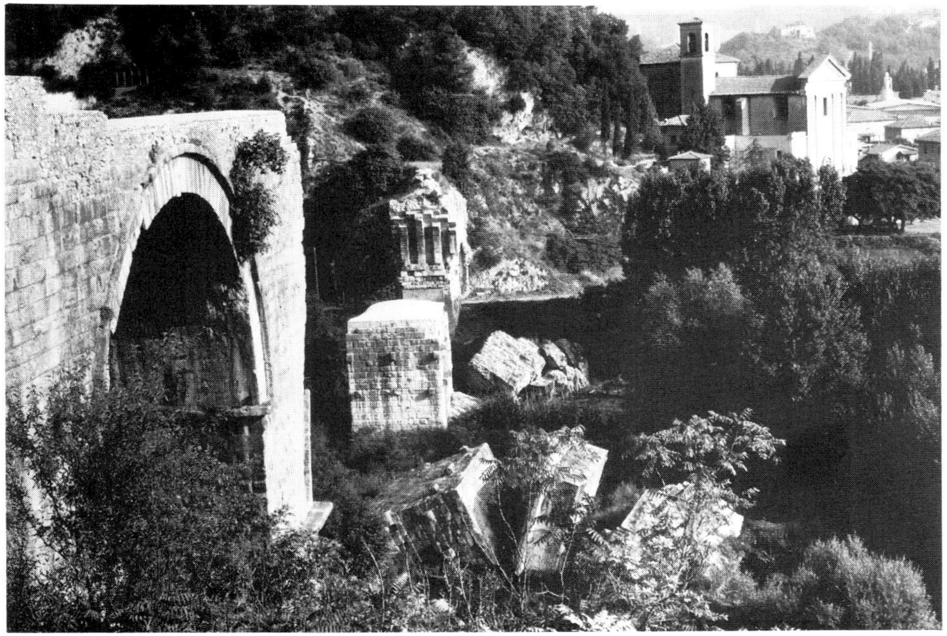

Fig. 37 General view of the remains of the Ponte d'Augusto at Narni, looking north (I87).

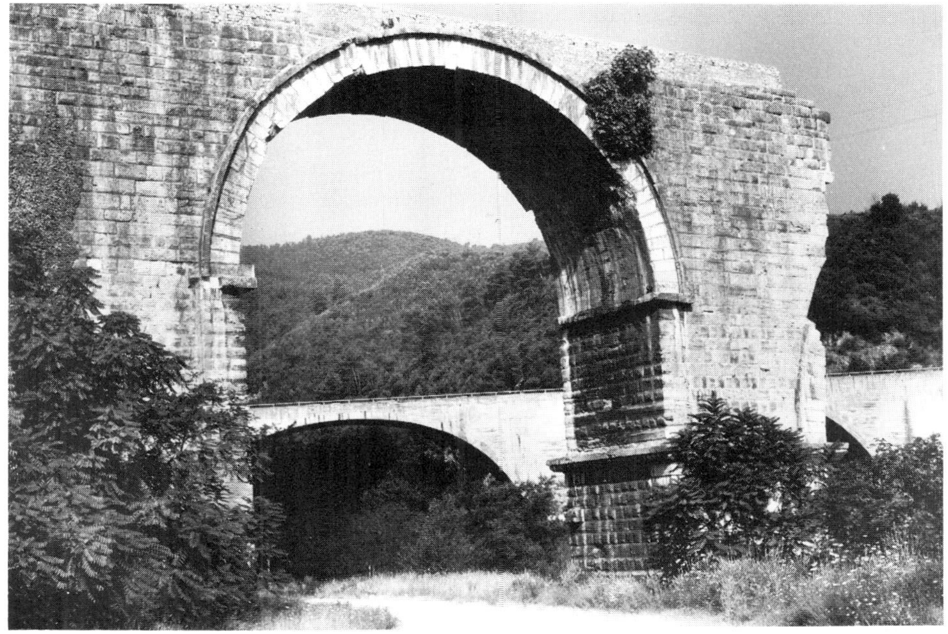

Fig. 38 The intact span 1 at the southern end of the Ponte d'Augusto at Narni (I87).

sum of 17, 18 and 10 m does not equal 43 m. More likely figures may be spans of 18 and 16 m (as given above), and a pier width of about 9 m. A span of 43 m would have been by far the largest built by the Romans – exceeding the 35.6 m span of the Pont St Martin (I155). Ballance does not accept this account.

An arch of so enormous a span could only be fitted into the scheme of things by assuming that it sprang from a point below the cornice of pier 2 . . . If the 43 m arch ever existed, it has left no traces, and, short of assuming that the whole of the bridge beyond the first pier belongs to a reconstruction, it cannot be made to fit into the extant remains.

The old photograph reproduced by Blake (1947, pl. 24) shows pier 2 before its collapse. The springing of span 3, although higher than that of span 2, is lower than in span 1, and this would imply normally that the rise and therefore the span of span 3 would have exceeded that of span 1. However, it is evident from the photograph that pier 2 had settled before it fell (see the springing levels of span 2). It is also likely that there was a slope in the deck of the bridge – a span of 18 m for span 3 could well be correct.

Two forms of construction are visible in the piers and spandrel walls. The lower levels are constructed both externally and internally of squared stone, with alternate courses of stretchers and headers in the external faces. The higher levels consist of concrete with stone facing. In this case the external walls contain more stretchers, with some headers used in alternate courses to bond the facing into the concrete core. In Figure 37, the boundary is at the springing level of pier 1 at the right, and a little above springing level at the left. Indeed, if pier 1 is viewed from the river, the upstream facing may be seen to protrude some distance, perhaps 1.5 m, beyond the concrete core in the region forming the spandrel of the missing span 2. Another unusual feature of the construction is the provision of a ring of stones to form a cornice around the extrados of the arch.

However, it is in span 4 that the most unusual construction detail occurs. As can be seen in Figure 38, the arch has four protruding ribs of stone, bonded into an adjoining vault of stone. The arrangement is asymmetric, with the upper vault extending to the outside of the left-hand rib, but not to the right. The purpose of the detail is unclear, unless perhaps it was intended to ease the centering problem. As far as is known, this arrangement is unique. Ballance dates this span as 'probably later than any other part of the bridge' (1951: 94).

It is also intriguing to speculate on the nature of the collapse. Until 1885 all three piers were standing, and the remains as they exist today show both pier 2 and span 4 fallen upstream. Span 2 was larger than the others, and it would be expected that any design fault would cause collapse of this span before the others. Also, if span 3 were to have gone before span 2, it would be expected that span 2 would have pushed pier 2 over to the north in the course of its collapse. The collapse is discussed further in Section 9.4, where it is suggested that the piers were undermined before collapse of the arches, and that the resulting horizontal shear forces in the arch rib, parallel to the stream flow, may have caused them to collapse first. The whole question is very difficult to resolve.

Both Blake and Ballance believe that the bridge was built in the time of Augustus, 31 BC–AD 14, possibly in 27 BC.

The next bridge along the more westerly branch of the Via Flaminia is **Ponte Calamone** (**I88**; Gazzola, 1963b, no. 105) about 2.5 km beyond Narni. It had two main arches, each with a span of 3.3 m, and an arched floodway of 1.4 m span set high in the substantial centre pier. The southern span and flood arch were destroyed in the Second World War and have since been replaced in brick (Ballance, 1951: 97); Figure 39 shows the interior of the span which remains. A further 2.5 km up the same road is the **Ponte Cardaro** (**I89**; Gazzola, 1963b, no. 106; see also no. 104). This had five spans of graduated size: 3.4, 5.5, 7.9, 5.5 and 3.4 m. The stonework was of travertine, which has survived well, but the central span again was blown up in the last war and replaced in brick. It is now much overgrown, but Figure 40 shows two of the smaller spans. The protruding string courses at springing level are characteristic of the bridge, as is the rustication in the face of the stonework. The illustration shows new brickwork at the upper right. The bridge had substantial retaining walls in the approaches, with two square buttresses in the wall to the south east. Both bridges appear to be Augustan.

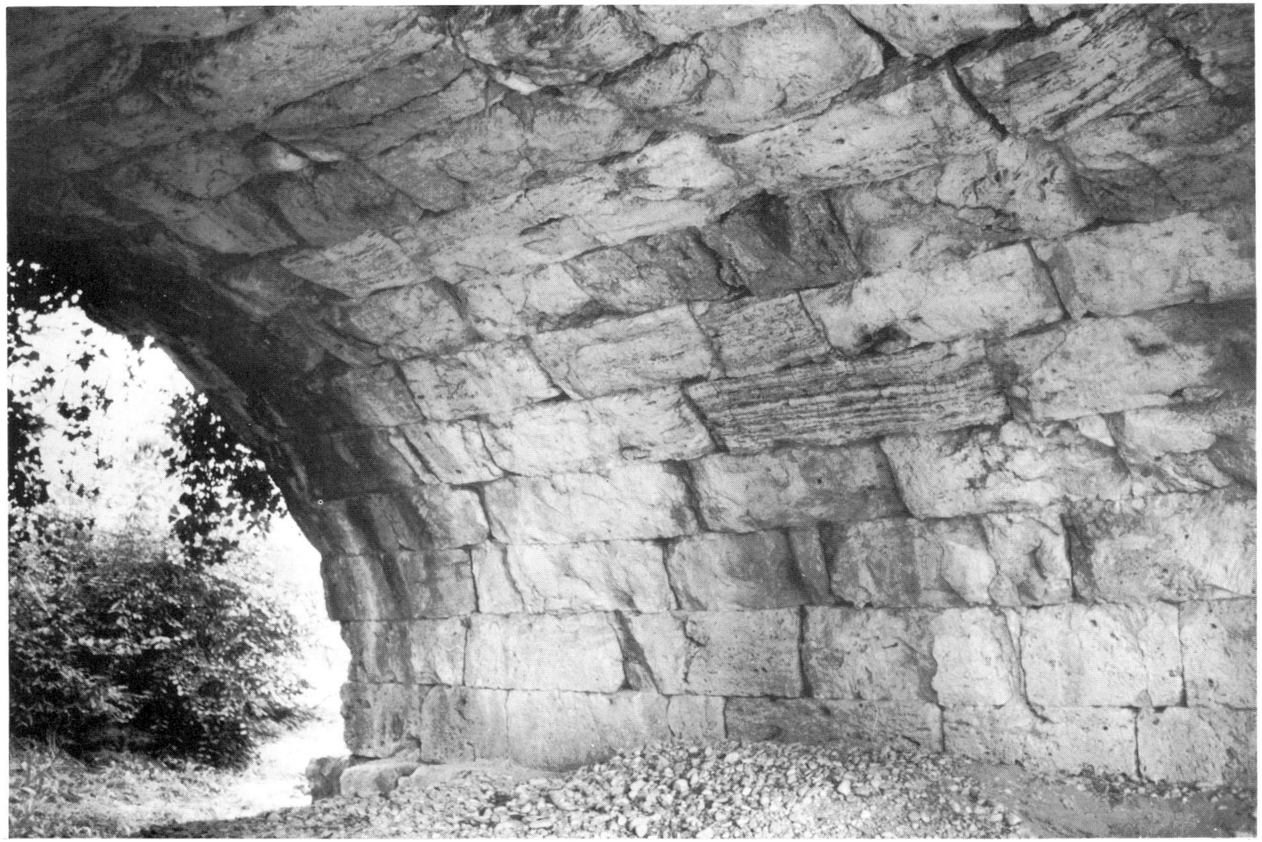

Fig. 39 Interior of Roman span of the Ponte Calamone near Narni. Note the stretcher bond (Section 8.5).

Further north on the same road are three other small bridges – I90–92. The second of these is the **Ponte Romano** or **Fonnaia** (**I91**; Gazzola, 1963b, no. 61). Although of only 3.4 m span, it is quite high, 8 m, and is notable for its high skew, with the arch axis at less than 60° to the road centreline, and for the great width of roadway (about 11 m). The third, the **Ponte del Diavolo** (**I92**; Gazzola, 1963b, no. 63), has the even greater width of 14.7 m.

The other branch of the Via Flaminia went through Terni and Spoleto to Foligno. Six kilometres beyond Terni it crossed the **Ponte Toro** (**I93**, Gazzola, 1963b, no. 97).

The northern approach from the plain to the steep and beautiful hill town of Spoleto is via the Porta Garibaldi.

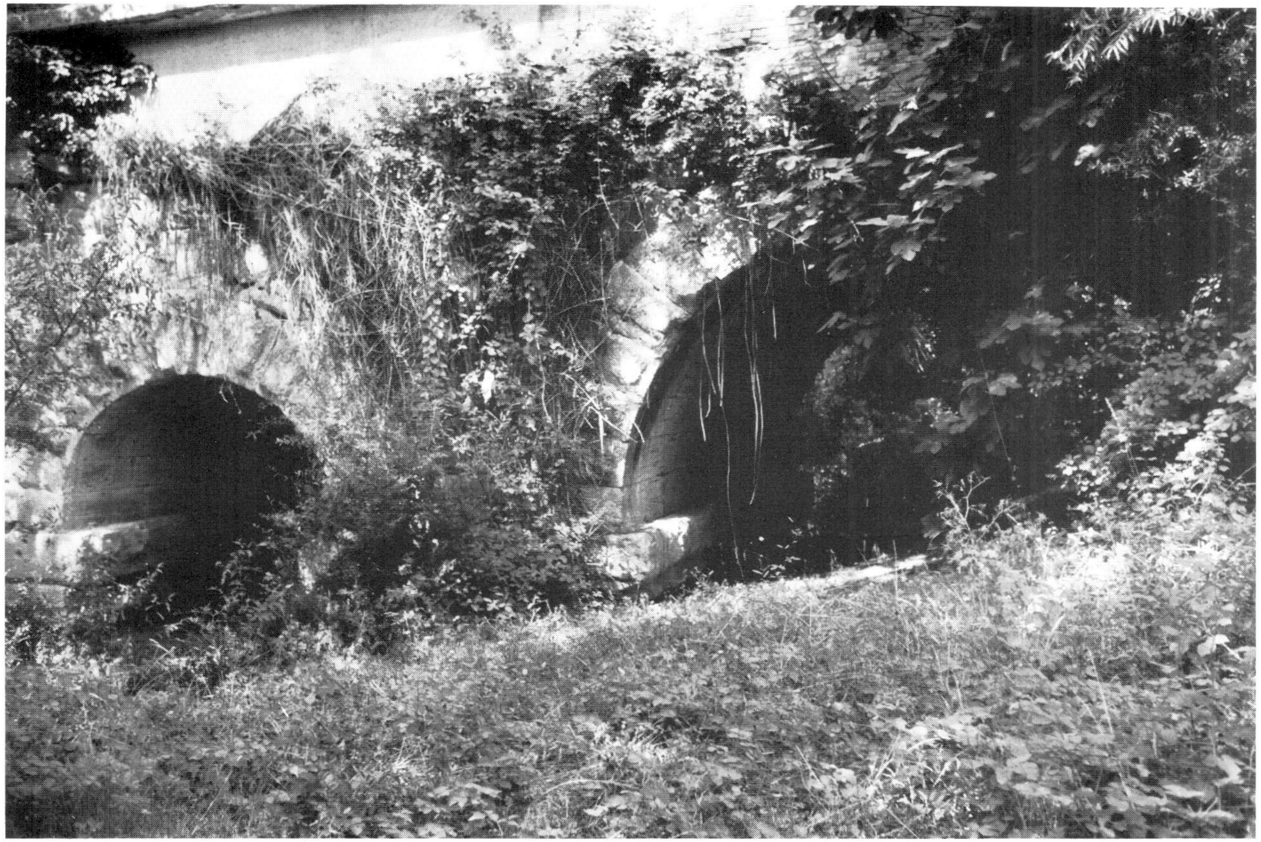

Fig. 40 Ponte Cardaro, north of Narni on the original Via Flaminia (I89).

Under the Piazza Garibaldi is still the **Ponte Sanguinario** (**I94**; Gazzola, 1963b, no. 53; Blake, 1947: 214, pl.25) which carried this eastern branch of the road across the Torrente Tessino, now running some 80 m further to the north. The bridge was excavated in 1817, re-excavated in 1843, and roofed over in 1848. There were probably four arches of travertine, of which two have been exposed, with spans of 6.2 and 6.8 m. Its width is only 4.5 m. The most unusual feature is the small pier width, scarcely greater than twice the arch thickness, with a single row of stones on the pier serving as springing for the two arches. Both Blake and Ballance believe it to be pre-Augustan, probably around the middle of the first century BC.

Gazzola lists four ancient bridges in Foligno (I95–8, Gazzola, 1963b, nos. 64–7). The first, the **Primo Ponte**, has four spans; the second had seven spans and is largely buried; the third is again at least partly buried, with five spans from 9 to 10 m; and the fourth, the **Quarto Ponte**, is similar, with three spans. They are not easily seen, and are not cited by Ashby, Blake or Ballance.

Beyond Foligno, the Via Flaminia starts to climb, and it is another 34 km before Gualdo Tadino, ancient Tadinum, is reached. Gazzola lists two bridges 'in this area'. His **Primo Ponte** (**I99**; Gazzola, 1963b, no. 110) is a fine high bridge with a single span in limestone. To judge from his photograph, it must have a clear height of the order of 7 m, and a span about 4.0–4.5 m. He also lists a 'Secondo Ponte of Gualdo Tadino' (Gazzola, 1963b, no. 111), but the photograph he uses to illustrate this is actually of the Ponte Spiano, which he also lists (Gazzola, 1963, no. 90), and it appears that there is some mistake.

Ponte Spiano (**I100**) carries the Via Flaminia across the River Fonturce some 6 km north of Sigillo (Ashby and Fell, 1921: 181). It is an excellent small bridge, with a single span of 3.25 m and a width of 10.6 m (Ballance, 1951: 104), built of a Lias limestone, known locally as *pietra corgnola*. The stonework is unusual for a Roman bridge. The arch of fifteen voussoirs springs from heavy cornice stones. The voussoirs themselves are cut with vertical and horizontal faces at the extrados to match the stones of the spandrel walls; in particular, the first stone above the springing extends well into the spandrel and this also is unusual. Blake (1947: 215) regards it as probably Augustan, but it may be later.

In this region, Gazzola also lists the **Ponte di Fossato di Vico** (**I101**; Gazzola, 1963b, no. 69). Fossato di Vico lies about 1.5 km east of the Via Flaminia, just south of the modern road which branches east to Fabriano; that is, south of Sigillo and south of Ponte Spiano. The bridge is small, with a single arch that contains Roman work. It crosses the Fosso Regalino. Further along the Via Flaminia, 1.5 km beyond Ponte Spiano, is the **Ponte Etrusco** (**I102**; Gazzola, 1963b, no. 222) with only the abutments remaining. It had a single span of 4 m built of *pietra grigna*, an 'intractable pinkish conglomerate' (Ballance 1951: 105–6) – see p. 82f.

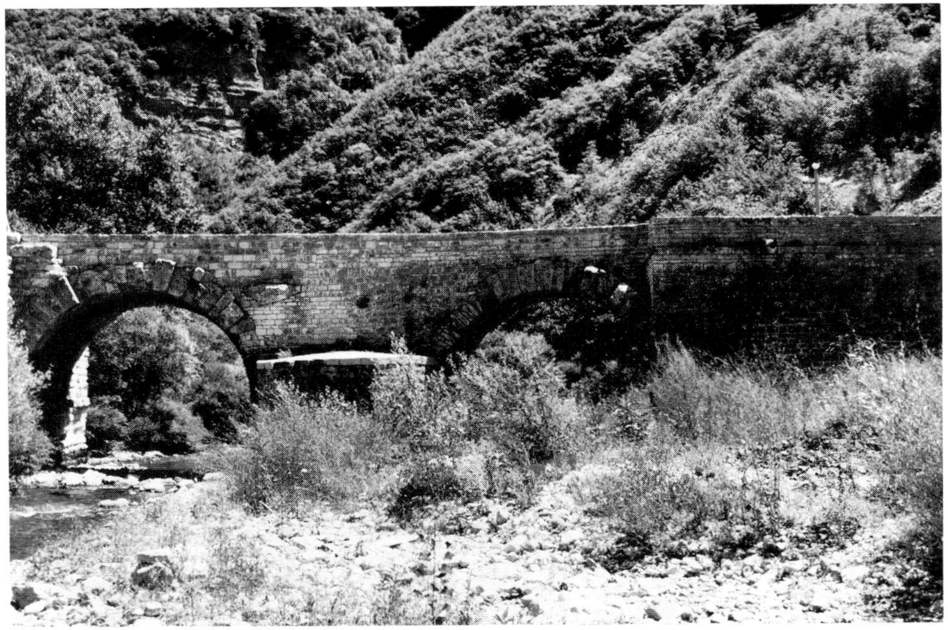

Fig. 41 The Ponte Grosso, in the gorge called the Foci di Cagli (I105).

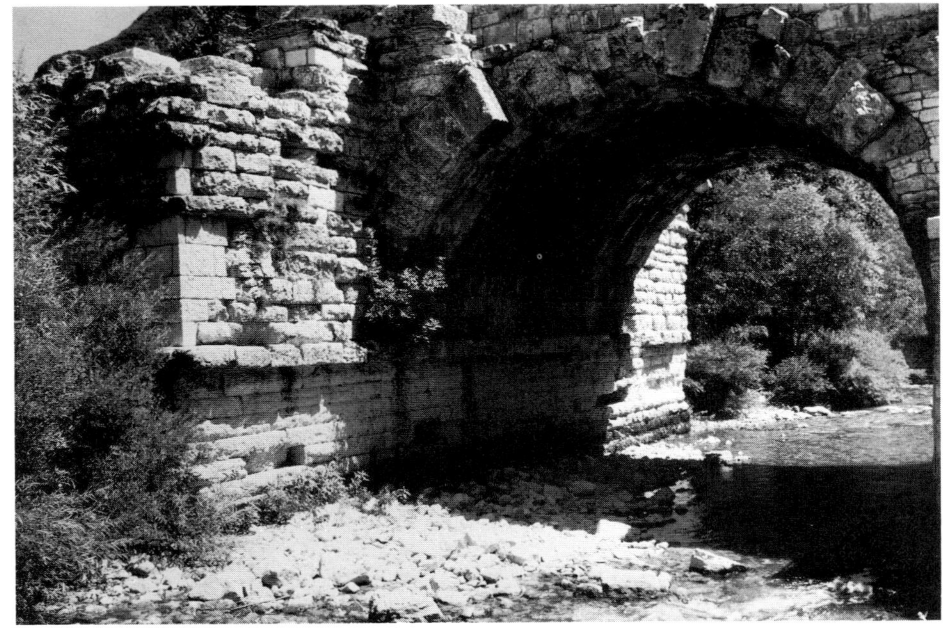

Fig. 42 Massive construction in the Ponte Grosso (I105).

The road now rises to the Scheggia Pass beyond Scheggia. Here, before the village of Pontericcioli, were two small bridges – **I103**, **I104** (Gazzola, 1963b, nos. 223, 95). Ballance was unable to find Roman remains (1951: 106).

However, beyond the pass, near Cantiano is **Ponte Grosso** in the Foci di Cagli (**I105**; Gazzola, 1963b, no. 98). To quote Ballance (1951: 107), 'In its magnificent setting and fine state of preservation Ponte Grosso is today by far the best extant Roman bridge between Foligno and Rimini.' It crosses the Burano River about 7 km south of Cagli, to the east of and immediately below a modern road bridge. Figures 41 and 42 show the upstream face of the bridge. The geometry is curious, for the spandrel wall on the western or upstream side is not straight, but bends to the west as it crosses the southern span. There are protruding voussoirs in both spans, but in the south span these extend to the western face of the abutment and it appears that this bend was present also in the original spandrel wall. The present cutwater is not original, although it is believed that there was a cutwater on the upstream face in the original design. The whole construction is massive – notably the widths of the central pier and abutments – leading to Ballance's comment, 'It was obviously intended to stand up to a river of unusual violence, and it is principally to this that it owes its exceptionally massive construction and its unusual plan.' It has two spans of about 7 m, a width of 6.4 m, and is built of a Lias limestone similar to that in the Ponte Spiano (*pietra corgnola*). The name, Ponte Grosso, is derived from the large size of the stones.

Between Ponte Grosso and Cagli are the scanty Roman remains of three bridges – **I106–8** (Gazzola, 1963b, nos. 224, 225, 93; Ballance, 1951: 108–9) – but just north of Cagli is another fine bridge, the **Ponte Manlio** (**I109**; Gazzola, 1963b, no. 46). It crosses the Torrente Bosso just west of the current road. Ballance regards it as 'the principal key to the dating of most of the bridges in the Apennine section of the Flaminia' (1951: 109–10).

The existing structure is of three well defined periods. To the first period belong the main arch with the walling of the spandrels, a massive string course, and a good deal of the parapet. Three embankment walls were subsequently added at an angle to the original bridge (the second stage). Finally three more walls were built parallel to the main axis of the

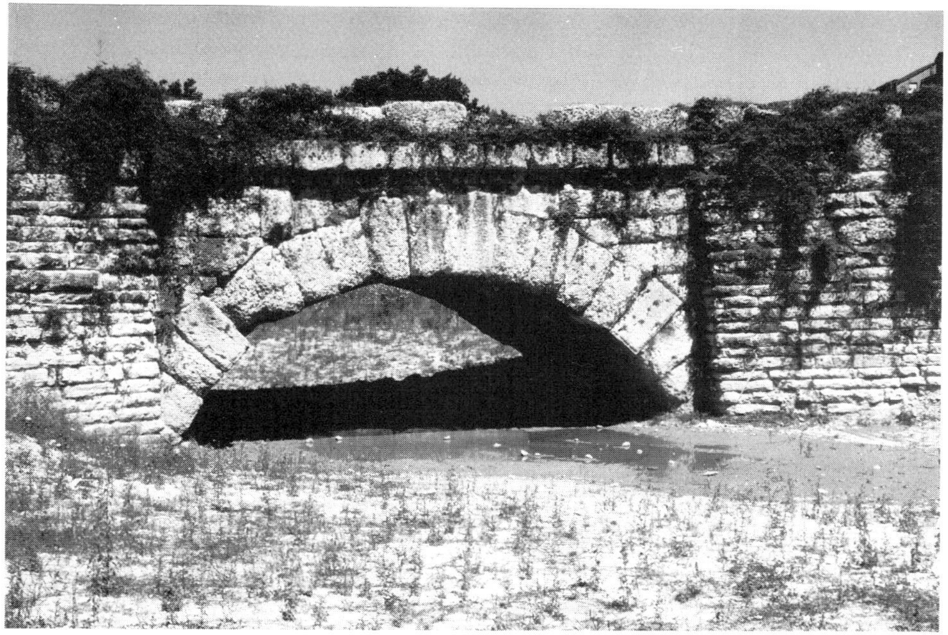

Fig. 43 The Ponte Manlio on the Via Flaminia north of Cagli is also of massive construction, with the heavy arch rib of the first period, and wing walls of the second period (I109).

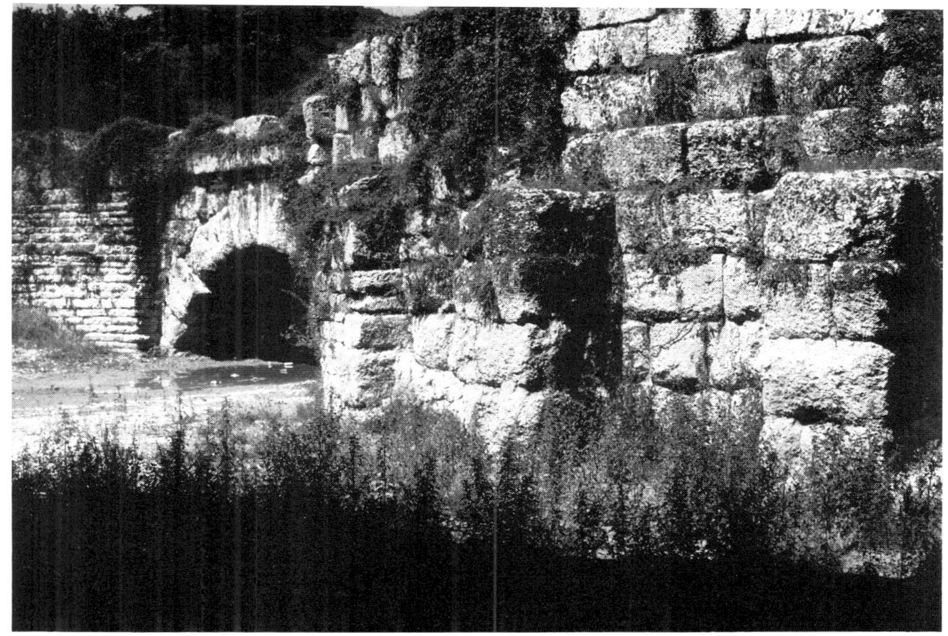

Fig. 44 General view of the Ponte Manlio, with masonry in the foreground described by Ballance (1951) as of the third period of construction (I109).

road and composed, like the arch, of enormous blocks of *pietra grigna*. A second and much smaller arch at the far northern end of the bridge belongs to the same period as the main arch.

Figure 43 shows the main arch. It has a span of 11.7 m, a width of 8.8 m, and is composed of massive stones of *pietra grigna* (pink conglomerate). The keystone is about 1.85 m high, and 1.15 m wide at the top; one of the spandrel stones (of the same material) is about 2.5 m long (see Fig. 134). The lower part of the arch is buried in the stream bed. Ballance quotes Mocchi as saying, on the basis of his excavations, that the arch is not semicircular, but circular, but this is surely unlikely. The smaller arch, possibly about 100 m north of the main arch, spans 3.4 m.

This figure also shows wing walls of the second period. They are built of *pietra corgnola* (Lias limestone), and run in front of the walls of the arch. Figure 44 looks south west, and shows the rough *pietra grigna* walls of the third period, reinforced with four square buttresses.

Ballance has classified other bridges on this part of the Via Flaminia by comparison with the masonry of these three stages. None have masonry which resembles the work of the first period. The *pietra corgnola* group, which resembles the second period, include the Ponte Grosso (I105). The third period includes the Ponte Etrusco (I102) and the second stage of the **Ponte dei Ciclopi** (**I106**). He dates the first period as late Republic (first century BC), the second as Augustan (*c.* 27 BC), and the third as from the repairs of Trajan (AD 98–117) or Hadrian (AD 117–38).

Between Ponte Manlio and the Furlo Pass were at least ten ancient bridges and culverts (Ballance, 1951: 111). The **Ponte di Smirra** (**I110**; Gazzola, 1963b, no. 92), 2 km south of Acqualanga, had a span of 3.8 m, and was destroyed in 1944. Two of the culverts were at **Case Nuove** (**I111**; Gazzola, 1963b, no. 91), with one 1.4 m span; and the **Ponticello dell' Abbazia** (**I112**; Gazzola, 1963b, no. 94), where a wall of *pietra corgnola*, some 3 m high and 50 m long, was pierced by two culverts, of 2.1 and 2.8 m span, spaced about 30 m apart.

The Burano River and the road turn north east near Acqualanga; some 6 km beyond the town they enter a steep and narrow gorge, the Furlo Pass, 1.5 km long. The road is cut into the left bank 30 m above the stream and at one point passes through a narrow tunnel curving through the face of a rocky outcrop. Between the present tunnel and the point is Vespasian's tunnel of AD 76–7, 38 m long (Fig. 4, Section 2.2.4). At the southern end was an even earlier tunnel 8 m long (Ashby and Fell, 1921: 185–6). The stream is now dammed at the outfall from the Furlo Pass, but the water level lies well below the old Roman road. Ashby described the scene, 'The cliffs are bare and precipitous on both sides, in places overhanging the road, and the scene is at once majestic and desolate', and speaks of Vespasian's tunnel as 'a remarkable achievement of engineering with the tools then available' – as indeed it was.

The road descends, and 2 km beyond the end of the pass reaches the Metaurus River, which comes in to join the Burano from the left. It is crossed by a bridge of three spans, 4.0, 11.3 and 11.3 m, the **Ponte di Traiano** (**I113**; Gazzola, 1963b, no. 178) dated AD 115 from an inscription on a neighbouring slab (Ballance, 1951: 112). This fine limestone bridge was destroyed during the Second World War. The road then reached the village of Calmazzo and turned right to follow the left side of the Metaurus valley some 5 km to Fossombrone. Half-way between was a branch road to the right which crossed the Metaurus River 150 m from the Via Flaminia. Photographs are all that remain of the **Ponte di Diocleziano** (**I114**; Gazzola, 1963b, no. 230), also destroyed in the Second World War. It had a considerable span, possibly as large as 20 m, springing directly from the cliffs that rose some 10 m vertically from the water. The arch rib was separated from the spandrel walls by a projecting cornice, as at Narni, but in no other respect was the work as fine.

The Via Flaminia reached the coast at Fano, and then turned north west to Rimini. The Metaurus joins the sea about 5 km south east of Fano. Gazzola records another bridge across the Metaurus, the **Ponte detto delle Piangole** (**I115**; Gazzola, 1963b, no. 78). Its location is unknown, but it appears to have been a fine bridge with four limestone spans. Up the coast in Pesaro was the **Ponte di Porta Rimini** (**I116**; Gazzola, 1963b, no. 84) with two segmental stone arch spans of 28.7 and 6.3 m, clearly late, but which are said to include some Roman remains.

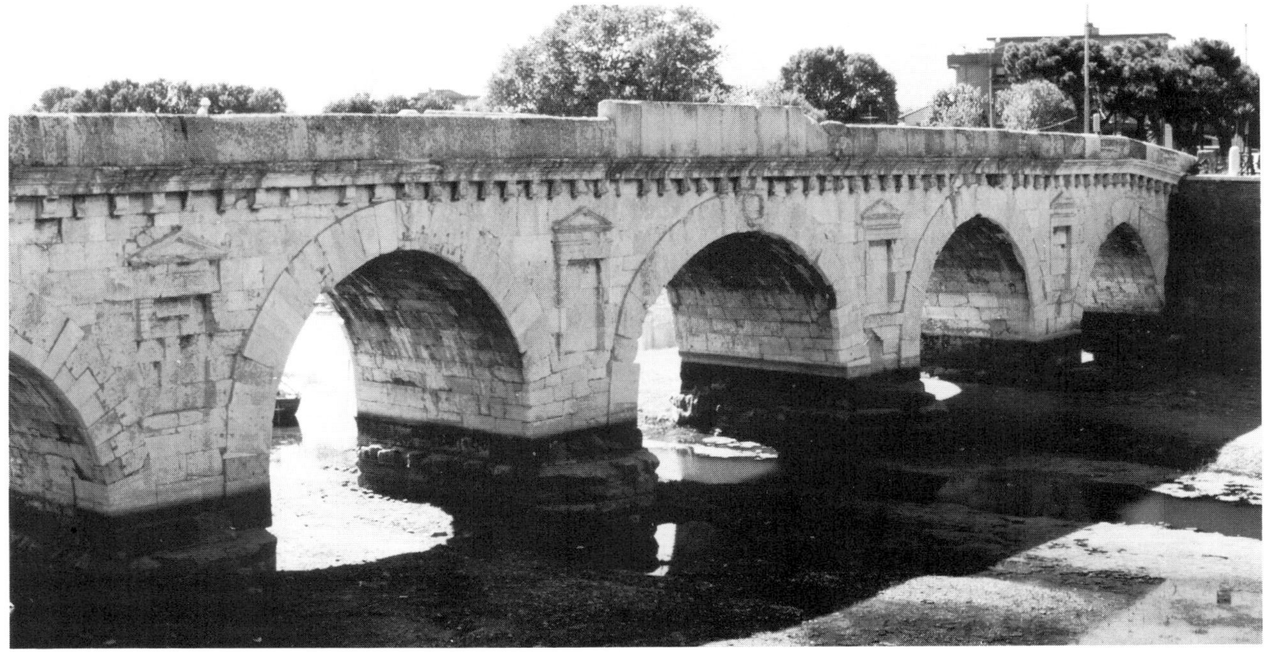

Fig. 45 Downstream face of the Ponte d'Augusto at Rimini (I117).

The end of the Via Flaminia was in Rimini, at the forum. The River Ariminus, modern Marecchia, flowed into the city past the northern walls of the town. The Via Aemilia, dating from 187 BC, travelled to the north west, up the valley of the Po, and Augustus, realising the importance of this road, started to build a bridge across the stream. It was finished in AD 20 by Tiberius (Ashby and Fell, 1921: 215) and carries an inscription recording it as 'given by both emperors' (Ashby and Fell, 1921: 190). It is usually called the **Ponte d'Augusto** (**I117**; Gazzola, 1963b, no. 83) but also the **Ponte di Tiberio**.

Figure 45 shows the bridge from the downstream side. The main flow in the river has now been diverted to another channel, leaving the old stream in use as a floodway, and as a boat harbour at its mouth. There are five spans: 8.7, 8.9, 10.6, 8.9 and 8.0 m. The shorter dimension of span 5 has resulted from inward movement of the northern abutment, and this has also caused disturbance to the alignment of the deck. Although the arches are notionally semicircular, there is a significant departure from this in the central, larger span, where the rise is less than half the span. Although it is difficult to say with certainty whether this is deliberate, or accidental and due to settlement, yet the impression given is that it is deliberate, and that the arch profile was designed as semi-elliptical. If this was so, then it is unusual. The overall length between abutments is 62.6 m, corresponding to an average pier width of about 4.4 m.

The appearance of the bridge has been carefully considered. The line of the deck has been emphasised by projecting cornices on ribbed supports and rises slightly to mid-span. Decorative niches with superimposed pediments have been built into the upstream and downstream faces of all piers. There are low cutwaters at the upstream faces, and probably also downstream. These are surmounted by what looks like ancient concrete at the upstream side of piers 2 and 4 (counting from the southern end).

The overall width of the bridge is 8.6 m; the width between spandrel walls is 7.4 m; and there is a 4.9 m wide roadway between footpaths and solid parapets. It still carries traffic.

There was a second ancient bridge in Rimini, the **Ponte sull'Ausa (I118**; Gazzola, 1963b, no. 82). The Torrente Ausa enters Rimini from the south and crosses the Via Flaminia near the Arch of Augustus, built by Augustus where the road entered the town. The bridge had two unequal spans, but only some elements of the ancient bridge are now included in the structure.

The next major road east of the Via Flaminia was the Via Salaria (Section 2.2.3). This left Rome up the Tiber valley to the east of the river, and then swung away to Rieti, Antrodoco and Ascoli Piceno. Between the two roads were a few bridges.

The **Ponte di Pioraco (I119**; Gazzola, 1963b, no. 85) was on a branch road which left the Via Flaminia to the east from Nocera Umbra through Pioraco, San Severino, Treia and Osimo to Ancona (Ashby and Fell, 1921: 179). It has a single arch, and is believed to be Augustan. South of Treia, at Pollenza, about 10 km south west of Macerata, was the **Ponte di Pollenza (I120**; Gazzola, 1963b, no. 86). The present bridge has two unequal spans on Roman foundations. **Ponte Genga (I121**; Gazzola, 1963b, no. 79) crosses the Sentino River in the gorge called the Gola di Frasassi not far from the cave of Frasassi, 16 km north east of Fabriano, on another branch road from the Via Flaminia to Ancona. The present bridge is largely medieval, with a single 8 m span in a picturesque site. It has Roman foundations believed to date from Augustus (31 BC to AD 14). Further north at Caudino, on the junction for Arcevia, is the **Ponte di Caudino (I122**; Gazzola, 1963b, no. 68), again with a single 8 m span.

The Via Salaria passed through Rieti. South of the town is the **Ponte Sambuco (I123**; Gazzola, 1963b, no. 96) a small single-span bridge that appears to be in reasonable condition, with substantial retaining walls of squared stone. Lugli (1957: 356) dates it 100–70 BC; Gazzola thinks it may be Augustan. In Rieti itself is the **Ponte Romano** over the River Velino (**I124**; Gazzola, 1963b, no. 284), with Roman origins.

On the Via Salaria approaching Ascoli Piceno, and in Ascoli Piceno itself, is a fine series of Roman bridges, of which four, I126–8 and I130, are really quite outstanding. First, the Ponte Romano at Acquasanta (**I125**; Gazzola, 1963b, no. 72), 32 km from Ascoli Piceno, is a robust single-span bridge of travertine beneath a modern seven-span stone arch bridge. It has a clear span of 10.5 m, a clear height of about 9 m, and supports one of the piers of the modern bridge at the centre of its span. Blake (1947: 216) believes it is Augustan. The modern bridge also is impressive with a total height from the bed of some 20 m. They cross the Torrente Garafo, a tributary of the River Tronto.

The **Ponte di Quintodecimo (I126**; Gazzola, 1963b, no. 75) is about 8 km towards Ascoli Piceno from Acqualanga (Fig. 46). It has a single main span of 17.1 m and two smaller high-set arches to carry passageways along the banks. It is built of travertine and has a width of 3.8 m over the parapets. A further 10 km along the Via Valeria is

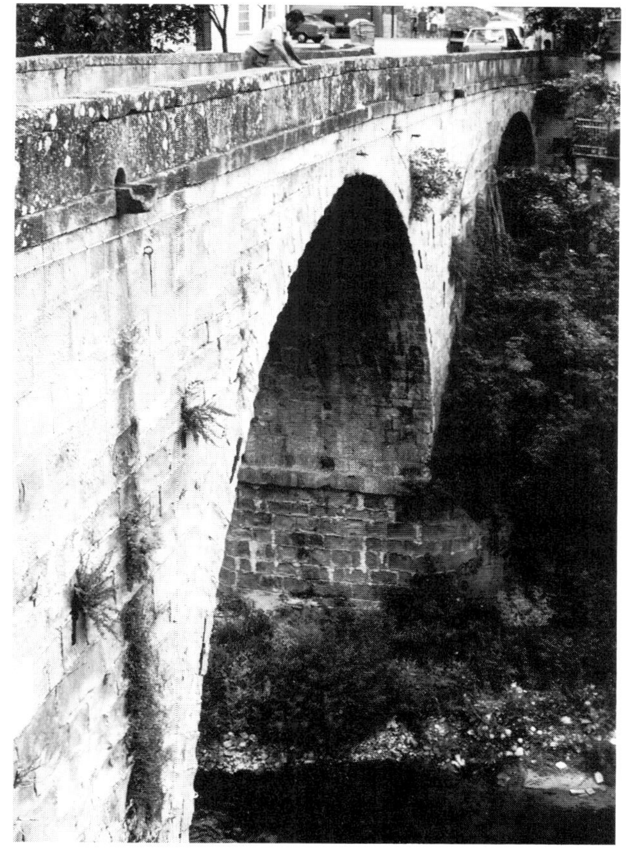

Fig. 46 The bridge at Quintodecimo (I126).

the **Ponte d'Arli** across the Tronto River (**I127**; Gazzola, 1963b, no. 76). Its span is 17 m and in this and other respects it resembles the bridge at Quintodecimo. However, there are no side spans, and the roadway, instead of being horizontal, rises to mid-span. As shown in Figure 47, it is another fine bridge in excellent condition. The arch rib is particularly slender (Section 8.3). Both are believed to be Augustan (Blake, 1947: 216).

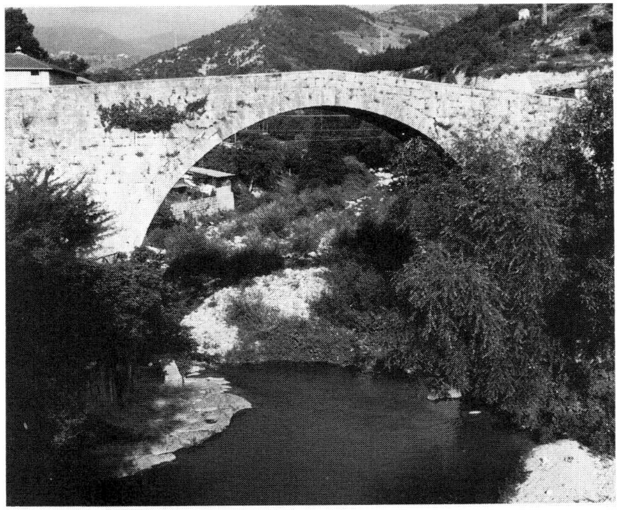

Fig. 47 Ponte d'Arli (I127), with an unusually slender arch rib (see Fig. 140).

The **Ponte di Porta Cappuccina (I128**; Gazzola, 1963b, no. 70) is an outstanding bridge. Figure 48 shows the bridge from the north, looking towards the city and the Porta Cappuccina which gives the bridge its name. The road continues as the Via di Solesta, entering the city from the north to the west of the Piazza del Popolo. The bridge is also called the **Ponte Romano** or **Ponte di Solesta**. It has a single arch of travertine, of 22.2 m span, with protruding cornices around the extrados, at the springings and at deck level. It has a height to the top of the parapets of 26 m, and a roadway width inside the parapets of 5.5 m. Blake (1947: 216) has identified points of resemblance between this bridge and that at Narni, including both the architectural treatment and the details of the stonework, and concludes that, 'though later than the bridge at Narni, it was not a late Augustan monument'. This would date the bridge in the last quarter of the first century BC.

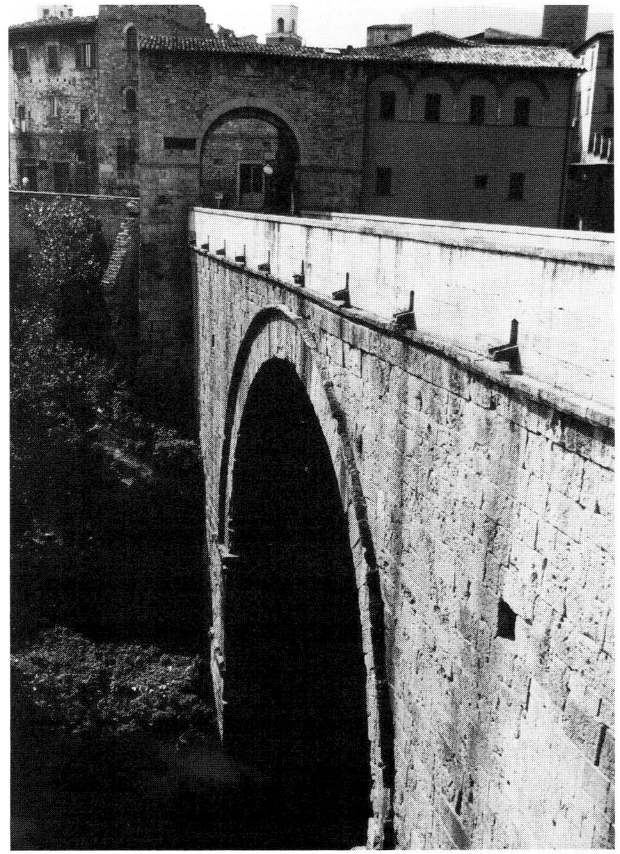

Fig. 48 The Ponte di Porta Cappuccina at Ascoli Piceno (I128) is in some respects similar to the Ponte d'Augusto at Narni – see in particular the protruding cornice.

Ascoli Piceno is on a narrow plateau or promontory running from west to east, bounded by the Rivers Tronto to the north and Castellano to the south and east, meeting to the north east of the town. The Tronto is the larger stream and is crossed by three fine stone bridges which, in order downstream, are the Ponte di Porta Cappuccina, the Ponte Nuova and the Ponte di Porta Tufilla. Figure 48 gives some idea of the depth from the city plateau to the stream, and the valley of the Castellano is similar.

The Ponte Nuova is an excellent stone bridge with three spans, but, as the name implies, it is modern. A short distance downstream to the east is the lower **Ponte di Porta Tufilla (I129**; Gazzola, 1963b, no. 77) (Fig. 49), carrying a narrow road which, to the south, turns sharply west, and climbs to enter the city through the Porta Tufilla. Its other name is the **Ponte San Antonio**. 'The foundations go back to the Roman age' and 'the construction of the walls exhibits a remarkable similarity to those at Quintodecimo' (Gazzola, 1963b). There are the remains of a pier below the southernmost of the two main spans; and there is a third small span at the southern end.

The **Ponte di Cecco (I130**; Gazzola, 1963b, no. 71), on the River Castellano, has a particularly fine appearance, with its chapel, the luxuriant undergrowth of its steep gorge, the waterfall below and the torrent falling from the eastern cliff above. Figure 50 is taken from the later Ponte Maggiore 100 m downstream at the eastern exit from the city (see also Fig. 139). It appears to show only a single arch; there are in fact two, for a smaller span is hidden behind the foliage to the left. The spans are 14.5 and 7.2 m; the height to the roadway is 24.8 m; and it carries a road of width 6.3 m. It is built of travertine. The heavy arch rib is unusual in that most voussoirs consist of two stones of unequal depth. The joints are staggered so that the stones interlock and are bonded together, rather than forming two separate rings, as is often the case (Blake, 1947: 212). The single pier has a prominent cutwater at water level, which tapers in width as it rises about half-way to the cornice at the springing level of the main arch. As with the Ponte di Porta Cappuccina, Blake (1947: 216) identifies some resemblances with the bridge at Narni, although these are less marked in the Ponte di Cecco. She dates both bridges as later than Narni but still early Augustan.

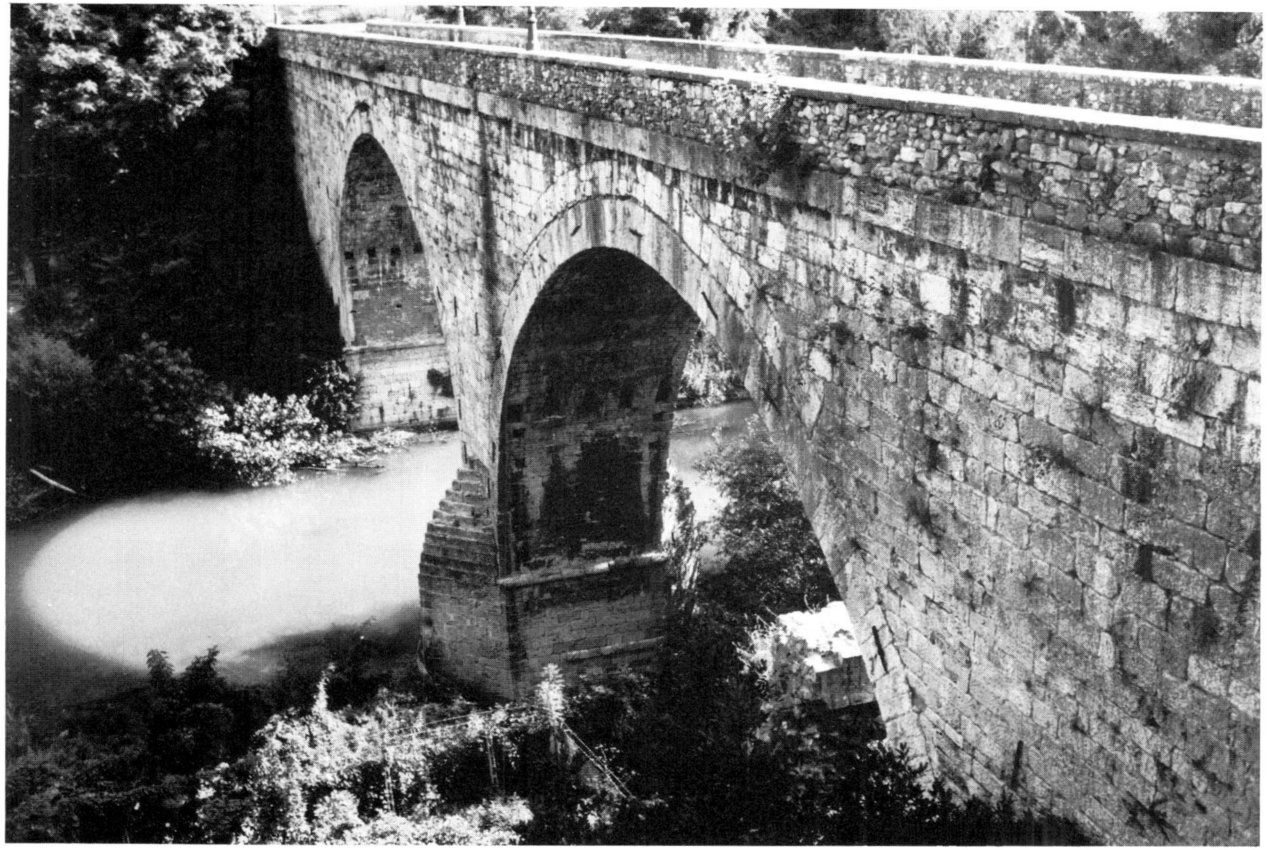

Fig. 49 Ponte di Porta Tufilla, Ascoli Piceno – with Roman foundations (I129).

There are two other small Roman bridges in the vicinity. **Ponte del Gran Caso (I131**; Gazzola, 1963b, no. 74) is 2 km east of Ascoli Piceno, immediately south of a new bridge over the Torrente Gran Caso. It is today scarcely visible through the surrounding vegetation, and serves only to carry a wood shed. Gazzola gives its span as 6 m, its width as 3.3 m and states that it is built of travertine. He has a drawing showing two relieving arches in the walls of one approach; the arch nearer the main span is segmental and springs from the ground to about the quarter point of the main arch rib. A similar feature is visible at Pioraco (I119), and is quite remarkable.

The location of the **Ponte della Scutella (I132**; Gazzola, 1963b, no. 73) is unclear. It crosses the Fosso Cavignano and it is believed that it may be between the Ponte di Cecco and the Ponte del Gran Caso. Gazzola describes it as choked with luxuriant undergrowth. It carries the Via Salaria and has a single 10.8 m span.

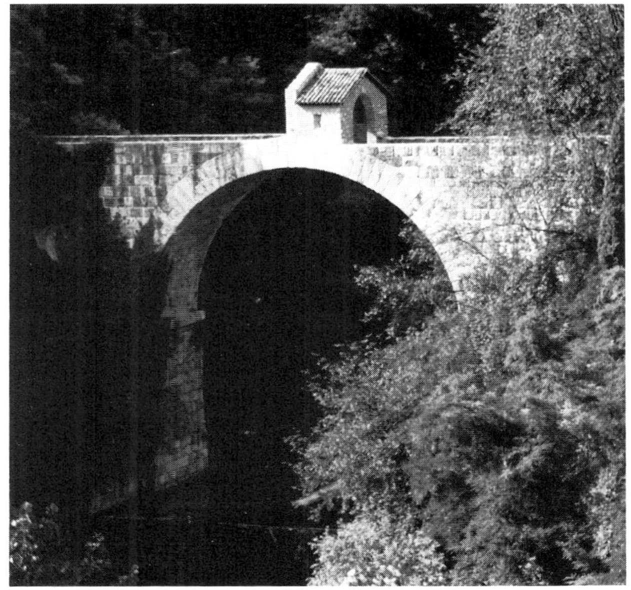

Fig. 50 The beautiful Ponte di Cecco in Ascoli Piceno (I130).

There is one final bridge to be described in this section. The Via Caecilia left the Via Salaria at Antrodoco and travelled east and north east to the Adriatic coast. **Ponte Nascosa (I133**; Gazzola, 1963b, no. 58) is on the road near Civitamassa, not far from San Vittorino. This is, in turn, about 9 km north west of L'Aquila. Only the vault and part of a retaining wall remain. Blake (1947: 212) believes that it may belong to the original construction of 117 BC. It had a single small span.

5.5 Northern Italy

This section considers bridges north of the Tuscan Apennines, excepting those in Rimini which have been discussed earlier with those on the Via Flaminia, in central Italy (Section 5.4). It follows first the great trunk roads (Map 1) – the Via Aurelia, and its successors, the

Viae Aemilia-Scauri and Julia Augusta, around the Mediterranean coast to France; then the Via Aemilia, which runs diagonally across the country from Rimini on the Adriatic to Piacenza, with its connecting roads to Milan, Aosta and the Passes of St Bernard; and the Viae Annia and Popillia around the Adriatic coast from Rimini to Aquileia. These are then linked by the Via Postumia from Genoa through Verona and Vicenza to Aquileia; the final discussion is of bridges in the north, in the foothills of the Alps on the roads to Austria (see Section 2.2.4).

The old Via Aurelia reached about as far as Populonia near Piombino south of Pisa (Wiseman, 1970: 133) by 241 BC. It was extended as the Via Aemilia-Scauri (or Aurelia) to about Savona in 107 BC, and then further to Arles in France as the Via Julia Augusta about the year 12 BC. The first bridge north of Vulci (see I65, Section 5.4) is the **Ponte Vecchio** at Rapallo (**I134**; Gazzola, 1963b, no. 150), a single-span bridge of considerable size which is believed to contain Roman elements.

It is not until after Savona that the next group of bridges is reached. In the region of Quiliano, about 6 km west of Savona, inland from Zinola in the Quazzola valley, were the remains of six bridges. The first is no longer visible. The second (**I135**; Gazzola, 1963b, no. 136), near the hamlet of Ricchini, is still in use. It has a single span of 6 m, with the arch formed of two rings of thin stones, founded directly on the solid rock. The spandrels are of small, roughly squared pieces of limestone laid in courses. The third, the **Terzo Ponte della Val Quazzola** (**I136**; Gazzola, 1963b, no. 137), is located 2 km beyond the second; it is generally similar, but the arches spring from stone walls protected by wing walls, all constructed of small coursed masonry. The fourth (**I137**; Gazzola, 1963b, no. 138) is a further 3 km on, in the Volta locality; the ruins of the fifth (**I138**, Gazzola, 1963b, no. 139) lie below the Colle sul Rio degli Scarroni; and the sixth (**I139**; Gazzola, 1963b, no. 140) is on the Rio del Gallo. Little is known of these bridges. Gazzola believes they may have been built by Hadrian at the time of his reconstruction of the Via Julia Augusta about AD 124.

The road follows the coast around Cape Vado and reaches the Val Ponci on its way to Finale Ligure. Here there are five Roman bridges, and it is from these that the valley gets its name. The arch of the **Primo Ponte di Val Ponci** (**I140**; Gazzola, 1963b, no. 141) has collapsed, but the abutments and part of the arch and spandrels still remain. The second bridge (**I141**; Gazzola, 1963b, no. 142), which is similar in form, still stands. Known as the **Ponte dell'Acqua**, it has a single span of limestone, with spandrels of squared stones of relatively small size, laid in mortar, in regular courses. The outer face of the arch ring has been removed. The third bridge (**I142**; Gazzola, 1963b, no. 143), called the **Ponte delle Voze** after a torrent of the same name which it crosses, is also standing, but not in use. It has a single span, possibly of the order of 4 to 5 m, with the arch formed of two rings of stone. The voussoirs are thin; about sixty were needed to form each vault. The combined thickness of the arch rib is large; indeed, one of the rings would have had a thickness not much less than that common for the complete vault of a bridge of this size. The spandrels are of the same construction as those of the first two bridges, with fairly small squared stones laid regularly in mortar. A few tens of metres further on, descending to the valley, are the ruins of the fourth bridge (**I143**; Gazzola, 1963b, no. 144); only part of the approaches remain. The fifth, also called the **Ponte delle Fate** (**I144**; Gazzola, 1963b, no. 145), has recently been restored and is still in use. The lower part of the side walls and the base of the main arch are of large squared stones. The upper part of the arch, which has a span of 5.7 m, is formed of two rings of thin stones, like the other bridges on this road. There is also a small arch, of similar construction, set low in one of the approaches. The upper part of the spandrel wall, and the inclined wing wall of one of the approaches, are of small squared and coursed stones. The whole is built of limestone. The third and fifth of these bridges are in good condition and form excellent examples of the form of construction that is characteristic of this road. Gazzola again believes that they may date from about AD 124.

Six kilometres beyond Finale Ligure, near Loano, are another two bridges. **Pontetto** (**I145**; Gazzola, 1963b, no. 146), which is in ruins but still standing, has a single small span of limestone across the Rio dei Gazzi. **Pontaccio** (**I146**; Gazzola, 1963b, no. 147), which appears to be partly buried, has a span of about 10 m, again with the characteristic double rings of stone.

The width of the Via Julia Augusta varies, from 5.5 m at the Ponte delle Voze (I142) to 6.2 m at the Pontaccio (I146), but, further on, at the **Ponte Lungo** near Albenga (**I147**; Gazzola, 1963b, no. 149), it is restricted to 3.5 m. This major structure carried the road for 148 m over ten spans varying from 6.6 to 8.4 m. It is partly buried and runs beside a modern road. It is either of Roman origin or contains elements of Roman work.

The final bridge on the Via Julia Augusta before the French border is the **Ponte sul Rio della Torre** at San Lorenzo al Mare (**I148**; Gazzola, 1963b, no. 148). Now in ruins, it had a single 4 m span.

The Via Aemilia was built from Rimini in 187 BC by M. Aemilius Lepidus (Section 2.2.4). About 10 km out, it crossed the Torrente Uso by two bridges. The first, the **Ponte San Vito** (**I149**; Gazzola, 1963b, no. 80), was near the church of San Vito. Its remains now stand on dry land, as the bed of the stream has moved. It had a single stone arch, with spandrel walls of small, coursed, rectangular stones, with a concrete core. The second is called the **Ponte di Sant'Arcangelo di Romagna** (**I150**; Gazzola, 1963b, no. 81), from the town of the same name. It probably had three arches, of which one remains. This has a vault of large stones, resting on lower courses of small rectangular stones, with a span of about 4 m. The arch thickness is large compared with the span.

The River Rubicon, made famous by Julius Caesar's decision to cross in the year 49 BC and march on Rome, lies about 16 km north west of Rimini. It was crossed by the Via Aemilia at Savignano and here, until the Second World War, was the **Ponte sul Rubicone** (**I151**; Gazzola, 1963b, no. 57). It had three 6.2 m spans of limestone, a total length of 26.2 m, and was probably Republican (Ballance, 1951: 115). It was blown up in 1944 and only part of the foundations remain.

Ponte sul Reno (**I152**; Gazzola, 1963b, no. 277), across the Reno River near Bologna, goes back to the Roman period. In Milan, the remains of a small bridge, with a span of 2.6 m, have recently been brought to light, beneath the Corso Vittorio Emanuele (**I153**; Gazzola, 1963b, no. 231), and are believed to date from the third or fourth century AD. Between were the crossings of the River Po with its many tributaries, and the surprising thing is that there are so few signs of bridges here or in similar locations (Brookes, 1974: 47–8).

Ivrea lies on the Dora Baltea at the entrance to the Aosta valley and the routes across the Alpine passes to Gaul. Although these routes were used from earliest times, the Roman occupation was relatively late. Aosta itself, Roman Augusta, was founded as a military camp by Augustus about 25 BC (Section 2.2.4). Blake (1947: 218) has discussed the dating of the fine series of Roman bridges in the valley. She concludes,

> It is my conviction that the engineers of the second century before Christ made the rock cuttings and built the original military roads. Traces of their work remain, but, in general, their knowledge of concrete was too rudimentary for the task. Consequently, at some time in the first century before Christ, probably prior to the foundation of Aosta, the road system was completely renovated. Whether the old roads were in such shape that Caesar could use them remains an open question. Possibly the restoration was made in anticipation of an eventual subjugation of Gaul. Finally, the entire road system was renovated by Augustus . . .

The first, but least important, in the series is the **Ponte**

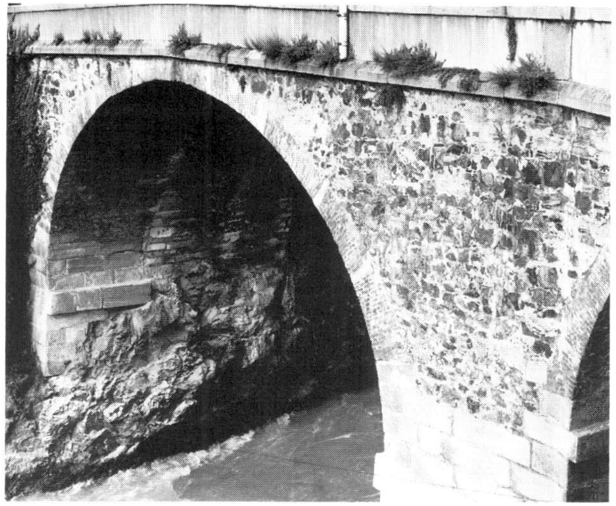

Fig. 51 Ponte sulla Dora in Ivrea (I154).

sulla Dora (**I154**; Gazzola, 1963b, no. 118). Known also as the **Ponte Vecchio** or **Ponte Romano**, it is the furthest upstream of the bridges in the town. As shown in Figure 51, the present structure is a fine bridge of three spans that has been widened at the right-hand or upstream side. However, it has been the subject of many repairs and restorations, one of which was in the time of Vittorio Amadeo II, and it is difficult to know how much of the present structure is original. Gazzola says, 'Of the primitive fabric there remains the lower part of the pier, faced with large squared blocks, and the larger arch, in accurate *opus latericium* [brickwork]'.

Seventeen kilometres upstream is the town of **Pont St Martin**, named after a bridge of that name (**I155**; Gazzola, 1963b, no. 119). The bridge, which crosses the Lys torrent on the left bank of the Dora Baltea River, has the largest span of all known Roman masonry arch bridges. It is shown in Figures 52 and 53. There is, unfortunately, some uncertainty as to what the span may be. Gazzola quotes the figure of 36.65 m, while Promis (1862) gives 35.64 m. The second figure is also given by Blake (1947: 217) and is preferred here; it exceeds the span of 34.2 m at Kâhta, Turkey (E36), and the 32.1 m of the collapsed span at Narni (I87).

Promis also gives some other dimensions: the width of the bridge is 5.8 m total, or 4.6 m between parapets; the height from the springing level to the top of the roadway is 12.0 m, and to the top of the parapet, 13.6 m. An unusual feature of the arch is that it is not a true semicircle. This is visible in Figure 52, and is apparent also from the dimensions (see also Fig. 138). The rise is in fact about one-third of the span, and the shape is slightly segmental.

Blake draws attention to a change in construction in the lower part of the bridge. 'The lower part of Ponte di San Martin exhibits construction which one might associate with the second half of the second century before Christ. The side walls for nine or ten courses were of blocks of local gneiss. They lack the uniformity of height which was almost universal at a later period.' She also refers to the tenth voussoir on one side which 'has been cut to form a firm basis for the rest of the span'. Both of these features can be seen in Figure 53. She concludes that 'It is conceivable that the original construction stops at this point. If so, the great span was too much for the indifferent concrete of the period. When it collapsed at some indeterminate time, it was replaced by the present span of seventy-one voussoirs.' Her reference to the second half of the second century before Christ is made in the context of a statement by Promis (1862: 87–9) who argued that the military roads in the vicinity of Aosta were constructed either in 141 or in 120 BC. She is suggesting, therefore, that the original construction dates from this time. By implication also (Blake, 1947: 218) she suggests that the rebuilding was in the time of Augustus – say about 25 BC. What is not so clear is whether the span was ever completed about 141–120 BC. The alternative is that the span was attempted but failed. The conclusion is,

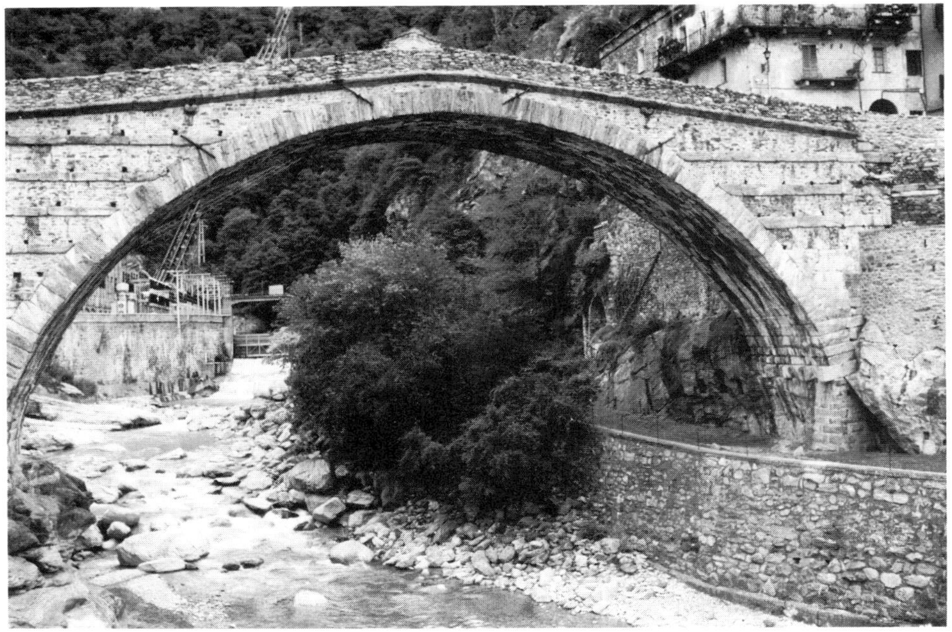

Fig. 52 General view of the Pont St Martin, believed to have the longest span (35.6 m) of all Roman masonry arches (I155).

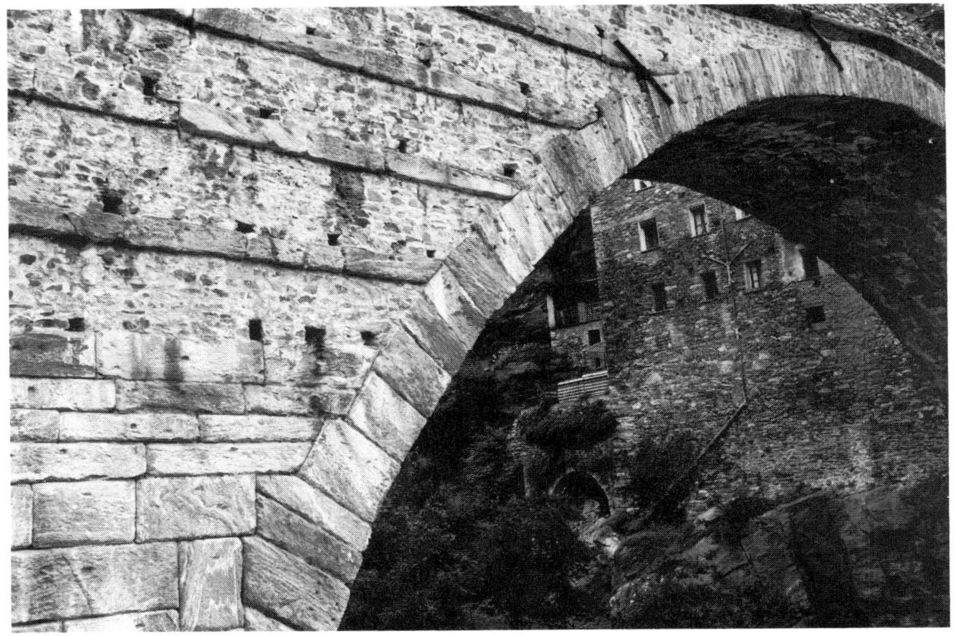

Fig. 53 Detail of stonework in the Pont St Martin (I155). Note both the spandrel face and the underside of the arch.

therefore, that a span of 35.6 m was achieved either about 141–120 BC or about 25 BC.

The **Ponte di St Vincent** (**I156**; Gazzola, 1963b, no. 122) is the next bridge on the road. It is located about 30 km beyond Pont St Martin, on the eastern outskirts of the town of St Vincent, across the Torrent Cillau. As shown in Figure 54, the bridge is unusual, with a main span over the stream, now collapsed, and a blind lateral arch to carry the roadway as it turns from the bridge around the cliff face. Promis (1862) has given the spans as 9.7 m for the main arch and 3.1 m for the lateral arch. The small pier in the background of the photograph carries a pipeline and is not related to the bridge. The stone is a calcareous *tufa*. Blake (1947: 217) dates the bridge in the early years of the first century AD.

About 3 or 4 km further on, at Chatillon, are the relics of a single span bridge across the Torrente Marmore (**I157**; Gazzola, 1963b, no. 123). Only part of the foundations and of the arch (Fig. 55) now remain, beneath a more recent bridge. The span of 15 m is quite large, as also is the height above a deep rocky crevice. Gazzola quotes Delbrueck as dating the bridge in the second century before Christ, but suggests that this date cannot be sustained, and that it is more likely to be Augustan.

Strategic Aosta was located at the junction of the routes west across the Little St Bernard Pass to France, and north across the Great St Bernard to Switzerland. Blake (1947: 113) describes it as 'presenting the most Augustan stonework at a single site in Italy outside Rome itself'. Its city walls enclosed a rectangle some 720 m long by 570 m wide. The eastern exit passes through the massive arches of the Porta Praetoria (Fig. 56) and then, 336 m to the east, reaches the Arch of Augustus (Fig. 57). A little further east the main road now crosses the Torrente Buthier. This has changed its course since Roman times, and further east again, about 300 m from the triumphal arch may be found the Roman **Pont de Pierre** (**I158**; Gazzola, 1963b, no. 121) (Fig. 58). Promis (1862) gives its span as 17.1 m and its width as 5.9 m. Blake (1947: 218) mentions the refinement of its stonework and the excellence of the concrete found within its spandrel walls. She ascribes it to 'a date somewhat later in the Augustan period, when the bridge furnishing access to an important city was afforded special treatment'. It is of *puddinga* or pudding stone.

The road up the Aosta valley has followed the left or

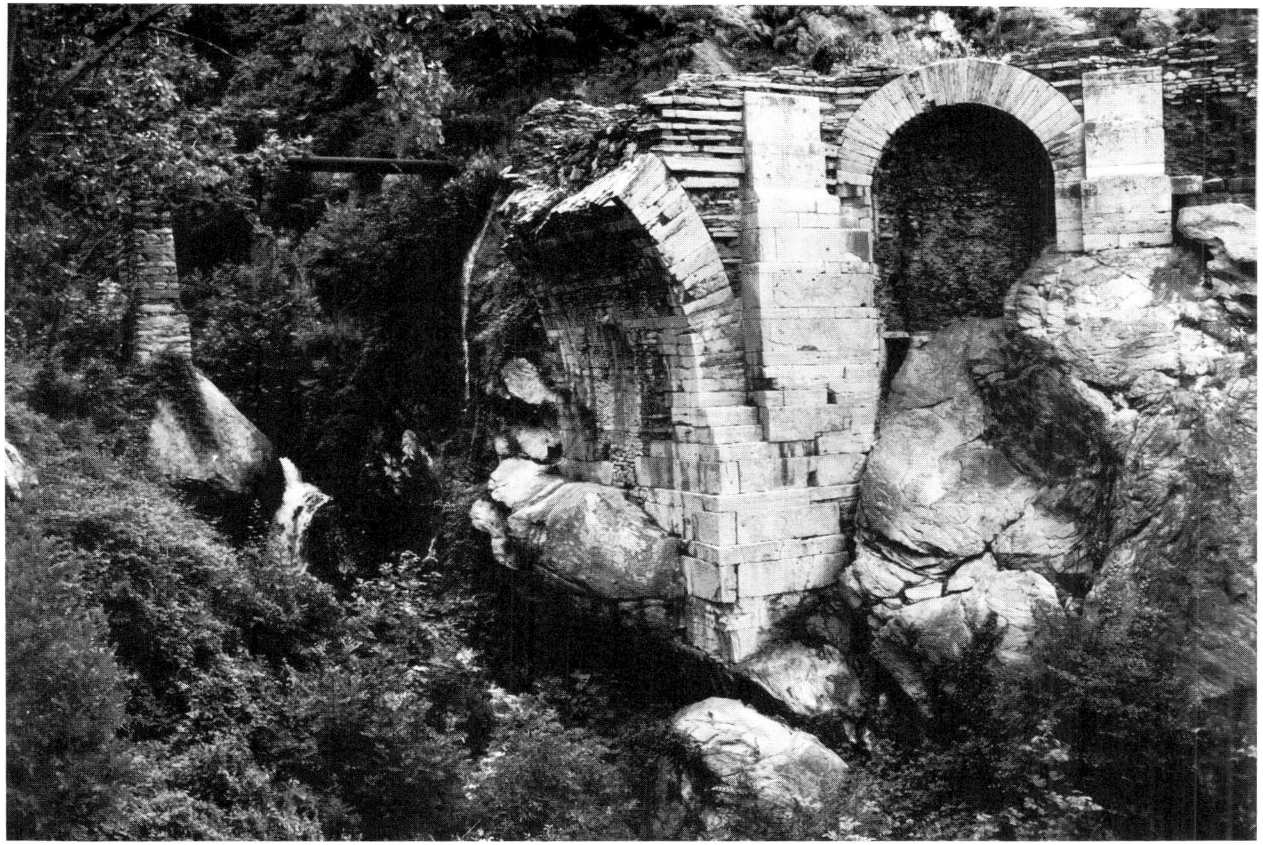

Fig. 54 The remains of the Ponte di St Vincent, in the Aosta valley. The half-arch at the right is distinctive and carries the roadway around a steep rock face (I156).

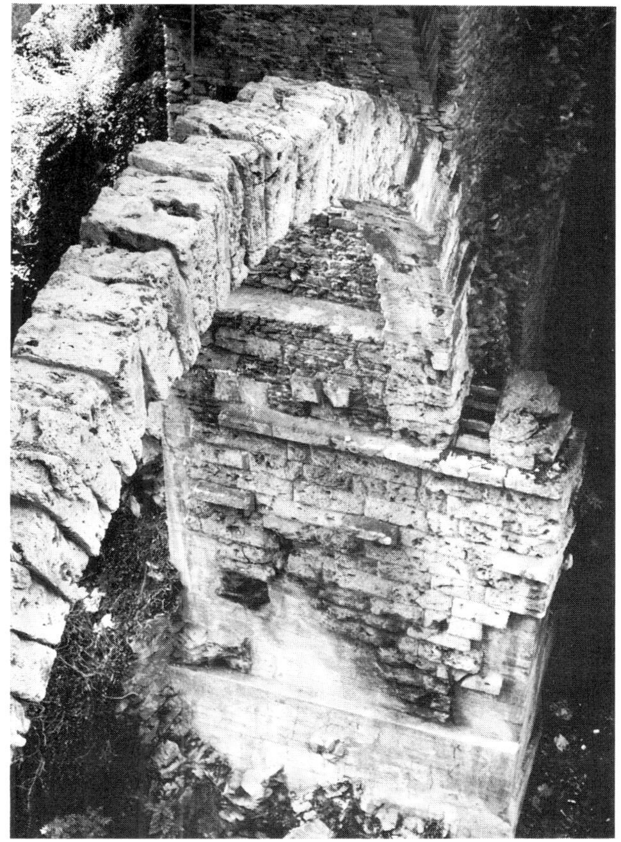

Fig. 55 Part of the arch rib of the Roman bridge at Chatillon (I157), near Aosta, beneath a modern bridge.

northern bank of the Dora Baltea River. Six kilometres beyond Aosta along the road to the Little St Bernard Pass and the Mont Blanc tunnel, a road branches to the left up the valley of the River Cogne through Aymavilles to Cogne itself. About 5 or 6 km along this road a minor road turns down to the right, to Pondel beside the stream. Here there is a remarkable bridge, the **Ponte del Pondel** (**I159**; Gazzola, 1963b, no. 120) (Fig. 59; see also Figure 135). It has three outstanding features. The first is its great height; from the top of the parapet to water-level is 40 to 45 m. Secondly, it carries an enclosed walkway. Figure 59 shows the great depth of the structure above the arch; the floor of the inner walkway is at the top of a protruding cornice 3.85 m above the crest; the walkway itself has a clear height of 3.88 m and is roofed with stone slabs 0.24 m thick; there is another open walkway above, protected with solid stone balustrade walls. The internal width of the walkway is 1.09 m, with side walls 0.59 m thick, giving an overall width of 2.27 m (Promis, 1862: 192–7). Its overall length is 50.3 m; it is entered by narrow doorways, 1.1 m wide, in the abutments on the downstream side, and is ventilated by a staggered system of air holes, some of which can be seen in Figure 59. It carries an inscription (*CIL* v(2).6899) – believed to be on the

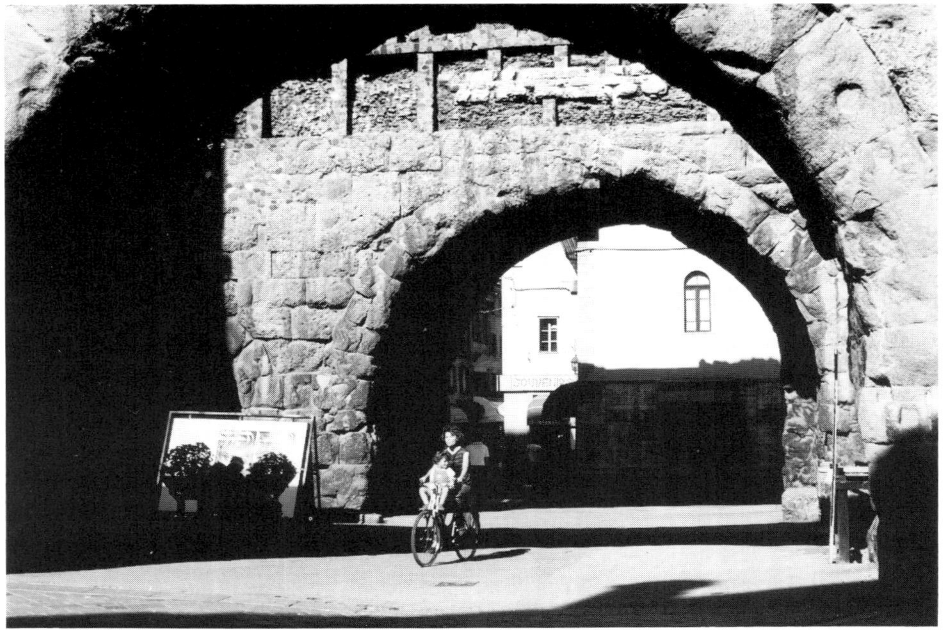

Fig. 56 Massive arches of the Porta Praetoria at Aosta.

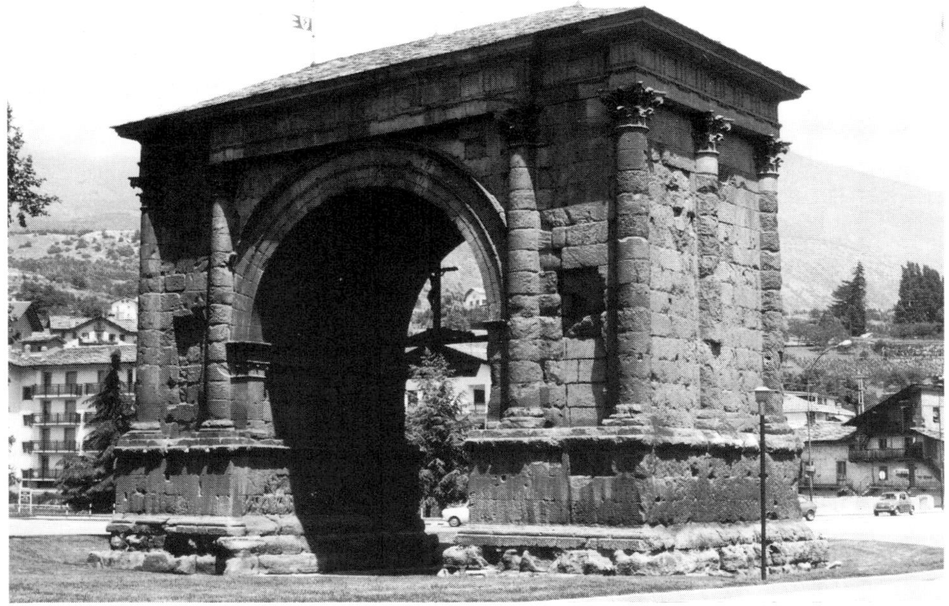

Fig. 57 Arch of Augustus at Aosta.

outside downstream face above the crest of the arch – to Caesar Augustus, attributing the construction to C. Avillius C. F. and C. Aimus Patavinus (see Section 3.2). Below this is another stone carrying the single word *privatum*. This is the third remarkable feature of the bridge and is interpreted as indicating that the bridge carried a private road (Blake, 1947: 218). In this respect it is worth noting that the bridge is well away from the natural route to the Little St Bernard Pass. The River Cogne has precipitous slopes, and the main road had no need to enter its gorge; in any case to do so would have required an unnecessary crossing of the larger Dora Baltea River. Promis (1862) adds the comment that the bridge was 'possibly for the transport of iron ore from a mine of which they were the owners'. The bridge has a clear span of 14.2 m (Promis, 1862). The inscription on the bridge carries further information which sets the date of construction as 3 BC. The material is a local gneiss.

From Rimini there were two routes around the Adriatic coast to Aquileia and the east. According to Wiseman (1970: 130), one of two Viae Anniae was built in 153 BC from Bologna through Padua (the other was from Capua to Rhegium or Reggio), and the Via Popillia in 132 BC from Rimini along the coast to join the Via Annia beyond Padua.

In 1900, the remains were found of a bridge on the Via Popillia south of Ravenna, a kilometre from San Apollinare (**I160**; Gazzola, 1963b, no. 278). It had three spans and was known locally as the **Ponte Petreo**.

Padua is believed to have had four Roman bridges across the eastern branch of the Bacchiglione or Edrone River, called in Roman times the Medoacus. The first was the **Ponte San Lorenzo (I161**; Gazzola, 1963b, no. 54), in the Via San Francesco near the University. Of its three spans, only the eastern span is visible, the others being hidden under the bank. Gazzola has a photograph of an inscription on the bridge which, he says, fixes the date of construction at about 47 to 30 BC. The bridge is remarkable both for the segmental shape of the arches and the slenderness of the piers. Gazzola gives the clear span lengths as 12.8, 14.4 and 12.5 m, and shows a rise of the order of 0.27 times the span. Concerning the piers, the arch ribs on either side are believed to spring from a common row of stones, with a pier width down to 1.72 m; that is, 0.12 times the central span. These matters are discussed in Chapter 8.

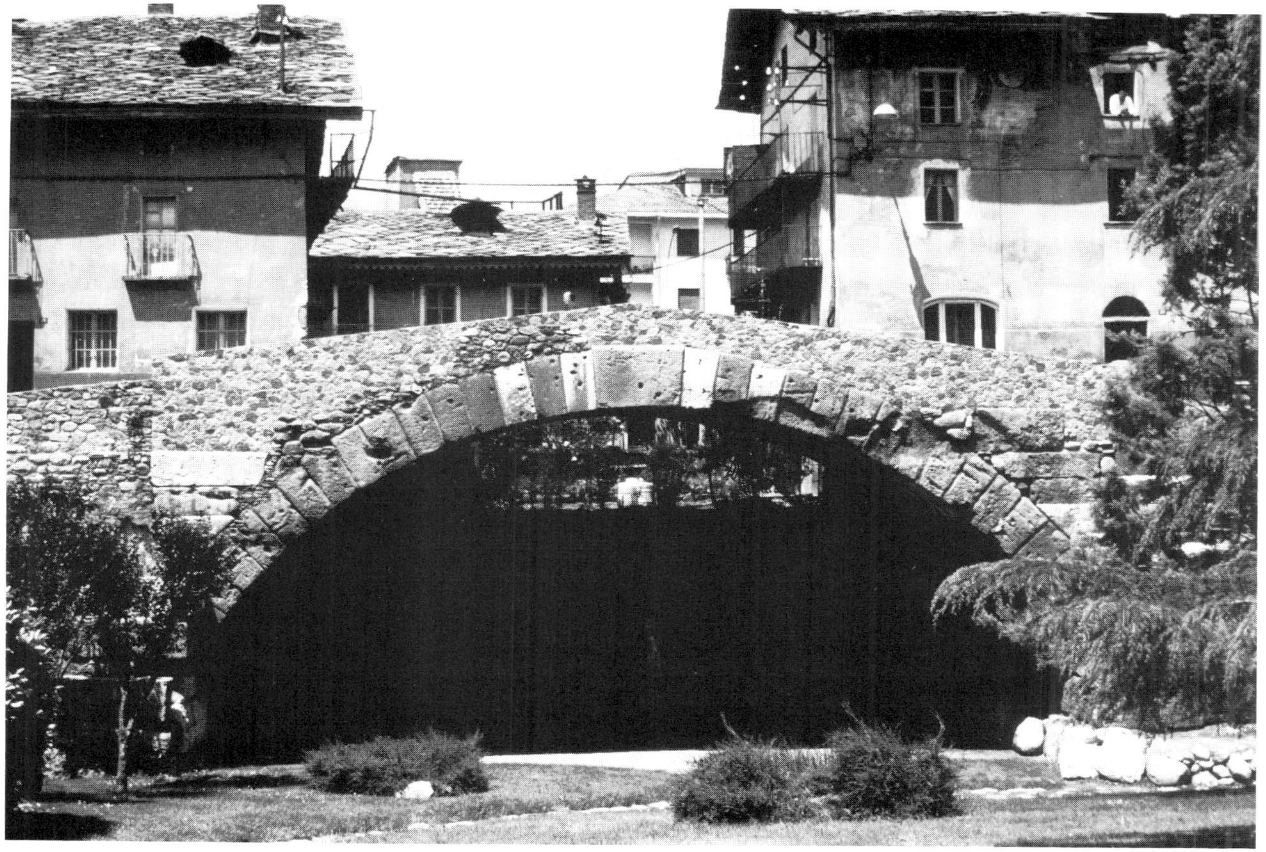

Fig. 58 Pont de Pierre, built originally to cross the Torrente Buthier at the eastern approach to Aosta (I158).

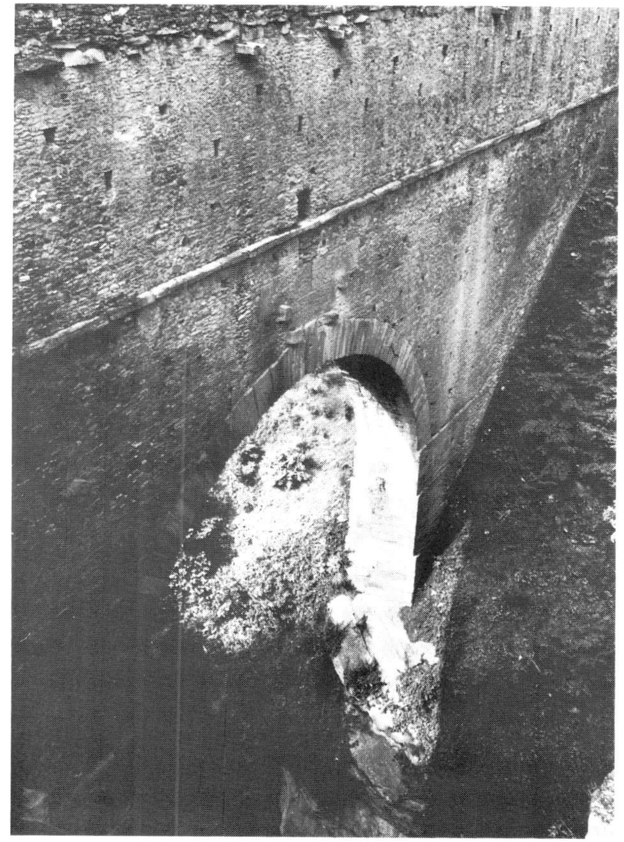

Fig. 59 The remarkable Ponte del Pondel, with its covered passageway above and great depth to the valley below (I159).

The **Ponte Corbo** (**I162**; Gazzola, 1963b, no. 55) is also partly buried. Its three spans have semicircular arch ribs. There is believed to be a third bridge, the **Ponte S. Matteo**, near the church of the same name (**I163**; Gazzola, 1963b, no. 56). The fourth, on the Via Altinate, is again only partly visible. Called the **Ponte Altinate** (**I164**; Gazzola, 1963b, no. 129), it has three semicircular limestone spans.

Brusin (1949–50) has recorded the remains of four bridges on or near the combined road, the Via Annia / Popillia, east of Venice (**I165–8**; Gazzola, 1963b, nos. 219, 220, 218, 221, in that order). In all cases there are only a few remains. The last, at Aquileia north of Grado, is believed to date from the second century AD.

The Via Postumia ran from the Mediterranean at Genoa to Aquileia at the head of the Adriatic, joining the Via Annia at Julia Concordia. It was built a little later than that road, about 148 BC (Wiseman, 1970: 140), and passed through Verona, where there was a fine Roman bridge, the **Ponte Pietra** (**I169**). Figure 60 shows the bridge as it is today. It crosses the Adige River at the north-eastern corner of the town, not far from the cathedral. It was destroyed in April 1945 and rebuilt between 1957 and 1959, with the old stones reused where possible. It has five spans. Figure 60 is taken looking towards the town

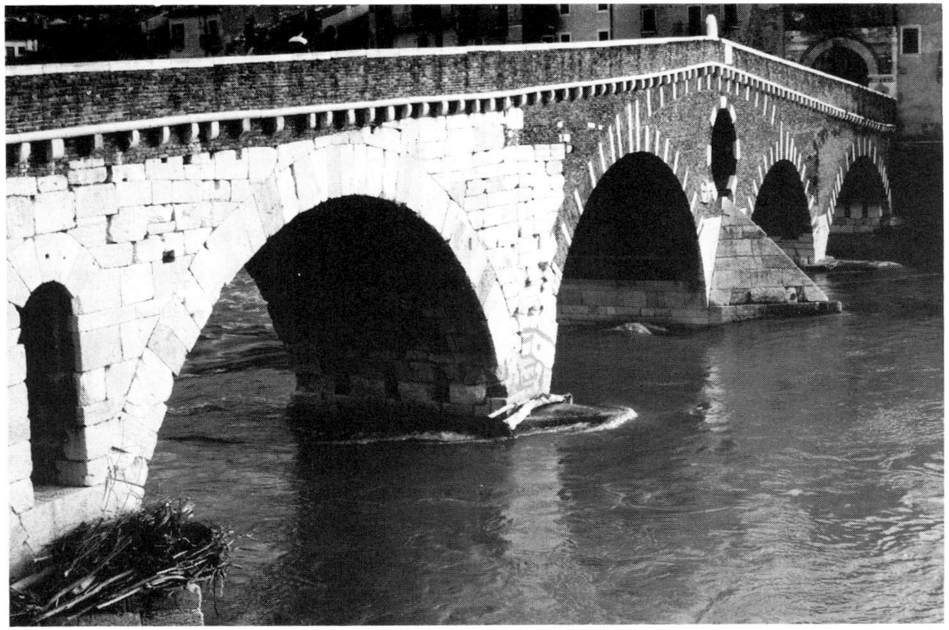

Fig. 60 The rebuilt Ponte Pietra in Verona; the near section is Roman (I169).

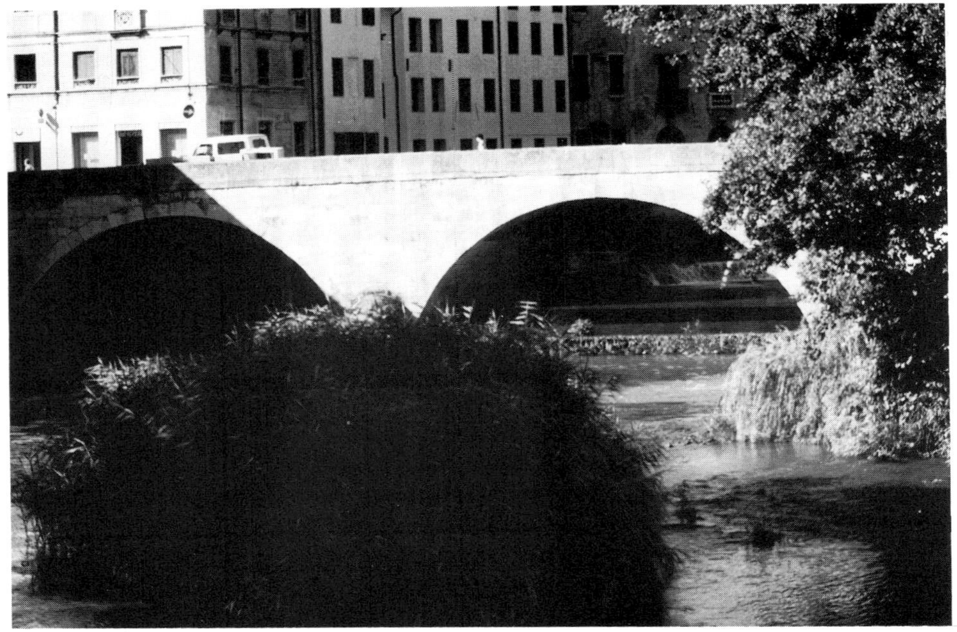

Fig. 61 Ponte Pusterla at Vicenza (I170).

and shows the Roman structure in the foreground. The pier with the floodway is narrow for a Roman bridge. The white stone is travertine. Blake (1947: 216–17) says that the bridge might be Augustan. Gazzola, in his other work (1963a), describes its reconstruction.

The Via Postumia also passed through Vicenza, at the confluence of the Bacchiglione and Retrone Rivers. **Ponte Pusterla** (**I170**; Gazzola, 1963b, no. 126) is on the Via Claudia Augusta. It has three spans and crosses the Bacchiglione River (Fig. 61). Gazzola doubts whether it is Roman. The **Ponte degli Angeli** (**I171**; Gazzola, 1963b, no. 124) crosses the Bacchiglione near the Chiesa degli Angeli. Although originally Augustan, the bulk of it was demolished in the thirteenth century, and reconstructed in the fifteenth. It was destroyed again in 1889 and then rebuilt in its present form. It had five spans.

However, the **Ponte Furo** (**I172**; Gazzola, 1963b, no. 125) in Vicenza appears to be original (see Fig. 62) excepting, of course, for the roadway and railing. It crosses the smaller Retrone River near the park called the Campo Marzio, towards the railway station. It has two spans, each of about 4.6 m, and what appears to be a rudimentary square floodway in the pier. It is a fine little bridge, possibly Augustan.

North east of Vicenza is the small town of Susegana on the River Piave with the small **Ponte di Susegana** (**I173**; Gazzola, 1963b, no. 127).

The remaining Roman bridges of northern Italy lie in the extreme north, above a line from Milan to Verona, in some cases approaching the southern slopes of the Swiss Alps.

Ponte de Arena (**I174**; Gazzola, 1963b, no. 232) was a small bridge with a span of 5.1 m in Monza, destroyed in 1842. Olginate is on the southern shores of the Lago di Garlate, south of Lake Como. Here there is one 4.75 m span remaining of what appears to have been a much larger bridge, possibly with 17 to 19 spans, over the River Adda (**I175**; Gazzola, 1963b, no. 280). In the same area, between Lecco and Bergamo, was the large **Ponte della Regina** (**I176**; Gazzola, 1963b, no. 180) over the River Brembo. To the east of Bergamo, over the River Serio, is the **Ponte Marzio** (**I177**; Gazzola, 1963b, no. 279), believed to be of Roman origin. Further north, the road from Chiavenna to St Moritz (Switzerland) passes through the little town of Promontogno in the Val Bregaglia. Near here is the **Ponte di Promontogno** (**I178**; Gazzola, 1963b, no. 290), possibly Roman, or at a Roman site.

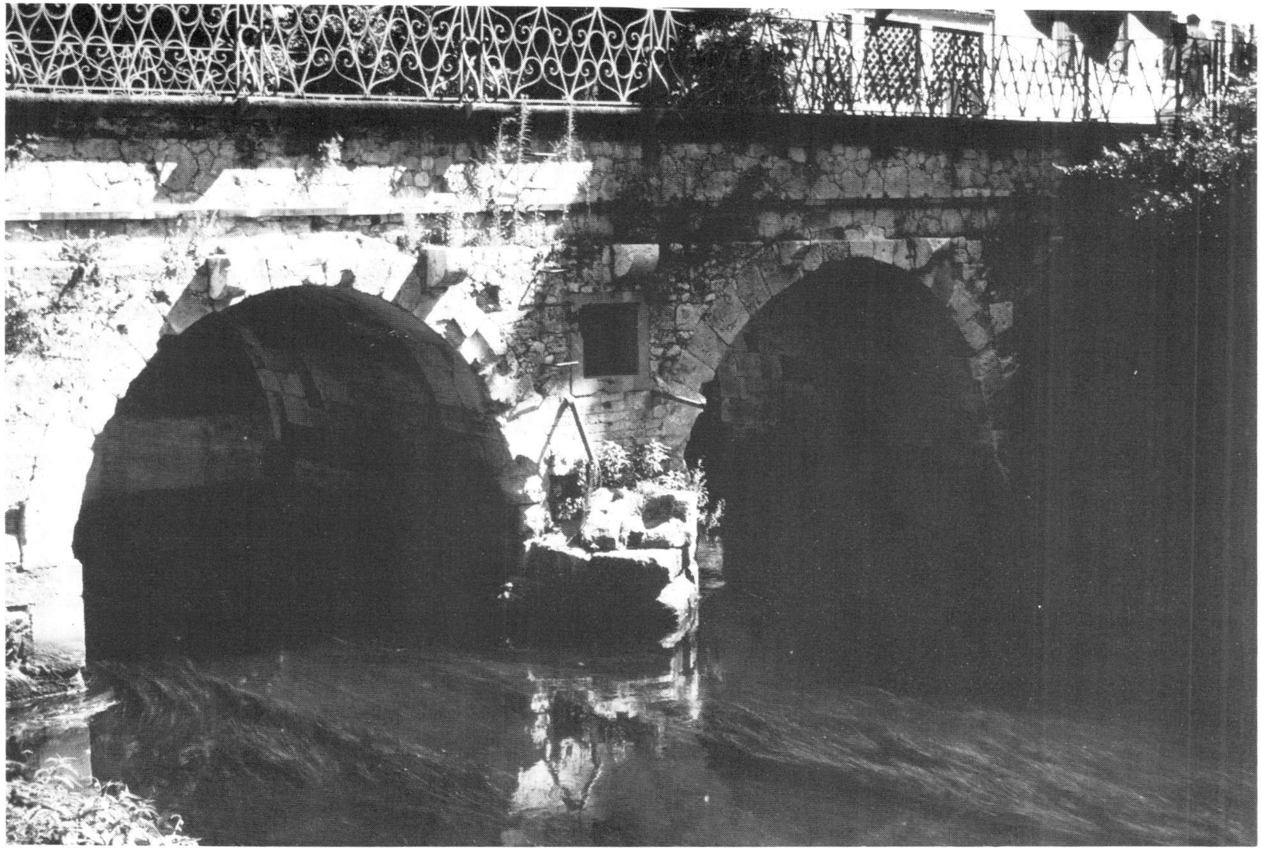

Fig. 62 The attractive Ponte Furo in Vicenza (I172).

Merano lies further east, on the western approach of the Brenner Pass to Austria. A medieval bridge here retains the name, the **Ponte Romano sul Passirio** (**I179**; Gazzola, 1963b, no. 281). Similarly at Franzenfeste (or Fortezza) the **Ponte Ladricio**, over the Isarco River, is reputed to be Roman (**I180**; Gazzola, 1963b, no. 128).

5.6 Gaul

It is convenient to include in this section the whole of the region of ancient Gaul. In effect this includes France, Belgium and Luxembourg. The bridge at Trier and Caesar's bridge across the Rhine are included with other timber bridges in Chapter 6.

Most of the bridges in France are located in the south, along the Viae Julia Augusta and Domitia and their tributaries (see Section 2.3). These roads will be considered first beginning at the point nearest Rome and moving outwards, and then more or less in a clockwise direction throughout the country (see Map 2). It should be remembered, however, that the spread of Roman influence in Gaul was determined as much by waterways as by roads. Rome's first contact with France was through Marseille or Massilia (Section 2.3), and it was reached by sea long before there was safe and easy access from Italy by road. Lyon, or Lugdunum, owed its importance not only to its central position but to its easy access to the valleys of the Saône and the Rhône, and from them to the Loire, Seine and Rhine. The Via Julia Augusta was not built until about 12 BC. It entered Roman Gaul at La Turbie, near Monte Carlo, left the coast at Fréjus and followed the valleys of the Argens and Arc through Aix-en-Provence to Arles. The Via Domitia was much older, having been built about 120 BC by the proconsul Domitius Ahenobarbus. It joined the Rhône to Spain, from Arles and Tarascon through Nîmes, Béziers and Narbonne to Le Perthus.

There were four major crossings of the Maritime Alps (see Sections 2.2.4 and 2.3): around the coast; from Turin and Susa past Mont Genèvre to Briançon; through the Little St Bernard Pass from Aosta to Moutiers, Albertville and Chambéry; and through the Great St Bernard Pass to Martigny and the Lake of Geneva (Lac Léman). The second of these, through the Cottian Alps, was particularly important. From Briançon there was a choice of routes: either north west through Grenoble to Vienne and Lyon; or south west down the Durance through Gap and Sisteron, across to the Coulon at Cereste and Apt, and then through Cavaillon to Arles.

The **Pont des Esclapes** (**G1**) is a small, three-span bridge, 2.5 km west of Fréjus, where the coastal Via Julia Augusta crosses a minor stream, le Béal, on its way up the Argen valley. Rivet (1988: 234, pl.48) believes it to be original.

The first major bridge on this road is the **Pont Flavien**

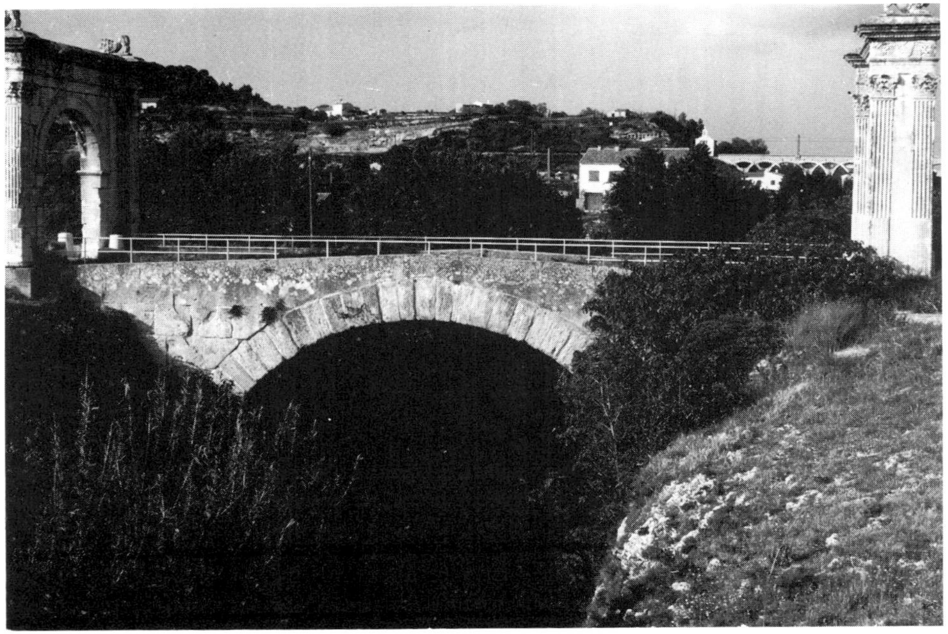

Fig. 63 The Roman arch at St Chamas, known as the Pont Flavien, with its fine triumphal arches (G2, France).

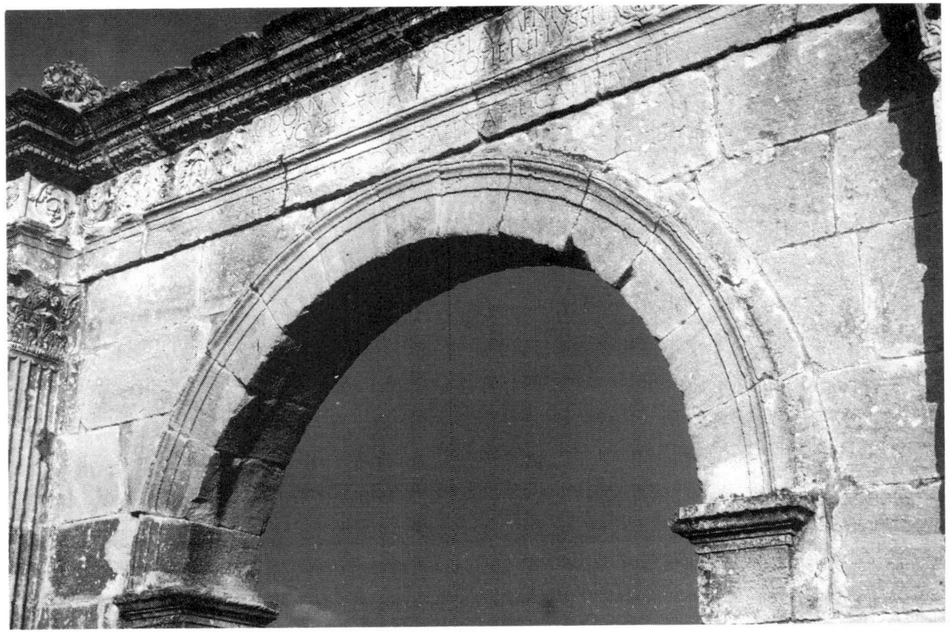

Fig. 64 Detail of one of the triumphal arches on the Pont Flavien (G2).

(**G2**; Gazzola, 1963b, no. 176; Rivet, 1988: 205, pl.35) near Miramas. The Via Julia Augusta on its way from Aix to Arles crossed the River Touloubre at St Chamas, near its entry into the Etang de Berre. The bridge, shown in Figure 63, is unique for its two fine triumphal arches. Emerson and Gromort (1925: 29) have detailed drawings of the bridge and the arches. The clear span is 12.3 m and the overall width 6 m, or 5.1 m between side walls. The triumphal arches have an overall height of 6.7 m exclusive of the carved lions; the Corinthian columns are 5.1 m high from base to top of capital; the archway is 3.6 m wide and the overall width of each arch at the base is 7.6 m. Their longitudinal spacing, centre to centre, is 23.5 m, giving an overall length of 25.0 m. The arch rib is 1 m thick. The bridge carries an inscription on the face of one of the triumphal arches to C. Donnius Flavos, a priest of Rome and of Augustus (see Section 3.2) (Fig. 64). It probably dates, then, from the time of construction of the road, from about 12 BC.

The inland road, north of Aix-en-Provence, crossed from the valley of the Durance to the valley of the Coulon. As it left Durance, it passed through the small town of Ganobie, over a small Roman bridge (**G3**) on the River Buès (Rivet, 1988: 278, 295; Chevallier, 1976: 99). Some 20 km further on, but away from the main road, 1 km west of Mane, is the **Pont sur Laye** (**G4**; Gazzola, 1963b, no. 175), crossing the Laye Torrent north of the modern road. The bridge has three spans, 3, 7 and 12 m, two of which are shown in Figure 65. The roadway rises sharply from the left bank of the stream to a crest above the main arch, and then drops to the higher bank on the other side. There are large triangular cutwaters on both the upstream and downstream faces of the main pier. The paved roadway is intact, but part of the balustrading has been removed. The overall length of the bridge, which is built of a local limestone, is 40 m, and its width 3.2 m. The arch rib is particularly slender, and appears to be partly segmental. Gazzola believes that it dates from either the end of the first or beginning of the second century AD.

About 38 km west along the same road, 7 km west of Apt, is the fine **Pont Julien** (**G5**; Gazzola, 1963b, no. 173; Rivet, 1988: 259, pl. 54). It is still in use, carrying a minor road across the River Coulon, to the south of the main road from Apt to Coustellet. Emerson and Gromort (1925:

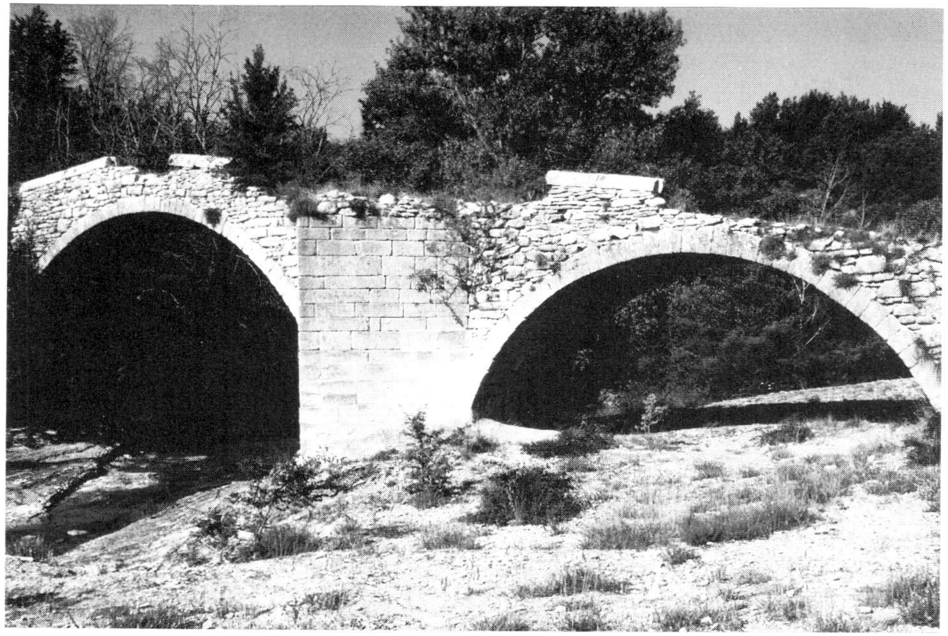

Fig. 65 The Pont sur Laye west of Mane in France (G4).

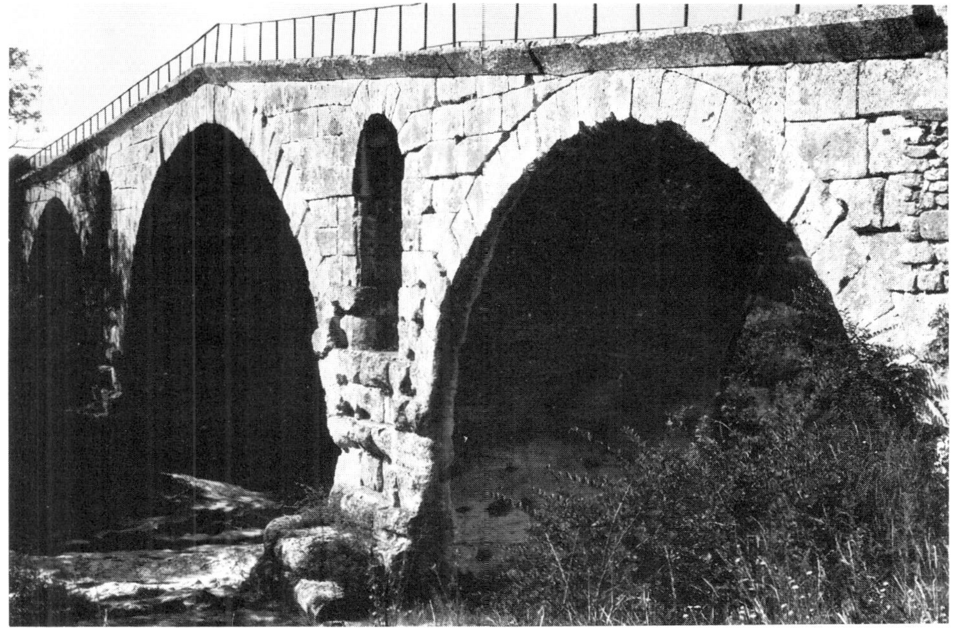

Fig. 66 The Pont Julien, 7 km west of Apt, France (G5).

30) again have an excellent scale drawing of the bridge. Figure 66 gives some idea of its excellent condition and proportions. It has three spans – 10.3, 16.3 and 10.3 m. The piers are 5.6 and 5.7 m wide, each carrying a flood arch 1.8 m wide by 3.4 m high. The roadway width is about 4 m, and the overall width excluding cornices is 5.5 m. The overall length is in excess of 85 m. It is built of limestone. Despite being named after Julius Caesar, it is believed to be of the same period as the Pont sur Laye, about the end of the first or beginning of the second century AD.

Gazzola (**G6**; 1963b, no. 267) mentions a bridge across the Rhône at Arles, built in the time of Constantine (AD 306–37), but little of this now remains. It was a floating or pontoon bridge with stone abutments (Rivet, 1988: 192–6; see Section 6.2). He also cites the **Pont de St Gabriel** (**G7**; Gazzola, 1963b, no. 265), a small bridge of 3 m span on the outskirts of Tarascon, near the chapel of the same name.

Having passed through Nîmes, the Via Domitia turned south west, and crossed the River Vidourle at ancient Ambrussium, now Ambroix, west of Gallargues, by the **Pont d'Ambroix** (**G8**; Gazzola, 1963b, no. 166). The bridge is now in ruins and access is difficult, but it was a substantial structure of five spans. In the thirteenth century a chapel to St Mary was added. This has now gone, and of the original structure there remain two arch vaults and part of the foundations of a third (Rivet, 1988: 172, pl. 19).

Further north, the same stream is crossed by the **Pont de Sommières** (**G9**; Gazzola, 1963b, no. 171), on the road from Nîmes to Lodève. This fine bridge is in excellent condition (Fig. 67). Seen from the river, it appears to have seven spans, but this is deceptive as there are in fact fifteen or seventeen. The far bank of Figure 67 is a wall, built to protect the town from floods. The original bridge continues beyond this wall through the entrance gateway (of which a glimpse can be seen) and on into the town. On either side it is lined by buildings, but their basements open into the space beneath the bridge. These spans can be seen from a square at the lower level and one of them is used as a subway beneath the road above. Emerson and Gromort (1925: 32) have an excellent drawing of the bridge. The central thirteen spans are each 8.5 m; the deck is horizontal over these thirteen arches, but slopes down at each end over smaller approach spans. The pier width is 3 m, giving a total

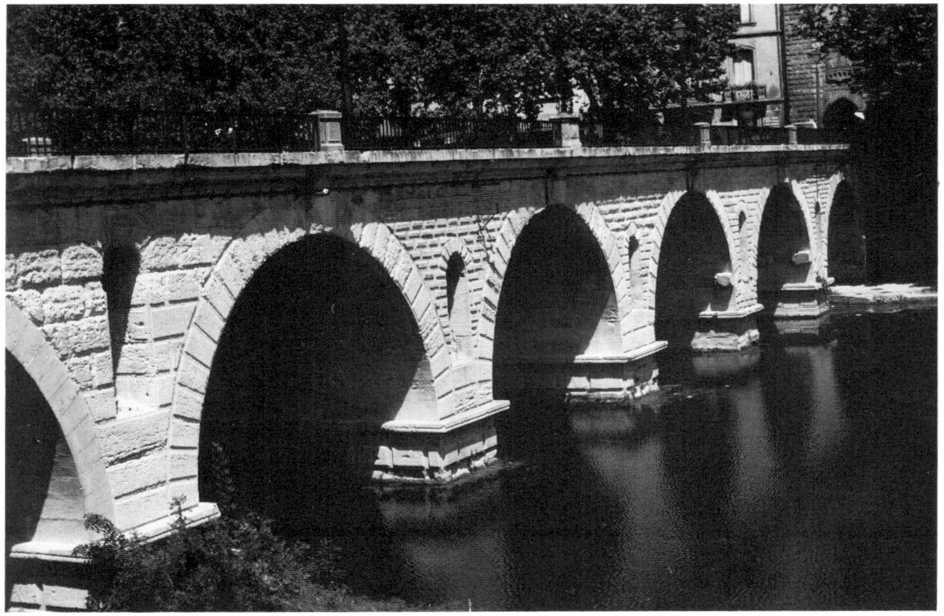

Fig. 67 Part of the Roman bridge at Sommières. It continues beyond the far training wall and gateway, with a total of seventeen spans (G9, France).

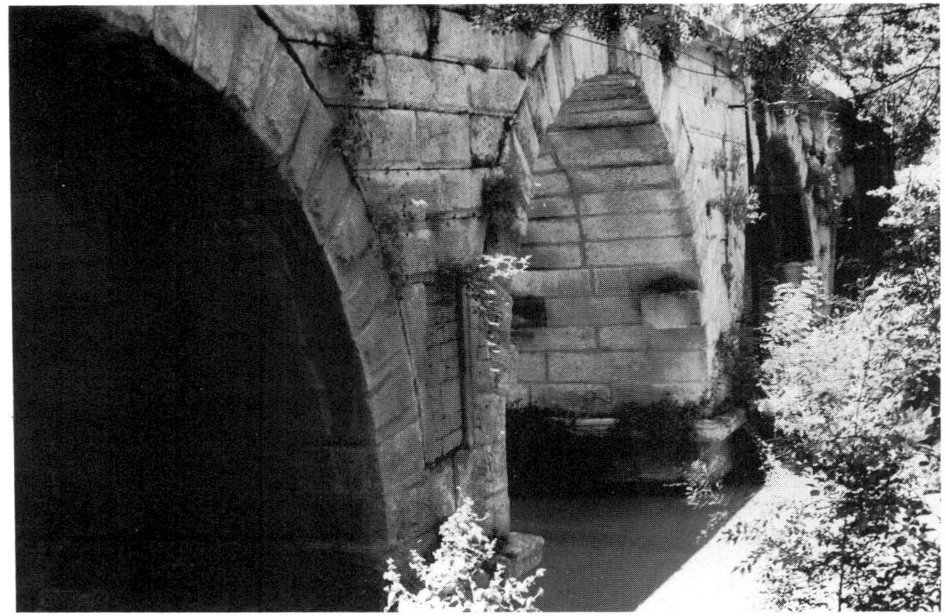

Fig. 68 The Pont de Boissezon, close to Sommières. The underside of the arch is clearly visible (G10).

length of about 190 m. Each pier is fitted with a low cutwater on the upstream face and carries a flood arch, 1 m wide by 3 m high. The arch width is 6.6 m, and is formed in plan by four parallel rings of stones, each 1.65 m wide, 0.9 m thick. The height from the roadway to the stream is 7.3 m. Gazzola dates the original construction to the period of Tiberius (AD 14–37). It has been restored at least twice, most recently in the eighteenth century.

The River Bénovie, Roman Venobia, enters the Vidourle from the west about 1 km below Sommières, and here at Boissezon is another excellent small bridge (**G10**; Gazzola, 1963b, no. 172). Figure 68 shows the eastern or downstream face; the bridge has been widened on the western side. It has five spans of graduated size – 6, 8.8, 9.4, 8.8 and 6 m – with the deck rising slightly to mid-span. The total length is about 50 m, and the width 3.5 m. Again, each arch vault is made of long voussoirs unbonded in plan, so as to form in this case two separate rings. This construction continues down towards springing level and then changes to normal stretcher-bonded construction in the pier. Another small point of interest is that rectangular flood channels were provided in the piers. At Sommières, the floodways were given arched openings but were roofed with slabs and left square in the interiors of the piers. The similarities between the bridges are such that it could be expected that they were built about the same time. The masonry also resembles that in the nearby Pont du Gard, to be described in Chapter 7. This is generally accredited to Marcus Vipsanius Agrippa, son-in-law to Augustus, about the year 19 BC, and so perhaps all three structures were built about that time.

Rivet makes no mention of this bridge, even though he describes the ruins of another over the Vidourle at Moulin de Boisseron, 1 km south (1988: 173, pl. 21).

The next major stream crossed by the Via Domitia was the Hérault River near St Thibéry (**G11**; Gazzola, 1963b, no. 169), 17 km east of Béziers; it can be approached south from Pézenas along the west bank of the river and lies south east of the town. There were originally nine spans, each from 10 to 12 m, with wide piers, substantial cutwaters both up and downstream, and arched floodways in the piers. The corresponding overall length would be about 150 m and this gives some idea of the substantial

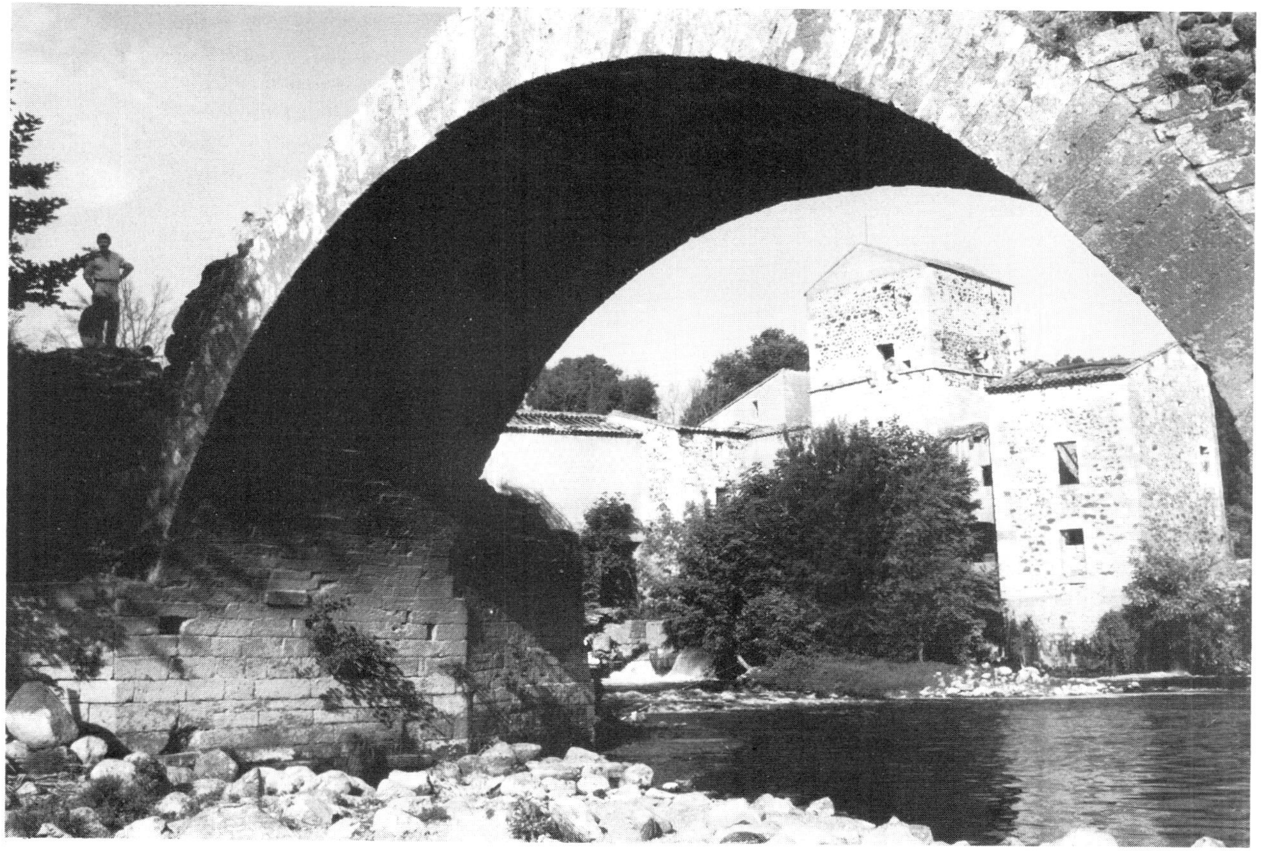

Fig. 69 One of the remaining arches of the Pont de St Thibéry (G11).

size of the bridge. Figure 69 shows one of the spans, with the large stone mill and millrace upstream. A close examination of the bridge confirms the impression given by the illustration, that the arch form is partly segmental, with a rise to span ratio of the order of 0.3, or possibly less. The road width would have been about 4 m. Gazzola (1963b) suggests that the bridge is Augustan (31 BC–AD 14), but Rivet (1988: 153) questions its Roman origin. The missing spans were destroyed by flood some time prior to 1536 (Société Languedocienne, 1905).

In the same district was the Roman **Pont Serme** (**G12**; Gazzola, 1963b, no. 170). West of Béziers, the Via Domitia crossed the River Orb and then the Etang de Capestang. This wide marshy area required the construction of a viaduct some 1500 m long (Brangwyn and Sparrow, 1915: 178), called the Pont Serme or Pons Selinus, and later the Pons Septimus. Little or nothing now remains except for the name, preserved as the name of a local village. Narbonne was an important city in Roman times. It lay on the Via Domitia, and also on the early road to the Garonne and the Atlantic through Toulouse. There is a Roman bridge here (**G13**; Chevallier, 1976: 99; von Hagen, 1967a, b; Rivet, 1988: 134, pl. 4) which serves as the foundation for a row of houses or shops. Its span is of the order of 15 m. Gazzola also records another bridge on this road, the **Pont sur Couesnon** (**G14**; Gazzola, 1963b, no. 167) at Vendelais, with a single span of 3 m. Its location is not known.

There are a few bridges in the general vicinity of the Via Domitia. Vaison-la-Romaine is 27 km north east of Orange. There is a fine Roman bridge here (**G15**; Gazzola, 1963b, no. 174; Rivet, 1988: 288, pl. 60) with a single span of 17 m across the Ouvèze River near the Hôtel-Dieu. Attributed to the early part of the first century AD, it was damaged in the last war and restored in 1954.

Viviers is on the west bank of the Rhône 10 km below Montélimar. Chevallier (1976: 99) records a bridge (**G16**) here across the River Escoutay, of which Rivet (1988: 186, pl.22) has a photograph. He believes that its foundations are probably Roman. Le Vigan is on a branch road from Nîmes to Millau, Roman Condatomagus. Gazzola lists the **Pont Vieux** (**G17**; Gazzola, 1963b, no. 291) as possibly a Roman bridge. There is in fact a fine two-span arch bridge here with that name, but it is scarcely Roman, and does not fit his description. Finally, far to the west is Gazzola's **Pont d'Ascain** (**G18**; Gazzola, 1963b, no. 263), reputed to be 'Roman by tradition'. It crosses the Nivella River about 8 km from the coast, 5 km from the western end of the Spanish border.

The neighbouring towns of Lyon and Vienne formed the centre of a system of radiating roads. Directly opposite Vienne, on the west bank of the Rhône, is Sainte Colombe les Viennes. Gazzola records a crossing here, the **Pont sur Rhône** (**G19**; Gazzola, 1963b, no. 262). He attributes it to Tiberius Sempronius Gracchus, consul in

177 and 163 BC. It was restored a number of times, particularly in the thirteenth century, and then destroyed by flood in 1651. Some ruins still remain (see also Rivet, 1988: 308–9).

One of the roads from Lyon went south west to Toulouse. Near this road is a crossing of the Loire, 8 km north of Le Puy at Lavoute (**G20**; Gazzola, 1963b, no. 264), where the remains of a bridge can be seen at periods of low flow. In the same area near Mondouilloux, was the **Pont de César** (**G21**; Gazzola, 1963b, no. 266), across the Velaj.

Another early road went due west from Lyon to Saintes, or Mediolanum Santonum (Chevallier, 1976: 160), near the Atlantic coast in the Department of Charente-Maritime. Chevallier refers to two bridges here, at Saintes, and at Champagne, 24 km to the north west.

Turning further north, the Saône River rises east of Langres, Roman Andematunnum, and runs south to join the Rhône at Lyon. Navigable for much of its length, it was followed also by a road from Lyon to Langres, and on to Metz, Trier and Mainz. Langres was an important crossroads, for it lay also on a direct road from the Great St Bernard Pass around the northern shores of Lake Geneva to Reims, Amiens and Boulogne, or through Reims to Bavay. Some 9 km north north west of Langres on the road to Reims is the **Pont de Rolampont** (**G22**; Gazzola, 1963b, no. 268). It lies about 1 to 1.5 km south east of the town, over the River Marne. An excellent small bridge in good condition, although somewhat overgrown (Fig. 70) it has three unequal spans, with the central span about twice the side spans, and the roadway rising to the centre between solid parapet walls. Gazzola says there is 'still some doubt about the Roman origin which is normally attributed to it'. There is some reason for this doubt. For much of the central arch, the voussoirs extend out radially across the spandrel to the underside of the cornice which defines the line of the deck. This and other features of the stonework do not appear to be Roman. However, it is on a Roman route, and may well be on a Roman site and contain Roman work.

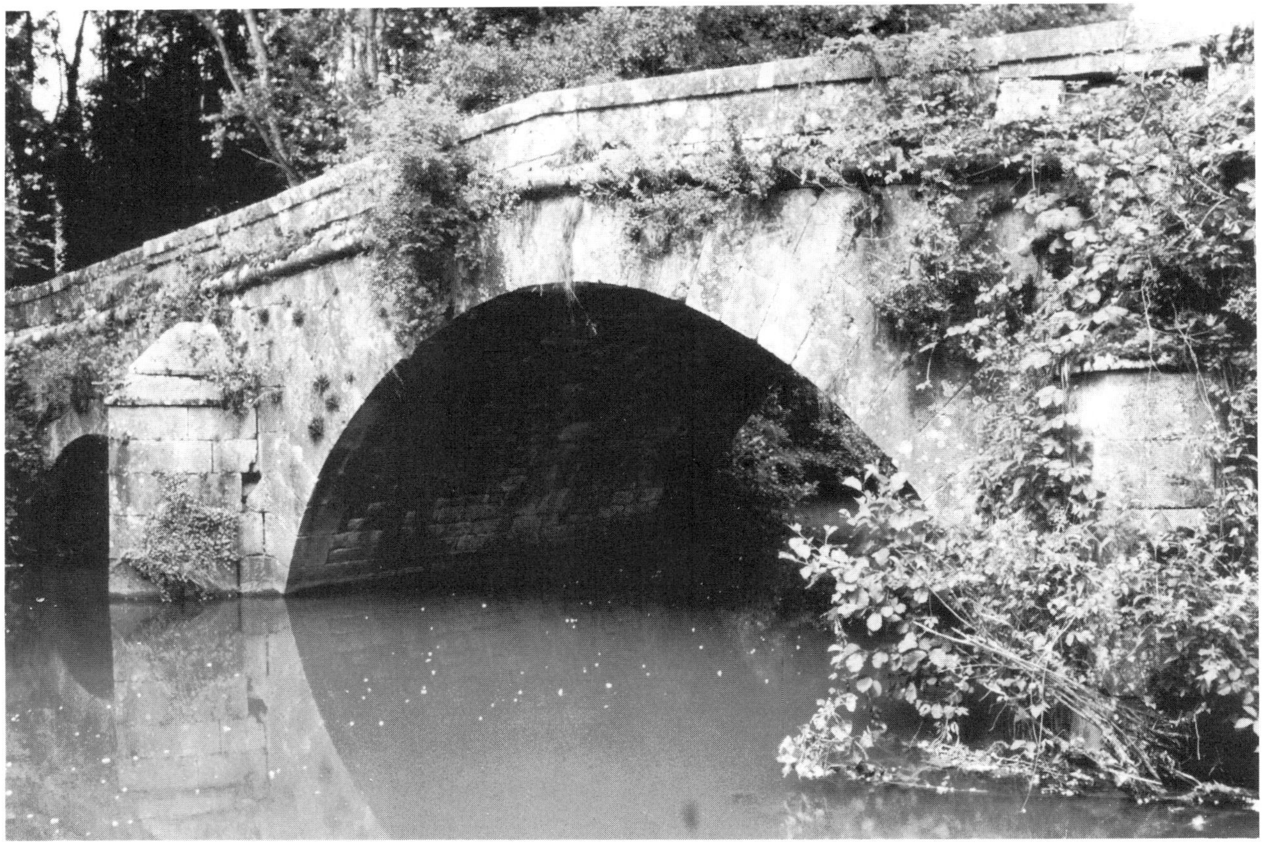

Fig. 70 The Pont de Rolampont lies north of Langres. Although reported as a Roman bridge, the stonework in the spandrels is scarcely Roman (G22).

To the north of Langres near Bourmont, still in the Department of Haute-Marne, is the **Pont d'Outremecourt** (**G23**; Gazzola, 1963b, no. 261) across the Mouzon. Known also as the **Pont Cinq Parts**, it provides a route between the five villages of Outremecourt, Vaudremecourt, Nijon, Soulacourt and Sommerecourt. It also has five spans – about 3.0, 4.7, 5.8, 4.7 and 3.0 m. The width of the piers is about 2 m, giving a length between abutments of about 29 m. In some respects it resembles the Pont de Rolampont – the treatment of the piers is similar, and in both cases the roadway height increases to mid-span as it passes over spans of graduated size. Both are included in a list of French monuments which go back to the Gallo-Roman period.

Gazzola also describes the **Pont de Domqueur** (**G24**;

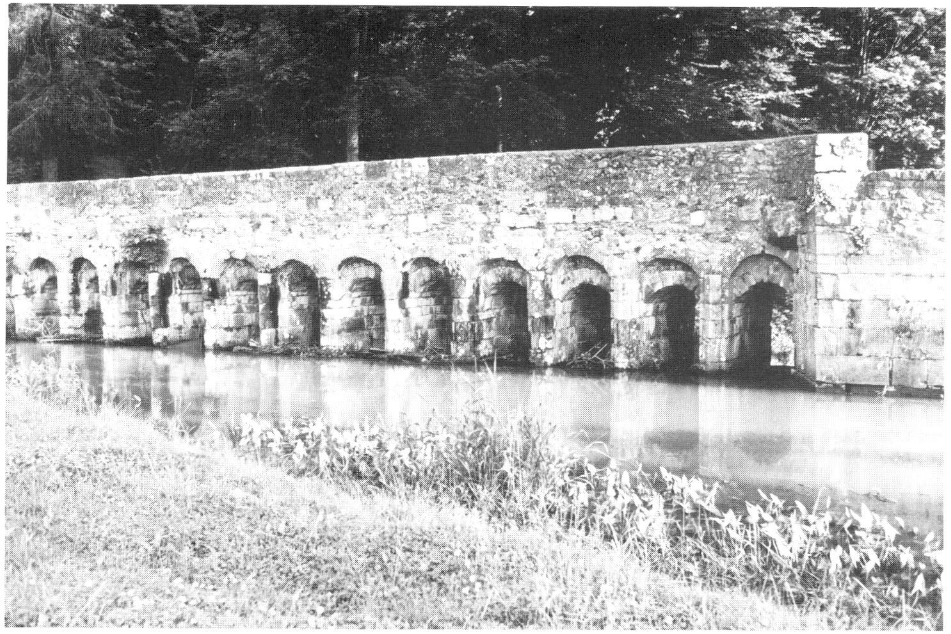

Fig. 71 The bridge at Montignies-St Christophe, just over the border from France into Belgium, has been reported as Roman but may be later (G25).

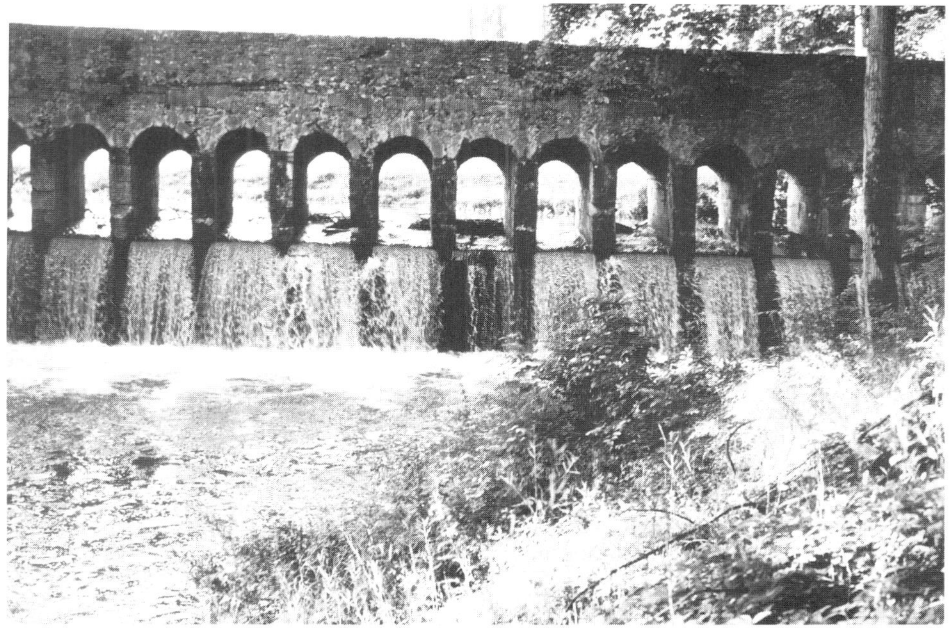

Fig. 72 Downstream side of the bridge at Montignies-St Christophe, showing its function as weir as well as bridge (G25).

Gazzola, 1963b, no. 227), presumably east of Abbeville.

Bavay, Roman Bagacum, was an important centre in Roman times (Chevallier, 1976: 171); it lies in France close to the Belgian border, between Valenciennes and Mauberge. A Roman road ran east from Bavay for some 90 km and then turned south east through Bastogne to Trier, where it met the road from Langres. In Belgium, just over the border, about 5 km north west of Beaumont, the road from Bavay crosses the River Hantes by the remarkable **Pont de Montignies-St Christophe (G25**; Gazzola, 1963b, no. 165). This bridge lies on a minor road, about 300 m south of the modern main road, 1 km north west of Montignies-St Christophe (Figs. 71 and 72). It has thirteen spans, each of about 1.5 m, and piers about 0.6 m thick, giving a length of about 27 m between abutments. The overall width is only 2.65 m. Both faces of the bridge are unusual: the upstream face (Fig. 71) has recessed double arches, with the spandrel walls carried on arches set higher than those which form the vault; and the downstream face (Fig. 72) displays the weir-type action of the substructure of the bridge. The piers of the bridge are in fact surprisingly slender and this, together with the substantial foundation, probably accounts for its survival. There are sharp cutwaters cut into the upstream faces of all piers. The stream turns sharply right as it passes through the bridge. The right bank upstream is parallel to the roadway and is retained by a vertical rough stone wall forming an extension of the abutment, which is constructed of heavier squared stones. These unusual features, and the lack of other bridges in the area, make the bridge difficult to assess. Gazzola (1963b) suggests a possible date ranging from pre-Roman to the time of Caesar; the first would be a brave assumption. Brogan (1953: 33–4) speaks of this 'handsome bridge ... now shown to belong to the eighteenth century'.

Further east was another road from Arlon to Trier. Where it crossed the River Semois are ruins of a Roman bridge, the **Pont d'Ombret (G26**; Gazzola, 1963b, no. 168), and there were two other bridges in Luxembourg, both destroyed in 1944 (**G27, 28**; Gazzola, 1963b, nos. 292–3).

5.7 Spain and Portugal

As far as the Romans were concerned, the early routes in Spain were established by its wars with Carthage (Sections 2.2.4, 2.4). Hamilcar founded New Carthage, modern Cartagena, in 227 BC and formed a treaty with Rome. The violation of this treaty in 218 BC led to the Second Punic War, in the course of which the Roman armies under the Scipios ranged not only down the whole east coast, but inland to the Guadalquivir River at Bailén, and downstream to Ilipa or Alcalá del Río near Seville. From about 206 BC the country was under Roman control although it was not until 18 BC that its subjugation was complete. Mining was also an important factor. The oldest roads (Map 3) were, therefore, from Le Perthus on the French border through Tarragona and Sagunto, north of Valencia, to Cartagena; from Tarragona to Huesca; and, from a point south of Sucronem at the mouth of the River Sucro or Júcar, west to Albacete, south west to Linares, and then down the Guadalquivir to Seville (Roman Hispalis) and Cádiz (Gades).

Spain possesses a remarkably fine collection of stone arch bridges, and it is often difficult to distinguish the Roman bridges from their later counterparts. The best treatment of them is in a compilation of papers by Casado, published about 1980 under the title *Historia del puente en España; puentes romanos* (Casado, n.d.). Its pages are not numbered. The page numbers quoted here begin with the author's prologue as page 1. A map of Roman roads is on page 3. As in the preceding sections, the order of treatment will be outwards and clockwise from Rome, following the major roads.

There have been only a few bridges preserved on the coastal section of the Via Augusta. At Besalu, 30 km north west of Gerona, is the twelfth-century **Puente Viejo (SP1**; Gazzola, 1963b, no. 287) built on the site of a Roman bridge.

Further south, inland from Barcelona at Martorell, is the remarkable **Puente del Diablo (SP2**; Gazzola, 1963b, no. 187; Casado, n.d.: 247–60). It crosses the River Llobregat at the eastern end of the town, close to a modern elevated highway bridge. In its present form it is one of the most spectacular of all ancient bridges (Fig. 73). It has a main clear span of 37.3 m, slightly pointed, with a stone chapel on top. There is also a triumphal arch (Fig. 74) at the eastern end, and the pathway – for it is used only for pedestrian traffic – climbs steeply to pass through the central chapel at a height of about 20 m above the stream. The secondary arch at the right bank, shown at the left of Figure 73, is recorded by Casado as having a clear span of 19.1 m; there is a minor opening above it, and a minor arch set high near the left bank with a span of about 6 m.

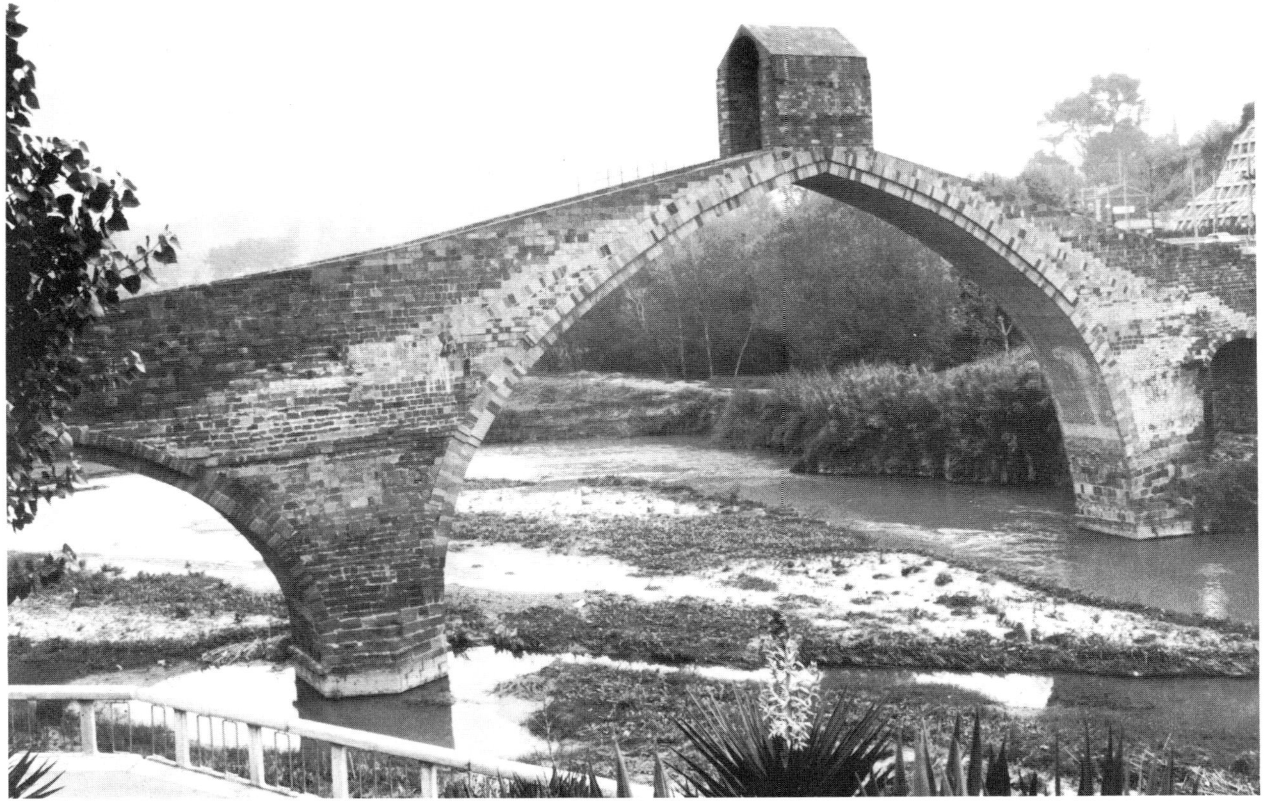

Fig. 73 The great Puente del Diablo at Martorell, Spain, originally Roman but rebuilt by the Moors (SP2).

The bridge is not, however, in its original form. It carries a stone which lists some of the important dates. The original construction is attributed to the second century, and this has given rise to speculation that this was the second century BC. Indeed, it has been said that

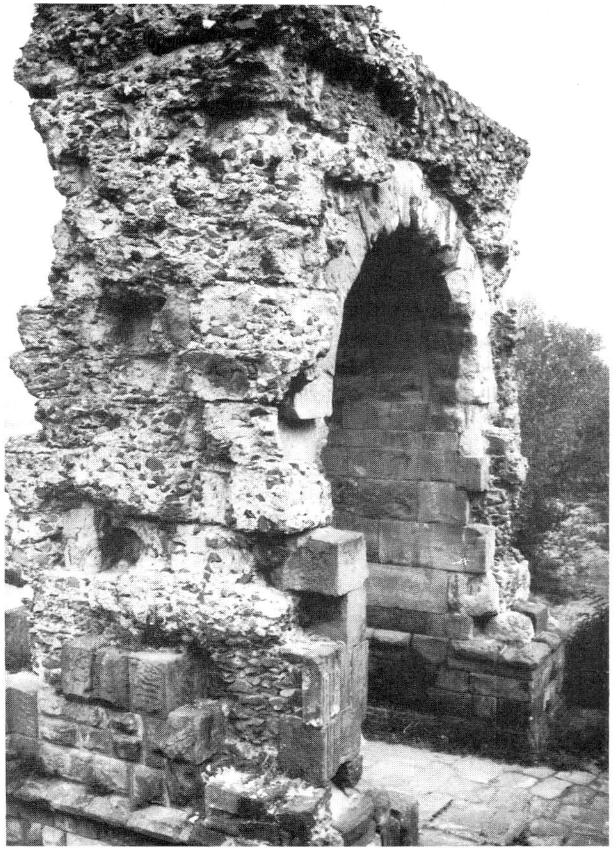

Fig. 74 Roman triumphal arch on the eastern abutment of the bridge at Martorell (SP2).

the bridge was erected by Hannibal in 218 BC, with a triumphal arch for Hannibal's father Hamilcar Barca, but this is hardly likely to be correct, and a date in the second century AD would appear to be more reasonable. In 1283, the bridge was rebuilt under Mohammed II with a Gothic arch; it was repaired in 1768, and again in 1928 and 1933. Destroyed in 1939 in the Civil War, it was then rebuilt in 1965. Its present form is generally similar to that which preceded this reconstruction.

The bridge contains Roman masonry in the triumphal arch and in its foundations. If this were to substantiate the existence of a Roman bridge of 37.3 m, then it would have been the largest span the Romans ever built. Casado suggests instead that the earlier bridge had four spans. The first of these corresponded roughly in location and size with the span on the right bank; the present central span was subdivided into two spans with an intervening pier; and there was a fourth span beneath the present minor arch near the left bank – giving four spans, each of about 15 to 16 m. This does not agree with Casado's figure of 19.1 m for the side span; indeed photographs, and a drawing by Gazzola, suggest that this span was somewhat less, possibly about 16 m. Casado's suggested reconstruction has a horizontal deck, at the level of the opening through the triumphal arch.

There is a series of fine medieval stone bridges further upstream on the River Llobregat – at Manresa, for example. Gazzola lists one of these, the **Puente de Monistrol** (**SP3**; Gazzola, 1963b, no. 188). It has four spans of 37, 27, 17 and 12 m, but does not appear to be Roman (Casado, n.d.: 281).

The next Roman bridges on or near the Via Augusta are in the valley of the Guadalquivir. The **Puente de Villa del Río** (**SP4**; Gazzola, 1963b, no. 160; Casado, n.d.: 119–23) is on the south bank, 2 km east of Villa del Río, which is in turn 10 km east of Montoro. It crosses the Río Salado de Porcuna and is an excellent small bridge with three major spans (Fig. 75). The arches and the lower parts of the spandrel walls are original; the upper stonework and the balustrades are later, as can be seen in the illustration. Casado gives the main spans as 3.0, 9.1 and 3.6 m. There are floodways 1.0 and 1.1 m wide in the piers and, on the far side, another small arched opening with a span of about 1.6 m. The construction of the side spans is unusual in that they spring directly from the voussoirs of the flood arches. The main span has been repaired in part with brick. An Australian will recognise the eucalypts, or gum trees, on its approach.

The north or right bank has two notable bridges. The **Puente del Jándular** (**SP5**; Casado, n.d.: 124, 240, 243) crosses the Jándular River near its junction with the Guadalquivir, about 5 km west of Andújar. Casado speaks of it as 'attributed to the Roman period, but probably from the sixteenth century'. It is 127 m long, with an unbalustraded deck horizontal over seven 5.1 m main spans, then sloping down at one end over two or three smaller spans. It has some Roman features – notably its semicircular arches and the general character of its masonry – yet on the whole its appearance is not convincing. Its piers have rounded ends, topped with a minor cornice which is extended through the arches beneath their springing in a way that does not appear Roman.

The River Guadalmellato joins the Guadalquivir further west, about 10 km east of Córdoba, and is crossed by the **Puente Mocha** (**SP6**; Casado, n.d.: 124–6). It has ten spans, each from 4.5 to 6.6 m, pier widths between 3 and 4.2 m, and a total length of about 100 m. The arches are semicircular, and the central pier has a floodway. Although the construction looks Roman, it probably is not.

In Córdoba itself is the **Puente Romano** (**SP7**; Gazzola, 1963b, no. 159; Casado, n.d.: 147–53, 191–205; Knapp, 1983). A very large bridge indeed, it has sixteen spans, listed by Casado as 12.8, 9.5, 9.8, 10.3, 10.5, 10.8, 11.1, 11.9, 11.9, 12.0, 10.2, 10.0, 10.0, 10.4, 10.4 and 7.9 m. The length between abutments is 274 m; for comparison, the sum of the spans is 169.5 m, or 0.62 of the total. The piers form an obstruction 105 m wide. It is not surprising, then, that the bridge has had a large effect on the stream flow. Casado has an instructive aerial photograph (p. 153) showing the river more than doubling its width as it approaches and then passes through the bridge, with a complex system of sandbanks and channels downstream. Conversely, the stream has done its best to destroy the bridge, and the detailed plan given by Casado (p. 147) shows the extensive protective works that have developed over the years.

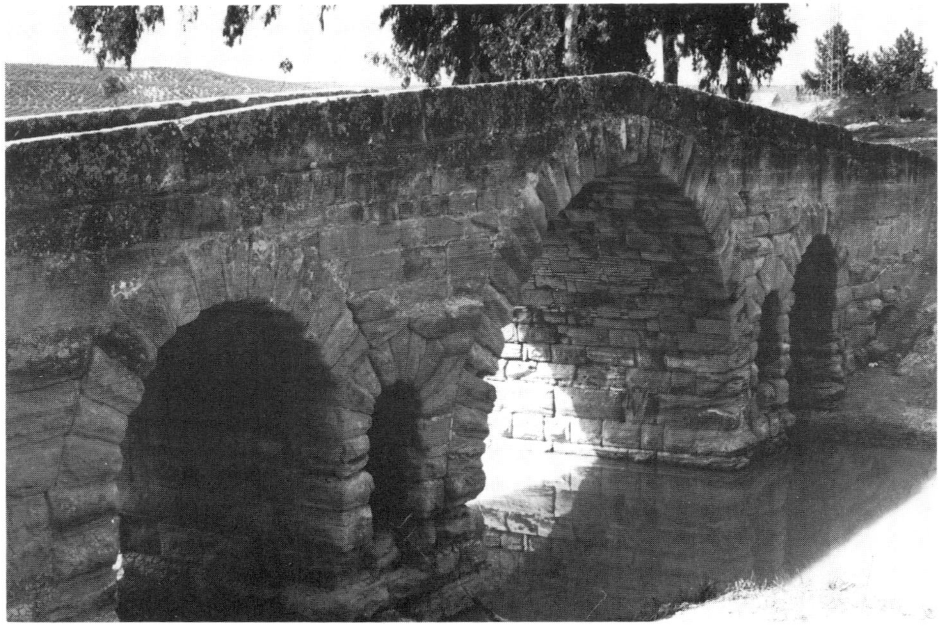

Fig. 75 Puente de Villa del Río, 12 km east of Montoro in Spain (SP4).

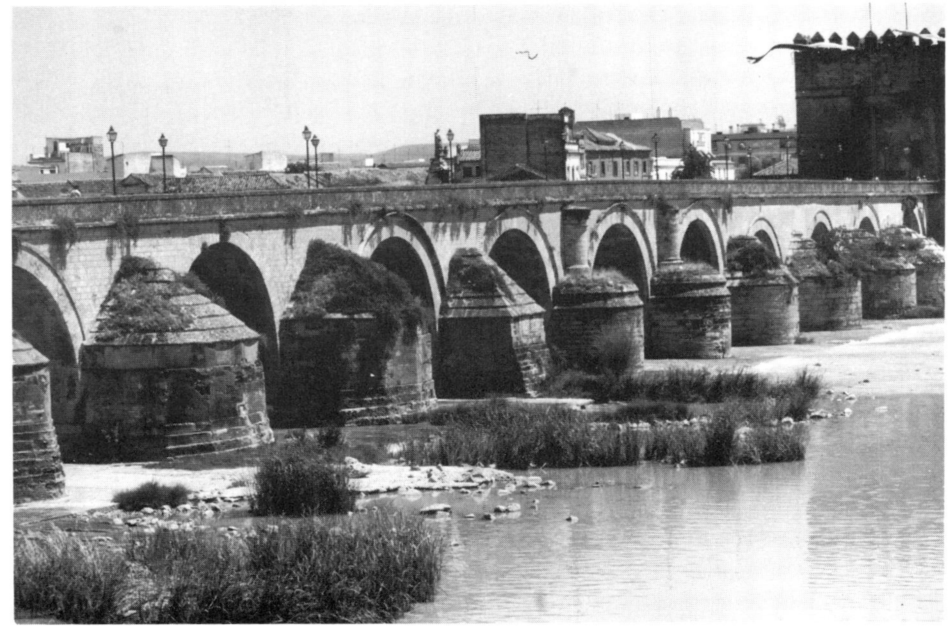

Fig. 76 Downstream face of the bridge at Córdoba (SP7).

Figure 76 shows part of the downstream face of the bridge, and is sufficient to indicate the alterations and restorations that have been made, many by the Moors. Gazzola suggests that the original construction may have been Augustan. It is still used by motor traffic and runs from the Torre de la Calahorra on the left bank to La Mezquita, the Islamic mosque, now the cathedral, in the Judería or Jewish quarter of the city.

Near Seville are two bridges that appear to be Roman. Carmona was a Roman town, 28 km from Seville, 14 km south of the river. The **Puente de Carmona** (**SP8**; Casado, n.d.: 227–30) is a five-span bridge, with spans of 4, 4, 5, 4 and 4 m, and both arches and spandrel walls made of bricks laid in thick mortar beds. The central span is distinctive in that the outer face is stepped back in three concentric rings, in contrast with the plain, flush finish of the other arches. The set back, and the larger span, make it appear higher than it is, higher than the other four spans.

The Puente de Alcantarilla (**SP9**; Gazzola, 1963b, no. 192; Casado, n.d.: 206–9) crosses the River Salado de Moran near Utrera 32 km south east of Seville, on the Roman road to Cádiz. It has two moderately spaced spans, of 5.4 and 5.7 m, and, with the exception of the modern balustrade, is built of squared stonework. An old drawing, of the year 1720, shows the bridge with a defensive tower at each end. It carries a stone with a three-line inscription, of which the first line reads, 'Augustus pontem'. It is attributed, therefore, to the period of Augustus.

Further west in southern Portugal is the old town of Mertola, Roman Myrtilis, on the lower reaches of the River Guadiana. Wiseman (1956: 183) reports the piers of a bridge (**SP10**) in the river here.

In the north, there was an important route across the country from Tarragona to Lérida, Zaragoza, Toledo, Mérida and beyond. **Puente de Piedra** was an old bridge across the River Segre at Lérida (**SP11**; Gazzola, 1963b, no. 158; Casado, n.d.: 288). It was badly damaged in the Civil War of 1936 to 1939, and has since been replaced. A bridge existed here at the time of the wars between

Caesar and Pompey, and it is believed that there were traces of Roman work in the foundations of the old bridge. Another **Puente de Piedra** (**SP12**) still crosses the River Ebro at Zaragoza; the view across the bridge to the town and its cathedrals is a famous one and has been painted by Velázquez and others. There has been a series of bridges here – one was destroyed in 827–8 and rebuilt in 839; one was here at the time of Alfonso I in 1118; and a bridge was reconstructed here in 1437 and 1659 (Casado, n.d.: 311–15). The present bridge is not Roman, but could well be on the site of a Roman bridge.

The road passed by Medinaceli and its triumphal arch, to the east of Madrid where there was no town in Roman times, and then to Toledo. The **Puente de Alcántara** across the River Tagus in Toledo, not to be confused with the bridge of the same name near the Portuguese border, incorporates Roman foundations (**SP13**; Gazzola, 1963b, no. 286; Casado, n.d.: 427–60). It was built about 1258 to take the place of another bridge, destroyed by flood in 1249, which in turn replaced an earlier Roman bridge, known to exist in 567, but destroyed by Mahomet Alameri, governor of Toledo, in 997. The present bridge has two spans, of 28.3 and 16 m, and a width overall of 5.4 m (Fig. 77). The roadway passes through a tower built in 1711 at the eastern end, across the bridge, and then through a larger defensive tower of 1484 before it climbs to enter the city immediately below the Alcázar. Figure 77 shows the heavy walling which forms part of the bridge on the east bank, beyond the major arch. Casado believes that the Roman bridge may have had a third span here, to match that on the west bank. The spans of this bridge would then have been 16, 28.3 and 16 m. The present pier is 8.4 m wide with the springing of the main arch not greatly above the river level. This suggests that the deck level of the Roman bridge was about the same as at present, 26 m above the river, with a total length of about 127 m. Casado also discusses the dating of the Roman bridge. The present structure carries a number of inscribed stones. One of these attributes the bridge to the time of Trajan (AD 98–117), and another to C. Julius Lacer, who built the other Spanish Alcántara bridge west of Cáceres about the year AD 104. This suggestion is not unreasonable.

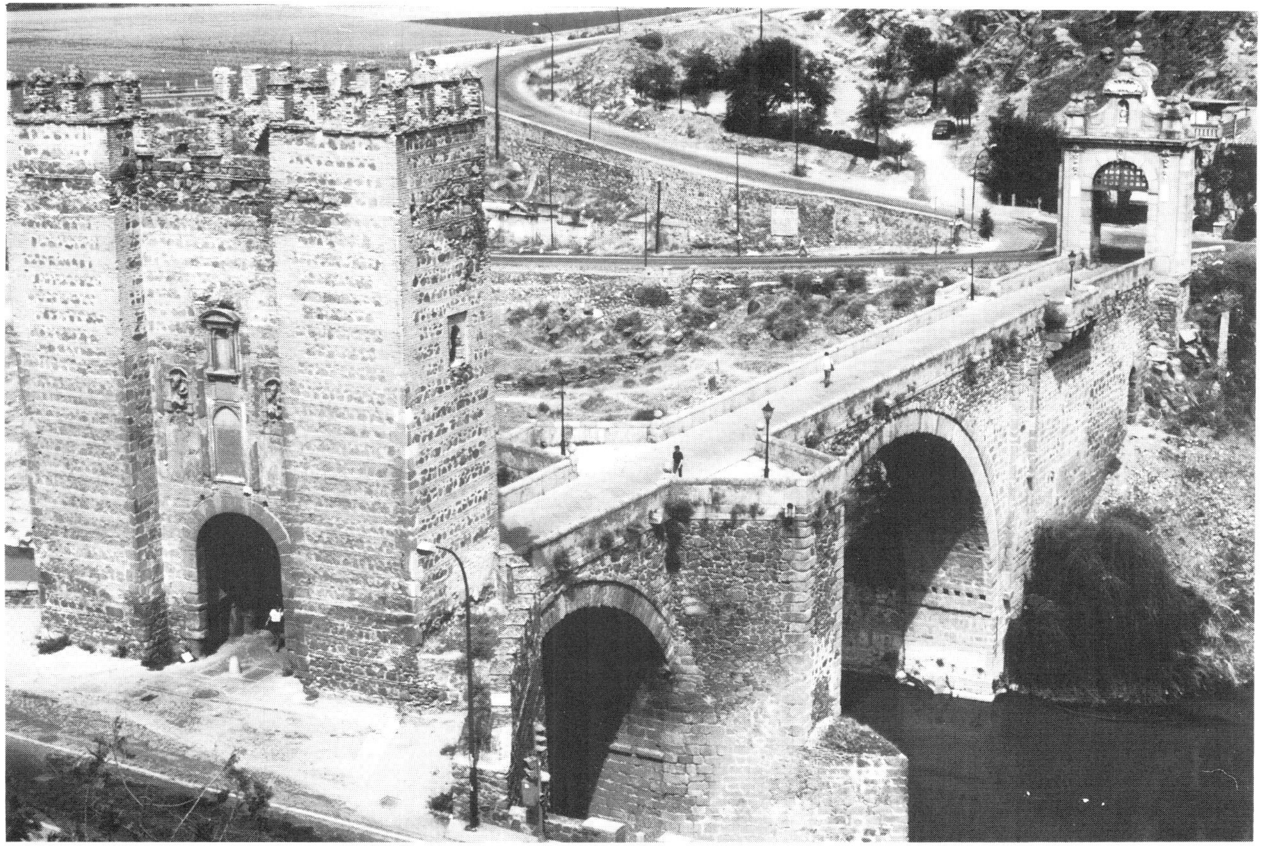

Fig. 77 The Puente de Alcántara in Toledo, SP13 (not to be confused with the other bridge, SP21, of the same name).

The remains of a Roman aqueduct (AQ67, Section 7.4) exist in Toledo about 300 m downstream of the Puente de Alcántara, beyond the Puente Nuevo. The Pont St Martin, at the western approach across the Tagus to the town, dates from about 1203 (Casado, n.d.: 457), and is not Roman.

The road through Toledo to Mérida was built by Vespasian. An earlier Augustan road (Chevallier, 1976: 157) joined Astorga in the north with Seville in the south, passing also through Mérida. There is a Roman bridge across the River Guadiana (Roman Anas) in Mérida itself. It is surprising, therefore, to find the remains of another Roman bridge (**SP14**) across the same stream, 35 km to

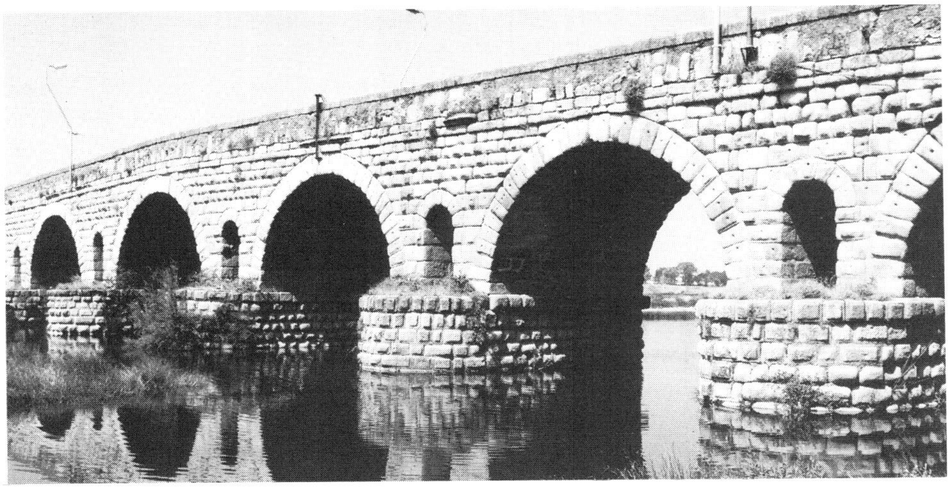

Fig. 78 The long Roman bridge at Mérida was built in three stages. This photograph shows bridge 1, closest to the north bank, but believed to have been constructed second (SP15).

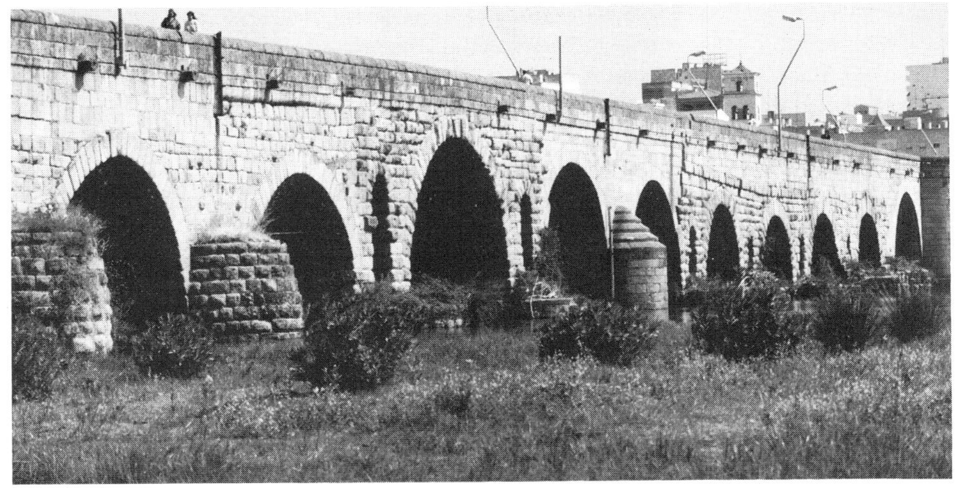

Fig. 79 Part of bridge 2 at Mérida, built first, but including some late spans (SP15).

the east at Medellín; and this has caused Casado (n.d.: 167) and Gazzola (1963b, no. 275) to suggest that the bridge at Medellín (SP14) may have been on the north–south road. Gazzola states that this bridge 'constituted, by its dimensions, the fourth of the great bridges of Lusitania'. Although a judgement like this depends on the criteria that are adopted, it clearly was a very large bridge. He suggests it had twenty-eight 12 m spans; the same span is given by Casado (n.d.: 145). Only the bases of the piers remain; these were 9.2 m long in the direction of the flow and 4.7 m wide, on stone foundations 10 m by 5 m. On these dimensions, the length of the bridge between abutments would have been about 465 m.

The **Puente Romano** (**SP15**; Gazzola, 1963b, no. 161; Casado, n.d.: 11–40, 107), still in service across the Guadiana River in Mérida itself, is the longest of all existing Roman bridges. Casado lists fifty-seven spans still in existence, with a total length between abutments of 721 m. Another three of the Roman spans are buried in the southern approach; five of the present spans (numbers 11–15) replace possibly seven earlier spans. This means that the original bridge had about sixty-two spans, covering in total about 755 m. The overall length including approaches is given by Gazzola as 790 m. It is of granite.

Casado divides the bridge into three parts. The 'first' Roman bridge consists of spans 1–10, running from Mérida on the north bank to the first ramp, which leaves the bridge at right angles on the upstream side and descends to a sandbank in the river. Figure 78 shows the upstream side of spans 2–6. As can be seen, the deck rises slightly to the centre of span 5, which is the largest span, and then falls to the south. These spans, in the order 1 to 10, are 7.3, 7.5, 8.8, 9.6, 10.5, 9.5, 8.5, 6.7, 6.7 and 6.7 m. The pier widths vary from 4.4 to 6.0 m, averaging 5.2 m. The total length is 128.8 m, of which the openings form 81.8 m, or 0.64 of the total. The piers in this part of the bridge are rounded on the upstream side, flush on the downstream side, and all carry flood arches.

The first ramp, with a total width of about 7.1 m, separates this 'first' Roman bridge from spans 11–15. These are also semicircular in elevation, with spans 11–14 of 15.5 m, span 15 of 12 m, and pier widths of 5.2 m. They date from the seventeenth century. The total length is 94.8 m, of which 74 m, or 0.781 of the total, is opening.

Spans 11–15 are separated from the 'second' Roman bridge by a prominent pier, 8.4 m wide, with semicircular ends, extending up to and including the balustrades, on both faces. This second bridge has 36 spans, of which spans 21–2 and 29–33 were replaced in the nineteenth century. The span lengths recorded by Casado are: spans 16–20, 11.3, 8.8, 9.6, 7.6 and 7.6 m; spans 21–2 (nineteenth century), 9.0, 10.0; spans 23–8, 11.6, 9.2, 9.2, 8.0, 7.6 and 8.0 m. Spans 29–33 (nineteenth century), average about 9 m; and 34–6 are about 6.5, 6.2 and 4.2 m. The overall length of this section is about 260 m, made up of 179.5 m of openings and 80.5 m of piers. The openings amount to 0.69 of the total. Some of these piers have floodways. Numbering pier 16 as that following or to the south of span 16, then there are floodways in piers 17, 18, 19, 20, 22, 23, 26, 33, 34 and 35. All piers have rounded cutwaters on the upstream face. The new spans are lighter in colour, with the stones finished flush rather than rusticated. There is a distinctive higher cutwater on the upstream side between the new spans 21 and 22. Figure 79 shows the upstream face of spans 16–25; the new spans 21 and 22 are visible, as is also span 23, the largest of the Roman spans (11.6 m). The line of the roadway rises from each end to above span 23.

The second ramp, 5.5 m wide, lies between spans 36 and 37, again on the upstream side. Spans 37–57 comprise the visible portion of the 'third' Roman bridge (spans 58–60 are buried in the abutment). Spans 37–51 are each about 5.6 m; spans 52–7 are 6.0, 6.7, 7.2, 7.8, 6.8 and 6.8 m. The overall length is 216 m, with openings amounting to 125.3 m or 0.58 of the total. There are no cutwaters or floodways in the piers, and the deck rises slightly to the centre of span 55. Figure 80 shows a typical span.

Summing these figures shows that the 721 m total length of the existing bridge divides into 461 m of openings, and 260 m of piers and other obstructions. Openings comprise only 0.639 of the total. It is not surprising, then, that the bridge was liable to flood. Casado (n.d.: 33) shows photographs of the bridge in the flood of 1942, which covered the roadway on the northern approach, and reached deck level at the first ramp. The Romans would have learnt, with other engineers, that it is difficult to fix the channel of a stream such as this, with large variations of flow in a wide sandy plain.

Casado (n.d.: 34) believes that the 'second' bridge was built first, either before or at the same time as the establishment of Roman Mérida, Emerita Augusta, by Augustus about 25 BC. It crossed the central main channel but by its very presence restricted the flow, raised river levels and caused overtopping of the approaches. These would then have been cut away, to create new channels bypassing the bridge. He believes the 'first' or northern bridge was built next, at about the same time as the castle or Alcázar. Finally the bridge was extended to the south – the 'third' bridge – possibly in the time of Trajan (AD 98–117), who was responsible for the reconstruction of many Spanish bridges.

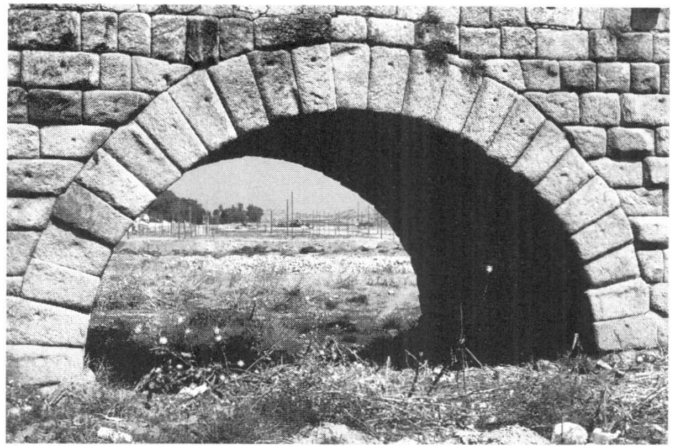

Fig. 80 A typical span of bridge 3 at Mérida (SP15).

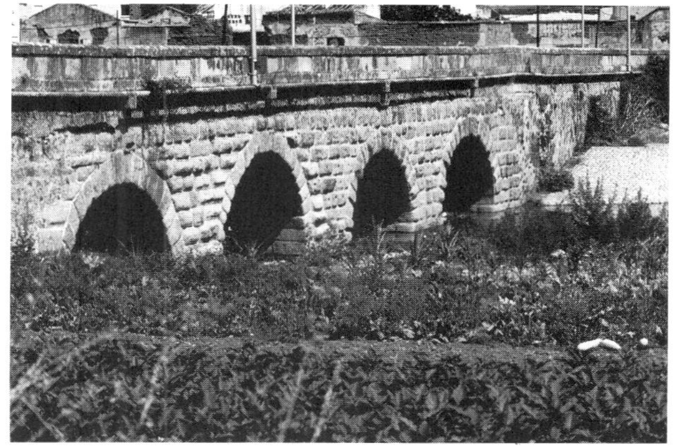

Fig. 81 The Puente de Albarregas at Mérida (SP16).

Apart from its aqueducts, which will be described in Chapter 7, Mérida has a second major Roman bridge. The smaller Albarregas River crosses to the north of the town, to join the Guadiana River downstream – that is, to the west. Approximately 600 m north of the Puente Romano, it is crossed by the **Puente de Albarregas** (**SP16**; Gazzola, 1963b, no. 186; Casado, n.d.: 139, 188–90). This granite bridge has four main spans, each 5 m, with two minor arch openings of 1.7 and 1.3 m in the long southern approach (Fig. 81). The overall length is 125 m; the height from the water to the roadway is 6.7 m; and it has an overall width of 7.9 m. As can be seen from the photograph, it is in excellent condition. It carries modern traffic on the road from Mérida to Cáceres.

West of Mérida, one road went directly to Lisbon, Roman Olisipo; another branched north west to join the road from Braga and enter Lisbon from the north. This latter road crossed the River Seda by the **Villa Formosa bridge** (**SP17**; Gazzola, 1963b, no. 163) (Fig. 82). It now carries the modern road from Alter do Chão to Ponte do

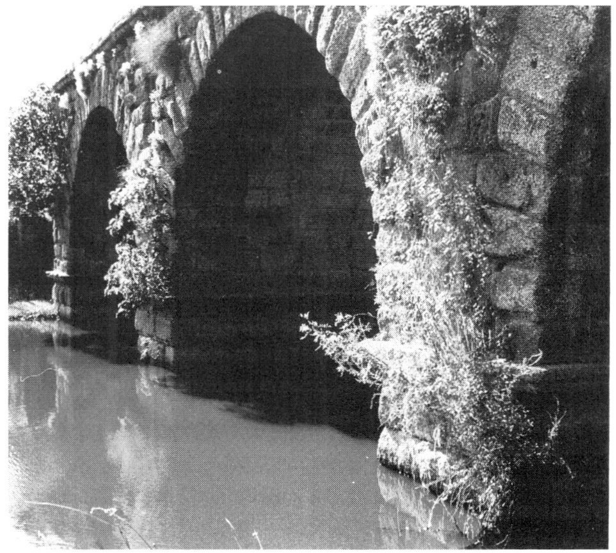

Fig. 82 Bridge at Villa Formosa, west of Alter do Chão in Portugal (SP17).

Sor, about 10 km west of Alter do Chão. It has six spans, each of about 9.6 m, spanning between piers about 2.8 m wide, each with a cornice at springing level, and a high set flood arch. The total width is 6.7 m. Wiseman (1956: 178) describes the bridge as the 'best preserved Roman bridge in Portugal'. This may be misleading, as there are two other excellent Roman bridges here – at Segura and Chaves. It is certainly true, however, to say that the bridge is a fine one, in excellent condition.

It is difficult to correlate with assurance the bridges and roads in the region north of Mérida, around Cáceres and Plasencia. What is certain is that there was a major north–south route which crossed the River Tagus at Alconétar and went north somewhat west of the present road through Plasencia to Salamanca and Astorga. But this does not include all of the bridge sites. There are also bridges at La Vera east of Plasencia along the right bank of the valley of the River Tiétar and it is not clear which route they served. There is some variation of opinion concerning the line taken by the road from Toledo to Mérida; this in turn seems to relate to some uncertainty as to the location of the Roman town of Augustobriga. Casado (n.d.: 3, 475) places this town south west of Toledo, and takes the road through here more or less on a straight line to Mérida. Chevallier (1976: 156) shows this road but also another, which leaves Toledo along the northern bank of the Tagus to Caesobriga at Talavera de la Reina, and then continues further west along the Tagus before turning almost due south to Medellín. Sitwell (1981: 54) places Augustobriga here, on the Tagus almost due north of Medellín, near Navalmoral de la Mata. Possibly, then, there was a road joining this point to the northern road near Plasencia, passing across the bridges of La Vera.

The problem is even more marked with the notable bridges of Alcántara and Segura, for neither Casado nor Sitwell shows a road which crosses them. The answer appears to be that given by Chevallier (1976: 156), that a road branched north west at Cáceres from the Mérida to Salamanca road, to cross the Tagus at Alcántara and the Elja at Segura, heading in the direction of Coimbra.

Both Gazzola (**SP18**; Gazzola, 1963b, no. 270) and Casado (n.d.: 157, 164) refer to the bridge at Vadillo, across the Tamuja River on the modern Cáceres to Trujillo road. Gazzola describes the bridge as 'undoubtedly late'. This is borne out by the single photograph available: the depth of the arch rib is slender and the general quality of the work does not appear to be Roman. A stone in the walls carries the inscription

L. TROBANO RUFO
AN. XXIV
LABEO A. UNCULUS

The site is about 30 km east of Cáceres, possibly on Sitwell's route from Augustobriga to Medellín.

The northern road crossed the River Tagus at Alconétar near the confluence of the Almonte. The site is now flooded, but Casado (n.d.: 174) has earlier drawings and photographs of the area. The old road from Mérida reached the Tagus immediately south of the Almonte. It could have crossed the Tagus directly; instead it turned right, and crossed the Almonte to reach the base of a rocky eminence. It turned sharply left and within the space of 100 m or so reached and then crossed the Tagus.

The bridge over the Almonte was called the **Puente de El Garro** (**SP19**; Gazzola, 1963b, no. 271; Casado, n.d.: 168). Even before the inundation there were only a few remains, including parts of the abutment and the retaining walls which supported the road as it followed the steep rocky bank. Casado has discussed the reasons for the chosen alignment, requiring two bridges instead of one, and suggests that it may have been for strategic reasons, to allow the fortification of the ridge which overlooked both bridges.

The **Puente de Alconétar** (**SP20**; Gazzola, 1963b, no. 183; Casado, n.d.: 145–6, 169–74, 468–70) has been moved to a site about 6 km north, just west of the junction of the Plasencia road with the branch road to Coria. Casado quotes Prieto Vives who gave the main spans as 7.3, 8.2, 9.0, 10.1, (11), (12), (13), (14), (15), (14), 13, 12, 11, (10), 9.3 and 9.1 m, in order from the north to the south bank. The figures in brackets are expected rather than measured values. Measured pier widths varied from 4.2 to 4.6 m. If an average of 4.4 m is assumed, then the total distance between abutments would have been 244 m, with the openings totalling 178 m or 0.73 times the total. There were additional openings in the north abutment, of 6.7 and 7.1 m. The relocated structure consists of the north abutment with these two approach arches; piers 1, 2 and 3, with parts of 4 and 6; and arch vaults corresponding to spans 1 and 3. Parts of other piers were visible at the original site before inundation. An unusual feature of the bridge is that all existing spans are segmental – Figure 83 shows span 3 from the upstream side, Figure 84 the downstream face of the second span in the northern abutment, and Figure 85 a general view looking south. It is unlikely that spans 1 and 3 are original, even if they have been transferred to this site without change. The underside has the form of a random rubble wall, without coursing. The arch profile too is irregular, being more parabolic than segmental. However, the spans of the north abutment look Roman in every respect other than the fact that they are segmental, and these segmental vaults are framed into the side walls in such a way as to make it almost certain that they are authentic. The stone supporting the extrados of the left-hand springing of Figure 84 is shaped with a reentrant corner in its upper face, and this looks quite deliberate. The conclusion appears to be inescapable that, in this case at least, the Romans used a segmental arch, and evidence in support of this can be adduced from at least one other bridge, near Limyra in Lycia (E23), along the Turkish Mediterranean coast. It will be remembered also that Ponte San Lorenzo in Padua (I161) is segmental, and the great Pont St Martin (I155) partly segmental. In this case, the ratio of

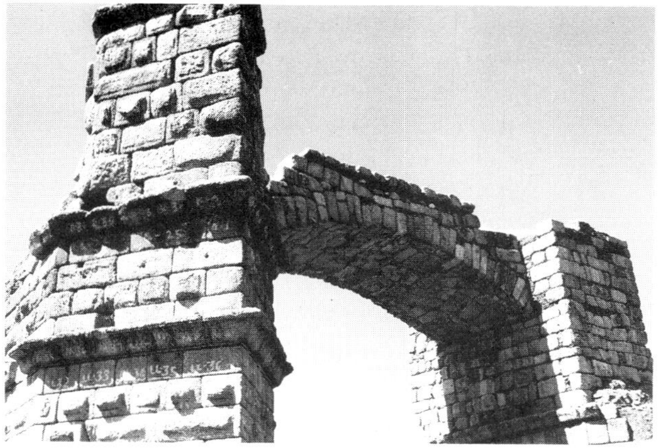

Fig. 83 The bridge of Alconétar (SP20) has been relocated at its present site. The span shown here is clearly new, although the piers appear to be original.

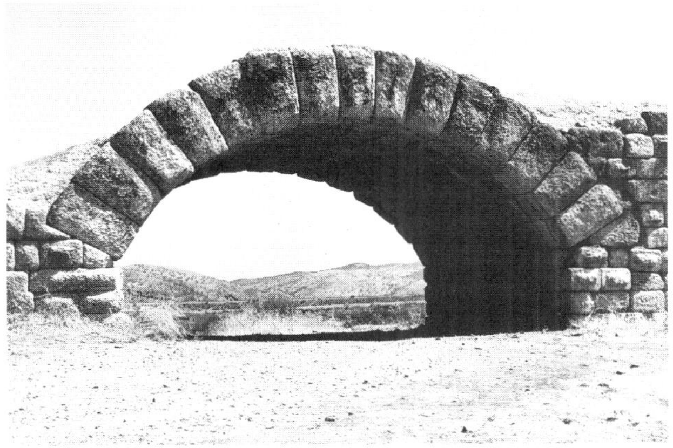

Fig. 84 This span in the abutment of the bridge at Alconétar is believed to be original, even though segmental (SP20).

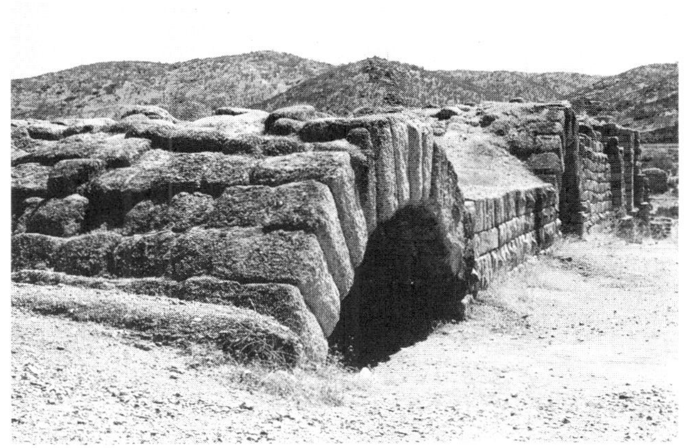

Fig. 85 General view of the bridge at Alconétar. Note the relatively rough finish of the upper side of the vault in the foreground (SP20).

rise to span is about 0.29; at Padua about 0.27; and in Lycia about 0.19.

This segmental profile has caused Casado (n.d.: 472–3) to speculate on whether the engineer might have been Apollodorus of Damascus, who built the great segmental timber arch bridge across the Danube for Trajan about AD 104–5. Although there is a great difference between the two bridges in other respects, it is not unreasonable to conjecture that, since the segmental arch is so unusual in Roman bridge design, the two bridges may be of about the same age. Furthermore, the adjacent Alcántara bridge is usually dated about AD 104 and this would, in any case, suggest a dating for Alconétar within the time of Apollodorus. Gazzola (1963b) says that there is no doubt that the original fabric goes back to the time of Trajan (AD 98–117) when the emperor was restoring the roads of the province. The material is granite, as at Alcántara.

To the west, at a distance of about 38 km downstream from Alconétar, is the **Puente de Alcántara** (**SP21**; Gazzola, 1963b, no. 181; Casado, n.d.: 41–80, 141–6, 454–5, 461–7), one of the greatest of all Roman bridges. The surprising nature of the site, away from the line of any of the great trunk roads, has already been mentioned. To the modern eye it would appear more reasonable to have built a road from Toledo parallel to the Tagus, possibly through Plasencia to Segura, but this may not sufficiently consider the lie of the land or the needs of the time. At Alconétar, the pressing need may have been strategic, and the same may have been the case at Alcántara. A Roman camp had been constructed at Cáceres in 79 BC by the consul Caecilius Metellus in his campaign against Sertorius and perhaps the major need was to deploy troops quickly from there into Portugal.

Whatever the need, it has given rise to a fine bridge. Figure 86 shows the upstream face from the southern end. In order from the northern end, spans 1–6, as given by Casado (p. 55) are 13.8, 21.9, 28.8, 27.4, 21.9 and 13.8 m. Gazzola's figures are different – 18.5, 24.3, 28.5, 27.3, 24.3 and 18.5 m; they agree with Casado's for the main spans but differ markedly for the smaller spans and do not appear to be correct. Casado's figures are preferred. The largest span, span 3 (28.8 m), crosses the stream immediately north of the triumphal arch. It is only slightly larger than span 4. If this difference is ignored, then the spans are symmetrical about the triumphal arch above pier 3; the stream, however, is not symmetrical, and pier 3 lies at the southern edge of the waterway, with piers 2 and 3 in the stream, and piers 1, 4 and 5 on the banks. The maximum height from the deck to the bed of the stream is about 48 m. The height above the water level depends on the state of the river but is about 40–2 m (scaled from photographs in Casado, n.d.: 56, 464). To the top of the triumphal arch is another 14 m. The lower parts of the main piers are 8.3 m wide in the direction of the bridge and 14.7 m long parallel to the stream, or, excluding the cutwater at the upstream face, 11.2 m long (Casado, n.d.: 146). The triumphal arch provides a clear opening 6 m wide, with a width between parapets of 6.7 m, and an overall width of 8 m.

There are massive stone retaining walls at each end of

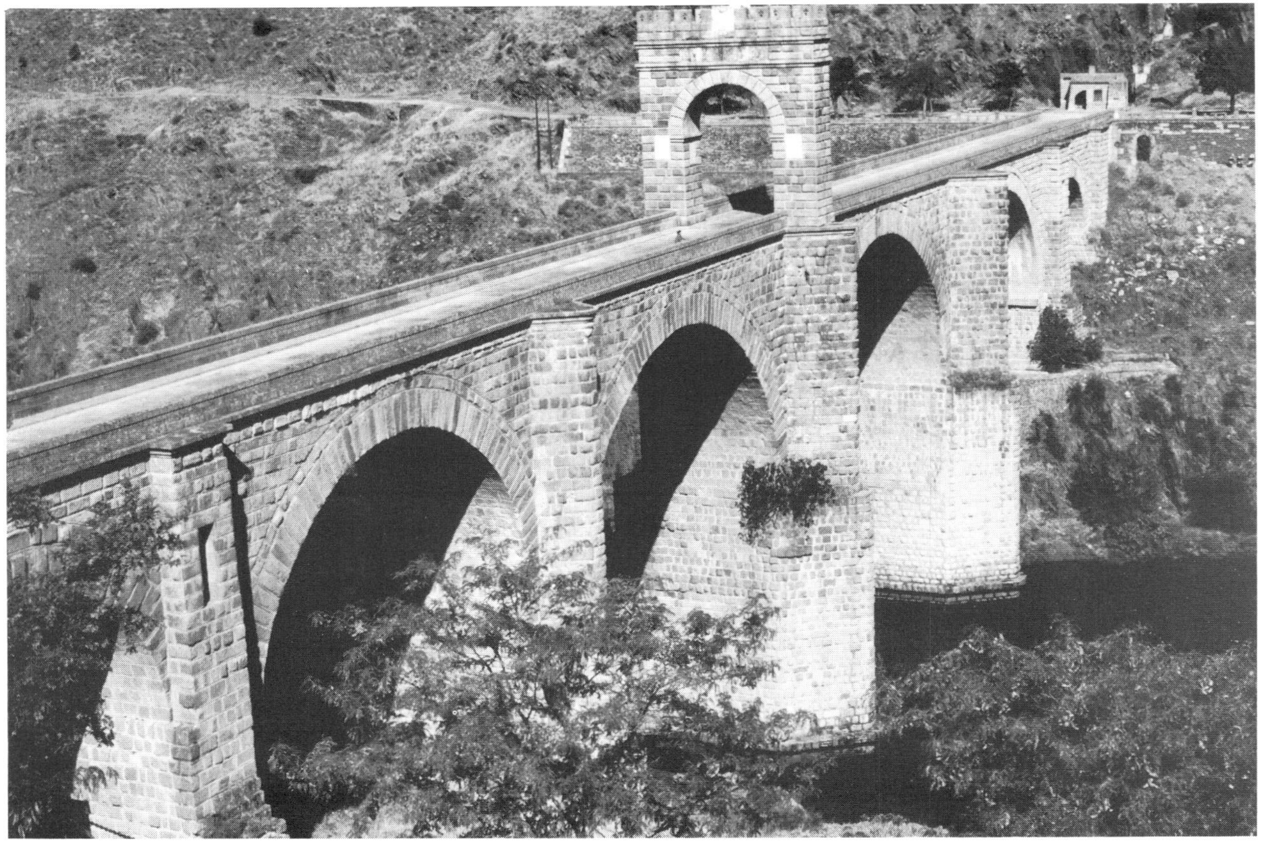

Fig. 86 The great bridge of Alcántara over the Tagus in Spain (SP21).

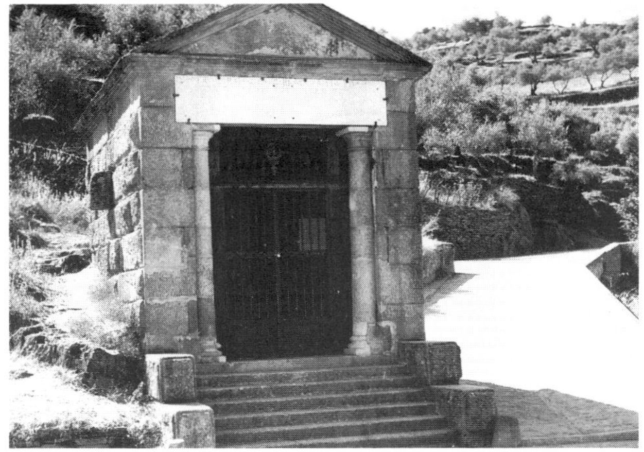

Fig. 87 Temple opposite the southern approach to the bridge at Alcántara (SP21). The tablet above the doorway includes the name of the builder, Lacer, and his words, 'I have built a bridge which will remain for ever.'

the bridge parallel to the flow. The length of the bridge between these walls is 194 m. At the northern end, the bridge joins a road running upstream and downstream along the bank at deck level. Opposite the bridge is a small building with two arched openings and a tower on the bank above. On the south side, the present road turns sharply left and follows the upstream bank. There is also another road which zigzags up the bank. Opposite the end of the bridge is a small votive temple (Fig. 87), which carries a nine-line inscription on a marble slab beneath the pediment and above the doorway facing the bridge. This begins (*CIL* II.761):

IMP. NERVAE. TRAIANO. CAESARI. AVGVSTO. GERMANICO. DACICO. SACRVM

and includes as its sixth line

PONTEM. PERPETVI. MANSVRVM. IN. SAECVLA. MVNDI. FECIT. DIVINA. NOBILIS. ARTE. LACER.

Although the stone has been replaced during the life of the bridge, it appears that most of the inscription is original, and this includes these words. The bridge, then, was built by Lacer in the time of Trajan. He adds the much quoted words, 'I have built a bridge which will remain for ever.' The bottom line of the inscription, which may be apocryphal, gives his name more fully as C. Julius Lacer.

There is also an inscription on the southern face of the triumphal arch. This also dedicates the bridge to Trajan, and refers to the fifth year of his reign. It is on this basis that the bridge is dated to AD 104.

The bridge has withstood many floods – Casado, for example (n.d.: 142), has photographs of a flood that reached well above the springing level of the arches. It has, however, suffered damage during wars, and was restored about 1543 by Carlos I, and again by Carlos III in 1778 following damage caused during the war of succession, in 1707. Span 2, which was damaged at this time, was completely destroyed in 1809 during the Napoleonic wars. It was rebuilt between 1858 and 1860, and the bridge was reopened to traffic on 4 February 1860. The name Alcántara was applied to the bridge at least as early as 1215; it simply means, in Arabic, 'the bridge'. The material is granite, taken from a quarry at Piedras Albas, 12 km north west of the bridge.

Beyond Piedras Albas, at **Segura**, over the River Elja which now forms the border between Portugal and Spain, is another fine Roman bridge (**SP22**; Gazzola, 1963b, no. 185; Casado, n.d.: 145–6, 161, 184–5) (Fig. 88). Casado describes it as the 'little brother' of Alcántara, and it was probably built at the same time. It appears to have carried a road from Norba (modern Cáceres) to Conimbriga, 5 km south of Coimbra.

The bridge has five spans – of 7, 7, 10, 7 and 7 m. The springing levels of the arches rise towards mid-span. When this is associated with the increased rise of the central span, it results in a marked increase in height at the crown of the central span relative to the side spans. The original road also followed this profile. The present road is horizontal and is retained by new stonework in the upper levels of the spandrel walls, as can be seen in Figure 88. There are cutwaters on the upstream faces only of the piers, which are 2.8 m wide. The overall length of the bridge is given by Casado as 78 m, with a width of 6.8 m.

From Alconétar, the northern road passed through Cáparra, Roman Capera, about 20 km north of Plasencia, west of the modern road. The **Puente de Cáparra** (**SP23**; Gazzola, 1963b, no. 189; Casado, n.d.: 186–7 crosses the River Ambroz here. It has two main spans, 8.7 and 9.2 m, separated by a pier 3.6 m wide, with a cutwater on the upstream face only. There is also a minor span in each of the approaches, one of which is at a considerable distance from the main spans. The overall length of the bridge is about 70 m. It is now in good condition, following some relatively minor work on the approaches.

Nearby, on the River Alagón is the **Puente de Guijo de la Granadilla** (**SP24**; Gazzola, 1963b, no. 288; Casado, n.d.: 90–1, 144). It has a single clear span of 19.1 m, resting directly on top of sheer walls of native rock 8 m high. The roadway rises quite steeply on either side to the crest, and the overall width is 5.6 m. Casado describes it as 'undoubtedly Roman'.

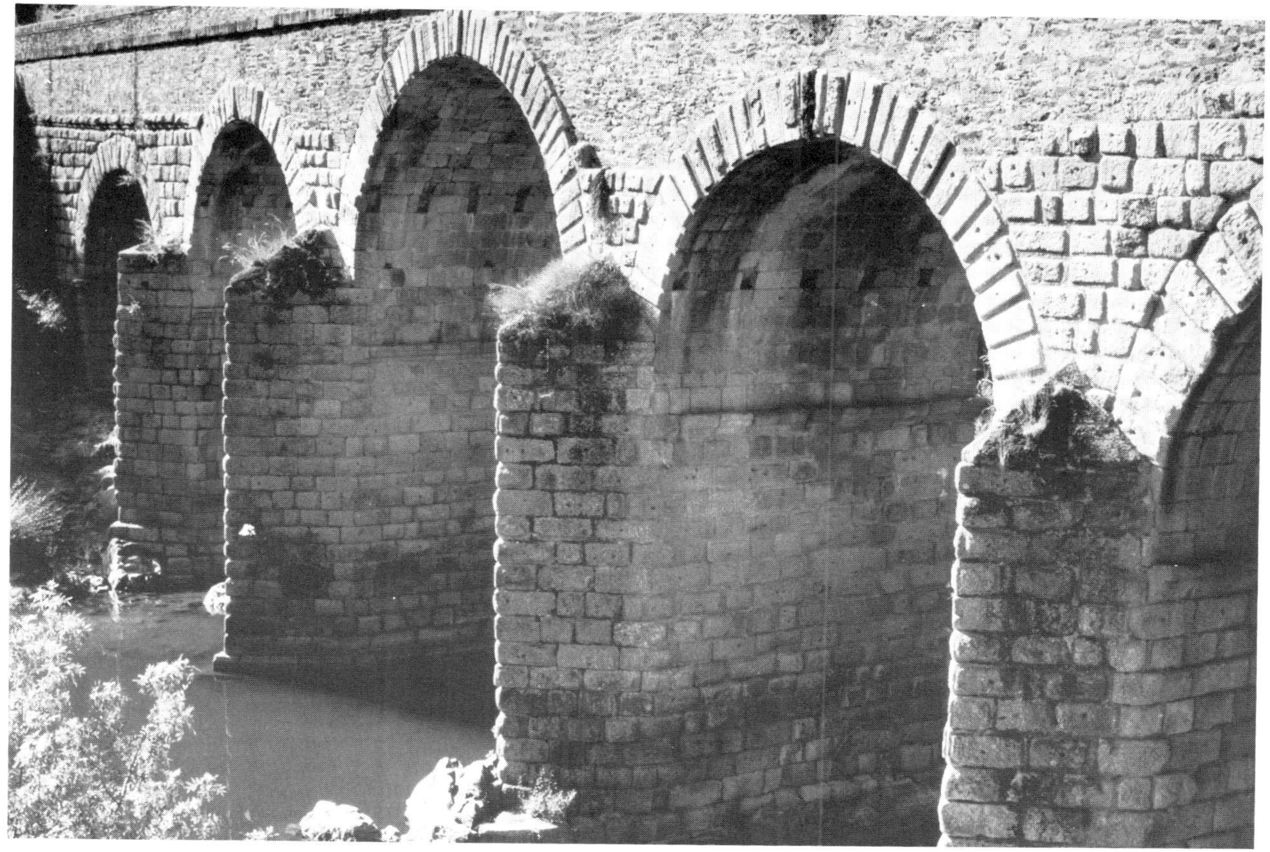

Fig. 88 The bridge at Segura, on the border between Spain and Portugal, west of Alcántara (SP22).

In the same area, also across the River Ambroz, are the remains of the **Puente de la Doncella** (**SP25**; Gazzola, 1963b, no. 274; Casado, n.d.: 163). It had a single span of 9.3 m; parts of the arch vault and spandrel walls still remain. Again, there is little doubt that it was Roman. Gazzola describes it as on the Via de la Plata, the later name given to the Salamanca to Mérida road, at the 93.4 km point – that is, presumably from Salamanca, about 39 km north of Plasencia.

There are two further bridges of interest in the region of La Vera, east of Plasencia on the northern side of the valley of the Tiétar River, between the river and the

Sierra de Gredos. In both cases the bridge-crossing is at a place where the stream is constrained to run between banks of native rock – a throat or *garganta* in Spanish. The first is the **Puente de Guallaminos** (**SP26**; Gazzola 1963b, no. 273; Casado, n.d.: 181) near Villanueva de la Vera, 75 km east of Plasencia. It has a main arch of 18.6 m and another smaller arch, possibly about 8 m span, in one of the approaches. The overall width is 6.2 m. The other is the **Puente de Alardos** (**SP27**; Gazzola, 1963b, no. 272; Casado, n.d.: 181). It is about 5 to 10 km further east, near Madrigal de la Vera, and has a single 22.8 m span. Gazzola believes that both are medieval on Roman foundations. The spans are, of course, quite large.

In the north, the main road entered Spain at its north-east corner, and travelled through Tarragona, Lérida and Zaragoza to Astorga. There were, in fact, two alternative routes, one through Soria (Roman Numantia) and the valley of the Duero, branching back north to Astorga and the west, and the other up the Ebro to Logroño and then across to Burgos and León. The older road from Astorga avoided the difficult country east of Orense by going north west to Lugo, and then west and south to Braga. Again, however, there were alternatives. One road followed the line of the River Sil to Orense and Valença, south of Vigo; another cut through a little further south directly to Braga; and a third went through Chaves, Roman Aquae Flaviae. These roads were also fed from the north. The Pyrenees were crossed from Gaul to Jaca and Pamplona; and there were roads to the coast near Santander, Oviedo, the River Navia and La Coruña.

The first bridge to be noted on the road crossing the Pyrenees from Gaul is in Andorra, across the Gran Valira River (**SP28**; Gazzola, 1963b, no. 228) near **Santa Coloma** about 3 km south of Andorra la Vella. The bridge is hump-backed, with two unequal spans. Gazzola says that the date is uncertain, and that the walls may be medieval or even modern restorations. However, Ganuza et al. (1981: 50) speak of a Roman bridge here, indicating at least a local tradition that part of the bridge is Roman.

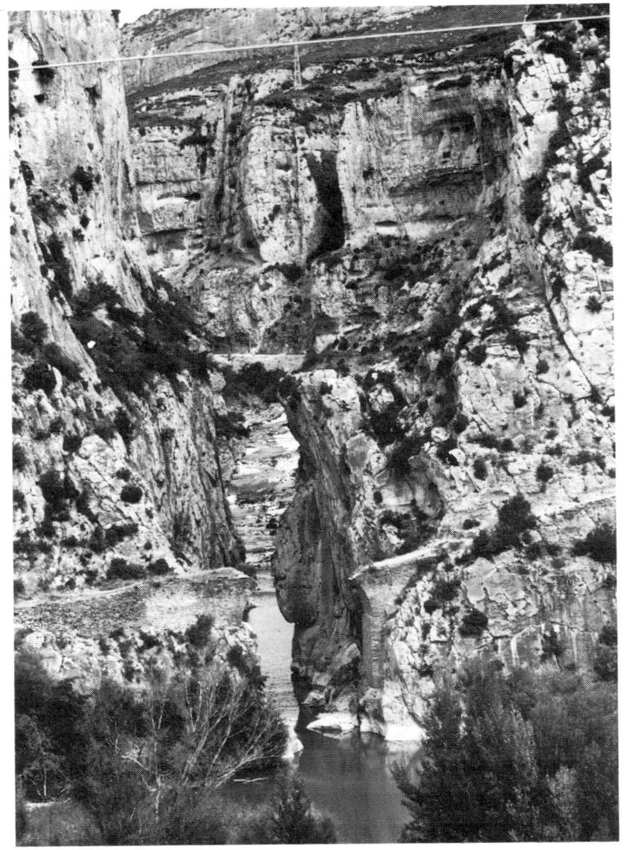

Fig. 89 The remains of the spectacular bridge at the base of the Hoz de Lumbier (SP29).

Chevallier (1976: 156) shows a road in this locality. Further west, the **Hoz de Lumbier** is a spectacular defile which carries the Irati River through the Sierra de Leyre between Jaca and Pamplona, a little west of Liédena. Here there was a single span bridge (**SP29**; Casado, n.d.: 308). The central portion of the span has now fallen, but the remains can be seen towards the bottom of Figure 89, carrying the old road around the cliff face and across the entrance to the gorge. Although Casado illustrates the bridge, he does not comment on its age. Von Hagen (1967b: 170) believes that it is Roman.

The old Roman town of Iruna is to be distinguished from the modern Pamplona-Irunea. It lies on the River Zadorra about 9 km west of Vitoria-Gasteiz. **The Puente de Tres Puentes de Iruna** (**SP30**; Gazzola, 1963b, no. 190; Casado, n.d.: 327) has thirteen spans, with the upstream cutwaters of the piers extended up and included in the balustrades (Fig. 90). It is believed to be Roman in origin and to include Roman materials. The same can be said of the **Puente de Villodas** (**SP31**; Gazzola, 1963b, no. 191; Casado, n.d.: 326). It has ten spans and is similar in many respects (Fig. 91). In this case some of the arches are pointed and are clearly medieval. It crosses the same stream a few kilometres downstream.

There is a very fine stone arch bridge at **Cangas de Onís** across the River Sella 66 km east of Oviedo (**SP32**; Gazzola, 1963b, no. 285; Casado, n.d.: 81–9, 337). Its three main spans are 7.7, 21.6 and 9.5 m, followed by three smaller spans of about 6.8, 4.3 and 3.6 m. The main piers 1 and 2 are 3.7 and 3.8 m wide; there are high-set floodways in piers 1 to 4; and the total length of the bridge is 92 m. The roadway rises steeply to the centre of the main span, where it is about 17 m above the stream. The picturesque appearance of this granite bridge is enhanced by the luxuriant creeper hanging from it, and the large metal crucifix suspended beneath the centre of the main span. It is clear that the bridge is not Roman in its present form, for all three main arches are pointed in profile. It is traditionally believed, however, to date originally from the second century AD; it was restored to its present shape at some time, destroyed in the war of 1936–9, and rebuilt in 1946.

Further west, the **Puente de Lugo** (**SP33**; Gazzola, 1963b, no. 251; Casado, n.d.: 130, 325) is clearly Roman, and presumably carried the main Roman road from Astorga. It crosses the River Jiloca with three main spans.

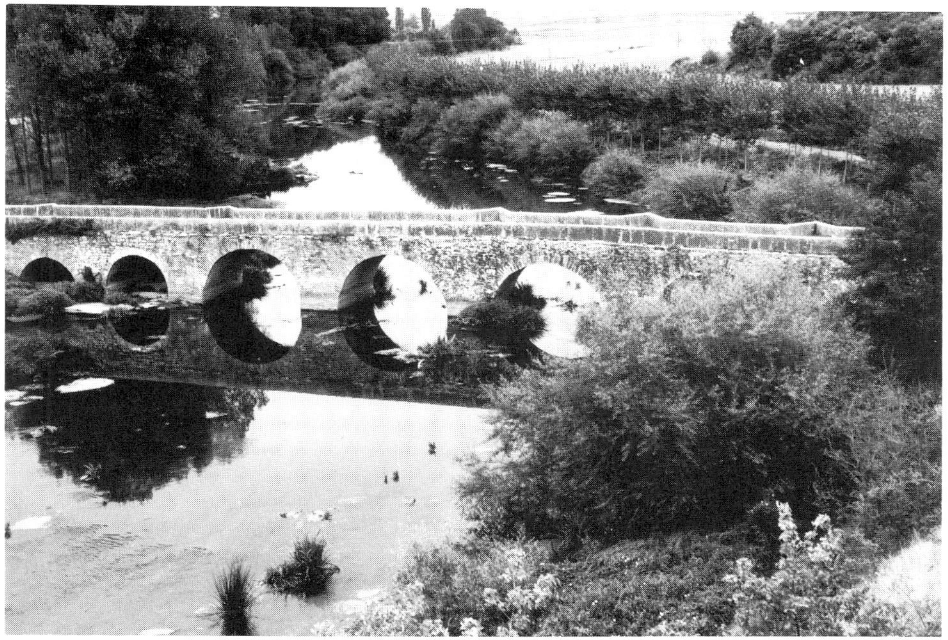

Fig. 90 The Puente de Tres Puentes at Roman Iruna (SP30).

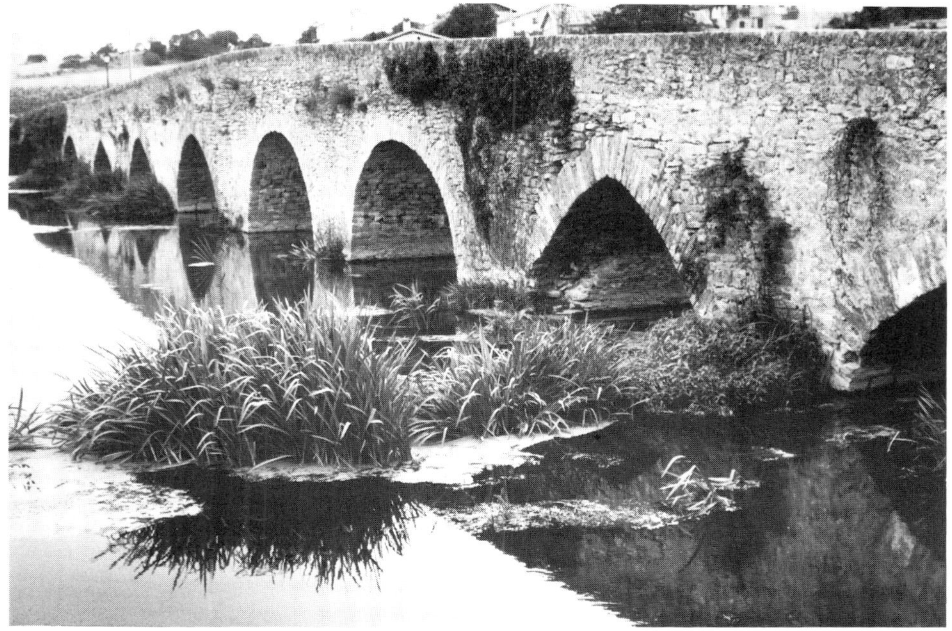

Fig. 91 The Puente de Villodas, a few kilometres downstream from Iruna (SP31).

The central span is 11.2 m, and the others each about 5.4 m; there are also arched floodways of about 1.7 m span in the two piers. There are now no balustrades, and the roadway climbs steeply over the bridge, just clearing the crowns of the arches. The bed of the stream has risen since Roman times, and has partly covered the bridge over its full length. The width is 3.6 m.

Gazzola also records the **Puente de Orbigo** on the Pilgrims' Way at Santiago de Compostela (**SP34**; Gazzola, 1963b, no. 289). It has more than five arches of graduated size; the present structure dates from the thirteenth century but has Roman foundations.

Instead of travelling from Astorga to Braga by Jugo, the Romans could also follow a route close to the Sil valley, crossing the Sil at Petín, about 60 km west of Ponferrada. Casado (n.d.: 357–8) gives the bridge the name **Puente de la Cigarrosa** (**SP35**), after the ancient capital of the Gigurro people in the Roman period. It has five unequal spans – 8, 10, 4, 19 and 10 m – all of which are pointed, except the large 19 m span (Fig. 92). Almost all of the bridge has been reconstructed, but, to quote Casado, there is work that is truly Roman in the lower part of one of the piers beside the main span, and there are also typically Roman details preserved in the reconstructed bridge. The level of the stream has been raised by the building of a dam downstream. When this is taken into account, the original bridge must have been of major proportions, in both length and height.

This can certainly be said of the bridge over the River Bibey (**SP36**, Casado, n.d.: 355–6) some 15 km to the south west. Having crossed the Sil, the road climbs over an intervening ridge and drops sharply to cross the **Bibey** about 5 km above its junction with the Sil. The bridge is shown in Figure 93. A nameplate at the eastern end of the southern balustrade gives the following information: spans 6.1, 18.5, 8.8 m; pier width 4.2 m; total length 75.2 m; clear height to the bed of the river, 23.3 m; width 7.1 m. Two inscribed columns have been mounted nearby (Fig. 94). One is a Roman milestone; the other dedicates the bridge to Trajan. A metal plate beneath the column records part of the inscription and indicates that it includes the words 'who reconstructed this bridge at his expense' (Section 3.2). The suggestion is, then, that the bridge predates Trajan. Casado, quoting Cea Bermúdez, says that it may have been built during the reign of Nerva and rebuilt by Trajan. It would date, then, from about AD 96–8. He also compares the style of the bridge with those

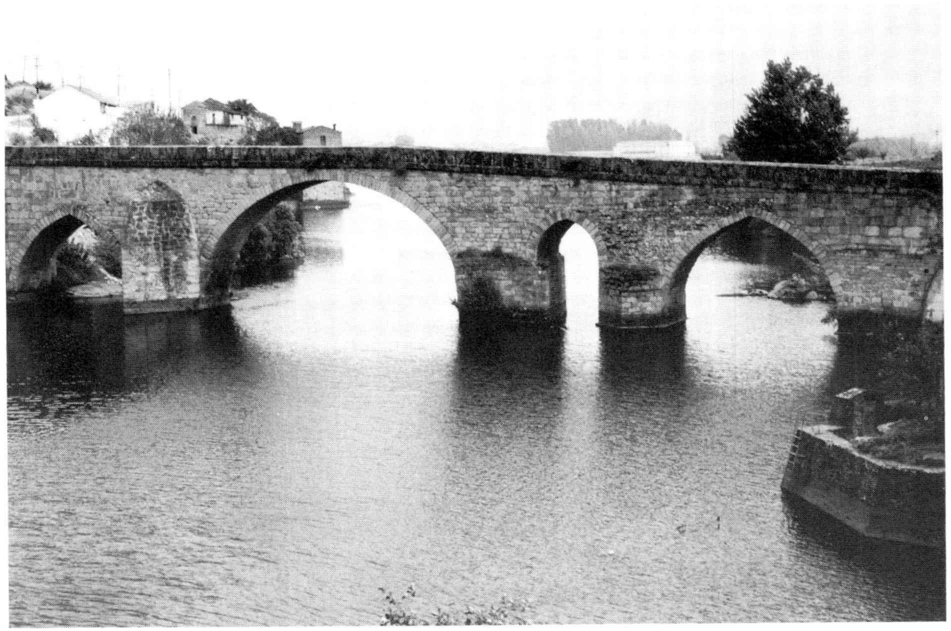

Fig. 92 The Puente de la Cigarrosa at Petín has been reconstructed, but is on Roman foundations (SP35).

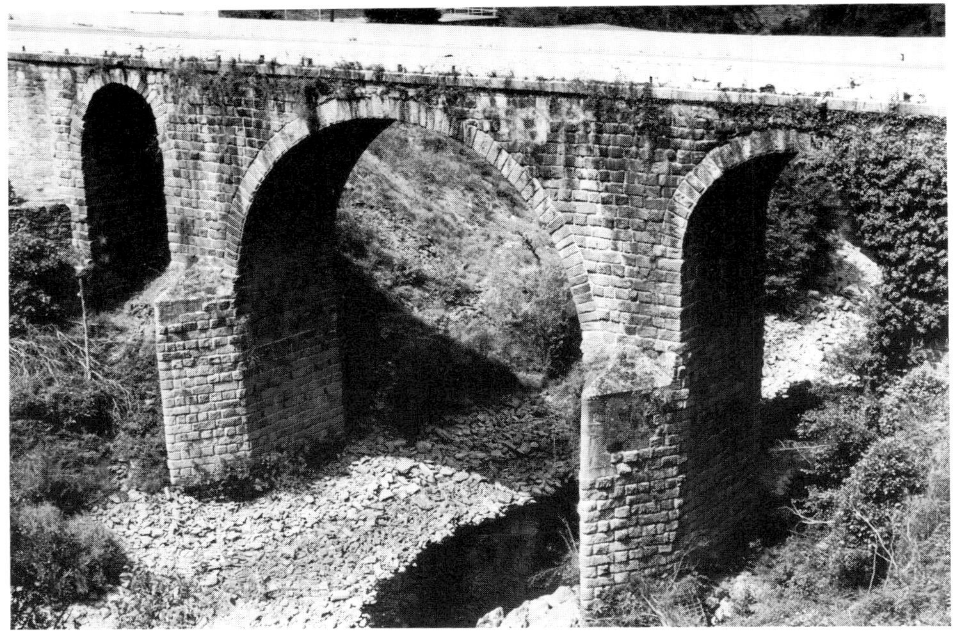

Fig. 93 For some reason, the fine Roman bridge at Bibey (SP36) is little known. It has a clear height to the river of about 23 m and a main span of 18.5 m (see also Fig. 137).

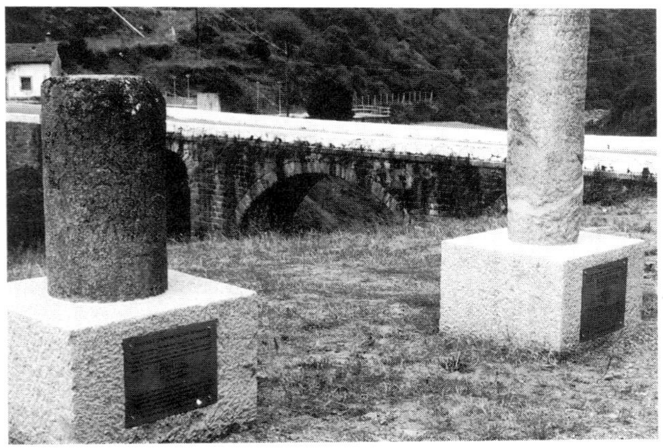

Fig. 94 Inscribed columns at Bibey (SP36).

at Alcántara and Segura, and believes that this alone would date the bridge no later than Trajan (AD 98–117). Like them, this bridge is built of granite.

It is interesting to note the difference between the dimensions of the side spans – 6.1 and 8.8 m. The western abutment (furthest from the camera in Fig. 93) shows clear signs of movement; it is believed that it has moved towards the stream causing slip between the voussoirs of that arch, and reducing the span (see also Fig. 137). It would be surprising, however, if this amounted to as much as 2.7 m, the difference between the span lengths. The piers have cutwaters on the upstream but not the downstream faces. The bridge as a whole is in good condition and is used by modern traffic.

Gazzola records another 'bridge over the River Bibey' (**SP37**; Gazzola, 1963b, no. 193). This may be the smaller bridge described by Casado (n.d.: 358) as the **Puente de Puebla de Trives** and located further west, near the town of that name, over a tributary of the Bibey, the Xares, close to the village of Mendoya, Roman Nemetobriga. Also built in granite, it has a single span of 7 m, and a width of 4.7 m.

This road proceeded south west to Bande, Roman Aquae Querquernae, 44 km south of Orense. The **Puente Linares** (**SP38**; Gazzola, 1963b, no. 162; Casado, n.d.: 362) crosses the River Limia between Bande and Porquera. It has four spans – two main spans, 7.5 and 7 m, and two much smaller side spans. Gazzola describes a two-span bridge here across the Limia, which he calls the Ponte Pedrina (1963b, no. 162). The two may not be the same, for there are other small bridges in this area believed to be Roman, as Casado has noted.

In Orense itself is the magnificent Puente Viejo (Casado, n.d.: 352–6; Brangwyn and Sparrow, 1915: 29). The present bridge is clearly medieval, with seven spans rising to a maximum variously recorded as 37.6 m and more. It has been suggested that there was a Roman bridge here, possibly from the time of Trajan.

The third road from Astorga to Braga passed through Chaves in Portugal, where there is a notable Roman bridge crossing the River Tamega (**SP39**; Gazzola, 1963b, no. 157). Figure 95 shows the downstream face, looking north towards the main part of the town. At the right is one of two widely spaced spans. There are then four spans to a protruding pier on which are mounted two inscribed Roman columns, one on each side of the roadway. A further three take the bridge to the present far bank, but there is then a succession of spans carrying the approach. The number of these is not clear, but possibly there may be four. There would then be a total of thirteen. However, there may be further buried spans both to the south and the north. Wiseman (1956: 143) mentions sixteen. Gazzola has two drawings, one of which shows nineteen spans and the other, incomplete, shows twelve. His text quotes eighteen. He also gives the width as 6.8 m, the overall length as 116.6 m and the clear span as 8.9 m. These data are inconsistent and unreliable. The actual pier width varies but an average would hardly be less than 2 m; a length of 116.6 m could then only accommodate ten or eleven. The actual total number is certainly not less than twelve.

The downstream face of the bridge differs markedly from the upstream. Apart from the pier carrying the Roman columns, the downstream face is flush. On the upstream side there are protruding rounded cutwaters, and these alternate between small cylindrical cutwaters of regular masonry and massive rounded cutwaters of rough construction, and even these vary in size. An

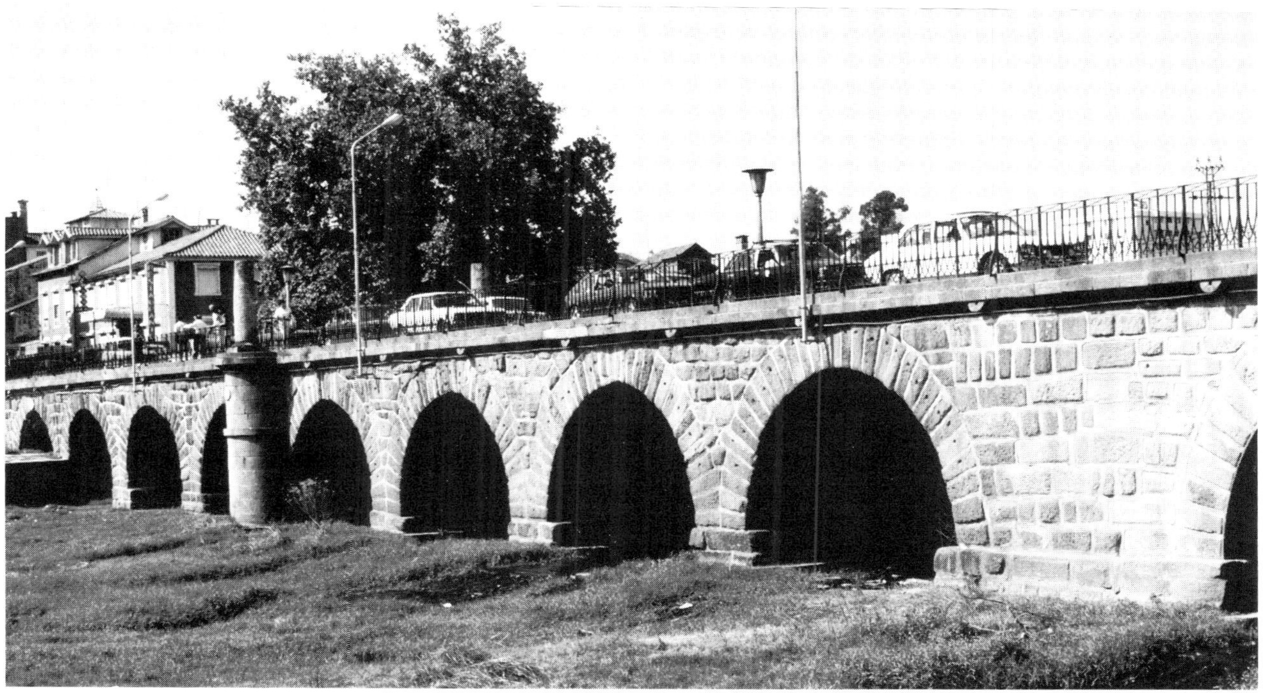

Fig. 95 Roman bridge at Chaves in Portugal. Note the inscribed columns on the bridge in the middle distance (SP39).

inscription dedicates the bridge to the three Flavian emperors, Vespasian, Titus and Domitian (Gazzola, 1963b). The reign of Domitian was from AD 81 to 96, so it appears that the bridge dates from about that time.

The final Spanish bridge to be described is the large and important bridge at Salamanca (**SP40**; Gazzola, 1963b, no. 182; Casado, n.d.: 140, 145–6, 175–80) on the Astorga to Mérida road. Called the **Puente Romano**, it crosses the River Tormes at the edge of the old town, south west of the cathedral. It divides into two parts, the Roman bridge closest to the town, followed by a seventeenth-century extension, the two being separated by a prominent pier. The Roman bridge has fifteen spans (1–15 from the north end), which are 9.5 (approx.), 9.5, 9.5, 9.4, 9.5, 9.7, 8 x 9.5, and 9.4 m (approx.). There is a northern approach with stone retaining walls 18.5 m long. The pier widths are: 1 to 6, 2.5 m; 7 and 8, 2.8 m; 9 to 14, 2.6 m; pier 15, 6.0 m. The length between the faces of the northern abutment and the end of the Roman work at pier 15 is 178.7 m, of which the openings amount to 142.5 m or 0.80 of the total. The overall length of the Roman bridge to the centreline of pier 15 is a little under 201 m.

The eleven spans of the later bridge (16–26) are 10.0 (approx.), 10.0, 9.5, 2 × 9.8, 9.9, 9.8, 9.5, 9.0, 8.4 and 5.5 m (approx.). The pier widths range from 3.2 to 3.9 m, averaging 3.4 m. The overall length between pier 15 and the abutments totals 135.2 m, of which 101.2 m, or 0.75, are openings. Including the southern approach, the total

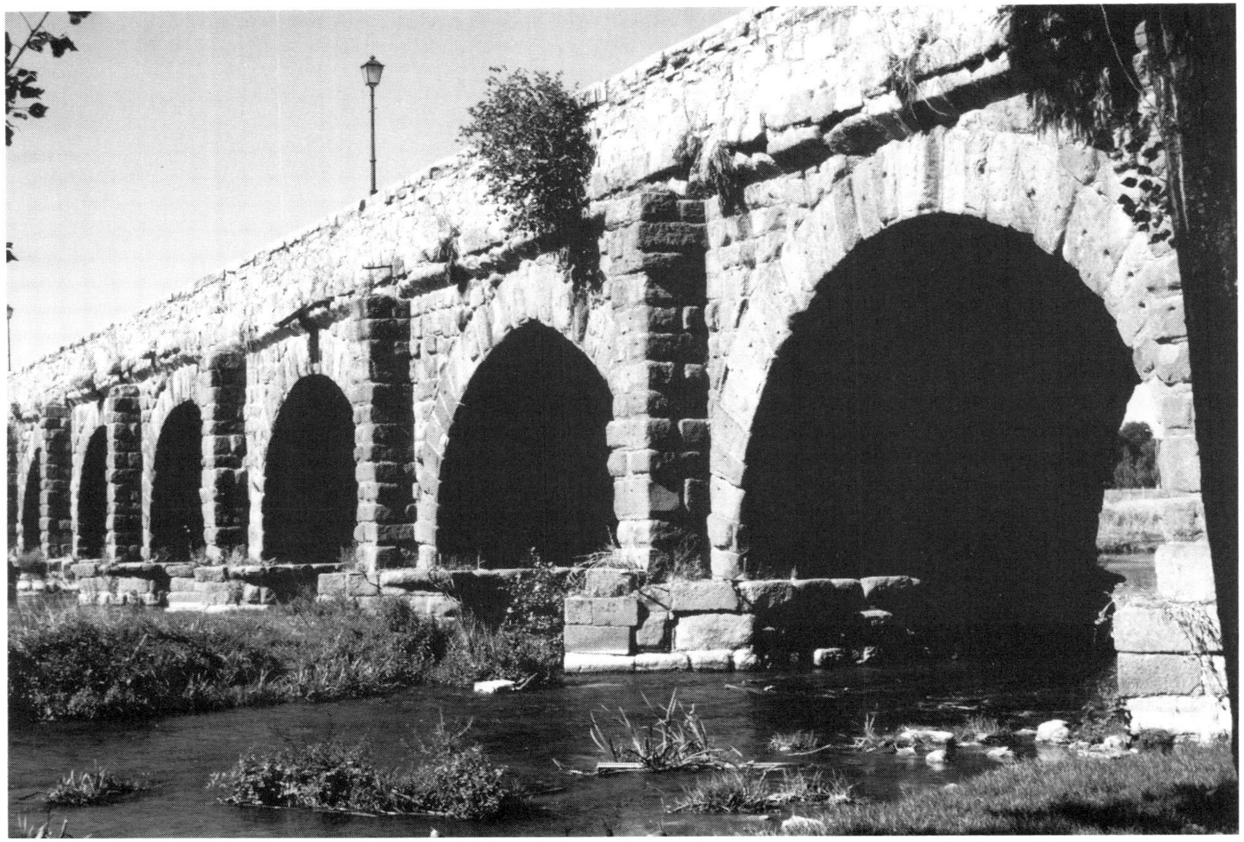

Fig. 96 Roman bridge at Salamanca (SP40).

length from the centreline of pier 15 is 155 m. The bridge as a whole therefore has twenty-six spans and is 356 m long.

The Roman bridge is built of granite and has a roadway 6.0 m wide between 0.65 m thick balustrade walls. It overhangs slightly and the arch width is a little less, about 6.6 m. Figure 96 shows part of the upstream face. There are no flood arches, but a square buttress is provided at each pier, resting on a shallow triangular cutwater. The downstream face is flush. By contrast, the newer bridge has prominent high cutwaters on most of the piers. The old bridge is commonly attributed to Trajan. It has been restored on a number of occasions, notably in 1627 by Philip IV. A further restoration was carried out in 1677 by Carlos II who built the eleven later spans.

5.8 Africa

The Africa of this section is the northern part of modern Africa, extending from the Red Sea and the Nile valley in the east to the Atlantic in the west. Roman occupation of this area was piecemeal and took some time: Carthage and most of modern Tunisia, the original Africa Vetus, after 146 BC; Cyrenaica after 96 BC; Tripoli and Numidia, including north-eastern Algeria, after 46 BC; Egypt after 30 BC; and coastal Mauretania, to Morocco, after AD 42. It reached its greatest extent only under Septimius Severus (AD 193–211) with the southward extension of the borders of Tripolitania and Mauretania, but this situation was short-lived, for by AD 235 much of this new territory had been abandoned (Section 2.5; Sitwell, 1981; Raven, 1969).

Roads extended around the Mediterranean coast from Carthage to Alexandria and then to the Red Sea and up the Nile; they crisscrossed the more fertile areas of Tunisia and Algeria; and probably reached as far west as Rabat, Roman Sala, and Volubilis (west of Fes) (Maps 4, 5, 11). The number of Roman bridges is relatively small. For this reason, it is not necessary to trace the Roman roads in detail here. However, the order of treatment will be the same as in other sections: from Rome, or in this case Carthage (near Tunis), and then clockwise. It starts, then, along the Mediterranean coast to the east, and ends along the Mediterranean coast to the west.

The Gulf of Tunis has as its southern end a deep bay; ancient Carthage lay on the north-western side of this bay, on a harbour that has since silted up. Tunis lies at the western end of the Lake of Tunis, a landlocked saltwater lake which opens into the Gulf of Tunis south of Carthage. It is bounded in turn to its south west by a completely enclosed salt lake, the Sebkhet Sejoumi. There were three routes available for the traveller to the south and east: across the Goulet, the channel from the Gulf of Tunis into the Lake of Tunis; around the western side of the Lake of Tunis and across the narrow neck of land between that lake and the Sebkhet Sejoumi; or further west again around the Sebkhet Sejoumi.

The most natural land route was the second of these; then overland to the region of modern Hammamet; around the coast to Sousse, Roman Hadrumetum; either across country or around the coast again to Sfax; and on to Gabès, Tripoli and Cyrenaica. The first of two Tunisian bridges on this route crosses the Oued Cherchar about 4 km north east of Bou Ficha, which in turn is 21 km north of Enfidaville. The ruins of the **Pont d'Oued Cherchar** (**A1**; Gazzola, 1963b, no. 197), referred to also as the Roman bridge at El Knatir (Babelon et al., 1893: 36.174), lie immediately west of the modern road.

This substantial bridge had originally at least fourteen spans, possibly sixteen, or even as many as nineteen, the number quoted by Gazzola. It is difficult to be sure without further measurement and study; the spans are graduated in size and some of the small end spans may be buried; some piers are also missing at the main crossing of the stream. Figure 97 shows one of the spans. The concrete core is clearly visible, but also some of the stone vault. Although by no means the largest, it is one of the larger spans, and forms part of the southern approach, which contains the bulk of the remains. Further south again the spandrel walls of the approach are in place and include an intact, small stone arch.

The second bridge is near Ksar Hellal (**A2**; Gazzola, 1963b, no. 207) on the coastal route across the Oued Bou-Zid about 30 km east of Sousse. There are only some remains of this limestone bridge, on the right bank of the stream.

The barren coastline of the Gulf of Syrte contains no Roman bridges, but there is one (**A3**) in Cyrenaica, at Tolemaide (Addirsiyah), ancient Ptolemais, across the Oued Ziuna east of Benghazi. Gazzola (1963b, no. 276) has a photograph and a drawing of a small two-span bridge here, with the stone vaults and part of the spandrel walls intact. His reference to Procopius (*Buildings* 6.6.9) is incorrect, although earlier (*Buildings* 6.2.9–11)

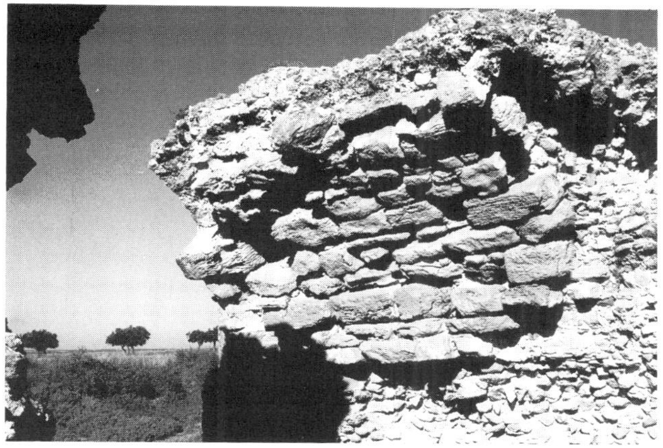

Fig. 97 Part of the ruins of the Pont d'Oued Cherchar, south of Hammamet in Tunisia (A1).

Procopius does refer to an aqueduct here. Further east, the only other remains of a North African bridge, possibly Roman, are at Aswan, on the left bank of the Nile (**A4**; Gazzola, 1963b, no. 236). The stonework, which could have formed the approach to a very large bridge, is part of an ancient construction known traditionally as the Baths of Cleopatra. The name may imply a construction date not later than the Roman conquest of Egypt in 30 BC.

From Tunis it was possible to reach Gabès, Roman Tacapae, and the Gulf of Syrte by an inland route, following the main road up the Mejerda River to Dougga, and then branching south to Sbeïtla. Between this route and the coastal road there are two bridges.

One of these is at Oudna 21 km south of Tunis. Raven (1969; end map) shows a local road from Carthage through Oudna to Thuburbo Maius, just north of el Fahs, meeting here a crossroad from the coast at Hammamet to Dougga. Oudna lies 5 km east of the magnificent Zaghouan to Carthage aqueduct, south east of Oudna station. The **Pont d'Oudna** (**A5**; Gazzola, 1963b, no. 215) crosses the Oued Oudna, a tributary of the Oued Miliana, about 200 m north to north east of the ancient town (Babelon et al., 1893: 28.48). It had three spans, of which the stone vaults now remain, partly buried by siltation of the stream.

The location of the second gives rise to a minor problem, for it is away from any of the normally recognised roads (Raven, 1969; BNI, 1945a; Sitwell, 1981). The bridge over the Oued Nebana (**A6**; Gazzola, 1963b, no. 212) lies about 36 km north of the holy Arab city of Kairouan, on the modern road from El Fahs. The presence of the bridge suggests that there was indeed a Roman road here, possibly from Thuburbo Maius near El Fahs to El Jem, Roman Thysdrus, and possibly joining a road from El Jem to Sbeïtla, Roman Sufetula, not far south of Kairouan (BNI, 1945a, fig. 28).

Gazzola records the original bridge as having eight spans; Babelon et al. (1893: 48.73) found seven spans standing in 1893; there are now either fourteen or fifteen, with the Roman bridge at the southern end, starting with span 1, and a later extension to the north. Figure 98 is taken from the south east and shows the downstream side of spans 3–8. Spans 3 and 4 are constructed of a lighter stone, and it is believed that they are later. It is easier to believe that these spans were breached at some time and later replaced, rather than the converse – that some circumstance led to the removal of spans 1–2 and 5–8, leaving 3 and 4 untouched. The corresponding view up the western face shows a change of construction after span 6 or 7, suggesting, with Babelon, that possibly seven of the original spans are incorporated as spans 1–7 of the present bridge. A notable feature is the provision of prominent cutwaters on both the upstream and downstream faces of the piers. A study of the stonework in the southern piers and spandrel walls suggests again that the darker, brownish stone is original.

Babelon gives some detailed dimensions of the bridge. The normal spacing between the centrelines of the piers is 5 m. The piers are 2.4 m wide, giving a clear span of 2.6 m. However, span 1 is smaller than the other spans; its crown is set lower and the width of pier 1 is greater than

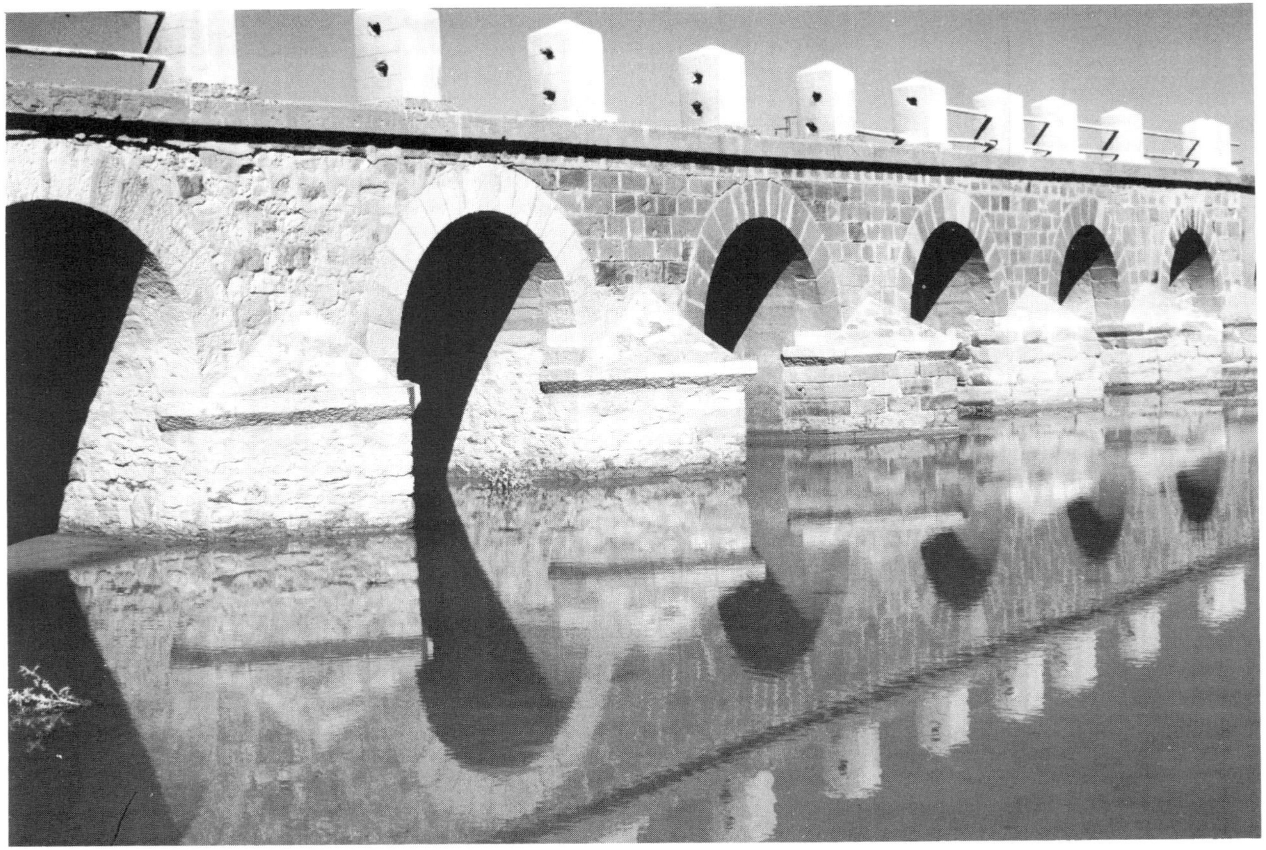

Fig. 98 Spans 3 to 8 of the bridge over the Oued Nebana in Tunisia, north of Kairouan (A6).

that of the central piers. It is likely, then, that the bridge did indeed have eight spans, as suggested by Gazzola, with a smaller span at each end. The overall width at deck level was 7.6 m; the length of the piers including cutwaters was 10.8 m; and the height to the deck, 3.6 m. BNI (1945a) has a photograph of the bridge (fig. 209) taken prior to 1945 with the caption, 'Restored Roman bridge over the Oued Nebana on track north from Kairouan', The 'track' is now a modern road. The bridge had already been extended to its present length.

The valleys of the Mejerda River and its tributaries provided convenient access from Carthage to the interior and the west. The importance of these routes was increased by the stationing of the Third Augustan Legion at Haïdra, Roman Ammaedara, about AD 6; it was moved some 75 years later to Tébessa (Algeria), 40 km to the south west (Sitwell, 1981; Raven, 1969), and then later, possibly in AD 128, to Lambèse (Section 2.5). Haïdra and Tébessa were the strategic centres of roads radiating out inland and to the coast, one of these being that built in AD 14 from Haïdra to the coast road at Gabès. There were two routes up the Mejerda. That on the right bank reached the river at Grich el Oued near Mejez el Bab and went then to Téboursouk, Dougga, Musti and Haïdra. There was a branch from Musti south to Sbeïtla and the Gulf.

The first bridge encountered on this latter route is the **Pont d'Adpertusa** (**A7**; Gazzola, 1963b, no. 209) across the Oued Chafrou, on the northern shores of the Sebkhet Sejoumi, south of Bardo, 5 km west of Tunis (Babelon et al., 1893: 20.43). It had a single span of 6.3 m, and a width of 6 m. Gazzola's illustration shows the vault of the arch standing but holed in two places, and with parts of the spandrel walls intact.

Grich el Oued, Roman Thisiduo, lay 6 km north east of Mejez el Bab on the south bank of the Mejerda River. There was a large multiple span bridge here (**A8**), about 500 m north of the old town. Gazzola (1963b, no. 195) includes this as one of three bridges across the Mejerda, the others crossing at Chemtou and Jedeida. Babelon et al. (1893: 27.28) speak of it as 'the Roman bridge on the bank of the Oued el Hamar. Following movement of the bed of the stream (*le torrent*), the bridge is now isolated on the right bank.' It is possible to read this as saying that the bridge crossed, not the Mejerda, but a tributary close to the main stream. It is not clear which is correct. Gazzola shows a photograph of the bridge (unfortunately somewhat indistinct), and there is every reason to believe that it is still there.

There were two adjacent parallel routes to Dougga: one through Mejez el Bab; the other a little further south. The **Pont d'Ain Younes** (**A9**; Gazzola, 1963b, no. 206; Babelon et al., 1893: 33.221) crossed the Oued el Kebir about 9 km east south east of Téboursouk. Babelon described it as being well preserved (in 1893). It had one span. A little further west, the **Pont d'Oued el Ahmar** (**A10**; Gazzola, 1963b, no. 205; Babelon et al., 1893: 33.213) was in ruins in Babelon's time. It was about 6.5 km east of Dougga. Dougga itself lies on the side of a ridge overlooking the valley of the Oued Kalled, a tributary of the Mejerda. Looking south from Dougga, part of ancient Agbia is visible, a few kilometres away, down in the valley. Agbia, or Ain Hedja, which is the modern name of the neighbouring community, has the remains of another bridge, the **Pont d'Ain Hedja** (**A11**; Gazzola, 1963b, no. 203; Babelon et al., 1893: 33.201).

A little further on is Musti, with its fine triumphal arch standing alone in the fields. The Roman road divided here, with the left branch passing to the west of Maktar down to Sbeïtla, Gafsa and Gabès. About 15 km east of Maktar was the **Pont sur l'Oued Djilf** (**A12**; Gazzola, 1963b, no. 199; Babelon et al., 1893: 30.250), north from el Kessera. It was originally high and long, with a total length estimated by Gazzola as between 100 and 120 m. Six arches now remain, going from one bank into the steep valley, up to 30 m high, 6 m wide, 48.9 m long, and with clear spans of 5.4 m.

Maktar itself lies among the headwaters of the Siliana River. Gazzola (1963b, no. 208) shows one of a series of small bridges or culverts (**A13**) in the valley of this stream.

Sbeïtla is another town with a Roman triumphal arch. It also has a fine bridge-aqueduct (**A14**; Gazzola, 1963b, no. 237; Babelon et al., 1893: 48.18). Babelon speaks of nine hydraulic structures built to supply water to the town, including a tunnel, aqueduct, bridge-aqueduct and canals. The implication is that the 'bridge' was in fact an elevated aqueduct, crossing the Oued Sbeïtla, about 300 m NNE of the town. However the structure is most unusual. Figure 99 is based on the excellent photograph facing p. 139 of BNI (1945a). The most notable features are the heavy piers with domed ends, unlike any other comparable Roman structure, but possibly later additions. The illustration shows four substantial arch spans, much wider than would be necessary to carry an aqueduct alone. It is possible then that the structure was dual-purpose, with a water conduit to one side of the road. Gazzola dates the structure from the third or fourth century AD, with some stonework from the second century AD.

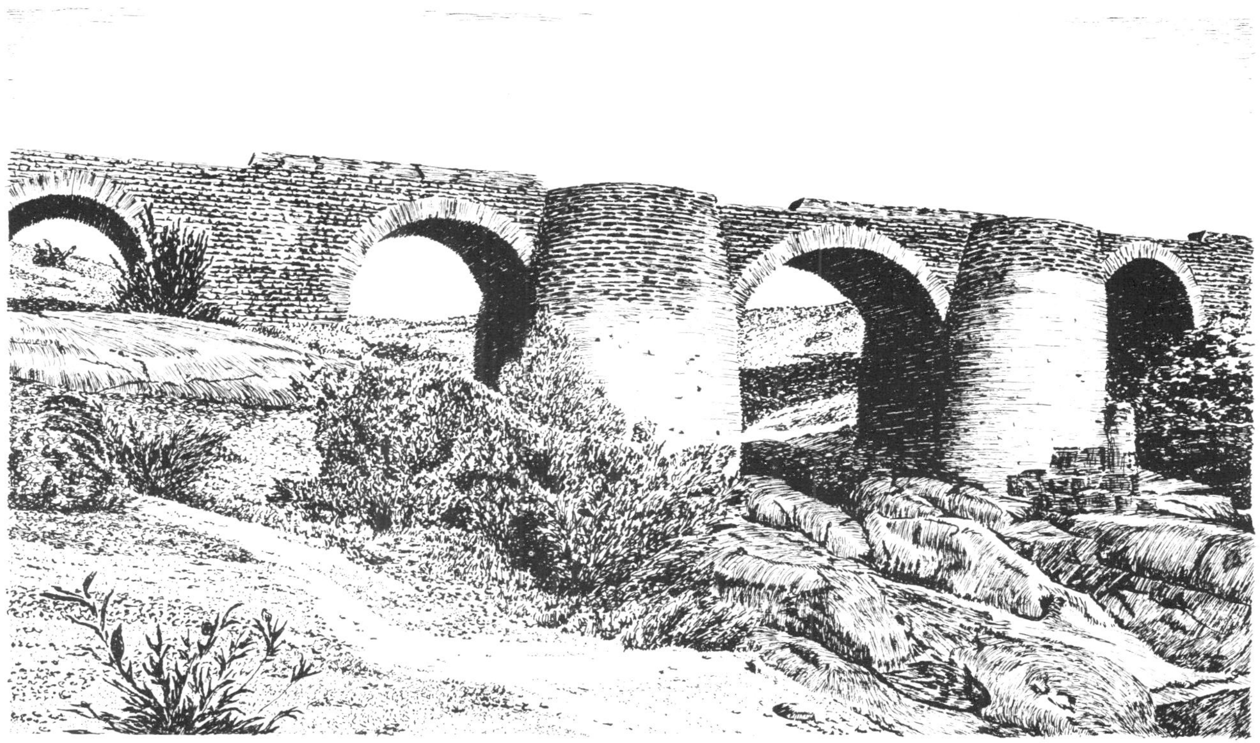

Fig. 99 Bridge-aqueduct at Sbeïtla in Tunisia (A14, based on BNI, 1945a, fig. 94).

The right branch from Musti led to the main road to Haïdra and Tébessa. Two and a half kilometres north east of Tébessa, in modern Algeria, is the **Pont de Tébessa** (**A15**; Gazzola, 1963b, no. 202), over a tributary of the Oued Meskiana. It consists of four arches, each with a clear span of about 3.1 m and a width of 5.1 m. The total length is 17.7 m, and the bridge is built of granite. It is believed to date from the time of Hadrian (AD 117–38). In the same area is the **Pont de Gastal** (**A16**; Gazzola, 1963b, no. 213), said to be on a road from Haïdra to Morsott (34 km north of Tébessa). It had a single span of 7.5 m and a width of 6 m. The vault and much of the spandrel walls appear to be intact.

Another road left the Tébessa road shortly after Musti, keeping to the right and running west through Le Kef and Constantine, Roman Cirta, as the main inland road to Mauretania. Having left the Mejerda before Téboursouk, it rejoined it at Souk-Ahras, Roman Thagaste, and followed it to Khémissa, 35 km to the south west. Here is another 'bridge-aqueduct' (**A17**; Gazzola, 1963b, no. 211), 2 km east of the town, Roman Thubursicu Numidarum. It has three arches, the central one having a span of

5.35 m. As at Sbeïtla, the width is quite large – in this case 3.1 m – suggesting that the structure carried a road. Gazzola thinks it was probably built in the time of Hadrian.

Constantine is a remarkable town, built on the northern edge of a high plateau, 585–665 m above sea level (BNI, 1944a: 70). To the east, it is bounded by the gorge of the Rhumel, which falls to the coastal plain by a series of waterfalls, with its bed 110–290 m below the town. To the west is another gorge of similar depth, leaving the town connected to the plateau only by a narrow isthmus to the south. The **Pont de Treviri (A18**; Gazzola, 1963b, no. 235) crossed the Oued Rhumel at a site believed to be to the south of the town, before the stream began its descent to the plain. The main crossing had three spans, a central span of 5.3 m at a height of 5.6 m, and two side spans each about 3 to 3.8 m, at a height of 4.6 to 4.8 m. There were another two minor arches, at a distance of 23.4 m from the nearest major opening, with spans of 2.6 and 1.8 m. The bridge as a whole was 60 m long and 7.5 m wide, with the road rising from low banks to a maximum height of 6.5 m. Believed to have been built in AD 335, it was still there in the twelfth century, but in ruins by the fourteenth.

A Roman road went south from Constantine to ancient Timgad. From there one could turn west to the vicinity of modern Batna, and then south west into the interior. From Batna, the road climbs the Massif de l'Aurès, and then, as it descends the other side, 65 km from Batna it enters the el Kantara gorge, 300 m long, with rocks 120 m high on either side (BNI, 1944a: 317, 406). Here is a notable Roman bridge, the **Pont d'el Kantara (A19**; Gazzola, 1963b, no. 253), across a stream also called the Oued Kantara. In excellent condition, the bridge is still in use and carries a branch road across the stream to join the main road which is along the far bank, both upstream and downstream of the bridge. This road and the approach from the other side are protected by massive stone retaining walls which flank the single 10 m span of the high arch. The bridge was built in the time of Caracalla (AD 211–17), and restored in 1862. The site is both spectacular and strategic, just where the road drops from the hills to enter the open wastes of the Sahara.

The next major Roman road from Carthage crossed the Mejerda River at Jedeida and continued up its left bank to Bulla Regia, north of Jendouba, and Chemtou, and then struck across to meet the coast at ancient Hippo Regius, later called Bône or Hippone, near Annaba. There is at present a bridge across the Mejerda in Jedeida at the eastern end of the town. Gazzola calls this bridge the **Pont de Jedeida (A20**; Gazzola, 1963b, no. 196), and there is an excellent photograph of the bridge facing p. 27 of BNI (1944a). Figure 100 shows one of the spans, taken from the downstream side, with the characteristic heavy rounded piers. There were originally nine such spans but, of these, at least two have been destroyed and replaced by a metal truss span. Gazzola describes the bridge as 'attributed to the time of Trajan' (AD 98–117), but adds that there is some valid controversy concerning its authenticity and that it may be medieval. BNI (1945a: 371) speaks of it as a restored Roman structure. It is not listed in Babelon.

There was a similar bridge (**A21**) 10 km further upstream, at Tébourba (BNI, 1945a facing p. 27) but this has evidently been rebuilt. The present bridge has a concrete deck on a series of at least sixteen small span modern brick arches, but these rest on heavy old stone piers of uncertain age. A further 35 km upstream is the bridge at Mejez el Bab (**A22**) (Fig. 101), a heavily restored bridge with seven major spans, and flood arches in the piers. BNI (1945a: 220) describes this bridge as built of Roman materials in 1677. There was a Roman bridge here – Babelon et al. (1893: 27.19) speak of a triumphal arch which formed the head of a bridge until it (the arch) was destroyed some forty years ago. Mejez el Bab means 'ford by the gate' and lies on the site of Roman Membressa. It seems likely that the ruins of this earlier bridge still remain.

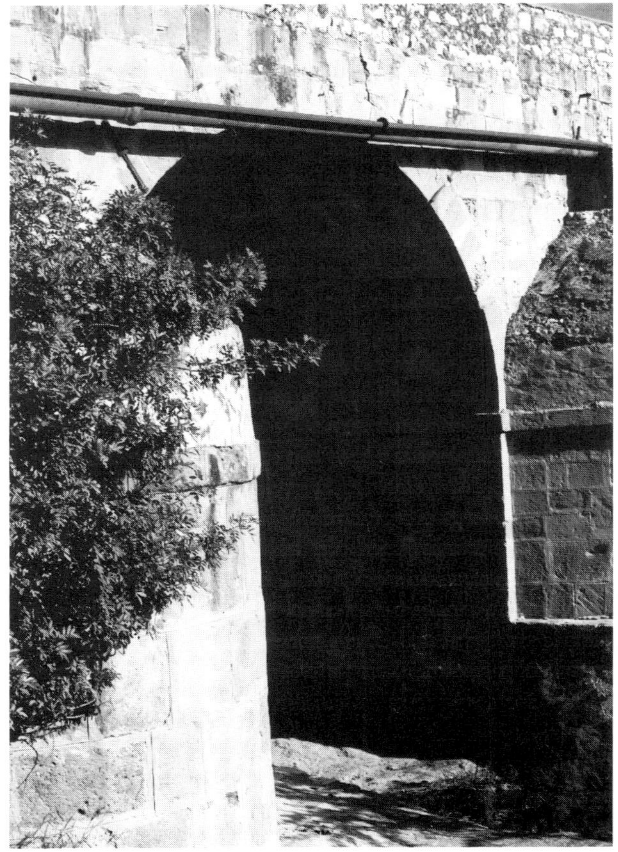

Fig. 100 Bridge at Jedeida (A20).

The Mejerda River is now dammed about 17 km west of Mejez el Bab. The village of Oued Zarga is 10 km north of the dam wall. There was a Roman bridge here (**A23**; Gazzola, 1963b, no. 214), with two spans of limestone. It is not clear whether it still exists. Babelon et al. (1893: 26.14) state that it was destroyed because of the construction of a railway bridge, and, in any case, the site may have been flooded by the dam.

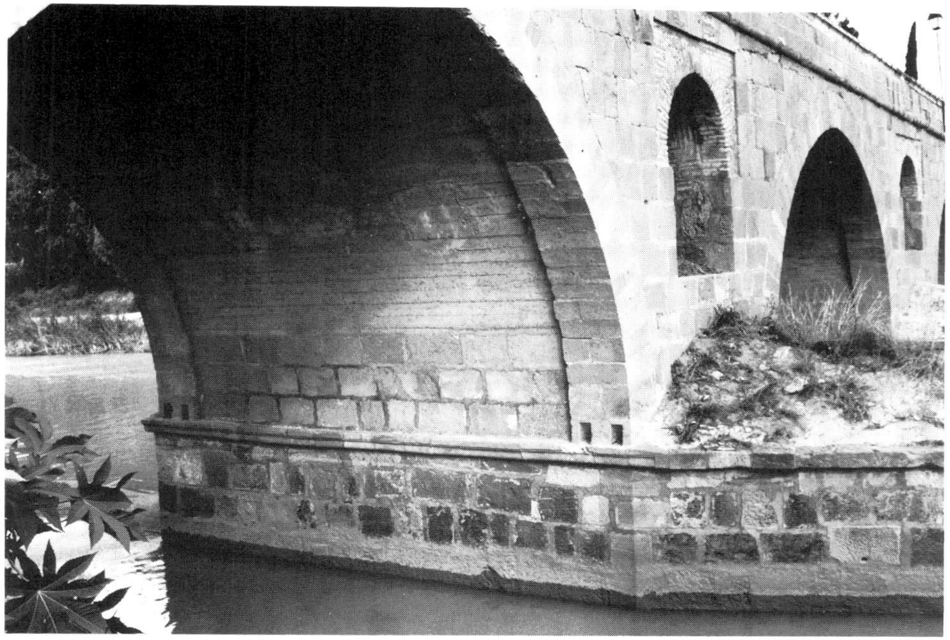

Fig. 101 Bridge at Mejez el Bab, heavily restored, possibly rebuilt with Roman materials on the site of an earlier Roman bridge (A22).

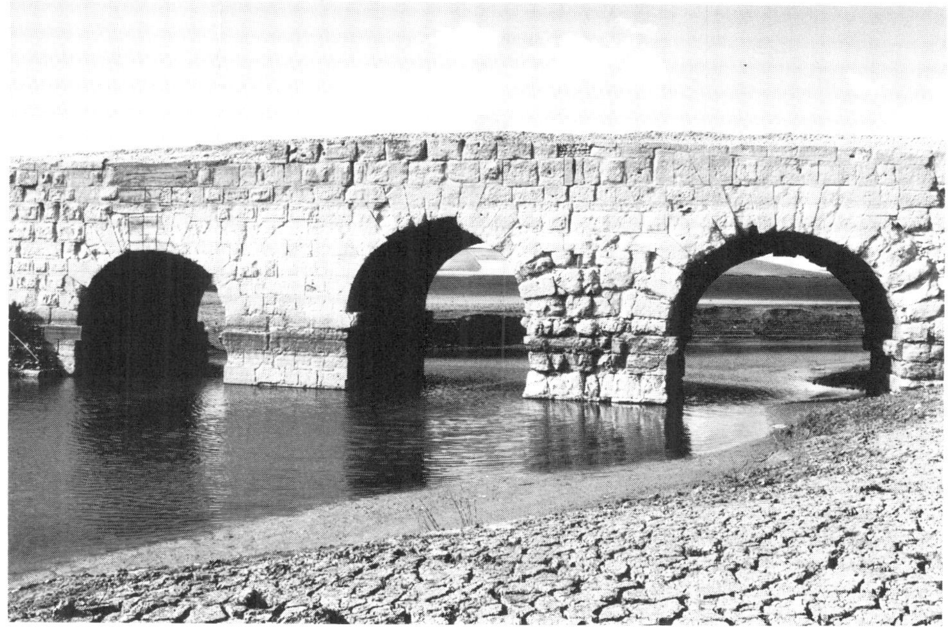

Fig. 102 The Pont de Trajan, a fine Roman bridge south of Béja in Tunisia (A24). It crosses the Oued Béja, a tributary of the Mejerda River which has been dammed downstream, so ponding the waters beneath the bridge.

Béja, Roman Vaga, lies above the western bank of the Oued Béja, on a branch road north of the main Roman road from Carthage to Bulla Regia. The **Pont de Trajan** (**A24**; Gazzola, 1963b, no. 151) crosses the Oued Béja 10 km south of Béja, just north of its junction with the Mejerda. This fine bridge is in its original condition (Fig. 102). Raven (1969: xi) describes it in these terms: 'A fine Roman bridge near Béja, which once carried the traffic from Carthage to Bulla Regia and the west, now carries men and donkeys from one ploughed field to another; of the highway which it served there is no sign.' She is not entirely correct, for it carries the occasional motor vehicle on a rough track, but it is certainly true that it lies by itself in the fields well away from signs of human habitation, excepting for the Béja to Bou Salem railway, scarcely visible to the west. It is reached by leaving the bypass road due south of Béja, travelling about 6 km to a junction and then west and south along a track that would be impassable in the wet. It has three spans, 5.8, 5.3 and 6 m; is 7.3 m wide, and 70 m long; and crosses the line of the Oued Béja after it reaches the lake. Despite its common name, both Babelon and Gazzola date its construction about AD 29, in the reign of Tiberius. Although not large, it is a most impressive bridge.

West of the Pont de Trajan is Mastouka; 2 or 3 km further west was a bridge over the Mejerda itself carrying a branch from the main road to Bulla Regia. It is given a variety of names. Gazzola (**A25**; 1963b, no. 210) refers to it as the **Ponte di Civitas Avensis**, near Souk el Khemis. Babelon et al. (1893: 25.107) identify it as near Souk el Khemis, which appears to be an older name for Bou Salem; the bridge lies 11 km east, with two spans.

Further west, Chemtou (**A26**) lies on the Mejerda, west of Bulla Regia, some 20 km west of Jendouba (Souk el Arba). Gazzola (1963b, no. 194) has an excellent drawing of the bridge as it was originally built. It was of granite, 47.6 m long, with five spans, and substantial river training walls of stone at each end. Now it lies in ruins – to quote Raven again (1969: xi), 'The great bridge over the Mejerda at Chemtou, where the much-prized henna red

Numidian marble used to be hewn from the quarries of ancient Simitthu, lies in ruins and has never been replaced; one must cross the river by a water-splash.' Babelon et al. (1893: 31.70) and Gazzola date the bridge from the time of Trajan (AD 98–117), but an earlier bridge may have gone back to Augustus (31 BC–AD 14).

From Chemtou, the road left the river and turned towards Annaba. Roman Thuburnica was 11 km to the north west and here a 10 m span, single-arch bridge crossed the Oued el Enja, a tributary of the Oued Meskiana (**A27**; Gazzola, 1963b, no. 200; Babelon et al., 1893: 31.7).

Annaba, in Algeria, is the present name for Bône. This in turn was derived from Hippone, or Hippo Regius, famous as the see of Augustine (AD 354–430), bishop of Hippo. To digress a little, his life tells something of the Africa of his times. Brought up at Thagaste (Map 4), he studied rhetoric at Carthage, and as a young man of 29 travelled to Rome to teach. He became dissatisfied with the quality of his life and in April AD 387 was converted (*Confessions* 8.12). He returned to Thagaste and in AD 395 was appointed bishop of Hippo. Writing from there, he was the most respected teacher of his day. He witnessed the fall of Rome to Alaric in AD 409 or 410 and this inspired his greatest work, *The City of God*, which compared the conduct in the city of Rome in those days (2.4 and 2.6) with that which it ought to have been (10.6). The Vandal leader Gaiseric went on in AD 428 to invade Africa, and Augustine died in AD 430 in Hippo, while it was under siege. The whole of Africa fell with Carthage in AD 439 (Jackson, 1914: 490–515; Bruce, 1958: 333–9; Ayerst and Fisher, 1971, vol. 1: 276–91).

Ancient Hippone was, therefore, a famous city in Roman times. It lies about 1 km south of the present town centre between the Oued Budjema and the Oued Seybouse. The first of these streams is crossed by a long stone bridge (**A28**; Gazzola, 1963b, no. 201), 98 m long, with eleven arches. The bridge is difficult to date, because of many restorations, but is believed to have been originally Roman. If so, it would have been the longest Roman bridge in Africa.

Much further west, the Roman town of Oppidum Novum lay near modern Duperré, 36 km west of Khemis Miliana, about 150 km south west of Algiers. There was a major Roman bridge here across the Oued Chelif, but today only the remains of two granite piers are to be seen (**A29**; Gazzola, 1963b, no. 198).

5.9 The east

This section considers masonry bridges to the north and east of Aquileia at the head of the Adriatic, through Austria, Yugoslavia and Turkey, to Mesopotamia, Syria and beyond. The bridges are summarised in Appendix table A.8; Roman roads are shown on Maps 7 to 11.

Although the Alps were a major obstacle, they were crossed by the Romans at many points, from Milan and Lecco north to the Rhine valley at Chur; through the Brenner Pass from Verona to Innsbruck; from Aquileia north to Villach and the Danube near Vienna; and east across the Julian Alps to Ljubljana, Roman Emona, and from there either down the valley of the Save, or across to Ptuj, Roman Poetovio, on the Drave (Sections 2.2.4, 2.7).

Innsbruck takes its name from an early Roman bridge, built by Claudius about AD 47 (von Hagen, 1967b: 127), but no trace of this now remains. The pavement of the Via Tiberia, earlier called the Via Julia Augusta (von Hagen, 1967b: 121), is still visible near Villach. Further north, the modern road to the Hohentauern Pass leaves Judenberg to the north west. Here, at Oberzeiring, 20 km from Judenberg, is a small bridge across the Blahbach Torrent (**E1**; Gazzola, 1963b, no. 269), believed to be Roman in origin. In the same general area, the Murufer Torrent between St Dionysen and Oberdorf (**E2**; Gazzola, 1963b, no. 233) is crossed by a Roman bridge curved in plan, with a single 12.3 m span.

To the east of Aquileia were the four provinces of the upper and lower Danube, separated by the Iron Gates of Orsova. One of the greatest of all Roman bridges, the bridge of Apollodorus, was built here for Trajan about AD 104–5 to join the towns of Pontes and Drobeta (Gazzola, 1963b, no. 184). It had a timber arch superstructure on stone piers, and for this reason will be reserved for a later chapter (T13 in Chapter 6).

Instead of crossing the mountains from Aquileia, one could go south east along the Dalmatian coast. Here, at Salona near Split is a bridge called the **Cinque Ponti de Salona** (**E3**; Gazzola, 1963b, no. 217). It has two small spans; although the roadway has been removed, the remaining stonework is in excellent condition. Further south, Mostar, 50 km inland, has a famous stone bridge, but it is not Roman.

Durrës in Albania, Roman Dyrrhachium, could be reached either by this road, or by sea from Brindisi and the Via Appia. The road from here east through Macedonia to Istanbul, Roman Byzantium, was called the Via Egnatia, and was a route of major importance in Roman times. It was this road that Paul followed on his journeys to Philippi, Thessalonica and Beroea (Acts 16:11–15; 17; 20:1–6). Hammond (1974) studied the western portion, down the Shkumbi River from Lake Ochrid to the coast, and mentions the remains of eight or nine bridges that may be Roman. One of these (**E4**) crossed the river near Shenjan at a point near Topçias south west of Elbasan. The river here is quite wide; the bridge had sixteen piers and a total length of 300 m. Hammond believes that these piers 'were made in the Hellenistic period' (1974: 191). He notes (footnote, p. 191) that 'the superstructure on the piers may have been of wood, because there is no sign of any arch ... There were some later repairs, and these may be the reason why [others] think that the bridge was of the Roman period.' He suggests there was another major crossing near Hadji Bektari (**E5**) some 15 to 20 km to the north east. The ruins of a third have been seen near Orakë (**E6**), just west of the pass above the lake, at a point which corresponds in distance with the **Pons Servili** of the Peutinger Table (p. 185).

There are no other recorded bridges on this road. In Istanbul itself there are some old stone bridges. One at least is surprisingly Roman in appearance – Odabaşi bridge (Tunç, 1978: 150, no. 180) has three unequal spans of 2.2, 4.7 and 2.2 m and a total length of 39 m. Tunç places it in the Osmanli period, which is quite late (*c.* thirteenth century). Another is Çobanceşme bridge (Tunç, 1978: 59, no. 67) with spans 3.3, 3.4, 3.6, 4.3, 4.1 and 3.7 and a total length of 39.3 m. It dates from the sixth century AD.

The references to Tunç (1978) here are to his book *Tas Köprülerimiz*, a systematic account of the older bridges of Turkey, which have been unfortunately generally neglected in the history of bridges. Like Gazzola (1963b), the format of the book is that each bridge is given its own section. The bridges are numbered in the table on pp. 205–215 at the end of the book, and in the map following (p. 222). These numbers have been cited here, together with the page number (but it should be noted that Tunç has one or two inconsistencies in numbering).

Turkey has many fine old stone bridges. Some are very large. The longest is at Uzunköprü in European Turkey, its 174 spans covering a total of 1266 m (Osmanli period). Büyükçekmece bridge near Istanbul is made up of a series of four distinct bridges in line, with a total length of 638 m and twenty-eight spans (Osmanli period). There are also large stone bridges in Asian Turkey, such as Kesik bridge to the west of Sivas, with nineteen spans totalling 326 m (Selçuk period; Tunç, 1978: 192, 39, 115, nos. 245, 41, 138).

It will be recalled that Constantine founded the city of Constantinople in AD 324, and in AD 330 made it the Roman centre of government. As early as AD 285, Diocletian had divided Roman rule, and by AD 395 the eastern and western portions of the empire were effectively distinct (Cary and Scullard, 1975: 550). Rome fell finally in AD 476, but the eastern Roman or Byzantine empire continued, and it was not until AD 1453 that Constantinople itself was taken by the Turks (Byron, 1964; Browning, 1980; Whitting, 1971; Runciman, 1933). The Turkoman or Turks invaded Asia Minor in AD 1071, with the Selçuks (or Seljuks) as their ruling dynasty. Turkish power passed to Osman I about AD 1281. When Tunç refers to the Selçuk period, he is referring then to the period AD 1071 to 1281, followed by Osmanli (or Ottoman) rule from AD 1281 onwards.

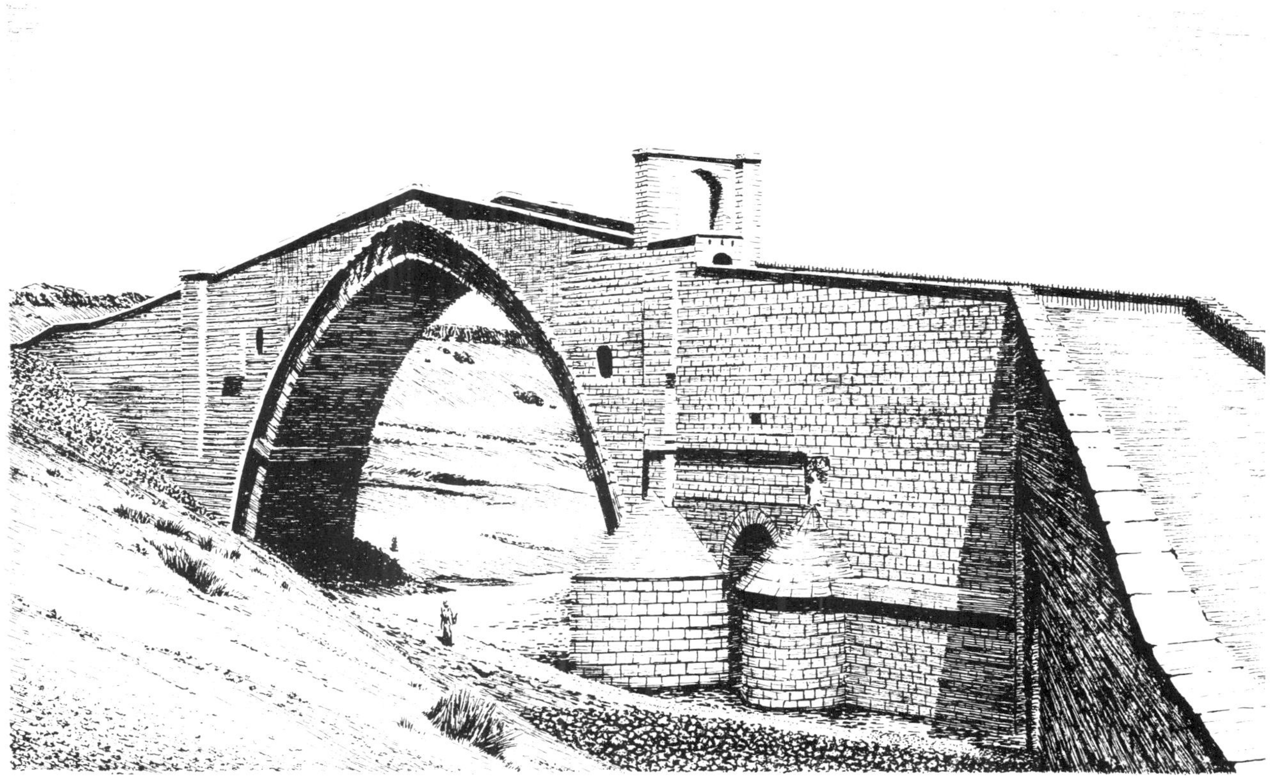

Fig. 103 Malabadi bridge, west of Diyarbakir in south-east Turkey, is a Moorish bridge with a main span quoted either as 38.6 or 41.0 m (based on Tunç, 1978: 132, 135). It has certain resemblances to the Moorish bridge at Martorell which was built on Roman foundations (see Fig. 73).

Malabadi bridge (Tunç, 1978: 130, no. 159) in the south east, west of Diyarbakir, has an extraordinary main span, quoted as either 38.6 or 41.0 m. It was completed in the Selçuk period about AD 1150. As shown in Figure 103, the bridge has some points of resemblance to the Puente del Diablo at Martorell in Spain (Section 5.7, SP2), reconstructed by the Moors in 1283 on a Roman foundation. Both have the characteristic pointed Gothic arch, and consequently a great height at mid-span. At Malabadi, the parapet has a maximum height approaching 25 m above stream level. Both are decorated with an arch over the roadway – in this case above the springing of the main span. The bridge is not straight in plan. The road climbs for about 80 m and then turns abruptly left, to climb a further 50 m to mid-span and then descend over a length of about 65 m to the higher bank on the far side. The initial ramp includes two smaller spans, of 5.0 and 5.8 m, and there is also another 4 m span before the main opening is reached. The main span, on either estimate, is greater than any achieved by the Romans.

These bridges are not Roman, but Turkey also has its fair share of Roman bridges, built either before the empire was divided, or as part of the eastern or Byzantine empire. There are other Roman bridges, or possible Roman bridges, in Iraq, Iran, Syria and Palestine. Their distribution is such that it is appropriate to consider them starting from Istanbul, and moving again in a clockwise direction, beginning within the bounds of modern Turkey to Antakya, Roman Antiochia or Antioch, at the north-east corner of the Mediterranean, and then again clockwise from there to the east and south.

Asia Minor could be reached by ferry from Istanbul across the Bosporus, or, for that matter, from any convenient point across the Sea of Marmora or the Hellespont to the west. If the crossing was made from Istanbul, it was possible to follow a road along the southern shores of the Black Sea to Trabzon or Trebizond, Roman Trapezus (Map 9; see also Calder and Bean, 1958), but there are no recorded bridges along this route.

An alternative route travelled south east from Istanbul to Izmit, Roman Nicomedia, and from there roughly parallel to the coast through Amasya, Roman Amasia, towards Erzerum. A short distance east of Izmit, the road crossed the ancient Sangarius River by the **bridge of Sakarya** (**E7**; Gazzola, 1963b, no. 257; Tunç, 1978: 156, no. 195; Texier, 1839; Whitby, 1985), the name now given to the stream. This fine large bridge, still in use, was built by Justinian about AD 560 (Procopius, *Buildings* 5.3.8–11). The main channel has moved since the bridge was built and this tendency for the stream to move may explain the stone breakwater (to follow Whitby) added later to protect the bank some distance from one end of the bridge. Whitby shows the original bridge with twelve spans – 3, 6, 9, 20, 4 × 24.5, 23, 19.5, 7 and 3 m; an overall length of 429 m and a width of 9.85 m. The first three spans are hidden by the later breakwater. The bridge piers are rounded upstream, with triangular extensions downstream. The site is close to the lake whose drainage was discussed by Pliny (Section 3.1) and this fact has caused some speculation concerning the detailed design of the bridge (Whitby, 1985).

Vermeule (1968: 500), refers to a bridge (**E8**) across the River Iris at Amasya which carried an inscription to Nerva, but nothing more is known of it. **Karas bridge** (**E9**; Tunç, 1978: 109, no. 126a) is a Byzantine structure 10 km north of Erzerum. It has at least four spans and possibly as many as eight, with semicircular arches and an attractive simple appearance, typically Roman in form.

Another road continued south from Izmit to Iznik, ancient Nicaea (Procopius, *Buildings* 5.3.4–6), then east to Ankara and south east to the Cilician Gates and Antioch. This was the Pilgrim's Road, an ancient route that became popular in later times (Winfield, 1977: 152). Alternatively, from Ankara one could travel east to Amasya or Erzerum. Macpherson (1954) mentions a number of bridges in the vicinity of Ankara. Some 100 km west, the Pilgrim's Road crossed the River Siberis at Sykeon (both Roman names). Anderson (1899: 65–7) has described the bridge here, with seven piers and a timber deck (see Chapter 6, T15). It was built by Justinian, presumably on the site of an earlier Roman bridge. Similarly, the Selçuk bridge at Akköprü on the western outskirts of Ankara (possibly Tunç, 1978: 13, no. 6) 'must have had its Roman prototype' (Macpherson, 1954: 111).

A more southerly alternative to this route passed from Iznik (Nicaea) to Dorylaeum, near Eskişehir, and then to Ankara. It crossed the headwaters of the Sakarya, or Sangarius, near Gordium. No trace of this bridge remains, but further west was another **Pons Sangarius** (**E10**; Calder and Bean, 1958) at Çandir Köprü (Macpherson, 1954: 112) on a road which ran directly from Dorylaeum to the south east. Some ruins of this bridge still remain, west of Roman Pessinus.

The eastern road from Ankara crossed the Kizilirmak or Halys River at Yahsihan. About 35 km south was another crossing, at Çeşnir Köprü, 90 km from Ankara. There is now a fine Selçuk bridge here (Tunç, 1978: 56, no. 63) with a high 18.2 m span over the river crossing, and a complex two-tiered pattern of minor arches.

To quote Akurgal (1978: 267), 'one of the best preserved archaeological sites in Turkey is located at Aezani, 54 km south west of Kütahya'. The modern name is Çavdarhisar on the Kemalposa River, the Roman Rhyndacus. Akurgal has a map of the site showing four Roman bridges; he goes on to say, 'Two very well preserved bridges over the River Rhyndacus, still in use today, together with the quay protecting the river against flooding, were built in Roman times.' The first of these (**E11**; Gazzola, 1963b, no. 239; Tunç, 1978: 51, no. 59) has five spans, from 3.9 (approx.) to 6.5 m, and a width of 4.1 m. It is believed to have been built in either the second or third century AD, and is very well preserved. The second is believed to be Gazzola's no. 250 (**E12**; Gazzola, 1963b). In Appendix table A.8 both are identified by the name **Rhyndacus River bridge** but Gazzola refers to no. 250 also as the Ponte di Kütahya. Aezani is in the Turkish province of Kütahya; there is a town of Kütahya, but it is not on the Rhyndacus. The bridge appears to have four spans and is constructed of particularly large stones. The arch ribs show vestiges of a cornice following the rib at the extrados, and are set back twice, to form three shadow lines on the rib, one at the cornice, and the other two internal. The spandrel wall over the piers is constructed of only four courses, of which the largest would appear to be about 1.2 m high. Gazzola suggests that it was built in the third or fourth century AD. There is no information available for the other two.

The main route through Turkey to Antioch passed

through Nicaea and Dorylaeum south east to Laodicea Combusta in Lycaonia and thence to the Cilician Gates. Aezani lay to the west of this route, about 85 km south west of Dorylaeum, and was surprisingly far away from any recognised major Roman roads. It is better to look for bridges down the west coast instead, to Bergama and Izmir, Roman Pergamum and Smyrna.

The River Granicus, now the Biga Çay, is the small stream made famous by the battle of Alexander the Great. Tunç describes a small bridge here, called the **Roman bridge** (**E13**; Tunç, 1978: 155, no. 192), built in the fourth century AD. Little is known of it.

Pergamum, the faithful city of the apocalypse of John (Revelation 2:12–17), is about 75 km due north of Izmir. The oldest part of the city lay on top of a steep hill between the ancient rivers of Selinus and Cetius. By Roman times, it had spread down the hillside to the Selinus and over the river to the west bank. Tunç describes four Roman bridges here (Tunç, 1978: 33, 192, 195, nos. 35, 36, 243, 247). Akurgal (1978: 34) speaks of 'the handsome remains of a stone bridge uncovered at Pergamum [which] date from Greek times. However, it is a modest work compared with the imposing examples of the Roman period.' It is not known which bridge this is. Gazzola (1963b, no. 204) has a photograph of one span of an excellent bridge of the time of Hadrian (AD 117–38). It appears that this may correspond to no. 36 in Tunç (1978: 33) in which case it has two spans of about 11.4 m (**E14**). Gazzola also refers to two other bridges, one of which 'presents the characteristic of having an exceptional width, several tens of metres'. This, no doubt, is the structure under the Temple of Serapis or Kizil Avlu (**E15**; Tunç, 1978: 33, no. 35). Both Akurgal (1978: 70, 104) and Bean (1966: 76, 82) have plans of this complex, described by Bean as the 'Red Courtyard, a monument of Roman grandeur remarkable equally for its layout and its vast size'. Akurgal speaks of it as 'the largest building in ancient Pergamum', consisting of a temple and the large courtyard in front of it covering an area of 260 by 100 m. 'Not the least remarkable feature is that this courtyard is built over the River Selinus, which flows obliquely across it in a double-vaulted tunnel which still serves its ancient purpose' (Bean, 1966: 81). Bean has an excellent photograph of the downstream end of the structure. The spans given by Tunç are 12.1 and 12.45 m. The ends of the vaults are aligned at their upstream face. At the downstream face the ends are staggered, each terminating as it leaves the side of the courtyard. As a result, the eastern vault extends some 14 m beyond the other, with the exposed face trimmed with a vertical stone wall. The tunnel lengths are about 195 and 181 m.

Roman Smyrna was another of John's churches (Revelation 2:8–11). Tyrrell (1911: 18) describes the **Caravan bridge** (**E16**) over the River Meles here, 'believed by archaeologists to be one of the very oldest in existence'. Bean (1966: 45–7) discusses the identity of the Meles. 'No fewer than six streams flow into the gulf between Bayrakh and Mt Pagus, but only three of them can reasonably come in question. The most considerable is the Caravan Bridge River [Kemer Cayi or Uzun Dere on the Turkish map] . . . This has acquired the name Meles in modern times, but is, in fact, the weakest candidate of the three.' He makes no mention of the bridge. He does, however, have a sketch of a 'joggled arch' in the theatre. An arch is formed of voussoirs, with their faces radiating from a centre. In this arch, each of the faces between voussoirs is stepped at mid-depth. In the keystone, this step is such that the extrados is significantly longer than the intrados, as if the stone was resting on the adjoining stones by means of the step. This orientation of the step is continued down each side of the arch, from the keystone to the springing. To use Bean's words, this form of arch is 'exceedingly rare'. It could conceivably assist the arch if its foundations were to move outwards. Bean describes the arch as belonging to a reconstruction following the calamitous earthquake of AD 178 and believes that the stepped construction was a precaution against further earthquake movements.

Laodicea is the third of John's cities to be mentioned here (Revelation 3:14–22). It could be reached in Roman times by travelling south from Izmir to Ephesus near modern Selçuk, and then about 150 km east to Laodicea in Phrygia; or alternatively directly along another road, east and south from Izmir. It lies close to Pamukkale, about 10 km north of Denizli, between the Asopus and Caprus streams which join the River Lycus about 2 km north of the city. Bean (1971: 251) speaks of 'a paved road leading from the Ephesus Gate to a bridge of Roman construction crossing the Asopus; this bridge still survives in part.' It is believed that this is the bridge described by Gazzola (**E17**; Gazzola, 1963b, no. 238), although he speaks of it as across the Lycus. Only the piers remain of a substantial bridge with spans 8.0, 12.3 and 7.1 m, and a width of 6.9 m. Tunç (1978: 156, no. 193) records another Roman bridge (**E18**) further to the east just beyond the south tip of Lake Eğridir, built in the second century AD. Calder and Bean (1958) show another (**E19**) to the east of Lake Beyşehir, Roman Caralis, near modern Kireli.

A road from Ephesus continued around the south-west coast to Lycia (Bean, 1978), and Pamphylia. Bean (1971, pl. 54) has an excellent photograph of the 'bridge at Incekemer' (**E20**), but, as he says in the text (p. 187), it is more likely to be part of the Alabanda aqueduct (see Section 7.6). Alabanda lies near modern Cine, and the structure crosses the River Marsyas 19 km to the south east on the Mugla road. It has three quite large spans. The island of Rhodes lies not far from the coast here. The **Ponte di Rodi** (**E21**; Gazzola, 1963b, no. 164) is an excellent small two-span bridge in fine condition. Unfortunately, Gazzola gives insufficient information to locate the bridge, but it may be in the city of Rhodes itself, for Torr (1885: 2) says that there is a bridge there.

From Laodicea, a road ran south to Cibyra west of Çavdir, with connecting roads to the coast at Kalkan,

Finike and Antalya. Wurster and Ganzert (1978: 304) have described a bridge at **Kemer** (**E22**), north of the Roman town of Xanthus, which in turn is north from Kalkan. Only part of one approach remains of a bridge which was evidently much larger. The present ruins are about 30 m long and rise from the bank over spans of 4.1, 4.5 and 4.0 m. The first two arches are segmental, like the bridge near Limyra, which will be described shortly. Another unusual feature is the inclusion of a chamber 1.5 m high, 3.5 m long and 3.2 m wide above the 4 m arch. The bridge itself is 4.5 m wide.

Wurster and Ganzert also describe the remarkable bridge near **Limyra** (**E23**), about 7.5 km north east of Finike, on a line between Limyra and modern Kumluca. The bridge as a whole is in excess of 330 m long, with twenty-six similar spans, and a further two smaller spans at one end, presumably replacing a single earlier twenty-seventh span. The outstanding feature of the bridge is that the arches are segmental (Fig. 104). Wurster and Ganzert (1978: 292) give the following typical dimensions: span centre to centre of piers, 12.7 m; clear span, 10.7 m; rise at the extrados, 2.0 m; distance from underside of crest to top of roadway, 1.25 m. The width averages about 3.6 m. The largest clear opening is in span 26, and is 13.0 m, corresponding to a distance, centre to centre of piers, of 15 m. Each arch rib is 0.8 m deep, made up of two levels of brick tiles. Allowing for the depth, the span at the centreline of the typical arch is 11.2 m; the radius is 8.5 m; and the effective rise, 2.1 m. This is only 0.189 times the span. It is believed that the bridge is Roman. If this is the case, then it is one of the few truly segmental Roman stone arch bridges, along with the Ponte San Lorenzo in Padua (I161), the Alconétar bridge in Spain (SP20) and the nearby Turkish bridge at Kemer (E22). The timber arch bridge of Apollodorus across the Danube (T13) was also segmental (see also Sections 6.3, 8.4). Bean (1968: 111, pl.51) describes another segmental arch nearby (**E24**), this time of stone, across the Arapsu River near its mouth, 5–6 km west of Antalya. 'The stream itself is crossed by an old bridge which is still sound and in regular use as a footbridge. Spratt calls it "ancient", but the shallow pointed arch betrays its post-Roman date.' The point in the arch is not marked, and the work at Limyra suggests that this conclusion should be reviewed. Spratt and Forbes (1847) identified the site with ancient Olbia.

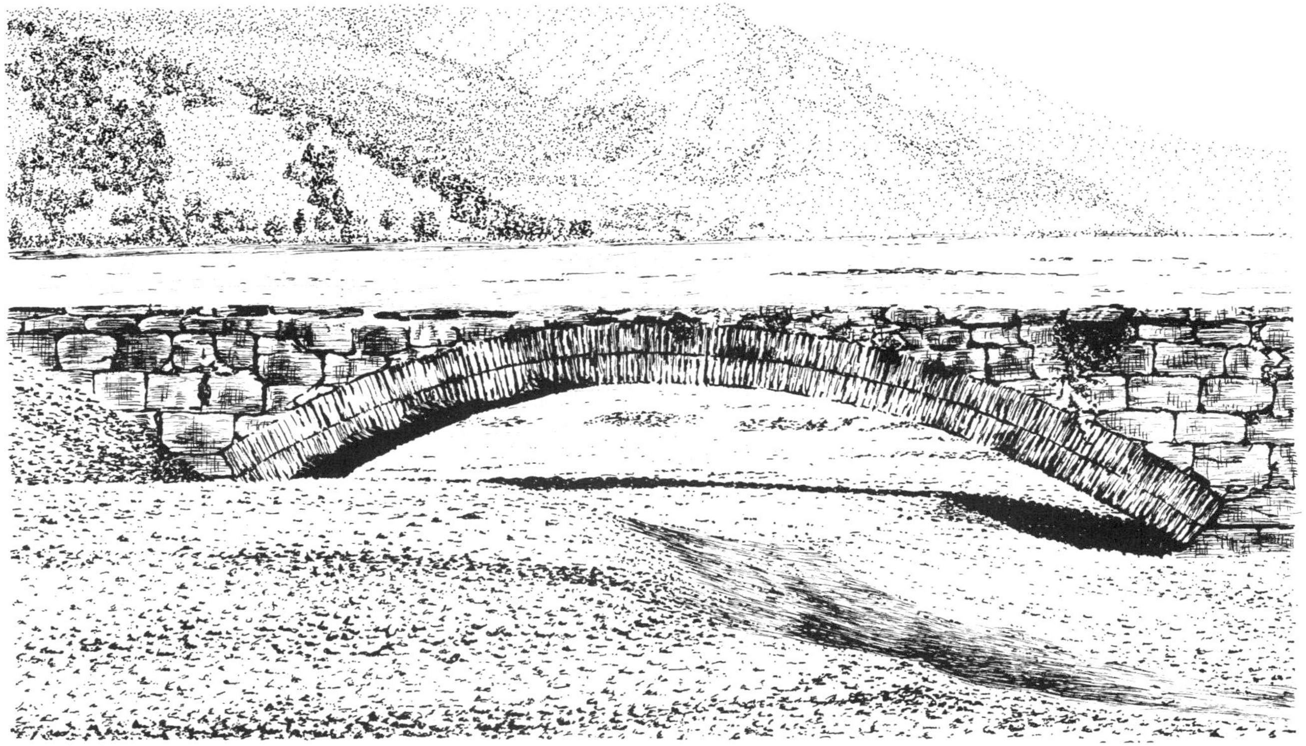

Fig. 104 Part of the 28-span bridge at Limyra, believed to be Roman (E23). The segmental shape is exceptional (sketch based on photograph by Wurster and Ganzert, 1978).

Further east, about 40 km beyond Antalya, the Roman Eurymedon is crossed by an excellent Turkish stone bridge. This is at the site of Aspendus. About 42 km north, on the road to Roman Selge, the same stream is crossed by a Roman bridge (**E25**). To quote Bean again (1968: 139), 'Beyond Beşkonak the road continues up the river for another three miles or so to a most attractive and excellently preserved Roman bridge; this has a single arch and is as sound today as when it was built.'

The road continues around Turkey's southern coast. Silifke, ancient Seleucia, lies at the mouth of the Göksu or Blue River, called in ancient times the Calycadnus. In AD 77 or 78, the governor, L. Octavius Memor, built a bridge here in honour of the emperor Vespasian and his two sons Titus and Domitian (Akurgal, 1978: 341). Akurgal describes the present bridge as modern, but it was erected on precisely the same site as the Roman bridge, and it seems likely that it includes Roman work (**E26**). It is illustrated by Tunç (Tunç, 1978: 168–71, no. 203 or 204), who also has a detailed drawing. The same bridge is shown by Sitwell (1981: 199). It has five main spans, three of them about 17.4 m and the other two a little less, followed by two much smaller spans, each of the order of 4.3 m. The total length is 120 m.

Ancient Tarsus, the home of the apostle Paul, lay about 38 km west of Adana. It was a strategic centre in Roman times, at the junction of the coast road and the major road coming down from the north west, through the pass in the Taurus Mountains known as the Cilician Gates. Tunç (1978: 26, no. 24) has a photograph of an ancient bridge in Tarsus (**E27**), built in the time of Justinian (AD 527–65), with three spans, now partly buried, and a total length of 101 m (see Procopius, *Buildings* 5.5.14–20).

In Adana itself there is a very old bridge, the **Taşköprü** (**E28**; Tunç, 1978: 165, no. 202), over the Seyhan or Sarus River (Procopius, *Buildings* 5.5.8–13). Its date of origin is uncertain. Tunç speaks of it as going back to Hadrian (AD 117–38) and Justinian (AD 527–65). Akurgal (1978: 345) refers to an inscription held in the museum at Adana (*CIG* 4440), which describes the bridge as built by Auxentius, 'who may have been the same as an Auxentius who lived and worked around AD 384 in Rome as a constructor of bridges'. The bridge is 310 m long, and originally consisted of twenty-one arches, some of which have disappeared. These arches differ – some are pointed, and the crest heights and spans are irregular.

A little further east is the **Misis bridge** over the Ceyhan or Pyramus River (**E29**; Tunç, 1978: 142, no. 169). This is again a fine stone bridge, 135 m long, with nine spans graduated in size and rising to mid-span. It is believed to have been built by Flavius Julius Constantius in the fourth century AD.

The Misis bridge was located at ancient Mopsuestia. There is some uncertainty as to the roads here. One led due east to the Euphrates at Zeugma (Calder and Bean, 1958), crossing another inland road south to Antioch. Chevallier (1976: 141, 144) shows a road directly around the coast through Issus. This reached Alexandria, at modern Iskenderun, and then struck across the mountains south to Antioch. Sitwell (1981: 187) shows here a tall, single-span bridge (**E30**), presumably ancient, high in a rocky pass. The **Asi bridge** (**E31**; Tunç, 1978: 21, no. 18) is in Antakya itself. It has four 10 m spans and is 48.2 m long, built early in the fourth century AD under Diocletian (AD 284–305). Now a city bridge, it is in service and in good condition, but has obviously been restored.

The Euphrates River rises east of Erzerum, not far below the Black Sea, and flows west and south across Turkey into Syria before turning south east to the Persian Gulf. It was of major strategic importance, and under Hadrian became the effective eastern boundary, the Cappadocian Limes of the Roman empire. Four legionary fortresses were established: at Trabzon, Roman Trapezus, on the Black Sea; Satala, some 105 km south; Melitene, modern Malatya, near the Euphrates; and Samosata or Samsat, on the river south east of Adiyaman (Mitford, 1974). A strategic road was established linking them along the right or west bank of the river. This could be reached from Antioch by going either east through Beroea, modern Haleb or Aleppo, or north east through Cyrrhus (now Horos) and Doliche (near Gaziantep) to Zeugma (Birecik). This strategic road contains one of the greatest of all Roman bridges, at Kâhta, which will shortly be described.

In Syria, north of Aleppo on the road to Cyrrhus, is a fine three-span bridge over the River Afrin (**E32**; Gazzola, 1963b, no. 258). Built of limestone, its spans are 11.5, 12.5 and 12 m. The road rises from either bank to mid-span, with a very shallow depth over the arches and no balustrades. The total length is 92 m, the width is 5.8 to 6 m, and the maximum height about 10 m. The appearance of the bridge is typically Roman and Gazzola dates it from the fourth century AD. It is still in use. Further north, 70 km from Aleppo, is another fine bridge, again still in use, over the Saboun-Souyou (**E33**; Gazzola, 1963b, no. 259), a tributary of the Afrin. It has six spans – 3.8, 8.4, 10, 9.8, 8.2 and 3.8 m – and a road only 4 m wide, 96 m long. The piers again have triangular, half-height cutwaters upstream, and are either rounded for half their height, or finished with triangular cutwaters downstream. There are no balustrades and, despite the narrow width, the bridge is still in use. Its similarity to the other bridge suggests that it was built about the same time. There is also a second bridge over the Saboun-Souyou on the Cyrrhus to Doliche road (**E34**; Gazzola, 1963b, no. 260). In poorer condition than the others, it has three spans, of 4.2, 9.3 and 4.7 m; its length is 48 m. Gazzola places it in the third or fourth century AD.

Calder and Bean (1958) show a Roman road following the right bank of the Euphrates River all the way upstream from ancient Nicephorium, modern Ar-Raqqah in Syria, to Zimara. A feature of this road was that it crossed all the right bank tributaries. One of these is the **Göksu** or Geuk Su (Anderson, 1897; Percy, 1901: 101) 25 km west of Samosata, south of Adiyaman. Gazzola (**E35**; 1963b, no. 255) has a brief account of a bridge over the Göksu with a single span of 30 m. Ballance (1951: 96, footnote) refers to the bridge at Kâhta (see below) and adds that there is 'a rather similar bridge over the Gök Su, with a span of over 30 m'. This span is very large. The name Göksu simply means Blue River, a common Turkish name, and there are three in this vicinity – one rising to the north west of Silifke and reaching the sea there, another to the north east of Adana which joins the Seyhan, and the present stream. It appears that it is the Göksu west of Samosata which is crossed by this large bridge, and Tunç also (1978: 87, no. 97) has a Roman bridge here. It has not been possible, however, to confirm the span (see also French, 1965).

The bridge at **Kâhta** (**E36**; Gazzola, 1963b, no. 254; Tunç, 1978: 43, no. 42; Dorner, 1981: 240, 329) has a single main span of 34.2 m, one of the largest ever built by the Romans. For comparison, the bridge at Pont St Martin (I155) has a span of 35.6 m, and the Ponte d'Augusto at

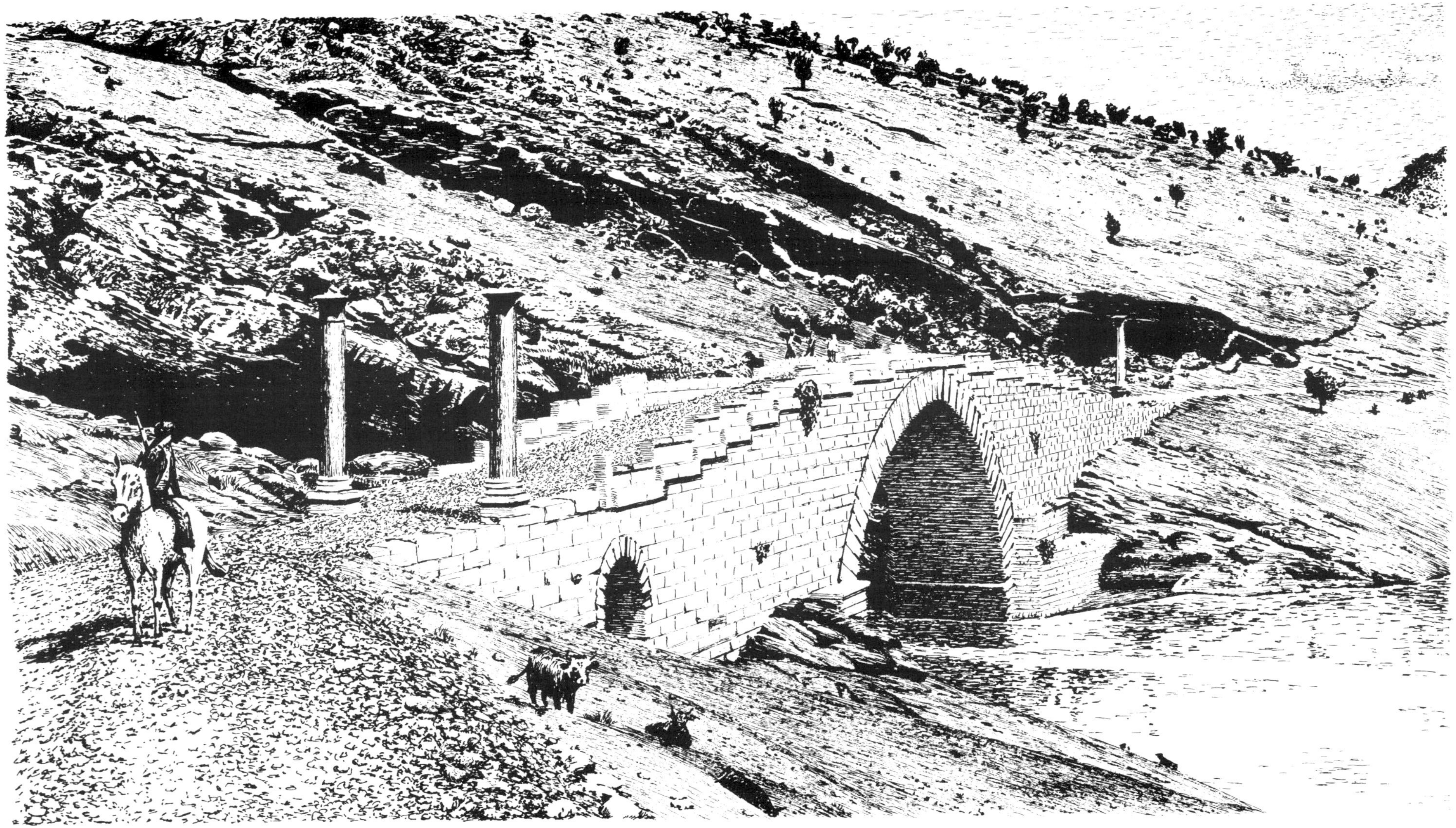

Fig. 105 The remarkable Roman bridge at Kâhta. Its main span of 34.2 m appears to be the second largest ever achieved by the Romans (E36, based on Mitford, 1974; Yorke, 1896; Tunç, 1978; Gazzola, 1963b).

Narni (I87), 32.1 m. Figure 105 is a drawing of this magnificent bridge. It was visited by Von Moltke in the last century and rediscovered by Sester in 1881. Yorke (1896) described the bridge in his account of a journey by horseback up the upper Euphrates in 1894. Percy (1901) has also visited the site, and his graphic description can hardly be improved:

In its wonderful state of preservation, the enormous span of its arch and the noble simplicity of the whole structure appeal the more forcibly to the imagination from the apparent incongruity and purposelessness of its situation. Were it, as it seems, the approach to some great capital strewn with the crumbling evidences of departed grandeur, we could understand the reasons for what must have been a prodigious expenditure of labour, time and money in this wild and remote region, on the far verge of the Empire. But not only is there no sign of a city; the very lie of the ground seems to disprove the previous existence of any road by which an army could have marched up through the main passes. There is the bridge and nothing more ... Behind the cliffs, springing abruptly from the water's edge, narrow as if they would meet across the chasm, while their walls, falling inwards towards the base, form a vault of inky blackness against which the massive fabric stands out clear and distinct like an arch of the purest white marble.

The site is variously spelt Kâhta or Kiakhta. The stream is called the Cendere Çay or Chabina by Akurgal (Akurgal, 1978: 347). It is also called the Bolam Su by Yorke (1896: 322; Percy, 1901: 102) and Gazzola, and joins another stream, the Kâhta Çay or Nymphaios, near the bridge. The ancient city of Arsameia lies on this stream, 2 or 3 km to the north west; it is 41 km east of Adiyaman, and can be reached from there by road.

The roadway of the bridge is 118 m long and rises from either side over the 34.2 m main span. There is, in addition, another small span set well back in one of the approaches. The roadway width is 7.6 m, and the height from the water to the summit of the arch is 17.1 m. The architecture of the bridge is simple, with no decoration apart from the distinctive stepped parapet. However, the bridge originally carried four dedicatory columns, two on either side of the approach, and of these three still remain. According to a Latin inscription on the columns, the bridge was built by four cities of Commagene, the region in which it is located, in honour of Septimius Severus, his wife Julia Domna and his sons Caracalla and Geta, and the four columns were erected in their honour. The one dedicated to Geta is missing, and was probably removed after he was murdered by his brother Caracalla (Akurgal, 1978: 347; Hogarth and Yorke, 1898: 315). The bridge dates, then, from the reign of Septimius Severus (AD 193–211), and is often referred to as the **Severus bridge** (see *CIL* III.6709, 6712–14).

Kâhta is about 25 km from the river, on the southern flanks of the Taurus Mountains. Within a few kilometres are high mountains, such as the Nemrud Dağ, 2150 m high. Travel to the north was limited to only a few practicable routes through the mountain chain. Calder and Bean (1958) show the road from Samosata to Melitene along the right bank, another more direct from Samosata to Melitene, and a connecting link through Kâhta. Mitford (1974: 161) shows only one road, across country from Samosata to Adiyaman, Kâhta and Melitene. Melitene was a crossroads, for here the Cappadocian Limes crossed the main road from Cappadocian Caesarea to the east.

About 15 km north is a 22-span Turkish bridge from the Osmanli period, the Kirkgöz bridge (Tunç, 1978: 118, no. 143), or, to quote Yorke (1896: 328), the bridge called Kirkgeuz-Keupri across the Tokhma Su. It is not Roman, and is mentioned here only to illustrate again the extraordinary range and size of old stone bridges in Turkey.

About 40 to 45 km north east of Malatya the Euphrates is joined by a major tributary from the east, the Murat. A little further upstream, another smaller tributary flows in from the west, called the Arabkir Çay by Mitford (1974), with the villages of Korpanik to the south and Ağin to the north. Yorke (1896: 330) calls this stream the Angu Chai and describes the remains of a Roman bridge (**E37**) here, 3 km east of Korpanik. 'The remains of this bridge, which was constructed in one bold arch with a span of about 100 feet (30 m) consist of two abutments ... of concrete faced with well-squared stones.' Again, the span, as with E35 over the Göksu, if it were correct, would be remarkable. He also speaks of another bridge further up the tributary. This may be the **Karamağara bridge** shown by Tunç (**E38**; Tunç, 1978: 108, no. 126), a Byzantine bridge of either the fifth or sixth century, with a single 17 m arch spanning between the cliffs of a rocky gorge.

As the road continues north, it reaches Zimara where the river turns to the east. Yorke (1896: 456) describes another tributary here, the Karabudak or Sabrina, and on the near bank, about 6 km from the Euphrates, the remains of another Roman bridge (**E39**). He saw an inscription near the bridge which Mitford (1974: 169) interprets as referring to Decius' restoration of about AD 249 to 251. The bridge was 30 m long.

Antakya (Antioch or Antiochia) was in Roman times a convenient point from which to reach Commagene, Mesopotamia, Syria and Palestine. It is convenient to use it also as a point of reference here, moving out and clockwise from Antakya to describe a final series of Roman stone bridges.

The first is one described by Gazzola (**E40**; 1963b, no. 252) as across the Tigris at Kakho (or Zakho) south of Lake Van; an excellent photograph by Wheeler may be found in Young (1980) following p.253. The bridge in its present form is not Roman. Young (p.220) suggests that it was built during the period of Abbasid rule (AD 750–1258), but according to Gazzola there is a tradition that its foundations date from the time of Alexander the Great. It is a major bridge with a central span flanked on each side by a smaller arch, and with at least one further arch in the approaches. Zakho is an island town in Iraq just south of its border with Turkey, about 30 km east of the point

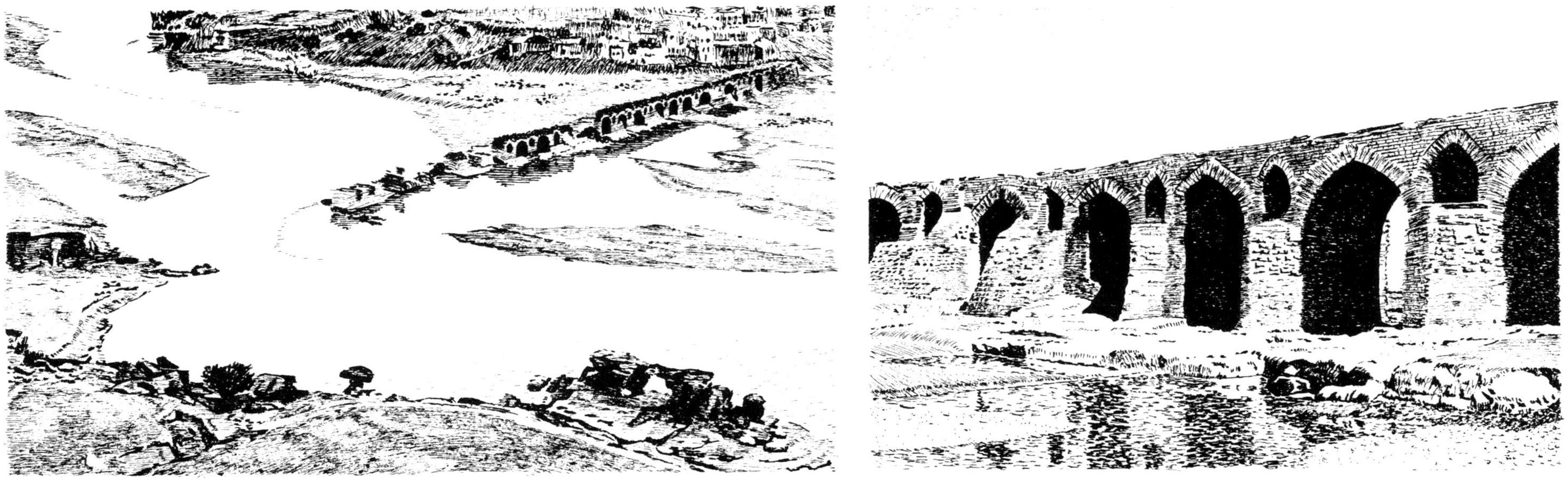

Fig. 106 The remains of the dam-bridge of Band-i-Kaisar on the Karun River at Shushtar or Susa in Iran (E42, based on Brice, 1966, pl. 10; BNI, 1945b, fig. 222).

where the borders of Turkey, Syria and Iraq converge. More correctly, the bridge crosses not the Tigris but a branch of the Khabur, a major tributary; it lies about 3 km upstream of the town (BNI, 1944b: 554; see also Lloyd, 1980: 30).

Mosul is on the Tigris, about 100 km south of Kakho. There are the remains of a Roman bridge near Eski Mosul or Old Mosul (**E41**; BNI, 1944b: 231). Only the arch vault remains, but it is completely intact and quite large, with a span that appears to be of the order of 15 m.

Shushtar is the name given now to ancient Susa in Iran. There is another remarkable bridge here over the Ab-i-Garghar branch of the Karun River (**E42**; Gazzola, 1963b, no. 229; Brice, 1966: 80; Matheson, 1972: 158), called the **Band-i-Kaisar** or the Dam-Bridge of Valerian (see Fig. 106). It is quite long, possibly of the order of 300 m, and had originally something of the order of thirty-five main spans, of which about twenty still remain. There are in addition small flood arches in the piers. Gazzola and Brice describe it as both a dam and a bridge.

BNI (1945b: 428) speaks of the

> extensive irrigation system [which] existed on the Karun between the third and thirteenth centuries AD, the rudiments of which probably go back to very much earlier epochs. The fully developed system was based on three barrages at Shushtar, Band-i-Kir and Ahwaz respectively. At Shushtar, the eastern channel (the Ab-i-Gargar) still branches off 500 yards above the ruins of a great dam and bridge, the Band-i-Kaisar.

This branch rejoins the main stream some 40 km south. The island between contained some 150, 000 hectares of irrigable land, serviced by the dam at Band-i-Kaisar. Despite this, the structure looks more like a bridge than a dam (Brice, 1966: 80; BNI, 1945b: photographs facing p. 420), and it is hardly likely that the impounded depth could have been greater than 7 or, at the most, 10 m. In plan, the bridge meanders a little as it crosses the wide bed of the stream. This may have been due to a desire to place the bridge on whatever rock outcrops were available, or alternatively the result of foundation movements resulting from shallow founding levels in alluvial material.

In AD 252–61, the Sassanid ruler Shapur I invaded the provinces of Syria, Cilicia and Cappadocia, and captured the emperor Valerian with some of his troops. It is said (Brice, 1966: 80; BNI, 1945b: 240) that, before killing Valerian, Shapur used him and his other captives to build the bridge. It fell into disrepair after the Mongol raids of the thirteenth century (see also Binnie, 1949/50).

From Antakya there was a choice of routes to the south

(Calder and Bean, 1958; Chevallier, 1976: 141, 144), one inland and the other along the Mediterranean coast. The inland route passed through Damascus to Bostra, south east of the Sea of Galilee and on the direct road from the Euphrates through Palmyra. There was a group of bridges here. The bridge over the Wadi Suweida (**E43**; Gazzola, 1963b, no. 247) is presumably near Suweida, north of Bostra, towards Kanatha (el Qanawat). It had a single span. The Wadi Zeidi runs west and then north west from Bostra to join the Yarmuk River, and is crossed by at least two Roman bridges. The first (**E44**; Gazzola, 1963b, no. 245) is 3.5 km from Bostra. It has three 3.8 m spans of granite, with piers 2.4 m wide and a height of 2.5 m to springing level. The second (**E45**; Gazzola, 1963b, no. 246) is near the village of el-Tali'a, between Der'a and Bostra, about 20 km from Bostra. It has two spans, each of about 5.6 m, and is intact up to crest height. Gazzola also refers to a third bridge across the Wadi Zeidi near Bostra, the bridge of Gemarrin (**E46**; Gazzola, 1963b, no. 256), but he also describes it as being in ancient Commagene, so its location is obscure. Perhaps it is on the road north from Bostra. It has three spans.

The Wadi Zeidi is a tributary of the River Yarmuk, which runs through a deep gorge to join the Jordan south of the Sea of Galilee (or Tiberias). The Nahr er Rukkad is another tributary which joins the Yarmuk below the Wadi Zeidi (Smith, 1935: pl. 7). There is a fourth bridge somewhere here (**E47**; Gazzola, 1963b, no. 244), with eight large arches of varying span. Another in the same vicinity, the bridge of Qodana (**E48**), has three spans.

These bridges lie either on or near the Roman road which ran west from Bostra to cross the Jordan below the Sea of Galilee; it then turned north to Tiberias on the western shore, and west again to Ptolemais, modern Acre, on the Mediterranean (Avi-Yonah, 1977: 187; BNI, 1943: 445). It is necessary to study a topographic map to appreciate the difficulties of the terrain; the Sea of Galilee is about 210 m below sea-level, while Der'a, only 50 km to the east, is at an elevation of 530 m (BNI, 1943: 508). It can only be assumed, therefore, that the road most probably descended by the gorge of the Yarmuk.

It was to this Yarmuk River that Lawrence of Arabia came, unsuccessfully and with great effort in November 1917, to blow up the bridge on the railway to Medina. In his words (1962: 430), 'Before us from a lip of blackness rose very loudly the rushing of the river which had long been dinning our ears. It was the edge of the Yarmuk gorge, and the bridge lay just under us to the right.' The Yarmuk River fell here into the valley of the Jordan. This was wide and flat, and it appears that there were two crossings of the Jordan here, one above and the other below the confluence with the Yarmuk. The more southerly crossing was by the bridge of **Jisr el Majami** (**E49**; Gazzola, 1963b, no. 243), or, to quote Avi-Yonah (1977: 138–9), 'the Roman bridge at Gesher'. This high bridge is located 10 km south of the lake and had two spans, of about 15 and 18 m; parts of the central pier still remain.

To the north of the Sea of Galilee was another Roman bridge at **Jisr Banat Ya'qub** (**E50**; Avi-Yonah, 1977: 186), or the bridge of Jacob's daughters. It lay on a direct Roman road from Damascus to Tiberias, which crossed the upper waters of the Jordan 4 km south of Lake Hula. A photograph of the bridge (BNI, 1943: facing p. 381) shows it with at least four spans, crossing the stream in two channels. It has, no doubt, often been restored.

The other road from Antakya to the south followed the coast. There are two bridges recorded on this route, one north of Beirut, and the other north of Tyre. Maameltein lies on the coast about 18 km north east of Beirut. The **bridge at Maameltein** (**E51**; Gazzola, 1963b, no. 240) crosses a stream of the same name. It has a single span of 12 m and is 4.4 m wide; it is believed to have been built in the third or fourth century AD. The vault and part of the approach walls remain. The second, the **bridge at Nahr Aboul-el-Assouad** (**E52**; Gazzola, 1963b, no. 241), crosses the River Kasniré, the ancient Leontes, about 10 km north of Tyre. Built about the same time, it also has a single span and, surprisingly, is segmental, with a rise to span ratio of about 0.32.

The final bridge to be mentioned is the **bridge of Jisr Jindas** (**E53**; Gazzola, 1963b, no. 242) at Lydda or Lod, 35 km north west of Jerusalem. The present bridge dates from AD 1273, but is believed to be on Roman foundations. There are, no doubt, other Roman bridges in Palestine and Transjordan (Gazzola, 1963b: 168, 170) but there is insufficient information to enable them to be documented here.

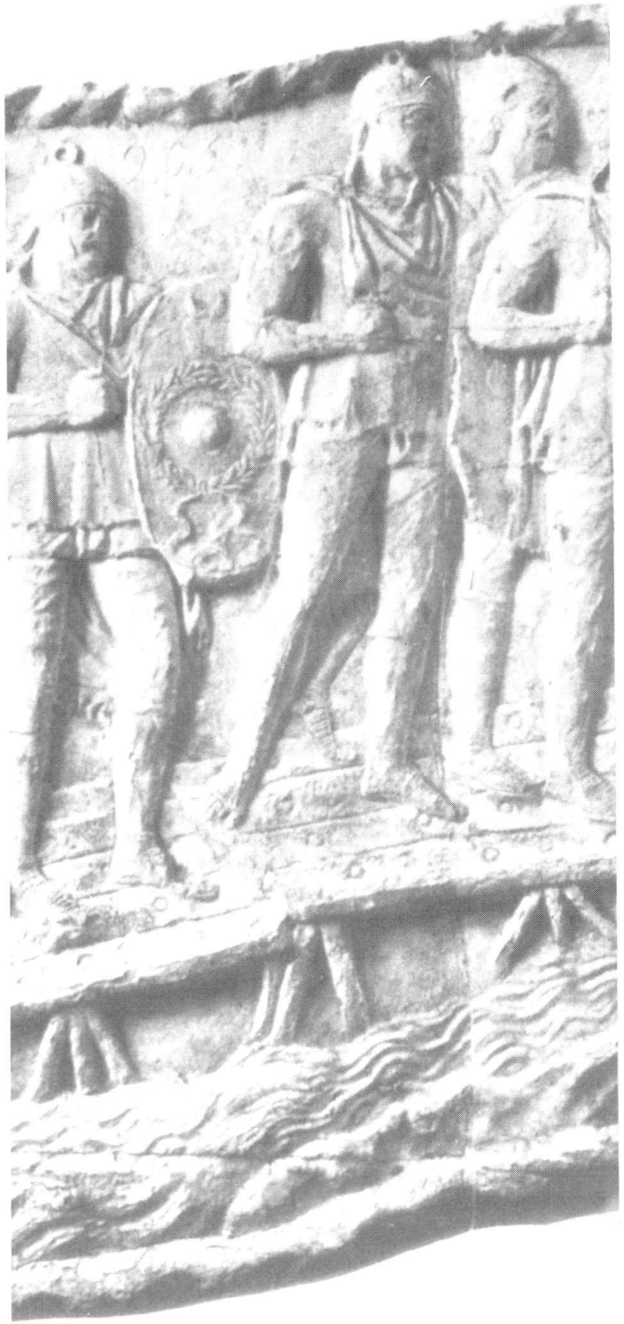

6

TIMBER BRIDGES

6.1 Introduction

A common view is that the typical Roman bridge was built of stone. This chapter will argue the contrary, that the use of timber was widespread but that the physical evidence for timber bridges has disappeared simply because timber has a relatively short life.

Reasons for this view include the following.

1 Suitable timber was available. Vitruvius (2.9) lists many useful building timbers, and it is reasonable to expect that the volumes of timber available in Roman times were much greater than today.
2 The Romans had the technology to cut and work timber (see Section 4.2).
3 A timber bridge is easier and quicker to build than a stone bridge.
4 Centering is required for the construction of a stone arch bridge. In Roman practice (see Chapter 8) this was built of timber; in effect, the Romans could not build a stone bridge without first constructing one of timber.
5 The columns of Trajan and Marcus Aurelius provide strong evidence that the Roman army often built bridges of timber.
6 Records of ancient timber bridges can be found.
7 In Great Britain in particular, the available evidence suggests that a great majority of the bridges built there had timber superstructures on stone piers.

What, therefore, were the forms of these timber bridges? By far the most common form of superstructure had timber decking on timber girders; the famous bridge of Apollodorus across the River Danube had timber arches; but the most intriguing question is whether the Romans had a knowledge of the timber truss. Evidence from Trajan's column suggests that they may have had, and leads to a study of ship-landing bridges, and the *corvus* of Polybius.

The existence of timber bridges offers a different perspective on stone bridge construction. It is erroneous to consider the Romans as advancing into an empty and unknown countryside, building stone bridges as they went. If they knew how to build timber bridges, then it is probable that others knew how to build them as well. As Roman culture spread, it advanced into areas that already had an established route system, and this system had at least some bridges. The Roman army built bridges of timber. The picture that emerges is not one of stone bridges being built with urgency as new routes were pioneered, but rather one where existing crossings – fords, ferries or timber bridges – were gradually improved by the building of bridges in stone.

On this view, therefore, stone bridges were likely to be second- or third-generation structures, built to replace earlier crossings, constructed in a stable environment, with access to adequate and known resources and a developed infrastructure. A list of timber bridges is given in Appendix table B.

6.2 The columns of Trajan and Marcus Aurelius

Rome possesses two carved columns, of similar proportions and style and built for similar reasons – the column of Trajan in the Forum of Trajan, close by the Piazza Venezia; and the column of Marcus Aurelius, or Antoninus, in the Piazza Colonna, off the Via del Corso.

Trajan's column has been described by Cichorius (1886–1900), Rossi (1971), Richmond (1982), Nash (1968) and others (Lepper and Frere, 1988; Florescu, 1969; Lehmann-Hartleben, 1926; Coarelli, 1974). About 100 feet (30 m) high, it consists essentially of a marble cylinder about 3 m in diameter, carved as a continuous spiral band, 0.9 m high and 200 m long. It was dedicated to Trajan in the year AD 113 and had the original purpose of showing the height of the Quirinal Hill, cut back to form the forum. Its dedication (Rossi, 1971) reads

> The Senate and the people of Rome to the Emperor Caesar, son of the divine Nerva, Nerva Trajan Augustus Germanicus Dacicus, pontifex maximus, at the time of his seventeenth tribunician power, of his sixth imperial salutation, and of his sixth consulship, the Father of the Country, to declare how high a hill and place have been excavated for these great works.

The chamber at the base of the column was used to contain his ashes.

Complete sets of photographs of the carvings may be found in Cichorius (1886–1900), of excellent quality, and in Rossi (1971), of much lesser quality. Lepper and Frere (1988) is a new edition of the Cichorius plates. The carvings form twenty-three spirals, with the first or bottom spiral starting above the centre of the doorway in the base. These are divided into 155 separate scenes. In Cichorius, they are reproduced in 113 plates, and are based upon 414 individual photographs. Richmond uses the scene numbers for reference, and this system is followed here. In Cichorius' reproductions, the scene numbers are placed above the carvings and the photograph (or cast) numbers below. Rossi identifies the locations of the scenes by reference to the spiral numbers. Lepper and Frere (1988) give the Cichorius scene numbers; the citations given here may readily be found in this later reference.

The purpose of the carvings has given rise to some conjecture. Rossi (1971) has argued that they give a chronological account of the events of the Dacian Wars. Richmond (1982: 2–5) rejects this – 'anyone of intelligence could have composed in a couple of hours a more telling continuous narrative of the wars . . .'. He argues that the chief purpose of the carvings was to show the army at work.

> It tells something indeed of history, but more of the labour by which history was made . . . Once the sense of flow is given to the various subjects which the form of the scroll so skilfully unites, there is no intimate connection between the successive scenes . . . They illustrate with a care and fullness never before attempted and never afterwards surpassed, the adage that the Roman army owed as much to the spade as to the sword.

Richmond studied the accuracy of the column by comparing its illustrations with objects found in the field in archaeological investigations. He believed that the carvings were copied by masons from an artist's sketchbook. In some cases misunderstandings arose; for example, a wall built from turfs is shown as a stone wall, even though adjacent scenes show the turfs being cut. However 'these muddles were quite exceptional'; generally the carvings are remarkably accurate.

The column of Marcus Aurelius is, in many respects, similar. Caprino et al. (1955) record its dimensions, and present photographs of its carvings. The column had an original height of 41.9 m. This consisted of a base 10.5 m high, a column of 29.6 m, and a further top stone 1.8 m high. The column height of 29.6 m included a square base and capital; the major length was cylindrical with a diameter varying from 3.8 to 3.7 m. It was formed by nineteen blocks, each with an average height of 1.6 m, but these were not made from a single stone. The carvings formed twenty-one spirals, each a little over 1 m high (perhaps about 1.2 m). The total length of carving in this case was about 246 m.

The column was completed in AD 193, after the death of Marcus Aurelius Antoninus in AD 180 (Nash, 1968). It depicts scenes from the wars against the Quadi, Marcomanni and Samartians (AD 172–5). Caprino et al. show the carvings in 139 figures (figs. 4–142), made up of 116 scenes. His figure numbers will be used here for reference.

The carvings on the column of Marcus Aurelius are somewhat more weathered than those on Trajan's column. They are generally similar, but perhaps not quite so good. The selection of subjects is different. Richmond (1982: 5) describes them as 'the scroll of an impressionist, interested rather in the tumult and honour of war than in the methods by which it was waged'. This is a comment on selection rather than style, for the carvings are accurately detailed. It is true that, in terms of engineering content, there are fewer scenes of interest than on Trajan's column, but they are still very valuable.

The first scenes on Trajan's column are clearly historical. They show the river bank, and then the Roman army crossing the River Danube by two bridges of boats, watched by Father Danube, and passing by the emperor in review (Richmond, 1982: 7). Figure 107 (Cichorius, 1886–1900, scene 4) is typical of the carvings on the column. The soldiers are accurately drawn, as are the boats, but there is a variation in scale. The crossing is clearly not as long as the width of the Danube, here possibly about 1100 m wide (Gazzola, 1963b: 137). Indeed, it would be impossible to show accurately both the soldiers and the bridge. Similarly, neither the boats themselves nor the railings of the bridge are shown to the same scale as the men.

The scene shows two bridges; in Figure 107 the second or more distant bridge can be seen at the right. It is not

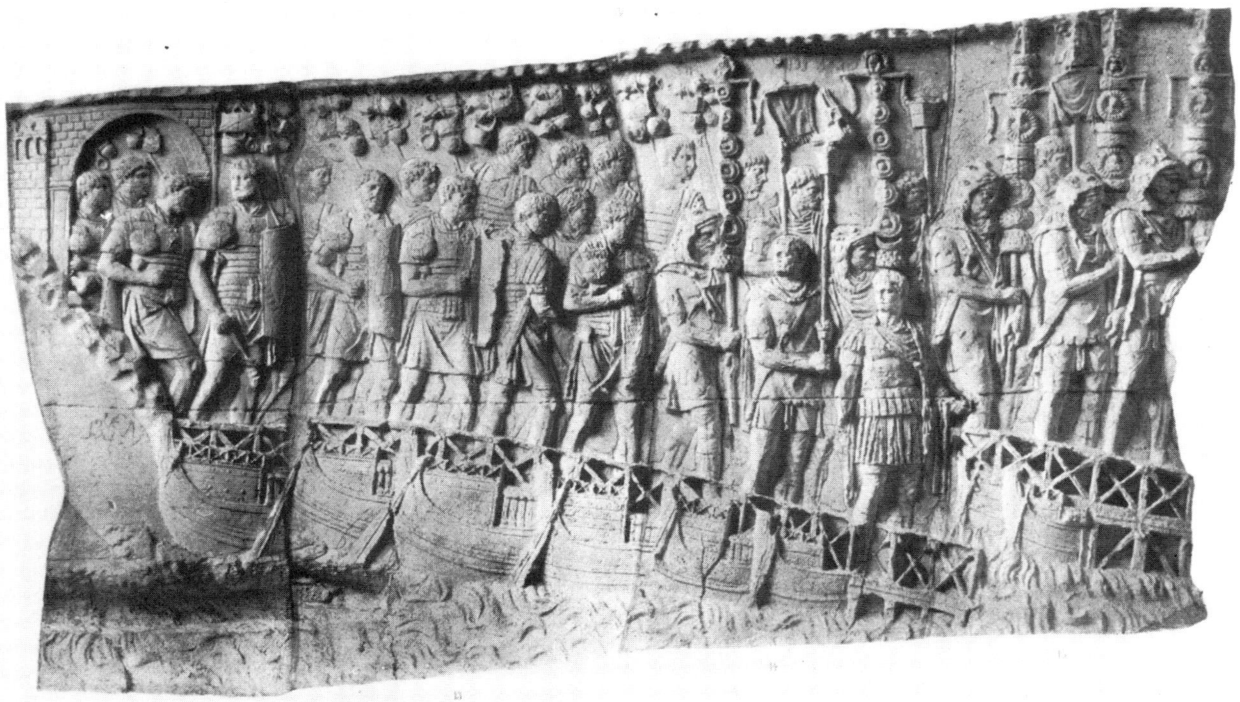

Fig. 107 Scene from Trajan's column – crossing the Danube (Cichorius, 1886–1900, scene 4).

known whether these bridges were adjacent. Sitwell (1981: 137) and Rossi (1971: 32) identify two sites, the first at Viminacium (**T1**, see Appendix table B), modern Kostolac, about 60 km east of Belgrade, and the second (**T2**) joining Pontes, on the south bank, and Drobeta (modern Turnu-Severin) on the north bank, some 100 km further east (see Map 7). Between, east of Orsova, lay the rapids of the Iron Gates, separating the upper and lower reaches of the river. To secure communication between the two sites Trajan built a road along the south bank which, as it passed through the gorge, was partly cut into the vertical cliff face, partly cantilevered in timber out over the water (see Section 2.7). The column either shows two adjacent bridges of boats at one of these sites, or, more likely, two widely separated bridges, with the distance between them greatly foreshortened.

The columns have at least six scenes with bridges of boats. Two are on Trajan's column (Cichorius, 1886–1900, scenes 4–6, 46–8). Of the four on the column of Marcus Aurelius (Caprino et al., 1955, figs. 8–10, 98, 104, 127), two are shown here in Figures 108 and 109 (ibid., figs. 9 and 104). The first is part of a scene similar to that on Trajan's column, with soldiers proceeding from a fort onto the bridge, about to be reviewed by the emperor, and with a river deity looking on. Despite these considerable superficial similarities, there are major differences of detail.

The Romans were not the first to build bridges of boats. Pannell (1964: 211) quotes as evidence an enclosed gold band now in the British Museum to indicate that the Assyrians built pontoon bridges as early as the ninth century BC. There was possibly a pontoon bridge across the Euphrates at Babylon about 700 BC: Darius built floating bridges (Hammond, 1988: 527), and in 480 BC Xerxes built his famous bridge across the Hellespont (Robins, 1948: 69–70; Steinman and Watson, 1941: 25–7; Tyrrell, 1911: 104–5). According to Herodotus (7.33–6), there were in fact two adjacent bridges, built with galleys and triremes lashed together, side by side, with 360 vessels in one bridge and 314 in the other. Hammond (1988: 527–32) has discussed these bridges in some detail, and includes a map (fig. 45) showing their location. Cables were laid along each bridge and pulled taut; the deck was formed by placing planks across the cables. Three gaps were left in the bridges to allow the passage of boats beneath the cables.

In the second half of the second century AD, Arrian wrote a history of Alexander the Great – his *Anabasis*, or 'Journey Up-Country', as he called it. In Book 5, he records Alexander's crossing of the Indus River, and later (5.7) says that, although he is not sure, he expects that it was by a bridge of boats, like those used by Xerxes, and by the Romans. He describes how the Romans built such a bridge.

At the word of command, the boats are allowed to float downstream, stern first, while their way is checked by a rowing boat which manoeuvres them into position. Once they are in position, pyramid-shaped wicker baskets filled with rough stones are lowered from the bows to hold them against the current. As soon as one vessel is securely moored, another is brought up alongside of it, bows-on to the current, at the proper distance to ensure a sufficiently strong base for the superstructure; then timbers are rapidly laid athwart

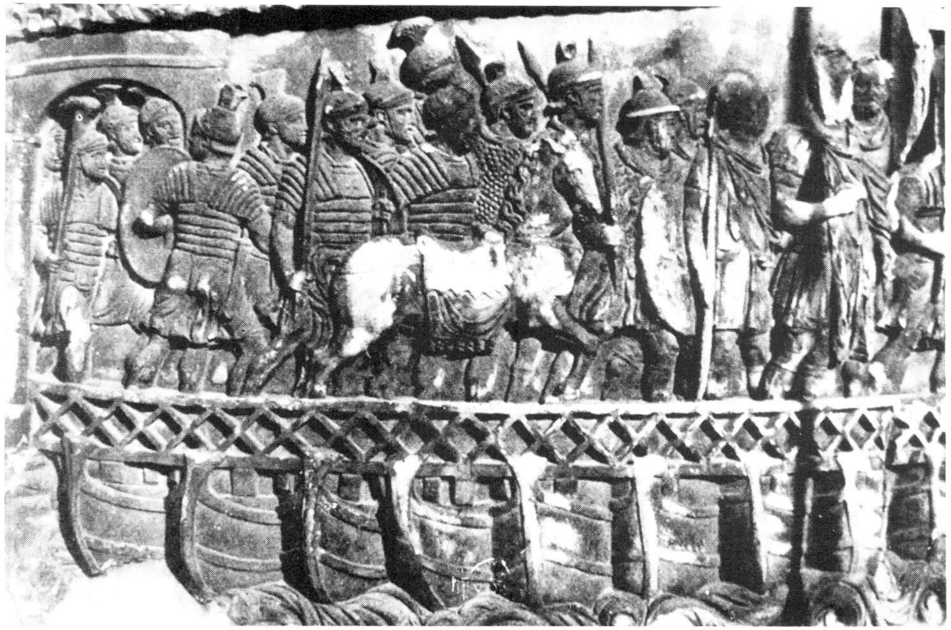

Fig. 108 Scene from column of Marcus Aurelius – a bridge of boats, with similarities to Figure 107 (Caprino et al., 1955, fig. 9).

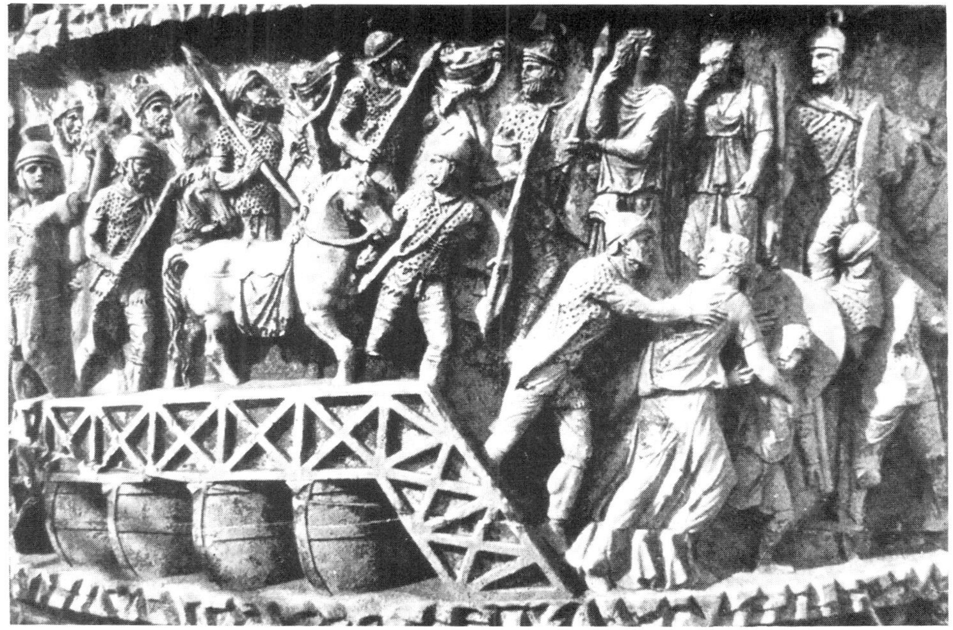

Fig. 109 Scene from column of Marcus Aurelius – another bridge of boats (Caprino et al., 1955, fig. 104).

ships from one vessel to the other with planks across them to make the structure rigid ... Such, then, is the traditional method employed by the Romans. Precisely how Alexander bridged the Indus I do not know ...

(Arrian, tr. by A. de Sélincourt, 1958: 169–70)

The Roman bridge across the Rhône at Arles (G6) was a floating bridge, pictured on a mosaic at Ostia (Rivet, 1988: 96). There is a late reference to a Roman bridge of boats across the Euphrates at Zeugma in AD 165 (Webster, 1979: 235), with another description of the method of construction in Dio Cassius (71.3), and, no doubt, there were many others.

In describing the scene on Trajan's column (Fig. 107), Richmond (1982: 7) draws attention to what he describes as a 'stout pier of short vertical logs held firm by horizontal slats', on each boat. He says, 'These are the main piers of the bridge; but, in addition, a pontoon floats between each pair of boats, and is built of closely fitted planks, heavily nailed.'

Concerning the first, Richmond is clearly in error, for other views, such as scene 34, show a bridge-boat in side view (Fig. 110). The 'pier' is in fact the seat placed in the rear of the boat and used by a steersman holding a steering oar. The 'piers' are omitted from the bridges on the other column, either because they were not present or, more likely, because the scene shows the bows of the boats. In one of the scenes (Fig. 109), the bridging units rest directly on the boats; in the other (Fig. 108), there appear to be baulks of timber resting on the boats and supporting the deck. In neither case are there 'pontoons' between the boats.

Whatever the meaning of the vertical slats shown beside the seats in Figure 107, it seems unlikely that they were pontoons. The boats themselves were the pontoons. Scene 133 of Trajan's column (Richmond, 1982: 30–1, pl. 10) shows legionary-artificers carving these boats out of solid tree trunks. Vegetius (1.10) indicates that army carpenters were regularly expected to make these boats, and the column of Marcus Aurelius (Caprino et al., 1955, fig. 132) shows such a boat being drawn by two oxen on a four-wheeled cart, with shields, helmets and spears stacked inside. Other illustrations suggest that the boats were dual-purpose, being used either for bridge-work, or for the ferrying of material across waterways.

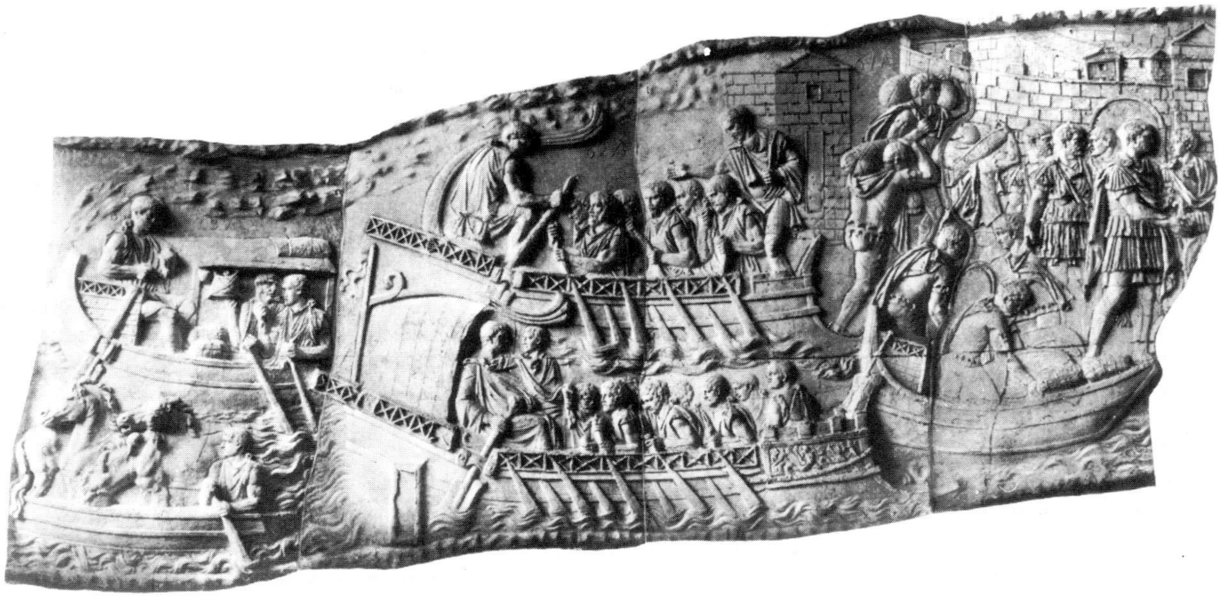

Fig. 110 Scene from Trajan's column – a bridge-boat in side view (Cichorius, 1886–1900, scenes 34 and 35).

The Roman army was therefore well equipped for amphibious operations. It is hard to believe that, having prepared themselves with a supply of boats, they did not carry decking also.

It has been mentioned already that the scale of these representations should be treated with care. The boats being shaped by soldiers (Richmond, 1982, pl. 10), for example, are clearly to a smaller scale than the men themselves. Figures 107, 108 and 109 appear to show the boats touching one another, side to side. Although this may have been the case, there are reasons to believe that some spacing would have been provided. A continuous raft would require a large number of boats. Even if the boats were side to side they would be unstable under pedestrian loads unless some form of longitudinal decking were present and this is suggested in the scenes. Figure 108, for example, indicates that the deck of the bridge, whatever its form might be, was spaced some distance above the boats. Any reasonable decking or superstructure could span some distance. An artist's view of the bridge would be foreshortened and the dominant impression would be that the boats were closely spaced. It is believed, therefore, that the actual boat spacing was somewhat more than that shown in the illustrations, and that a properly designed system of decking was provided.

All illustrations of the boat bridges show a substantial truss-like structure spanning between the boats. The common view is that this was a railing; for example, Richmond (1982: 7) says, 'The roadway structure is of three thicknesses of timber' (see the right hand side of Fig. 107), 'but the detail is not clearly discernible. Stout railings are fixed to uprights and diagonal cross-spars with very large nails. Where the bridge comes to land, it is carried on vertical struts, braced by cross-spars parallel with the roadway.' A better suggestion could be that the 'three thicknesses of timber' consist of longitudinal girders, transverse planking and the bottom longitudinal member of the railing, similar to the bridge shown in Figure 112 (Cichorius, 1886–1900, scene 21). There is no doubt that railings of this kind are shown elsewhere when their sole function is to serve as railings (Cichorius, 1886–1900, scenes 1, 14, etc.). Indeed a railing of this form, with verticals, horizontals and double diagonals, appears to be characteristic of Roman construction.

However, it is possible that these railings could have a considerable capacity to span longitudinally as trusses, and this raises the questions of whether the Romans were aware of this capacity and, if so, whether they deliberately used it. An alternative superstructure for the bridge-boats would consist of two longitudinal timber trusses instead of girders, with transverse decking attached beneath the lower chords.

For this to be a reasonable hypothesis, it would be necessary for the members of the trusses to be properly joined. A close examination of the original of Figure 107 shows the circular end of a large fastener at the centre of each joint – Richmond calls it a large nail. Alternatively, it could have been a heavy metal rod or bolt, passed through washers at its ends and riveted over. This would be effective only if it passed through the members. The truss of Figure 107 has four members at each joint – the horizontal, vertical, and two diagonals. It would not be difficult to pass the vertical member into the horizontal with a mortice and tenon joint. To join the diagonals in the same fashion would be more difficult. Irrespective of the nature of the loading, only one compression diagonal would be required at a time: that is, one for some loadings, and the second for others. If both diagonals were closely fitted to the junction of the vertical and horizontal members, an effective joint would be achieved. The diagonals would have to be kept in place in some way, perhaps by skew nails, but this would be the case also if the truss served only as a railing.

Figure 108 shows a somewhat different geometry, with horizontals and diagonals but no verticals. This geometry would be somewhat less suitable as a railing. If it was used as a truss, all three members at a joint would have to be shaped so as to engage the joint. Another geometry, with single diagonals, is shown in scene 92 of Cichorius (1886–1900). An underslung deck would also cause difficulties with attachments, in view of the fact that the Romans probably did not have screwed metal bolts and nuts (Section 4.2).

The erection problem differs between the two systems. With a system of girders, transverse decking and railings, the girders would be erected first, then the decking and finally the railings. In the other system, the trusses would need to be prefabricated. They would be erected first, followed by the decking.

All things considered, it seems likely that the Romans did not use trusses for their bridges of boats. The most compelling argument is that one would expect that, if prefabricated trusses formed part of the Roman plan for amphibious operations, they also would have been shown on the columns. However, this still does not remove the possibility that the Romans used the timber truss. The two cases that arise are in the ship landing bridges, and as a stiffening truss in the timber arch bridge of Apollodorus.

The columns have many views of boats; one of them, taken from Trajan's column (Cichorius, 1886–1900, scenes 34, 35), is shown in Figure 110. The boats at the left and right are bridge-boats, with the steersman sitting in a decorated chair, as described previously. The others, at the centre of the picture, are biremes with two rows of oars. They form another example of the lack of scale – or rather the change in scale – in these carvings, for the men are clearly too big for the boats. The man at the left is the steersman, for he is holding a large steering oar. He is sitting then in the stern of the boat, whose forward motion is from left to right, and is in front of a removable

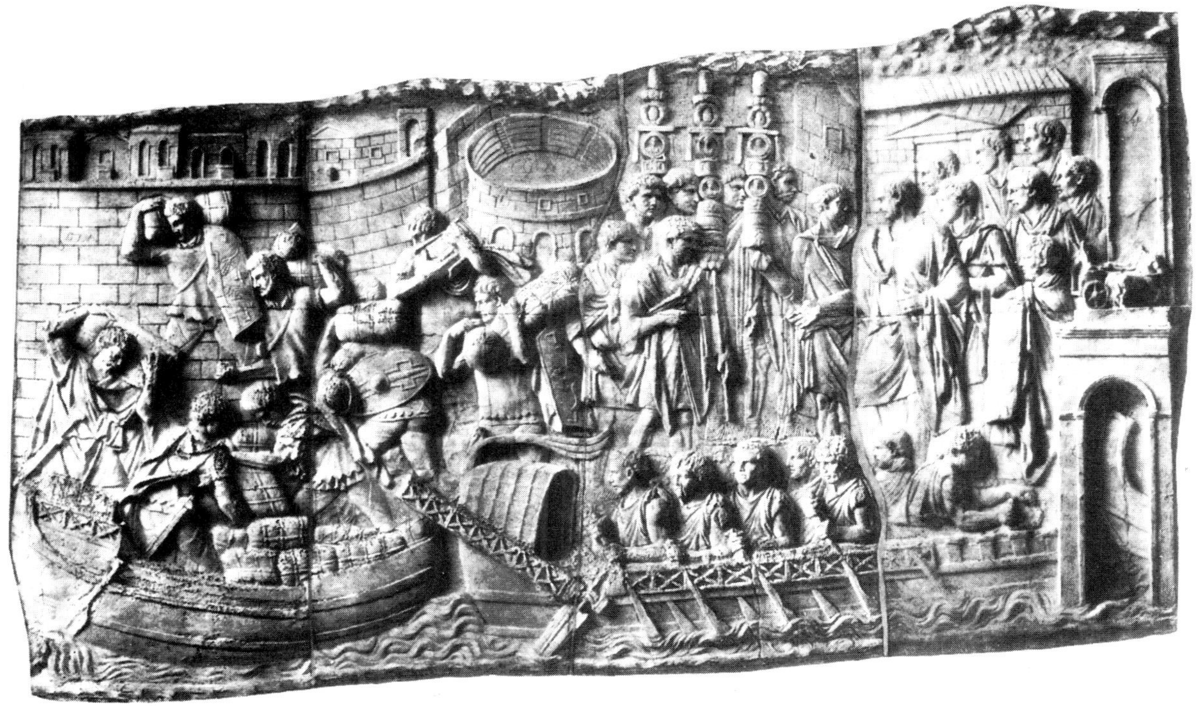

Fig. 111 Scene from Trajan's column: boat with pivoted landing-bridge (Cichorius, 1886–1900, scene 33).

tent or cover slung from an extension to the stern post.

Each boat has two sections of railing, with horizontals, verticals and double diagonals, as described before. The amidships section appears to serve simply as a railing. The other section forms part of a landing bridge. It is shown more clearly in Figure 111 (Cichorius, 1886–1900, scene 33), where its pivoted base is visible. Figure 110 has the bridge in its stored position, whereas in Figure 111 it has been lifted to span to the shore, and is being climbed by a man with a load. It is not clear how the man has gone past the covered area near the stern post, but possibly the bridge is in two parts, one on each side of the tent.

Earlier in Roman history, Polybius (1.22) describes the *corvus* used briefly by Rome in her naval battles with the Carthaginians (264–201 BC). The standard fighting ship in Greek and Roman times (Landells, 1978; White, 1984) relied on its speed to attack the enemy. It was oar-driven, and Landels has estimated (p. 146) that it could achieve a maximum speed of about 9.5 knots. Its chief weapon was a heavy bronze pointed ram, mounted in the bows below the water-line, used to disable the enemy ship and put it out of action. The Romans, at least for a short time, changed the strategy. They fitted their ships with a boarding bridge, enabling their soldiers to board and overcome the enemy. This boarding bridge was called a *corvus* or crow (or raven); it has been described by

Polybius and reconstructed by Thiel (1954: 432–47, 101–13).

Polybius' description of the *corvus*, slightly abbreviated, is as follows

On the prow stood a round pole four fathoms (i.e. about 24 ft or 7 m) in height and three palms (10 in or 0.25 m) in diameter. This pole had a pulley at the summit and round it was put a gangway made of cross planks attached by nails, 4 feet in width and 6 fathoms (about 36 ft or 10.5 m) in length. In this gangway was an oblong hole, and it went round the pole at a distance of 2 fathoms (about 12 ft or 3.5 m) from its near end. The gangway also had a railing on each of its long sides as high as a man's knee. At its extremity was fastened an iron object like a pestle pointed at one end and with a ring at the other end, so that the whole looked like the machine for pounding corn. To this ring was attached a rope with which, when the ship charged an enemy, they raised the ravens by means of the pulley on the pole and let them down on the enemy's deck, sometimes from the prow and sometimes bringing them round when the ships collided broadsides. Once the ravens were fixed in the planks of the enemy's deck and grappled the ships together, if they were broadside on, they boarded from all sides, but if they had charged with the prow they attacked by passing over the gangway of the raven itself two abreast.

(Book 1, pp. 60–3, tr. by W. R. Paton)

The *corvus* received its name from the heavy iron beak, driven into the enemy's deck. Thiel (1954) points out that the boarding bridge vanishes after the battles of Mylae and Ecnomos. Neither Polybius himself nor any other ancient authority ever hints at its being employed after those battles. It was a surprise weapon which enabled the Romans to use their land-based power to defeat the seafaring Carthaginians. Although there is some doubt as to its actual form, Thiel offers a credible model based on suggestions by Wallinga and Bolkestein.

Now the landing bridges shown on Trajan's column differ from the *corvus* – they are mounted on merchant rather than fighting ships, on the stern rather than the bow, and cannot be rotated about a vertical axis. However, the similarities are sufficiently marked to suggest a historical link. It is then that Polybius' description becomes important – 'a gangway made of cross-planks attached by nails' or, to quote another translation (Chambers) 'nailed together of criss-cross planks'. This description fits some form of timber truss. While it is difficult to conclude that the superstructure of the bridge of boats made deliberate use of the principle of the truss, it is easier to reach this conclusion about the *corvus* and the landing bridge. If it was employed in the landing bridge, then it may have been used elsewhere in locations where its use was appropriate. One such possibility was in the timber arch bridge of Apollodorus (see Section 6.3).

There is no doubt that the most common type of timber bridge built by the Romans was the timber girder bridge. As many as perhaps thirteen are shown on the columns – twelve on Trajan's column (Cichorius, 1886–1900, scenes 12, 14–15 (three bridges), 17, 19 (possibly two bridges), 21, 56, 58, 101, 131) and one on the column of Marcus Aurelius (Caprino et al., 1955, fig. 108). The columns, by contrast, only rarely show stone arch bridges, and in these cases there is no indication that the army was involved in their construction. There is, for example, a long 34-span stone arch bridge in scene 86 of Trajan's column (Cichorius, 1886–1900) and a single-span stone arch at the entrance to a city in Figure 119 of Caprino et al. (1955).

Three typical timber girder bridges are shown in Figures 112, 113 and 114. In all cases, the problem of scale again arises. Figure 112 (Cichorius, 1886–1900, scene 21) shows a four-span bridge. The braced piers have vertical timber piles and a timber crosshead. They support longitudinal girders, on top of which is transverse planking, with the ends clearly shown. The whole is surmounted by the characteristic timber railing. Figure 113 is also from Trajan's column (Cichorius, 1886–1900, scene 131). It is described here as a bridge, but could just as well be a

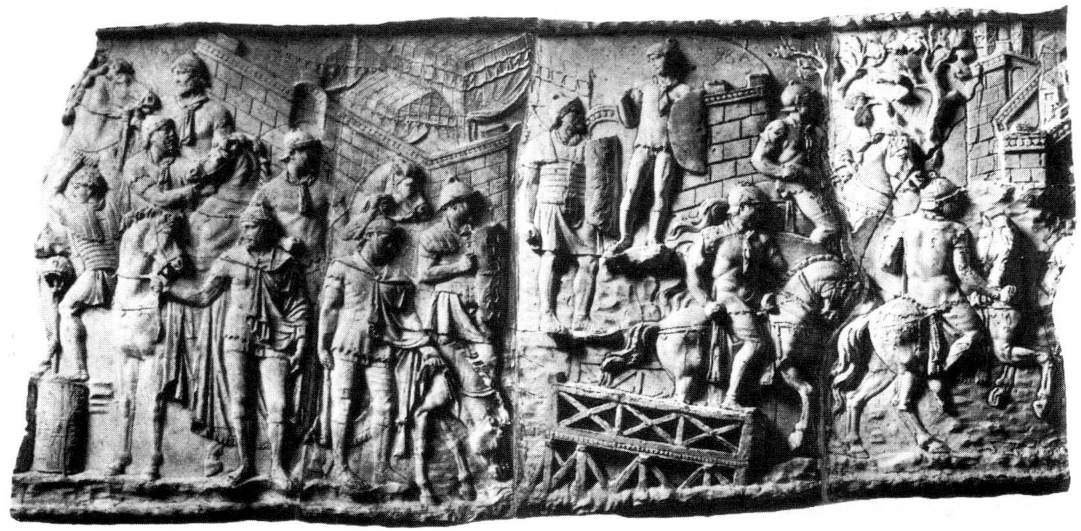

Fig. 112 Scene from Trajan's column showing timber girder bridge with railings (Cichorius, 1886–1900, scene 21).

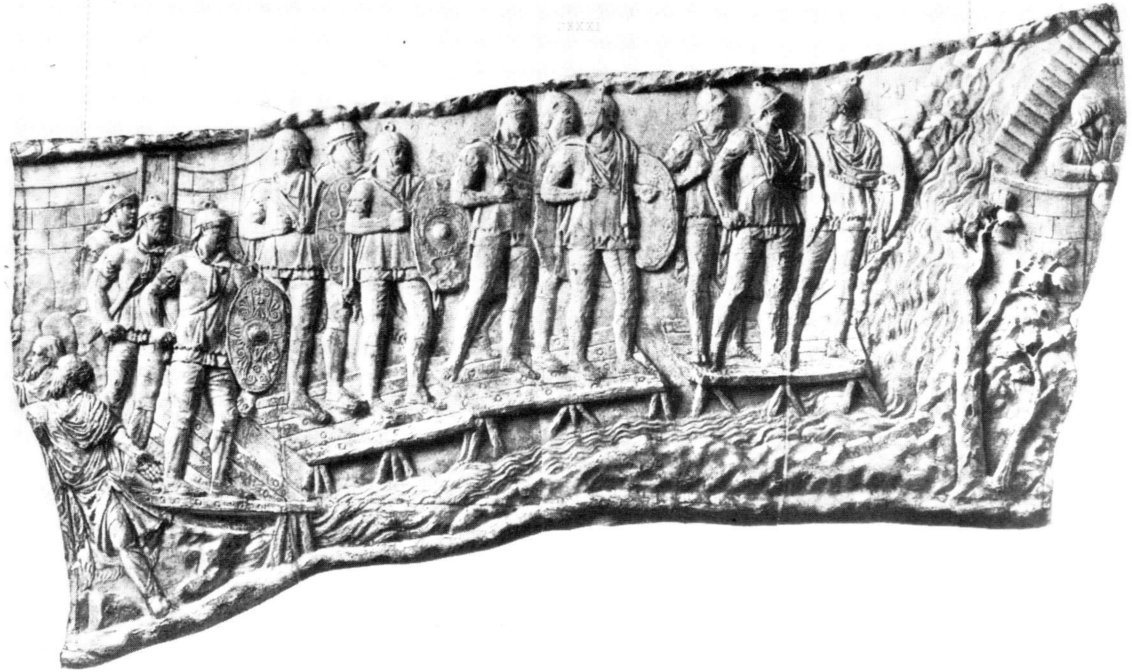

Fig. 113 Scene from Trajan's column – bridge or wharf (Cichorius, 1886–1900, scene 131).

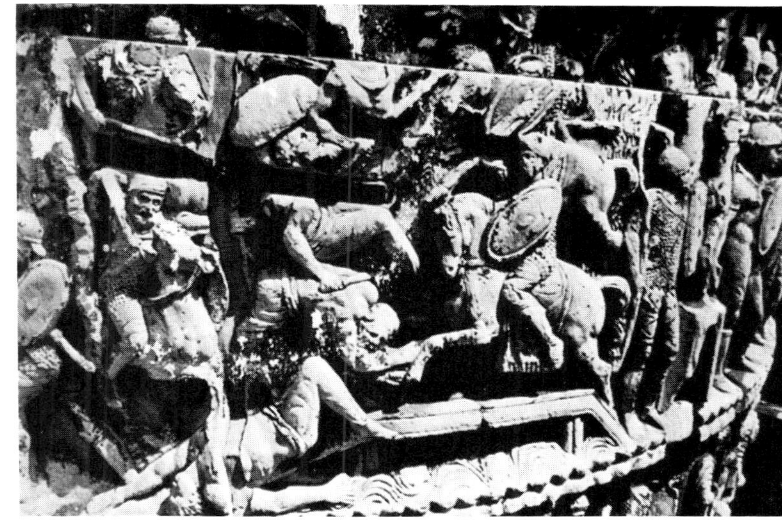

Fig. 114 Scene from column of Marcus Aurelius – timber girder bridge with single main span and sloping approaches (Caprino et al., 1955, fig. 108).

timber wharf. Again it has braced timber piers; the piles and bracing are shown, and the crosshead is implied. The deck consists of longitudinal planking spanning between the piers. There is no railing. The last, Figure 114, from the column of Marcus Aurelius (Caprino et al., 1955, fig. 108) is simpler again with one main span and two inclined approach spans. The form of the structure is unclear.

The final bridge shown on the columns is the timber arch bridge designed by Apollodorus for Trajan, and built across the Danube between Pontes and Drobeta, to replace the downstream bridge of boats. The description of it will be reserved for Section 6.3.

What conclusions then may be drawn from this study?

1 The Roman army built many bridges.
2 These were commonly of timber.
3 The most common army bridge was the timber girder bridge.
4 The army was equipped for amphibious operations and carried with it boats for the construction of boat bridges.
5 The decking of these boat bridges probably consisted of planking on timber girders. However, the railings of these bridges were truss-like in form and suggest that the Romans may have made either inadvertent or deliberate use of truss action.
6 Ship-landing bridges shown on the columns appear to be related to the *corvus* of Polybius, and may also rely on truss action.

6.3 Early timber bridges

The earliest recorded major bridge is a timber bridge built by the Chaldean king Nabopolassar over the Euphrates River at Babylon. Father of the mighty Nebuchadnezzar, who succeeded him in the year 605 BC, he built a bridge said to be 115 m long, with timber girders on seven piers of baked brick, timber and stone, 9 m by 20 m in plan (de Camp, 1977: 75). The corresponding pier spacing or girder span would have been about 7 m; of the total length of 115 m, only about 56 m or 48 per cent was left as waterway. The bridge was described by Herodotus (1.186) and Diodorus (2.8.1–3) and for centuries was the only structure of its kind (de Camp, 1977). BNI (1944b: 631) describes it in these words, 'Along the river front were great quays, later built up into fortification walls, and the river was crossed by a brick bridge, the most ancient bridge of which traces survive in the world.' Presumably the reference to 'brick' is in regard to the piers.

Hammond (1988: 526–7) speaks of bridges built by the Persians as part of their road system. To quote him, 'It is possible that bridges were built on wooden piles driven into the clay bed of the river. This was the method employed *c.* 425 BC at Nine Ways [or Amphipolis] in Macedonia.'

A detailed description of a Roman timber bridge (**T3**) has been provided by Caesar:

Caesar had determined to cross the Rhine, but a crossing by means of boats seemed to him both too risky, and beneath his dignity as a Roman commander. Therefore, although the construction of a bridge presented very great difficulties on account of the breadth, depth, and swiftness of the stream, he decided that he must either attempt it or give up the idea of crossing. The method he adopted in building the bridge was as follows. He took a pair of piles a foot and a half thick, slightly pointed at the lower ends and of a length adapted to the varying depth of the river and fastened them together 2 feet apart. These he lowered into the river with appropriate tackle, placed them in position at right angles to the bank, and drove them home with pile-drivers, not vertically, as piles are generally fixed, but obliquely, inclined in the direction of the current. Opposite these, 40 feet lower down the river, another pair of piles was planted, similarly fixed together, and inclined in the opposite direction to the current. The two pairs were then joined by a beam 2 feet wide, whose ends fitted exactly into the spaces between the two piles forming each pair. The upper pair was kept at the right distance from the lower pair by means of iron braces, one of which was used to fasten each pile to the end of the beam. The pairs of piles being thus held apart, and each pair individually strengthened by a diagonal tie between the two piles, the whole structure was so rigid that, in accordance with the laws of physics, the greater the force of the current, the more tightly were the piles held in position. A series of these piles and transverse beams was carried right across the stream and connected by lengths of timber running in the direction of the bridge; on these were laid poles and bundles of sticks. In spite of the strength of the structure, additional piles were fixed obliquely to each pair of the original piles along the whole length of the downstream side of the bridge, holding them up like a buttress and opposing the force of the current. Others were fixed also a little above the bridge, so that if the natives tried to demolish it by floating down trunks or beams, those buffers would break the force of the impact and preserve the bridge from injury.

(Caesar, *Gallic War* 4.17, tr. by S. A. Handford, 1951)

Caesar goes on to say, 'Ten days after the collection of the timber had begun, the work was completed, and the army crossed over . . . After spending a total of eighteen days across the Rhine . . . Caesar recrossed the bridge into Gaul and destroyed it behind him.'

Webster (1979: 235) suggests that the purpose of the bridge was at least partly to demonstrate the speed and skills of Caesar's army – 'to overawe the German tribes'. Chevallier (1976: 219, n. 97) makes a similar comment, 'The technical achievement . . . was outweighed by its importance politically (it was aimed at the Barbarians' morale) and as a religious symbol (victory over the river god).'

A reconstruction of the bridge is given in Figure 115. Others may be found, for example, on page 201 of Edwards' translation of Caesar's *Gallic War*, Hamey and Hamey (1981: 29) and Connolly (1981: 240). All are based on Caesar's description, but with some variations concerning the deck width, the diagonal ties and the form of the bracing piles.

The incident took place in 55 BC. The precise location of the bridge is not known, but it appears to have been near the junction of the Moselle and the Rhine. Handford (p. 266) suggests Coblenz. There was a Roman fort and bridgehead at Roman Moguntiacum or Mainz (Brogan, 1953: 117) but this bridge (**T4**) appears to have been later, possibly at the time of Trajan (Gazzola, 1963b, no. 177). If Caesar's bridge was at Coblenz, then its length would have been at least 260 m. Assuming a pier spacing of 10 m, the number of piers would have been about twenty-six. To complete the bridge in ten days, Caesar would have had to average almost three spans a day. Despite the daunting logistics, it is difficult to say that this would have been impossible. Equipment would have been his major problem – in particular for lifting and pile driving. His description implies that his piles were driven in pairs; that the two piles were attached together 2 feet apart and then driven into the river bed. They were driven at an angle, and it would have been necessary to achieve at least sufficient penetration for them to stand without falling over after driving was complete. His task therefore required the driving of about six of these main pile-pairs each day. Again, it is difficult to conclude that this was impossible, but if the account is true, then Caesar's achievement was indeed remarkable.

The 40-foot pile-spacing is presumably at the base of the piles. They were inclined and this base spacing is not inconsistent with a deck width of about 25 feet (8 m). The account does not imply that the timbers were squared. There is no reason why untrimmed round members could not have been used, and this would have saved time (see also Bundgard, 1965).

Caesar refers also to other bridges, for example across the Loire (**T5**; *Gallic War* 8.27), and a boat bridge (**T6**) built by his general Labienus (7.58) across the Seine near Paris. Some of these references imply the existence of bridges before the Roman occupation. Paragraph 11 of *Gallic War* Book 7 describes Caesar's attack on Cenabum (Orléans) on the Loire in 52 BC (Handford's translation).

Caesar reached this town in two days, but by the time he had encamped in front of it, it was too late to attack that day. He therefore directed the necessary preparations to be made for

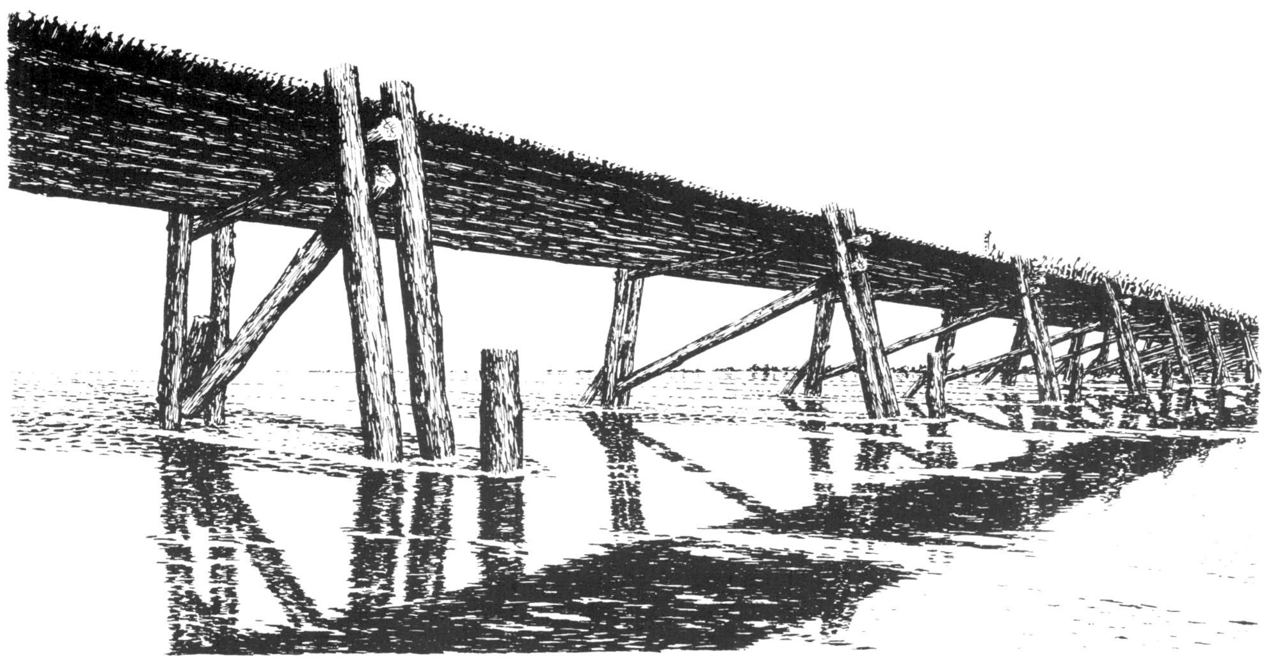

Fig. 115 Possible reconstruction of Caesar's bridge over the Rhine (T3, reconstruction by the author).

an assault on the next day, and as there was a bridge over the Loire right under the town walls, and he was afraid that the inhabitants might escape under cover of night, he ordered two legions to remain under arms all night. Shortly before midnight the people of Cenabum moved silently out of town and began to cross the river.

Caesar attacked 'and all but a very few of the enemy [were] taken prisoner – for the narrow streets and bridge were blocked by the crowd of fugitives' (**T7**; see also Cary and Scullard, 1975: 211).

Similarly, Dio Cassius (60.20.6), in describing Caesar's defeat of the Britons near London, seems to imply the pre-existence of a bridge (**T8**) across the Thames (in 54 BC). Another was at Geneva (**T9**; Caesar 1.7). In 58 BC the Helvetii sought to leave their country (Switzerland) and attack Gaul. They had two possible routes – one between the Jura Mountains and the Rhône, the other south across the Rhône and then west through the land of the Allobroges (Transalpine Gaul), recently subdued by the Romans. Caesar writes, 'The frontier town of the Allobroges, nearest to Switzerland and connected with it by a bridge, is Geneva.' Caesar closed the southern route, and when the Helvetii went west along the other, he followed and defeated them. 'After this success Caesar wanted to pursue the remaining forces of the Helvetii. He therefore built a bridge over the Saône and led his army across.' There is mention then of two bridges, a pre-Roman bridge at Geneva, and a bridge built by Caesar across the Saône (**T10**), which was possibly another bridge of boats.

At Trier – Roman Treves – there is today a stone arch bridge across the Moselle. As mentioned in Section 4.3, this bridge has a Roman foundation. Cüppers (1969), followed by White (1984: 99) and Chevallier (1976: 100), has described the stages in the development of this bridge. The first bridge (**T11**), completed about AD 44, or at least prior to AD 70 (Tacitus, *Histories* 4.77), had stone piers on timber piles, supporting an understrutted timber girder deck. This was reconstructed about AD 140 (Chevallier, 1976: 101), again with a timber deck, and survived in this form until about 1400, when the deck was replaced by stone arches on the same foundations.

The principle of the understrut is well illustrated by the restored old bridge at Thouars, France, Figure 116 (Emerson and Gromart, 1925: 46–8). One span of this originally thirteenth-century bridge has timber main girders supported by understruts; that is, by diagonal timber members beneath the main girders, running from the piers to about the third points of the span. A similar system was used at Trier.

To some, the best known of all Roman bridges would be that defended by stout Horatius and his two friends.

> Four hundred trumpets sounded
> A peal of warlike glee,
> As that great host with measured tread,
> And spears advanced, and ensigns spread,
> Rolled slowly towards the bridge's head,
> Where stood the dauntless three.
>
> (Macaulay, 1894)

The bridge was called the **Pons Sublicius** (**T12**) and crossed the River Tiber in Rome downstream of the Isola Tiberina, the island in the Tiber, and the Pons Aemilius (or Rotto), upstream of the modern bridge with the same name (Chevallier, 1976: 95). The original bridge dates from about 600 BC. The incident described by Macaulay

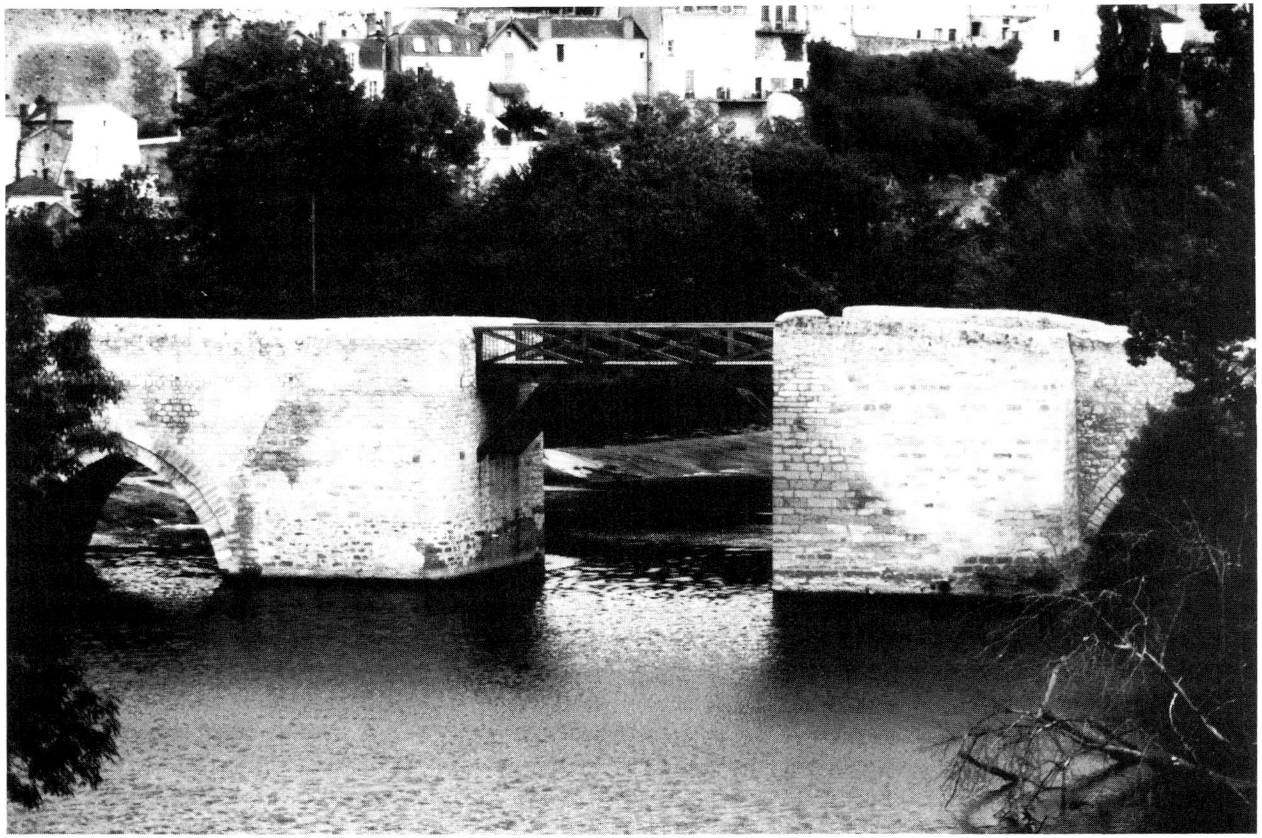

Fig. 116 The restored thirteenth-century timber bridge at Thouars, showing the principle of the understrut.

arose when the small city-state of Rome was ruled by the Etruscans. Rome rebelled and drove out Tarquinius, the Etruscan king. Tarquinius gained the support of Porsenna and returned to attack Rome, which was defended by Horatius Cocles and his two friends on the Pons Sublicius, or so the legend says (Section 2.2.1; Cary and Scullard, 1975: 55).

The name of the bridge derives from the Latin word *sublica*, pile or timber, and, as the name implies, the bridge was built of timber. It was destroyed during the defence by Horatius and subsequently rebuilt and destroyed by flood a number of times. There was a bridge at the site until about the fourth century AD. The bridge appears to have had a religious significance (Chevallier, 1976: 95; de Camp, 1977: 192). It has been attributed to Ancus Martius (Cary and Scullard, 1975: 40; Livy, 1.33.6); he built it entirely of timber, without metal fittings, and placed it in the care of the high priest, leading possibly to the title of *pontifex*, or *pontifices*, or *pontifex maximus*, later assumed by the emperors (see also Chapter 1, and p. 133).

One of the greatest of all Roman bridges was a timber arch bridge, that of Apollodorus across the Danube (**T13**). Shown on Trajan's column (see Fig. 117; Cichorius, 1886–1900, scenes 98, 99), it was built for Trajan about AD 104–5 (Gazzola, 1963b, no. 184) by Apollodorus the engineer-architect of Damascus (Sitwell, 1981: 137; Procopius, 4.6; see also Section 3.2). It crossed the Danube between Pontes, modern Turnu-Severin, and Drobeta, replacing the more easterly of Trajan's boat bridges (Rossi, 1971: 32; see Section 6.2). It was made unusable by Hadrian (AD 117–138), but later restored by Constantine about AD 328 (Gazzola, 1963b: 137). At least one of the piers remains.

The bridge is notable for two reasons – its great length, and its unusual form. Dio Cassius (58.13) cites the bridge as having twenty piers, each 150 feet high, 60 feet wide, and 170 feet apart. Taking one Roman foot as 0.294 m (Section 4.7), these dimensions become 44.1 m, 17.6 m and 50 m. If, as seems probable, the 170 foot spacing is the distance centre to centre of the piers, then the clear span would be 110 feet or 32.3 m. The sum of twenty-one such spans and twenty piers gives a length between abutments of 3510 feet or 1032 m. The illustration (Fig. 117) shows in addition a stone arch at one end of the bridge, so the total length would be somewhat greater. Forbes (in Singer et al., 1954–84, vol. 2: 512) quotes 1070 m. Gazzola gives the slightly higher figure of more than 1100 m, but he also quotes somewhat larger span lengths, of from 35 to 38 m clear. Hopkins (1970: 109) has an unsourced drawing of the approaches. This includes one timber arch of span 12.6 m. He also quotes a height from water-level to the underside of the deck of 10.2 m, and a roadway width of 6.6 m.

Although some of the dimensions given by Dio Cassius must be viewed with caution – the pier height in particular is difficult to accept – the span dimensions should be more reliable, as the total length has to accord with the known width of the river at the site. A bridge of length

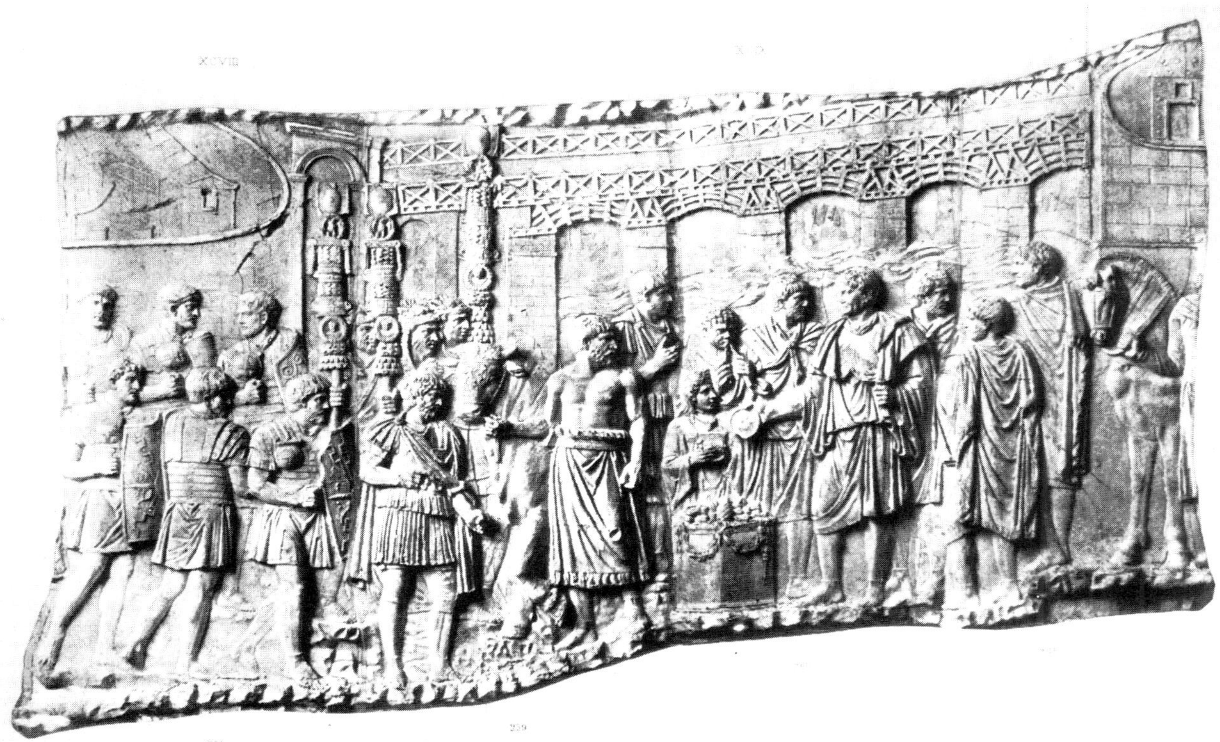

Fig. 117 Scene from Trajan's column – the great timber arch bridge of Apollodorus over the Danube, believed to be the longest, and in many respects the greatest, of all Roman bridges (T13, Cichorius, 1886–1900, scenes 98 and 99).

1070 to 1100 m would almost certainly be the longest ever built by the Romans. For comparison, the longest existing bridge is the stone arch bridge at Mérida, with an overall length in excess of 720 m (Section 5.7).

The timber arch spans are unique among known Roman bridges – although J. Collingwood Bruce has speculated that a similar form of construction may have been used for the Pons Aelius at Newcastle (see Section 6.4). The following discussion is based on the representation of Trajan's column, and an assumed clear span of 32.3 m. A suggested reconstruction is given in Figure 118.

The timber arch shown in the column is segmental, with a rise to span ratio of about 1:7. The rise would then become 4.6 m. The left-hand span shown on the column differs from the other spans in two respects – the lower ends of the radial members extend below the bottom of the lowest timber arch, and the timber arches proceed down to the top face of the abutment. Those differences may result solely from the fact that it was an end span or may represent some accidental variability in the carving.

All spans show connections to radial members at the quarter points of the arches. There is no necessity to regard this as accurate – just as there is no need to regard the representation of four bays of railing per span as accurate. However, if the arches are assumed to be interconnected at their quarter points, then, for the lower arch, the distance between these connection points becomes 8.5 m. The simplest form of construction would have each arch rib consisting of four members, each about 8.5 m long. They would be carefully fitted at their ends, cut so as to bear on one another over their full depth and held in line by suitable splice details, probably composed of iron plates. The connection difficulty discussed in Section 4.2 would arise – the Romans would not have been able to use nuts and bolts, but could have used plain metal rods, with their ends riveted over as they passed through the splice plates. These timber members would be straight; the resulting arch would not be circular, but would consist of four straight segments. If it were desired to have an arch that was more nearly circular, then it would be necessary to form the arch of more than four members, with more splices, and more radial interconnections.

An arch like this needs to be analysed for two load cases: under maximum load, possibly with a load spread over the full length of the bridge; and under non-uniform loading – either a heavy local load or a distributed load spread only over about one half of the span, from about the support to mid-span. In Section 4.6, a timber girder bridge was analysed under the action of its self weight plus a distributed live load equivalent to 250 kg per square metre. Consider first this load applied over the whole span. If it is assumed that each arch is composed of three members 450 mm square in cross-section; that the horizontal spacing between arches is 1.5 m; that the decking is 100 mm thick; and that the timber is oak; then the total thrust in one arch is 224 kN or about 75 kN per member. The nominal axial stress is 0.37 kN. Each member is, in effect, a column, supported at its ends, with a length to width ratio of 18.9. It is apparent from the allowable stresses quoted in Section 4.6 (which

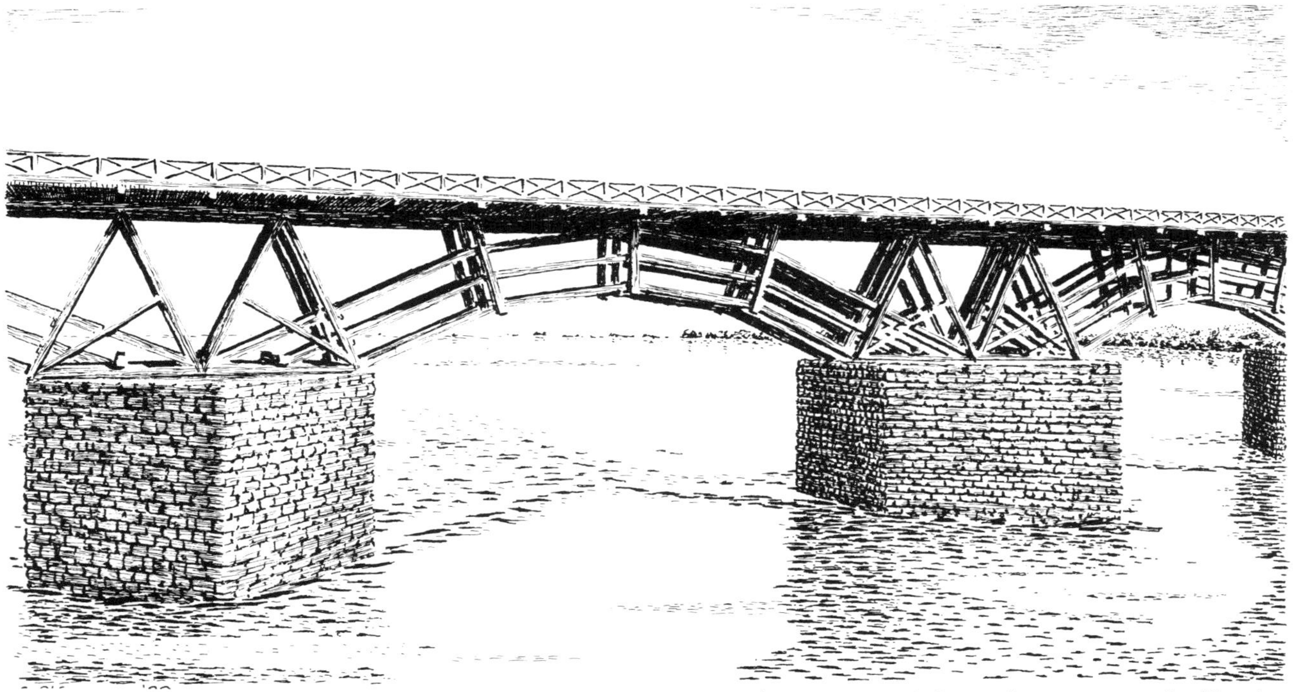

Fig. 118 Possible reconstruction of the great bridge of Apollodorus, with twenty-one timber arch spans across the Danube (T13, reconstruction by the author).

are for bending, not for axial loads) that the arch members could support these loads. Indeed much smaller members would be strong enough.

This assumes that the arch members are effectively braced at the quarter-points. It would also be desirable for the connection details between successive members to be such as to achieve full continuity through the quarter points. It is most unlikely that this could be done. Under these conditions, to take a limiting case, if there were no effective bracing at the quarter points, and no moment continuity at all between successive arch ribs, then the arch would buckle and collapse; in any case it would be unable to support local or asymmetric loads.

A close view of Figure 117 shows two additional features. The deck is shown with longitudinal planking on transverse deck beams; and each railing is in the form of a truss. It is possible, therefore, that the bridge of Apollodorus behaved as a braced arch, with at least some of the bracing stiffness provided by the railings (see also Section 6.2).

The width of the deck is not known. Trajan's column shows a considerable width, in excess of the span, but it would be unwise to rely on its accuracy for relative dimensions of this kind. More likely the deck width would have been of the order of 5 to 6 m, sufficient to allow vehicles to pass – a necessity on such a long bridge. This width is much less than the span; the transverse deck beams should then be sufficient to transfer bracing loads from the railings to any interior arches. If three sets of arches were provided, then each would support about 2 m of deck, a little greater than was used in the above calculations. Arch members of oak 450 mm square would still be more than adequate. It would be necessary to provide transverse connections between the ribs.

If the bridge does indeed behave as a braced arch, with the bracing provided by the railings, then the forces applied to the railings would be considerable. The railings would need to be substantial, with ample member sizes and proper connections.

There are other possibilities – accurate cutting of the member ends associated with suitable end fastenings could provide at least partial continuity between arch members at the quarter-points; the radial interconnecting members could be used to provide some shear connection between the component arch ribs; the use of deep members for the main ribs could add to their bending stiffness and strength; and, most importantly, the pier detail could provide end fixity to the arch as a whole.

It has been mentioned previously that the representation of the end span differs from the interior spans – the end arch members are continued down to reach the top face of the pier, whereas at the intermediate piers, the arches appear to stop at the face of a triangular timber structure mounted on the pier top. It is suggested that this difference may be more apparent than real. If the arches were to stop at the face of a timber structure, then the forces on this structure would be large and difficult to support. A more efficient detail would carry these forces directly to the pier top, either by providing members to act as extensions to the arch ribs, or by extending the arch ribs themselves. If the latter device were adopted, then the triangular frames could be used to provide end fixity for the arch ribs. These frames would be placed on either side of each rib, and would partly obscure the ribs from view.

With this arrangement, the upper members of the arch would start at the pier top, pass through the bracing frame and cantilever forwards to the first quarter-point. If this upper member is considered as a whole, it would notionally form a three-pinned arch, with equivalent pins at the three interior quarter-points; that is, in the limiting case when the bending stiffness of the splice is ignored. The overall stability of the complete arch would be improved, and the erection problem would be eased. The length of this outer member from the pier top to the first quarter-point would become about 16.5 m. Although longer than the normal member length of 8.5 m, it would still be procurable. From the engineering point of view, this literal interpretation, with the arch having only four bays, is preferable to other modern reconstructions showing a multiplicity of bays.

As mentioned above, the stability of the arch ribs could have been improved by truss action in the railings. Other authors have also made this suggestion. Hopkins (1970: 109), who was Professor of Civil Engineering at the University of Canterbury, New Zealand, says, 'the detail on Trajan's column, stylised though it may be, of Apollodorus' bridge over the Danube shows an efficiently triangulated lattice girder supported by timber arch ribs.' De Camp's observations (1977: 221) are similar: 'if the sculptor who carved the relief was accurate, Apollodorus used diagonal bracing in the wooden part of the bridge. As far as I know, this is the first definite example of the truss, which depends for its strength on the rigidity of three beams fastened together to form a triangle.' Both go somewhat further than the facts suggest; although it is possible that truss action was used, it is difficult to be sure.

Gazzola (1963b: 136) has a photograph of a model of the bridge. This also carries the arch ribs down to the pier tops. It includes radial frames at the tenth points of the span.

It is also notable that a segmental arch was used; that is, the arch did not form a full semicircle. Richmond (1982: 36) says, 'the methods of building economically in wood dictate the segmental form of the arches. These methods no doubt inspired the military architects who built the fine stone-arched bridge at Alconétar [Section 5.7] and the segmental stone arches in the amphitheatre at Mérida, in Spain; but the Romans did not carry them far in application to masonry.'

Brookes (1974) has described underwater surveys carried out at Minturnae where the Via Appia crossed the Garigliano River over the **Pons Tirenus (T14)**, mentioned by Cicero (16.13.1). The Via Appia reached the river in 312 BC. Possibly a bridge was built then, but if so it was relocated when a *castrum* was constructed in 295 BC. The relocated bridge was almost entirely of wood, apart from the eastern pier, which appears to be a later addition, possibly in the second century BC, and was of masonry. The coins found in the stream next to the bridge indicate that it continued in service until at least AD 548. On the more conservative estimate, a timber bridge continued in use at the site, with regular maintenance, from about 295 BC to AD 548, or for about 840 years.

Brookes enquires why the bridge was not replaced by a masonry structure, and suggests two possible reasons. The river at Minturnae maintains a constant, year-long flow with a depth of about 9.5 m. There is no period of low water, and this, together with the great depth, would have made underwater construction difficult. The underlying rock dips steeply from the western bank, and continues at this slope beneath the river. The slope and the great depth to bedrock may together have caused the Romans to prefer timber.

Brookes suggests that, if the locations of surviving masonry bridges are plotted, they are either in rocky, mountainous terrain or alternatively in places where rivers run very low or dry at some time during the year. Roman bridges are rarely found along coastal stretches of the Viae Appia, Aurelia, Domitia or Tiburtina; there are no traces of bridges, for example, at crossings of the River Po. He concludes, 'The bridge remains at Minturnae would appear to demonstrate that the Romans were willing to use wooden bridges where masonry bridges were unsuitable.'

The final example of a timber bridge is **Justinian's bridge (T15)** across the River Siberis or Ala Dagh Su at Roman Sykeon in Turkey, about 100 km west of Ankara on the Pilgrim's Road (see also Section 5.9). Anderson discovered the bridge in 1898; his description (1899: 67) reads, 'In the bed of the river are seen great stone piers, varying slightly in size and in distance from one another; on these a level wooden bridge was laid, similar to that which is used at the present day.' Scaling from his drawing indicates that the clear spans varied from about 5.4 to 9.6 m, with an average of about 6.8 m. The roadway width appears to have been about 3.8 m. Macpherson (1954) has a later photograph of the piers, still surmounted by a timber deck. Procopius (*Buildings* 5.4) has a brief description of the bridge. It was built by Justinian and must date therefore from the period AD 527–65.

6.4 British bridges

Other compelling evidence for the widespread use of timber in bridge construction, at least for the superstructure, comes from British bridges. The substantial and well established system of Roman roads – Stane Street, Watling Street, Ermine Street, Fosse Way, Dere Street and the rest (Map 6) – would have required many bridges; the surprising thing is that there are so few substantial remains. The reason is that most, if not all, of these bridges were of timber. Collingwood and Richmond (1969: 2), for example, say, 'All attested Roman road-bridges in this country had stone piers and timber superstructures, like that of Trajan across the Danube. There are numerous examples . . .' Dymond (1961: 145) says, 'It is certain that the bridges built by the army during the conquest and pacification were of timber, for the sake of speed and destructibility . . . After pacification, when time and labour were available, bridges originally of timber were perhaps rebuilt in stone, and gradually new bridges of both timber and stone must have replaced the original fords.'

Dymond (1961) lists 87 bridges, divided into four categories: 12 bridges of category A, where it is assumed that a bridge has existed, but there are now no traces of it; 13 bridges of category B, where there is physical evidence of an old bridge, but it is not certain that it is Roman; 55 bridges of category C, where either physical remains exist now, or have been known to exist in the past, and it is reasonably certain that these are Roman; and, category D, another 7 bridges where the place name or documentary evidence implies the presence of a bridge.

It is possible that when Caesar's armies reached the Thames about 54 BC there was a bridge there (Dio Cassius 60.20.6), a little upstream of London. There are firmer grounds for believing that the Romans built a bridge across the Thames (Jackson, 1971), but even here there is some doubt; Hopkins (1970: 36) suggests there was a ford: 'There would be less obscurity regarding Roman roads in London if it were known whether a bridge was here in Roman times.' A smaller timber bridge left Roman Londinium at Newgate (**T16**; Singer et al., 1954–84, vol. 2: 509).

Most of the bridges in Dymond's category C, with physical remains believed to be Roman, are in the north. The following descriptions are based largely on his account, and are selected from his lists (Dymond, 1961). Not all bridges are included.

Cirencester, Roman Corinium, 15 miles south east of Gloucester, was the second largest town of Roman Britain. In front of the town gate, on Akeman Street (National Grid ref. SP027021), is the massive abutment of a bridge across the River Churn (**T17**). Further north is Wroxeter, Roman Viroconium, 5 miles south east of Shrewsbury, an important centre linked to London by Watling Street. A pier and scattered lumps of masonry are the remains of a bridge (**T18**) across a tributary of the Severn (map ref. SJ562083). The Severn itself is a little further west, and it is believed that there may have been a bridge here in Roman times.

Further north again is Wallasey, opposite Liverpool. Here (**T19**; map ref. SJ322894) were found the remains of a bridge 100 feet long and 24 feet wide (30 m × 7.3 m) buried deeply under silt. The bridge had stone piers, two of which have been found, and timber girders, some of which also have been found. The description given by Dymond (1961: 155) states that each span was supported on four compound girders each 33 feet (10.1 m) long, consisting of three well squared beams 'which carried the transverse rafters'. The girder spacing would have been no more than 2.4 m, possibly less, and it would have been possible to support timber decking directly on the timber girders; perhaps this is meant by 'rafters'. Alternatively, there may have been transverse deck beams and longitudinal decking. The description goes on to say that the timbers had mortice holes, some vertical and some oblique, for the railings. The carvings on the columns of Trajan and Marcus Aurelius (see Section 6.2) seem to imply that the typical Roman railing had a lower horizontal member into which were framed diagonal and vertical members, the whole being attached to the top of the decking and held against sideways collapse by external sloping braces. The presence of mortice holes in the main members would not then necessarily be expected, but the description of them is, in other respects, consistent with the distinctive form of the Roman railing. They therefore offer some support for the belief that the structure is Roman. If this is so – and Dymond thinks that it is – then the find is of major importance and has disclosed the only known existing timber bridge girders of Roman origin.

The main Roman road from London to the north was Ermine Street, which crossed the River Nene at Durobrivae, 'the walled town (or fort) by the bridges' (Sitwell, 1981: 45); from Water Newton to Castor (map ref. TL117976), 5 miles west of Peterborough. Here stone piers have been found (**T20**). Further north, Fosse Way joined Ermine Street at Lincoln. Some 11 miles south west, a branch road left Fosse Way near Brough, and crossed the River Trent between Collingham and Cromwell (map ref. SK808613). A timber raft has been found here, similar to that described by Vitruvius (5.12.3) and quoted in Section 4.2, of the kind that would be used as an underwater bridge foundation (**T21**).

The major relics of Roman bridge construction in Britain are to be found on Dere Street, between Piercebridge and Elishaw, and on Hadrian's Wall.

Piercebridge is on the River Tees, 5 miles west of Darlington (map ref. NZ213156). Stone piers were visible here until 1771 (**T22**). More recently, detached stones and thirty-two timber piles have been found (see also Bidwell and Holbrook, 1989: 110–12). Further downstream, south east of Darlington at Pounteys Bridge (map ref. NZ346121), also on the River Tees, are said to be vast quantities of squared stone in the river, and metal cramps (**T23**).

The structures at Piercebridge have given rise to the book *The Piercebridge Formula* by Selkirk (1983) and a review article by Lewis (1984). Lewis refers to the bridge carrying Dere Street (T22) and some further Roman stonework 500 yards downstream. This stonework is believed by Selkirk to have been a dam, bypassed by a lock, and forming a basin upstream. Piercebridge is 47 km beyond the tidal limit and 53 m above sea-level. Selkirk suggests that there was a series of locks downstream allowing boats or barges to be drawn inland as far as Piercebridge, and that there were similar engineering works on other streams. The purpose of these works would have been to provide low cost water-borne transport to the forts, but if this was available, it could be used also to carry the materials required for bridge construction. (Smith, 1977–8, also discusses Roman canals.)

North of Piercebridge, Dere Street crosses the River Wear between Binchester and Hunwick (map ref. NZ204318), a mile and a half north of Bishop Auckland. Dymond has found here four rusticated stones; the river

is 24 m wide and he suggests that the bridge (**T24**) had either stone piers and a timber superstructure, or stone abutments and timber piers. He believes there was also a timber bridge a little further north west, at Hunwick Gill (**T25**; map ref. NZ194326).

Corbridge, Roman Corstopitum, lies 3 miles south of Hadrian's Wall, 16 miles west of Newcastle. The River Tyne is crossed here by a fine, seventeenth-century stone arch bridge 100 m long. Dere Street crossed the river about half a mile upstream (at map ref. NY983646). The masonry substructure of the bridge which carried it (**T26**) is partly visible at low water; there are between five and ten stone piers and two stone abutments; the piers have cutwaters at the upstream ends and are 7.5 m long, 5.9 to 6.0 m wide, at a clear spacing of about 7.5 m (Bidwell and Holbrook, 1989: 104). The total length of the bridge is probably about 74 m (for six spans), or alternatively 132 m (for eleven spans). The form of the superstructure is unknown.

To the north, along Dere Street, there are remains of two other bridges. At Risingham (map ref. NY887862) over the River Rede, 2 miles north of Ridsdale, masonry is visible in the river (**T27**, Bidwell and Holbrook, 1989: 114). The road crossed the River Rede again 5.5 miles further north at Elishaw (map ref. NY854953) where there are stones and the remains of a timber raft (**T28**).

Hadrian's Wall (Bruce, 1978; Birley, 1963; Harrison, 1962; Breeze and Dobson, 1987) crossed Britain from the River Tyne at Wallsend to about Bowness on the Solway Firth. It started at a bridge, and crossed three major streams, the Rivers North Tyne, Irthing and Eden, in its 115 km length. Built about AD 120, it was garrisoned by some 10,000 men (Sitwell, 1981: 40). Its many forts were joined to allow the rapid deployment of troops along its length.

The eastern end of Hadrian's Wall is at Wallsend, about 3 miles east of Newcastle, on the north bank of the Tyne. Bruce (1978) suggests that it was originally planned to end at the **Pons Aelius**, a bridge built by Hadrian across the river at Newcastle (**T29**). Work on the present Swing Bridge was commenced about 1872 and disclosed the massive timber raft and piles that formed the foundation of one of Hadrian's piers. The raft was 10.4 m long and 4.9 m wide, with cutwaters at each end. Bruce believed that the superstructure may have consisted of timber arches like the Danube River bridge of Apollodorus (Section 6.3), but there is no evidence that this was so. Dymond (1961: 158) suggests that one of the timbers that was found may have been a bridge girder like that at Wallasey. The raft was two baulks deep, each 300 mm × 150 mm, attached to piles 2 m long. The roadway width may have been about 5 to 6 m (see also Smiles, 1862, vol. 1: 240; Bidwell and Holbrook, 1989: 100).

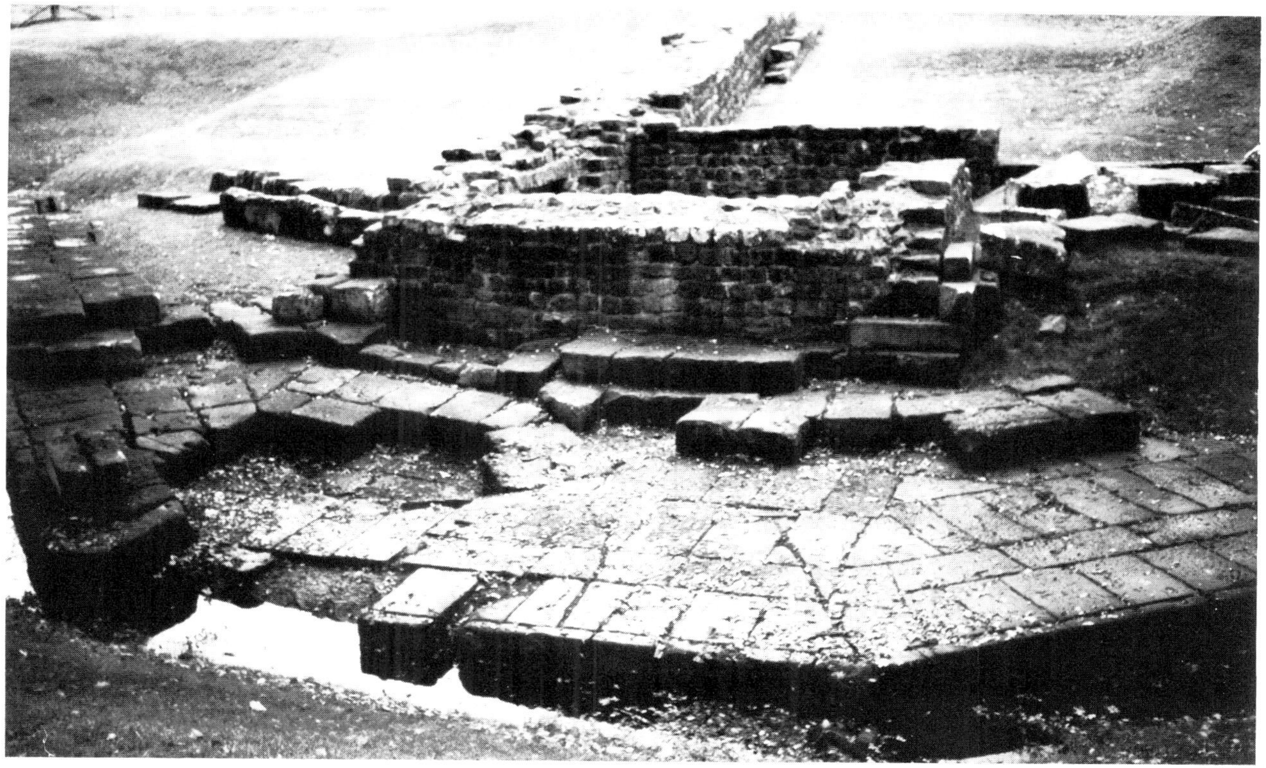

Fig. 119 Hadrian's Wall showing abutment of bridge at Chesters, including a lozenge-shaped earlier pier at left (T30).

The next bridge to the west was across the North Tyne River at Chesters (**T30**; map ref. NY913701), Roman Cilurnum; it may best be approached down the east bank of the river from Chollerford. Figure 119 shows the abutment. The river has since moved its course somewhat to the west, and is retained by an earth bank. The photograph is taken from the earth bank looking east – that is, away from the river – and shows the east abutment, the base of a square tower, and the remains of the wall running off into the distance. The straight face of the abutment, excluding wing walls, is about 6.7 m long. Exploration has shown another pier underneath the earth bank and a further two in the stream, visible at low flows (Bruce, 1978: 80–3; Mothersole, 1929; Bidwell and Holbrook, 1989: 16). These piers are 4.6 to 4.9 m wide and are at a clear spacing of about 10 to 11 m. The western abutment has not been located, but it appears likely that the bridge had a total of four spans, with a length of about 58 m between abutments. The piers have a cutwater at the upstream face only.

The abutment displays an unusual feature, visible in Figure 119. There is a recess in the top face at the mid-left caused by the removal of some stones in the top course. These stones, together with some adjacent stones, come from an earlier lozenge-shaped pier, with cutwaters at both ends, built into the present abutment. It is believed, therefore, that there was an earlier, narrower bridge at the site. Bruce thinks that this earlier bridge had stone arches, based on the presence of some reused, voussoir-shaped stones in the later structure; the later bridge, however, he suggests carried a timber superstructure. Dymond concludes that both bridges had decks of timber. The recent thorough study by Bidwell and Holbrook (1989) suggests that both had stone arches.

It is common to assume that the bridges carried the wall and a walkway or roadway across the stream. This wall could hardly have been of stone, as the resulting loading would be excessive applied to a timber girder bridge. It is also difficult to see what would have been achieved by extending the wall across, as the waterway itself would still have to be guarded. There is a large cylindrical stone, 2 m long and 0.5 m in diameter, now lying on the downstream wing wall (see Fig. 14). Bruce believed that this might have been a bollard 'to carry gratings, chains or hooks, slung across the river downstream from the bridge to catch anyone passing below it'. This also is conjecture, as it would be possible, if desired, to sling these obstructions from the bridge itself. Bidwell and Holbrook (1989) suggest that this was one of a number of stone columns built into the balustrades of the second stone arch bridge, and, indeed, base their argument for a stone arch superstructure largely on the existence of stones which they call cornice stones – stones at the edge of the roadway used to support both a stone balustrade and these columns. Their figure 91 is a reconstruction drawing of part of the parapet, and shows the balustrade as a vertical stone slab, about 800 mm high, 160 mm thick, extending a further 90 mm down into a groove in the cornice stone (see also their fig. 30 and table 3). This detail is not entirely convincing from the engineering point of view, as the balustrade slab would scarcely be stable, and the groove in the cornice block, unless it fitted closely, would be largely ineffective. Their reconstruction of the complete bridge is shown on their front cover, and in their plate 8. A point of interest is that

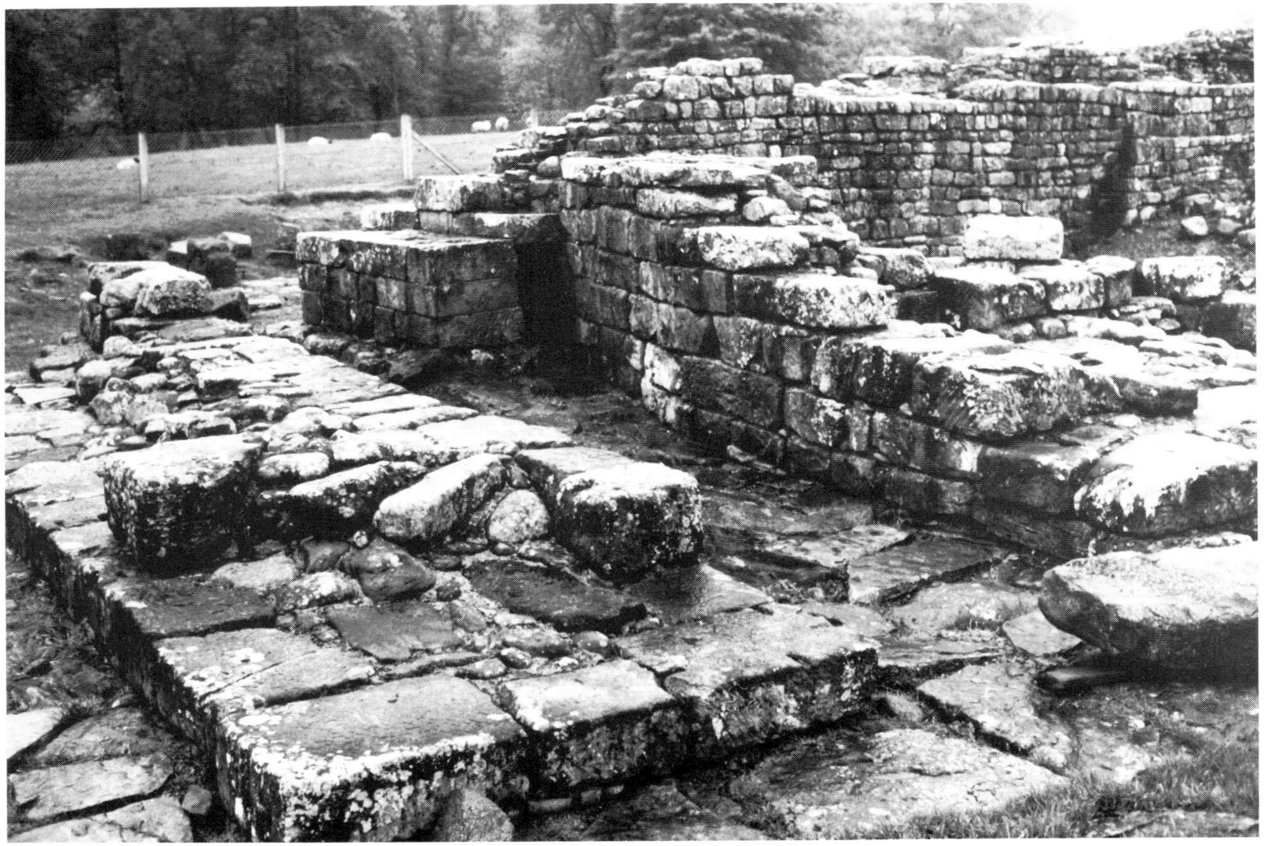

Fig. 120 The complex abutment of the bridge which carried Hadrian's Wall over the River Irthing near Willowford. The wall continues on to the right; the stream flowed to the left of the abutment (T31).

a mill-race was carried across the rear of the abutment, through the base of the tower, which probably served also to house a mill.

The next bridge, across the River Irthing (**T31**; map ref. NY622664) west of Willowford and Gilsland, three-quarters of a mile east of Birdoswald, is also believed to incorporate successive modifications. The sign at the site reads:

> The river has changed its course since Roman times, leaving the eastern abutments of a series of bridges, built and rebuilt over many years, high and dry. The west abutment has probably been swept away by the river. The bridge was originally planned to spring from a splayed abutment which was surmounted by a turret, of which only slight traces can be seen. At a later stage this was built over, and the bridge abutment moved westwards ... The abutment now incorporated a large platform of masonry ... and a series of channels, possibly a mill-race, but more likely to be flood channels. A new tower was added ... At a later stage, a new abutment and pier was built out westward, partially blocking the earlier flood channels.

Figure 120 shows the complexity of the existing remains. The final abutment or pier is at the left, the wall is at the far right, and one of the intermediate abutments is at the middle right. The stream, which would be to the left in the photograph, has a width similar to that at Chesters.

Dymond (1961), Bruce (1969: 160–2) and Bidwell and Holbrook (1989: 78, 94) all believe that the original bridge had stone arches, and was replaced probably by a timber bridge on stone piers, this conclusion being based on the presence of voussoir-shaped stones in the existing works. Bidwell and Holbrook (1989, pl. 7) show a possible reconstruction of the first bridge.

The last of the bridges in Hadrian's Wall crossed the River Eden at Hyssop Holme Well (**T32**; map ref. NY397568), immediately north west of Luguvalium or Carlisle (Dymond, 1961; Bruce, 1978: 155; Bidwell and Holbrook, 1989: 108). Some 80–90 stone blocks have been found here, of an average size 860 × 480 × 300 mm, with no voussoirs, but with Lewis and cramp holes as at Chesters and Willowford, suggesting again a timber bridge on stone piers.

The Romans later built another line of defence, possibly in AD 139 (Section 2.6; Sitwell, 1981: 41), from Duntocher at the mouth of the Clyde to Bo'ness on the Forth. Built of turf rather than stone, the Antonine Wall was about 55 km long. There are remains of a bridge near the western end, across the River Kelvin near Summerston and Balmuildy (**T33**; map ref. NS580719). It had stone abutments and piers, with no voussoirs. There are the remains of some timberwork. Robertson (1969) speaks of another bridge near the eastern end of the wall, over the River Avon at Inveravon, south of Grangemouth.

Finally, a coin issued by Caracalla (AD 211–17) depicts a pontoon bridge (**T34**) thrown across the Forth during the Caledonian campaign of Septimius Severus. It shows three boats, and the distinctive criss-crossed parapets.

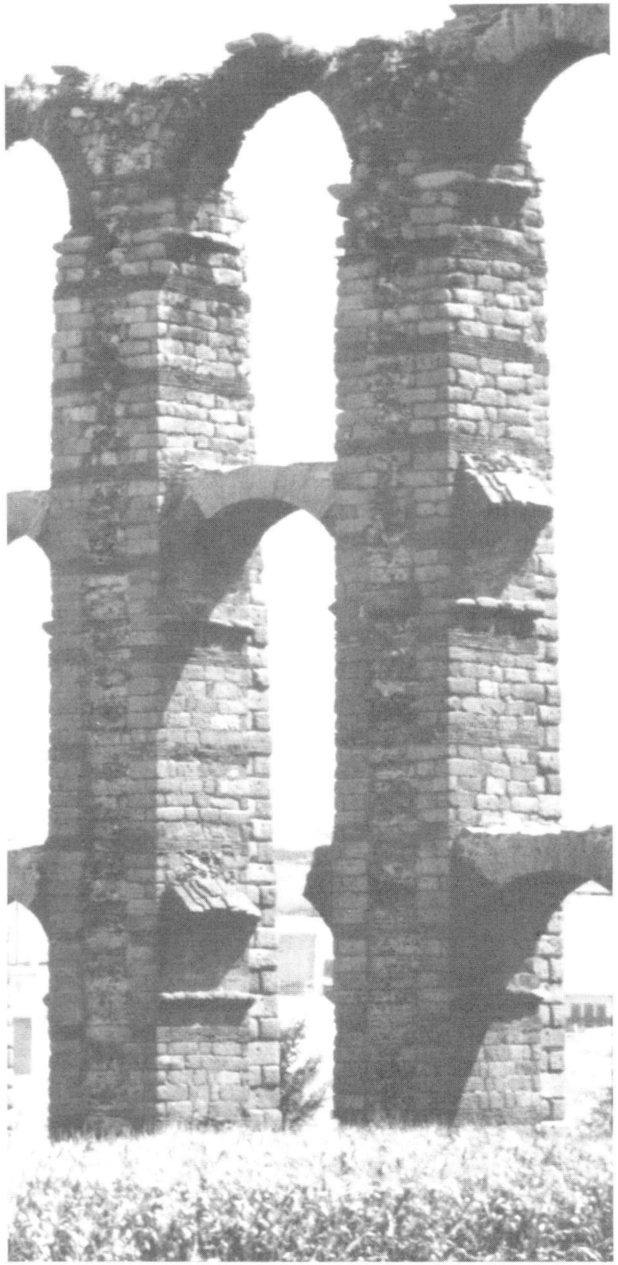

7

Roman Aqueducts

7.1 Introduction

An aqueduct is simply a channel constructed to carry water, and may be on the ground, below ground, or above ground. In the last case it becomes a bridge, but it differs from other bridges in certain respects – notably it is narrow, but it may also be long and high. Its method of construction may also differ – the use of *opus reticulatum*, for example, is not uncommon in aqueducts but is unknown in bridges.

The purpose of this book is not primarily to deal with aqueducts, but it is impossible to speak of bridges without referring to some extent to aqueducts also. The most famous of all 'bridges' is, in fact, an aqueduct – the Pont du Gard. In a later chapter the proportions of arch bridges will be examined in an attempt to elucidate the Roman principles of design and it is appropriate there to include also the aqueducts. For these reasons, this chapter needs to describe some Roman aqueducts.

A thorough description and listing of aqueducts is not, however, readily available. The pioneering works of Ashby (1935, 1925) and van Deman (1934) deal properly with the ancient aqueducts of the city of Rome, but a similar treatment has not been extended generally to structures elsewhere. Smith (1976b: 86), for example, says, 'Public water-supply was a feature of every part of the Empire in which Rome exerted her urbanizing influence . . . to list all Roman water-supply systems would require a book in itself . . . the twin criteria of interest and significance must now be applied.' The same comment applies here – not all aqueducts are described, and for an exhaustive treatment one must look elsewhere. However, because of this lack of access to a listing it has been decided to refer briefly to those aqueducts for which data have been obtained. The resulting list, although fortuitous and incomplete, may at least be of some benefit to those wishing to do further study (see also Appendix table C). (As mentioned in the Introduction, Hodge's *Roman Aqueducts and Water Supply* (1992) was not yet published when this was written. Nor have I seen the book by Belgrand, 1875, *Les Aqueducs romains*.)

7.2 Italy

Smith (1976b: 81–6) has an interesting summary of the aqueducts of Rome. Eleven were built between the years 312 BC and AD 226, with a total length of 507 km, of which 434 km was underground, 15 km on the surface, and 59 km, or 11.6 per cent, on bridges. Of the length on bridges, a considerable part was made up of multiple channels on a single bridge. When this is allowed for, 'only about 5 per cent of Rome's aqueduct-mileage was carried on bridges'. To quote Smith again, 'So far as was practicable, the lines of aqueducts followed a steady grade at or below ground level. The use of tunnels, long arcades, high bridges over rivers, or siphons across deep depressions was a last resort when difficult terrain could not be avoided.' (See also Blackman, 1978, 9.)

There is no doubt that this is an important point, but it still leaves the parallel observation – that, although aqueduct bridges formed a small proportion of the total, yet their length was considerable. On his figures the following aqueducts included the longest bridges.

Marcia	10 km	144–140 BC	(AQ1)
Tepula	9 km	126 BC	(AQ2)
Julia	10 km	33 BC	(AQ3)
Claudia	14 km	AD 38–52	(AQ4)
Anio Novus	11 km	AD 38–52	(AQ5)
Alexandrina	2 km	AD 226	(AQ6)

Of these, about 9 km of each of the Tepula and the Julia were carried on the Aqua Marcia, with the *specus* or water channel of the Julia placed on that of the Tepula and the Tepula on the Marcia. Winslow (1963: 174) gives the somewhat higher length of 14 km of bridges on the Aqua Anio Novus; of this, some 10 km were carried on the Aqua Claudia. It follows, then, that the long aqueduct bridges are to be found particularly on the lines of the Aqua Marcia, Aqua Claudia and Aqua Alexandrina, but there are other important bridges also on the Anio Vetus and Anio Novus.

The Anio Vetus (272 BC), Marcia, Claudia and Anio Novus received water either from springs in the valley of the Anio River, or from the Anio itself, east of Tivoli. Having reached Tivoli, they went south to the vicinity of Gallicano, then west south west and finally north west into Rome. The Aqua Tepula rose in the Alban Hills about 18 km south east of Rome, and joined the others in their final journey. The Aqua Julia started a little further east, but followed essentially the same route. The Alexandrina started about 20 km east of Rome and ran west to Rome some distance north of the others. The remaining four were the Aquae Appia (312 BC) and Virgo (21 BC), which also entered from the east, and the Alsietina (2 BC) and Traiana (AD 109) which came from the north west.

Sextus Julius Frontinus (see also Section 3.2) wrote the first account of the aqueducts of Rome about the year AD 97. Herschel (1913), a hydraulic engineer, has published a translation of this work with explanatory chapters. Ashby (1935) and van Deman (1934) described the aqueducts as they existed earlier this century; the account by Winslow (1963) is more recent and less detailed and refers particularly to aqueduct bridges. It identifies three areas of interest – east and north east of Tivoli; around Gallicano; and the final approaches, within about 10 km east and south east of Rome. Locations of the bridges are given by both Winslow and Ashby (1935).

Castelmadama is about 5 km east north east of Tivoli. Although there are minor bridges on the Aquae Claudia and Anio Novus about 3 km to the north, the first major bridges are on the Claudia within 1.5 km of the town – **Ponte Majuro** on the Fossa della Vallana with a single major span and seven small arches; the bridge in the Fosso della Noce with eleven spans still remaining; and another near Fonte Luca with a major span of about 9 m and twenty-six smaller arches (Winslow, 1963: 152–60; van Deman, 1934: 80–1; Ashby, 1935: 204–8).

An even finer set lies in the Valle d'Empiglione and the Valle degli Arci or Valley of the Arches, 2 to 4 km east of Tivoli. The most easterly lie on either side of the Via Empolitana, on the Anio Novus after it divides into two loops, with a low eleven-span bridge on the Tivoli loop to the north of the road, and two fine groups of multiple arches on the Hadrianic loop to the south. Further west, on the Tivoli loop of the Aqua Anio Novus, is a bridge with ten remaining spans, made up of a single arch over the roadway and another group of nine, at the **Ponte degli Arci**; and a little further west, the seven-span bridge over the Fosso Arcese (Winslow, 1963: 136–7, 142–9; van Deman, 1934: 84, 294–8; Ashby, 1935: 267–72; 1925: 578).

Some 2 km south of Tivoli is the two-span **Ponte Arcinelli** on the Anio Novus (Winslow, 1963: 149–51; van Deman, 1934: 304; Ashby, 1935: 279); but it is not until the vicinity of Gallicano, about 9 km south of Tivoli, that the next group of major aqueduct bridges is reached. To this point, the channels have followed the valley of the Anio and its tributaries, then turned south along the eastern flank of the Valle Pussiano, but here they veer south west across the grain of the country, crossing a succession of steep gorges falling to the north west. The older aqueducts were built to the lowest elevations, and followed the contours of the surface with irregular alignments to avoid the need for bridges. The route of the Aqua Anio Vetus is generally a little over 2 km north west of Gallicano, and has only two major bridges – the **Ponte San Gregorio** (Winslow, 1963: 134–5; van Deman, 1934: 45–7; Ashby, 1935: 69), a fine bridge 155 m long, up to 24 m high, with about twenty-one spans, some double-tiered, on a later diversion north of the Via di Poli; and the **Ponte Taulella** (Winslow, 1963: 162–3; van Deman, 1934: 48–9; Ashby, 1935: 72–3), with a single span over the Rio Secco.

The Aqua Marcia, a little closer to Gallicano, has a fine series of four bridges. To the north of the Via di Poli, 2 km north east of Gallicano, is the **Ponte San Pietro** (Gazzola, 1963b, no. 112), originally with a single span of about 16 m, and a clear height of about 19 m, both now greatly reduced (Winslow, 1963: 132–3; van Deman, 1934: 92–4; Ashby, 1935: 116). The **Ponte Lupo** has been described as the greatest of all the aqueduct bridges of Rome (Winslow, 1963: 126–30; van Deman, 1934: 95–100; Ashby, 1935: 117–21). Although immediately south of the Via di Poli, it is not easily seen, being partly hidden from the road by trees and shrubs. Also listed by Gazzola (1963b, no. 114), it is of massive construction with an overall length of 110 m, a height of 30 m, and two arched openings. The third of these bridges is the **Ponte Caipoli** (Winslow, 1963: 163–4; van Deman, 1934: 101–3; Ashby, 1935: 112–13; Gazzola, 1963b, no. 115), a tall, one-span, two-tiered bridge, 1.5 km north west of Gallicano. The fourth is the smaller **Ponte della Bullica** (Winslow, 1963: 161–2; van Deman, 1934: 103; Ashby, 1935: 123; Gazzola, 1963b, no. 116) about 0.5 km to the south.

The last series of aqueduct bridges in the vicinity of

Gallicano is on the Aqua Anio Novus. The first, the **Ponte San Antonio** (Winslow, 1963: 139–41; van Deman: 1934: 306–8; Ashby, 1935: 282–4, 1925: 578; Gazzola, 1963b, no. 113), is the furthest of this entire group from Gallicano, about 3.5 km to the north, across the Fosso dell'Acqua. With the nearby Ponte San Gregorio and the Ponte Lupo, it is one of the most imposing of these bridges. The entire structure is about 120 m long, 36 m high and has a magnificent central arch, still in its original condition, with a span of 10.4 m and a clear height of 32.3 m. Although the arch was at one time filled with concrete, this infill has since fallen away, leaving exposed the original lofty stone-built arch.

The next bridge still standing is the **Ponte dell'Inferno** (Winslow, 1963: 132–3; van Deman, 1934: 311; Ashby, 1935: 285–6), a much smaller single-span bridge just north of San Giovanni in Campo Orazio on the Via di Poli. To the south west of Gallicano, near the Via Praenestina is the Ponte Amato (I26) described in Section 5.2. Built originally to carry the Aquae Claudia and Anio Novus, it was later widened and rebuilt as a road bridge (Winslow, 1963: 160–1; van Deman, 1934: 225; Ashby, 1935: 215; Gazzola, 1963b, no. 17). Another bridge for the two aqueducts is about 1.5 km to the west – the **Ponte Barucelli** (Winslow, 1963: 169–70; van Deman, 1934: 226–7, 314–15; Ashby, 1935: 216), also called the **Ponte Diruto** or Ruined Bridge, with four or five spans.

The distinctive feature of the bridges near Gallicano is that generally they are not long, but cross deep and narrow gorges. The opposite may be said of the elevated aqueducts near Rome – the higher and later aqueducts, the Claudia, Anio Novus and Alexandrina, marched across the plains to the city on extensive arcades. The most famous and impressive of these carries the Aqua Claudia, with the Anio Novus superimposed (Fig. 121), from a point where they emerge from the ground, 600 m north of the railway station of Capannelle, for a distance of about 2.5 km north west past the Casale di Roma Vecchia towards the Tor Fiscale (Winslow, 1963: 99–103; van Deman, 1934: 234–40; Ashby, 1935: 228–32). It consists of an initial series of 163 arches, 1375 m long, followed by groups of 1, 4, 2, 6 and 6 arches, then a fine group of 14 arches 27.4 m high, and a final single arch. The original total length from the point of emergence to Rome was over 10 km, with all but 0.8 km on arches; the total number of arches was at least 1000.

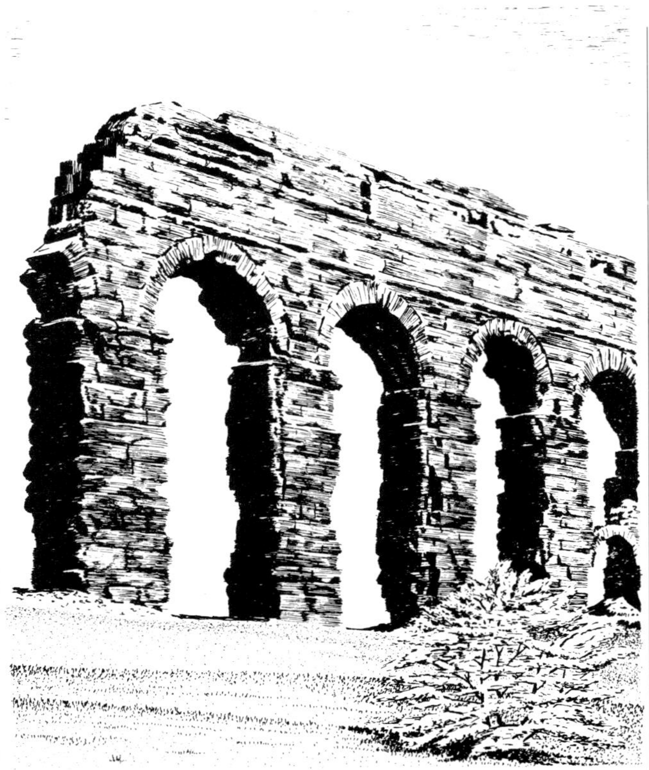

Fig. 121 Aqueduct bridge, carrying the Aqua Claudia with the Anio Novus superimposed, from Capannelle north to Rome (AQ4, AQ5, based on Ashby, 1935, pl. 13a).

The Aqua Marcia runs parallel to the Claudia in this vicinity, along a line about 100 m to the north east of the first long arcade. Its level was lower, and so, although it is above ground, it has fewer and lower arches, with a group of ten and another group of eighteen still standing. The Tepula and Julia are superimposed (Winslow, 1963: 103–7; van Deman, 1934: 105; Ashby, 1935: 130–1). Some distance upstream, about 5 to 6 km east of Capannelle, is another arcade of the Aqua Marcia, with four arches still standing (Winslow, 1963: 107–8; Ashby, 1935: 126–7). There are also some elevated remains of branch aqueducts from the Claudia and Anio Novus in the region of Capannelle.

The Aqua Alexandrina runs from east to west about 5 km north of the great arcade of the Claudia at Capannelle. Here, about 0.5 km south east of Centocelle is a group of 62 arches, with 48 forming a continuous arcade (Winslow, 1963: 113; van Deman, 1934: 352–3; Ashby, 1935: 314–15). This is the largest of a number of bridges on the aqueduct, lying between Centocelle and a point some 5 km to the east. Sections of the aqueducts may also be found above ground in Rome.

Away from Rome, records of Roman aqueduct bridges are less easy to obtain – to quote Robins (1946: 68), 'there are few remains of Roman aqueducts in Italy, outside the vicinity of Rome itself'.

Blake refers to aqueducts at Minturno (AQ7), Syracuse (AQ8), Taormina (AQ9), Turin (AQ10), Venafro (AQ11), and Viterbo (AQ12; Blake, 1947: 237, 311, 289–91, 220, 240); and also at Anzio (AQ13), Civitavecchia (AQ14), Termini Imerese (AQ15) and Terracina (AQ16; Blake and Bishop, 1973: 280–2, 275, 280). Only a few of these incorporate bridges: at Minturno; Venafro – 'for the most part subterranean, but where it emerges to cross a stream it shows an unusual type of bridge'; Civitavecchia, with the remains of piers; Termini Imerese; and Terracina, whose aqueduct was 54 km long, containing some well-constructed bridges over stream beds. According to Robins (1946: 68), there is an aqueduct at Catania (AQ17), and 'some traces at Pompeii' (AQ18). Another, at Bologna (AQ19), was reconstructed in 1878.

There are major aqueducts still standing at Spoleto (AQ20) and near Tarquinia (AQ21). The **Ponte delle Torri** at Spoleto has ten spans up to 81 m high, with a total length of about 230 m, but it dates from a reconstruction in the fourteenth century; it is believed to be based on an earlier, late Roman aqueduct (Robins, 1946: 68; Sandstrom, 1970: 100). The road from Monte Romano to the coast, and then north to Tarquinia, follows a long, stone-arched aqueduct, but its date is unknown. Smith (1976a: 54, 66, 70) refers also to an inverted siphon of 134 BC at Alatri (AQ22).

Gazzola describes a high bridge near the Via Valeria (AQ23; 1963b, no. 42). Blake illustrates the same bridge with the remark, 'A superb specimen, **Ponte Funicchio**, is still extant near Ferento, but I know nothing of its history' (Blake, 1947: 212). Ballance (1951: 89) adds, 'It appears to be an aqueduct', and the height and narrow width suggest that this conclusion is correct. It has at least two spans, probably many more.

7.3 Gaul

Grenier (1931, vol. 4) has a list of Roman aqueducts in France. In the district of Narbonne he names – Fréjus (AQ24; p. 41), Cimiez (AQ25; p. 55), Antibes (AQ26; p. 60), Saint Rémy-de-Provence (AQ27; p. 63), Aix-en-Provence (AQ28; p. 68), Arles (AQ29; p. 75), Nîmes (AQ30; p. 88), Pézenas (AQ31; p. 101), Vaison (AQ32; p. 104), Die (AQ33; p. 106), Vienne (AQ34; p. 111). His list for the Trois Gaules is Lyon (AQ35–38; p. 118), Saintes (AQ39; p. 143), Rodez (AQ40; p. 151), Périgueux (AQ41; p. 154), Poitiers (AQ42; p. 164), Sens (AQ43; p. 171), Lutèce (AQ44; to Paris, p. 180), Evreux (AQ45; p. 191), Metz (AQ46; p. 199) and Bavay (AQ47; p. 213). Smith (1976b: p. 86) adds another three: at Bordeaux (AQ48), Marseille (AQ49) and Orange (AQ50). There are also aqueducts, believed to be Roman, at Luynes (AQ51; Boussard, 1957: 13) and St Bertrand de Comminges (AQ52; MacKendrick, 1972: 92, 98). The resulting total of twenty-nine indicates how extensive was Roman aqueduct construction, and it may be expected that this was the case also in the other colonies.

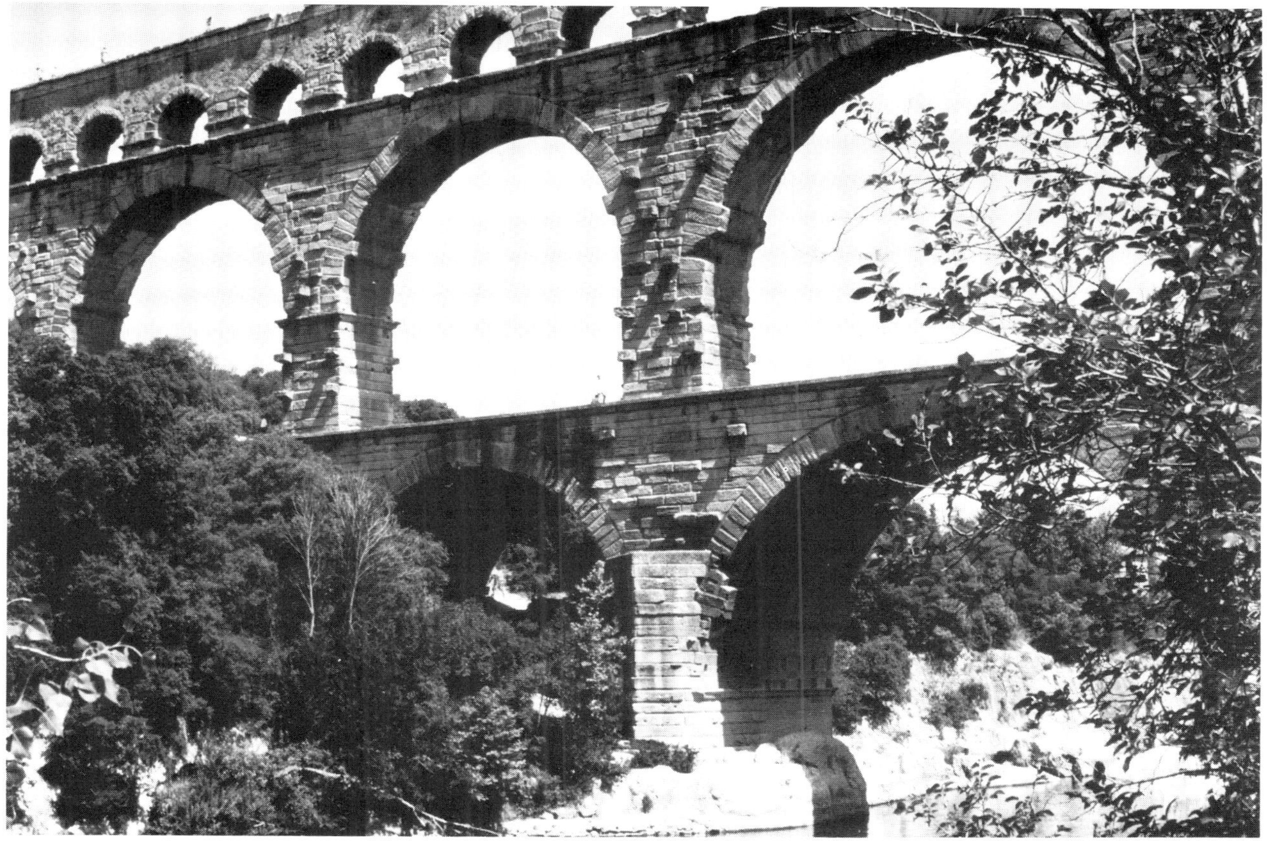

Fig. 122 The Pont du Gard, carrying the aqueduct to Nîmes, France (AQ30).

Of these, the more important aqueduct bridges are at Nîmes, Lyon, Fréjus and Metz, with other substantial remains at Antibes, Arles (to the east, near Barbegal; Rivet, 1988, fig. 23), Luynes (north west of Tours), and St Bertrand de Comminges (in the foothills of the Pyrenees, south west of Toulouse, near Barbazan).

The **Pont du Gard** (AQ30), Figure 122, is probably the most famous of all Roman bridges (Emerson and Gromort, 1925: 25–8; Grenier, 1931, vol. 4: 88–101; Robins, 1946: 66; Brogan, 1953: 82–3; de Camp, 1977: 195, 212, 214). It forms part of an aqueduct which carries the waters of the Airon and the Eure from Uzès to Nîmes, Roman Nemausus, a distance of some 50 km with a fall of 17 m. The bridge itself crosses the River Gard near Remoulins, 18 km north east of Nîmes. Although there is some doubt concerning its origin, the commonly accepted view is that it was built by Marcus Vipsanius Agrippa about the year 18 BC. Rivet (1988) places it later. Smith (1990–1) presents Rivet's view, but still prefers an early date, about the beginning of the first century AD.

The aqueduct is constructed in three tiers, with the two lower tiers – the first and the second – having equal spans (see Fig. 122). Emerson and Gromort (1925) have an accurate drawing with detailed dimensions. Tier 2 has eleven spans which are, from north to south, 11.9, 19.2, 18.9, 19.2, 24.4, 19.1, 19.1, 18.1, 15.7, 15.7 and 15.7 m. The lower tier has six spans, corresponding to the fourth to the ninth of these, from 19.2 to 15.7 m, including the large 24.4 m span over the stream. The pier widths are typically 4.5 m, but vary from 4.4 to 4.7 m, the larger dimensions corresponding to the piers on either side of the 24.4 m span. The length of the first tier is thus 139 m and that of the second 243 m. The top tier is arranged so that the centreline of a pier always coincides with the centreline of a pier of tier 2. There are four spans in the top tier over the 24.4 m span, two over the short northern 11.9 m span, and three over every other span. The northern end of the top tier is broken above the northern abutment of tier 2; to the south the top tier extends two spans beyond the end of tier 2. It follows then that there are thirty-five existing spans in tier 3. If the northern end were intact, there would probably be a further three spans, or thirty-eight in all. The lengths of these upper spans vary, but are typically of the order of 4.7 m clear, with pier widths about 3.1 m. The present length of the top tier is about 270 m, but if all spans were intact it would be about 287 m. There is some variation in the overall lengths given for the tiers – Grenier, for example, quotes 142, 242 and 275 m – but this probably results partly from differences of opinion over what constitutes the end of a tier. The figures quoted here of 139, 243 and 287 (or 270 m), are from face to face of the opening. The successive heights from water-level in the Gard to the top of the masonry in each tier are given by Emerson and Gromort as 20.5, 19.5 and 7.4 m (or 47.4 m total). Grenier gives the somewhat greater figure of 48.7 m, presumably resulting from a variation in river height. On either figure, the Pont du Gard was the highest bridge ever constructed by the Romans, exceeding the 40 to 42 m at Alcántara (SP13, Section 5.7) and the 30 m height at Narni (I87; Section 5.4).

The width of the structure varies, being about 6.0 m in tier 1, 4.6 m in tier 2, and 3.1 m in tier 3. The water conduit is 1.25 m wide and 1.8 m high. It is roofed with stone. The main structure was built without mortar, excepting for that necessary to waterproof the conduit at the top.

It is of interest to notice the variability of the span lengths, particularly in view of the appearance of great regularity. Three top-tier spans of 4.75 m clear, with pier widths of 3.1 m are precisely compatible with lower tiers of 19.1 m clear span, with 4.45 m piers. In fact, the corresponding lower tiers vary from 15.7 to 19.2 m. The main span of 24.4 m is flanked by larger 4.7 m piers, giving a dimension of 29.1 m centre to centre of piers. This gives an average length of 7.3 m between the centres of the piers in the top tier, somewhat less than the figure of 7.8 m (i.e. 4.7 + 3.1 m) that is typical for the remainder of the structure. The conclusion is, then, that although the basic structure is regular in form, the detailed dimensions vary substantially. The range of lower tier spans from 15.7 to 19.2 m corresponds to a variation that is close to ±10 per cent about the mean. Similarly, the module of 7.3 m adopted for the top tier in the large span is about 7 per cent less than that used elsewhere. A variation of the order of 10 per cent is therefore not unusual.

The structure was damaged in the fifth century and again in the seventeenth. The initial widths of 6.0 m for tier 1 and 4.6 m for tier 2 left a passage about 0.7 m wide – possibly a little larger – that could be used by pedestrians. This was increased by cutting into the side of tier 2 and corbelling out from tier 1. As a result, the entire structure was in danger of collapse. It was restored in 1669, and again in 1855. In 1747 a stone-arched road bridge was added to the downstream side, level with the top of tier 1; the designer was Henri Pitot, famous in another context for the invention of the Pitot tube, the device still used to measure the velocity of fluids, or of aircraft relative to the surrounding atmosphere.

Lyon had no less than four separate aqueduct systems: Mont d'Or (AQ35) and Brevenne (AQ36) from the north west; Craponne (AQ37) from the west; and the Gier aqueduct (AQ38) from the south west (Hodge, 1983, 1985; Smith, 1976a, 1976b: 89–93; de Montauzan, 1908). These contained nine inverted siphons. In the normal aqueduct, the water flowed with a free surface, at atmospheric pressure. In the inverted siphon, the water was collected in an open basin or header tank, from which it was led into closed pipes and taken down and across the valley, and then up the other side to another open collecting basin with a free surface a little below the level of the water in the header tank. The pressure in the intervening pipes was well above atmospheric pressure and the pipes had to be built of a material that could withstand the pressure without leaking – the Romans chose lead.

Smith, followed by Hodge, has argued that the use of the inverted siphon was not uncommon in Roman practice, and he cites examples at 11 Roman towns (Smith, 1976a: 54), 4 in Italy, 3 in France, 2 in Spain, 1 in Britain and 1 in Turkey. Of these the most important were at Lyon in France and Aspendos in Turkey. He argues further that the question of whether to use an aqueduct bridge or siphon was resolved on the basis of the depth of the valley and relative economics. The highest Roman bridge was the Pont du Gard, with its height above stream-level quoted variously as 47.4 to 48.7 m. The bridge at Alcántara (SP13, Section 5.7) had a height of about 40 to 42 m; that at Narni (I87, Section 5.4) was lower, about 30 m. Smith gives somewhat different heights – 48.7, 47.3, and 36.6 m – but his point remains the same. These are the highest of all Roman bridges; only a few exceed 30.5 m (100 ft) and the highest is only a little over 45.7 m (150 ft).

On the other hand, the depths of the valleys with

inverted siphons were substantially greater: for the nine Lyon siphons he gives values from 25.0 to 123.5 m, with seven in excess of 45.7 m. He concludes, therefore, that the Romans used aqueduct bridges for depths up to 45.7 m or so; for greater depths they used inverted siphons.

The longest of the Lyon siphons is on the Craponne aqueduct (AQ37), built by Augustus (31 BC–AD 14), with a length of 4800 m and a depth of 114 m. The deepest is the Beaunant siphon, 2610 m long and 123.5 m deep, on the Gier aqueduct (AQ38), built in the time of Hadrian (AD 117–38). This aqueduct rises 43 km south west of Lyon with an actual length of the order of 75 km (Brogan, 1953: 83–4). It contains four inverted siphons: at St Genis, 890 m long and 82 m deep; Soucieu, 1200 m by 93 m; Beaunant, 2610 m by 124 m; and St Irénée, 610 m long by 47 m deep.

Figures 123, 124 and 125 show the Gier aqueduct near Beaunant. The first view is of the aqueduct before the siphon, beside Route D50 near the village of Plat-de-l'Air. It shows the concrete core, *opus reticulatum* facing interrupted by brick courses, and the *bipedales* arches that are characteristic of the structure. The second is at the point where the aqueduct begins its steep descent to the valley below. The arch at the top of the ramp formed part of a header tank that ran across the full width and was fitted with an arched roof vented to the atmosphere (Smith 1976a: 62; Hodge, 1985: 97). The channel or *specus* of the aqueduct entered from the far side. On the near side, the water was carried down the slope in lead pipes, possibly ten in number (Smith, 1976a: 60). Figure 124 shows an opening from the right hand side of the header tank. The pipes were sealed into a series of such openings, placed on the ramp, and then, it is believed, encased in concrete. The substantial width of the ramp may be noted: the ramp and the header tank were considerably wider than the aqueduct bridge (Fig. 123) which entered at the far side.

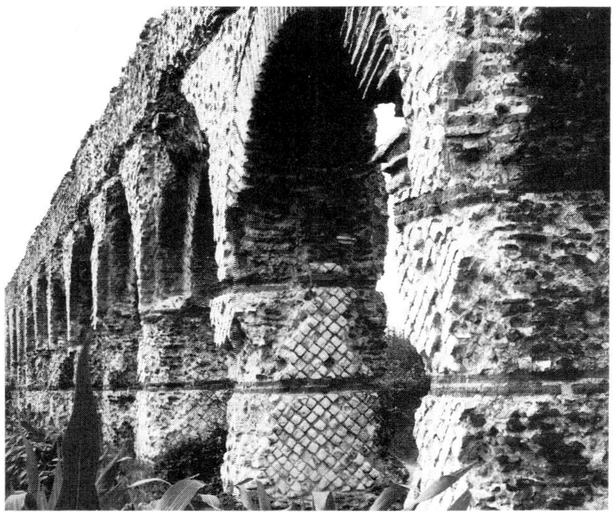

Fig. 123 Upper portion of Gier aqueduct at Beaunant near Lyon (AQ38).

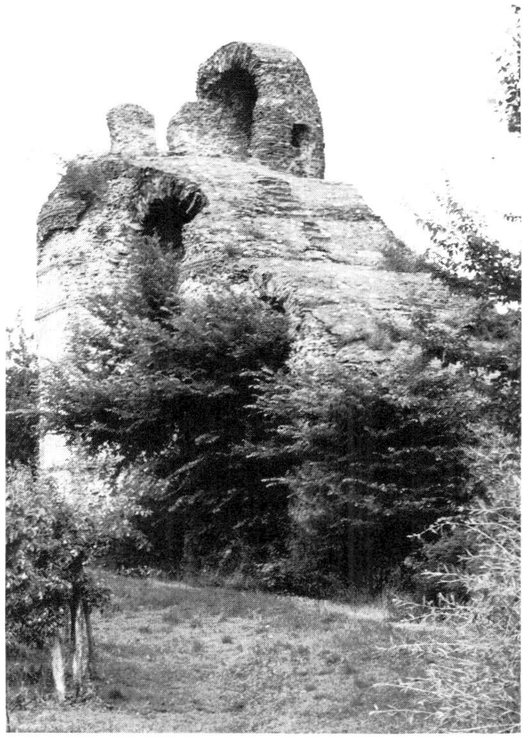

Fig. 124 Beginning of the inverted siphon in the Gier aqueduct, Lyon (AQ38).

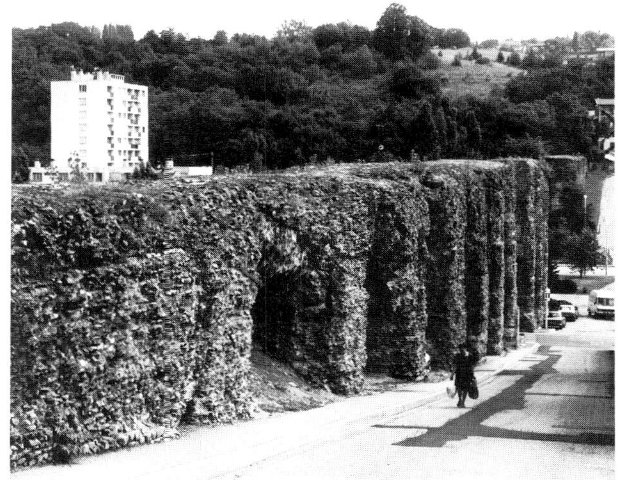

Fig. 125 *Venter* at the base of the inverted siphon in the Gier aqueduct, Lyon (AQ38).

At the bottom of the valley, the pipes were taken across an arched bridge or *venter*. Figure 125 shows the *venter* looking in the direction of Lyon; the great width of the valley and the depth of *venter* below the height of the surrounding hills are clearly shown. A *venter* was not included in all inverted siphons. The pipes could be placed in the bed of the valley, or alternatively on an embankment or wall rather than an arched bridge. However, a crossing of the stream had to be made and the use of a *venter* bridge served three purposes: it provided the necessary waterway for the stream, it reduced the hydraulic pressure in the lead pipes; and it gave a firm base on which the pipes could be laid. The depth quoted by Smith of 123.5 m for the Beaunant siphon is either the distance of the *venter* below the hydraulic gradient, or below the upstream water-level. There was a head loss of 9.3 m across the siphon; that is, the downstream water-level was 9.3 m below the upstream level. The *venter* itself had a maximum height of 17.4 m; it was 269 m long and consisted of thirty arches of 6 m span (Grenier, 1931, vol. 4: 118–43).

The width of the *venter* at Beaunant is 7.35 m (Hodge, 1985: 95), wider than that of the ramp, and provided both access and space for the pipes which carried the water. The remains are insufficient to indicate the number of pipes, but it is known that upstream at Soucieu there were nine and Smith (1976a: 60) suggests that there may have been about ten at Beaunant. The external diameter of each pipe was probably about 250 mm – as at St Genis, Soucieu and St Irénée – with a wall thickness of the order of 20 mm. The total length of the nine siphons was 16.7 km, or, if it is assumed that on the average there were nine pipes in each siphon, the total length of pipe was about 150 km. This represents about 15,000 t of lead or, allowing for uncertainty, about 10,000 to 15,000 t, which is de Montauzan's estimate as quoted by Smith. The procurement of this quantity of lead in Roman times was a major exercise, and the weight of the lead would have made its handling and transport difficult. In many applications it would have been easier and cheaper for the Romans to build an aqueduct bridge in stone – the twin factors of limiting bridge height and the cost of lead would have determined the decision between bridge and inverted siphon.

Hodge (1985: 97) indicates the typical form of the pipes. They were formed from sheets about 3 m long and with a width slightly greater than the required circumference. After forming around a wooden core, the joint was lapped, butted or folded, and then soldered to form the pipe. Hodge points out that the Romans were aware of the danger of lead poisoning, but suggests that the short retention time and the build-up of a lining of calcium carbonate in the pipes meant that this was not a problem. Vitruvius (8.6.4) describes the manufacture of lead pipe and the principle of the inverted siphon, and discusses the effect of lead on health (8.6.10–11).

Brogan (1953: 83) has described the aqueduct at Fréjus (AQ24) on the French Riviera as one of the finest still surviving. Some 36 km long (Grenier, 1931, vol. 4: 41–53), it rises at Neissoun, 6 km north north west of Mons, about 30 km north of Fréjus. Arches survive at Bouteillière, where there are three; at La Molte, with six arches 60 m long; Gargalon, where there were fourteen arches, 130 m long, 6 to 14 m high with spans of 5.2 m, but only six of these remain; and near Fréjus itself, where there is a series of 5.2 m spans. Illustrations may be found in Grenier (1931, vol. 4), Boussard (1957), and Rivet (pls. 45–7). Rivet also has a map of its course (1988: 130).

Another fine aqueduct is at Metz (AQ46) in the north (Grenier, 1931, vol. 4: 199–213; Robins, 1946: 69; Brogan, 1953: 84). It carried water from a large reservoir at Gorze, 22 km south west of the town, initially underground, but then across the River Moselle at Jouy-aux-Arches, 10 km from Metz, by a large bridge. Grenier describes the remains of 15 arches on the left bank and another 16 to 20 on the right. There was at least one span of 12 m in the original crossing, with the others at 5.6 m, and a maximum height of about 20 m. Many of the spans still remain.

Roman Moguntiacum, or modern Mainz in Germany, also had a Roman aqueduct (AQ53) of which some portions still remain (Brogan, 1953: 117; Robins, 1946: 66).

7.4 Spain

The best account of the aqueducts of Spain is given in Casado's book *Acueductos romanos en España* (1972). As the pages are not numbered, the page numbers given here begin with the author's Prologue. He lists sixteen towns with aqueduct remains – the more important are given below; the others are at Granada (AQ54), Bellone Claudia (AQ55) west of Tarifa, Gades or ancient Cádiz (AQ56), Sadaba (AQ57) on the road from Astorga to Zaragoza, and Chelva (AQ58) inland from Valencia.

Tarragona (AQ59)	Augustus (31 BC–AD 14)
Segovia (AQ60)	Claudius (AD 41–54)
Mérida (AQ61–3)	Hadrian (AD 117–38)
Seville (and Italica) (AQ64, AQ65)	First century AD
Almuñécar (AQ66)	First century AD
Toledo (AQ67)	Second century AD
Barcelona (AQ68)	First and second centuries AD
Alcanadre (AQ69)	Second century AD
Valencia de Alcántara (AQ70)	First century AD

The aqueduct bridge at Tarragona (AQ59), commonly called the **Puente de las Ferreras** or **Puente del Diablo** (Casado, 1972: 16–32; Wiseman, 1956: 122), lies about 5 km north west of the town, a few hundred metres east of the road to Valls and Lérida. Intact and in excellent condition (Fig. 126), it is constructed in two tiers, with eleven arches in the lower tier, and twenty-five in the upper. The maximum height is 26 m; the overall length 200 m; and the typical span is 5.9 m ± 0.15 m clear, corresponding to 20 Roman feet. The distance centre to centre of piers is 7.9 m. Although the width of the structure is 1.86 m for much of the height, the lower piers are stepped in both directions, from 3.8 m (longitudinal) by 3.1 m at the base, to 2.1 by 1.9 m below the rectangular cornice which marks the springing of the lower arches. The water channel is uncovered, with a width of about 0.7 m and a height of 1 m. The aqueduct of which the bridge forms a part is about 35 km long.

As stated above, Casado regards the structure as Augustan. It was restored in the period 1781 to 1790, and again between 1855 and 1856. Like some other Spanish aqueduct bridges, it is remarkably complete, and characterised by simplicity and elegance.

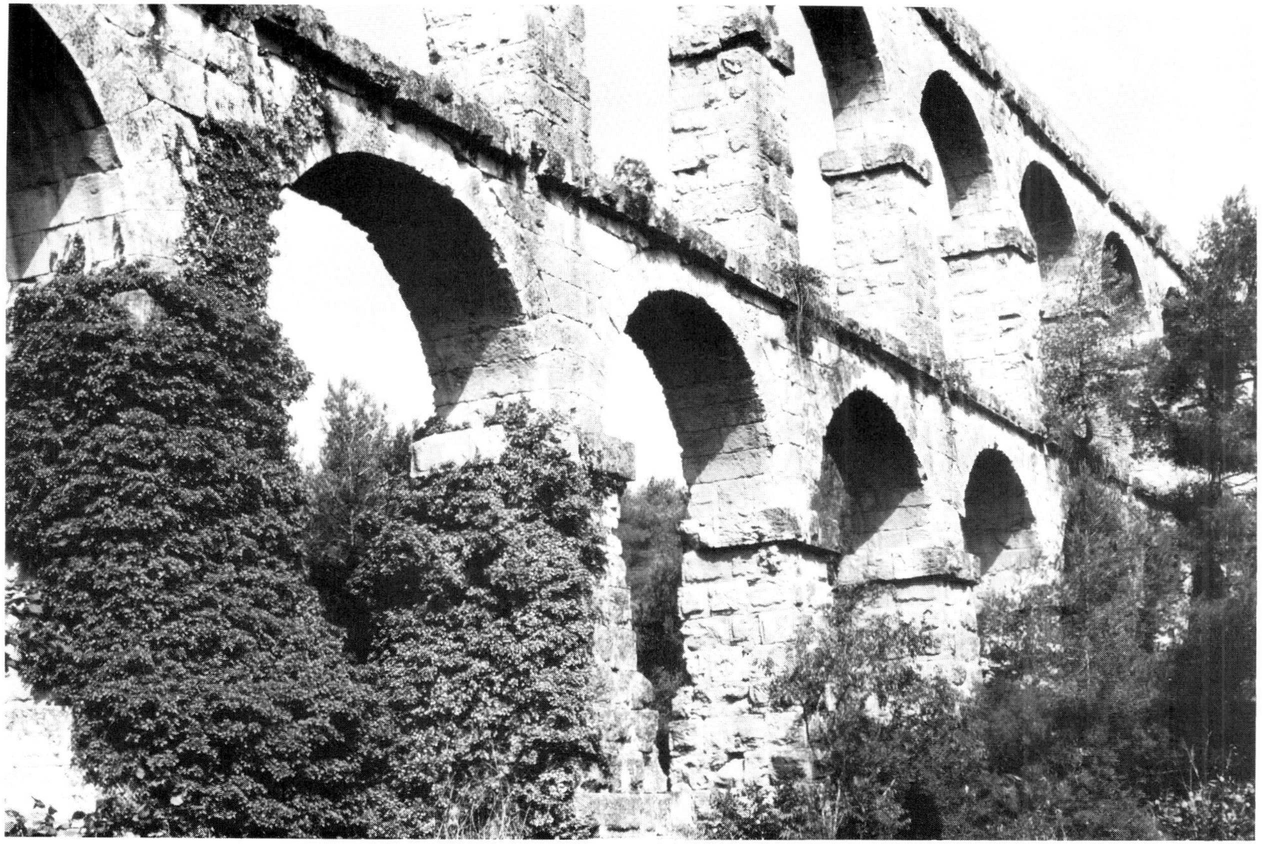

Fig. 126 Aqueduct at Tarragona, Spain (AQ59).

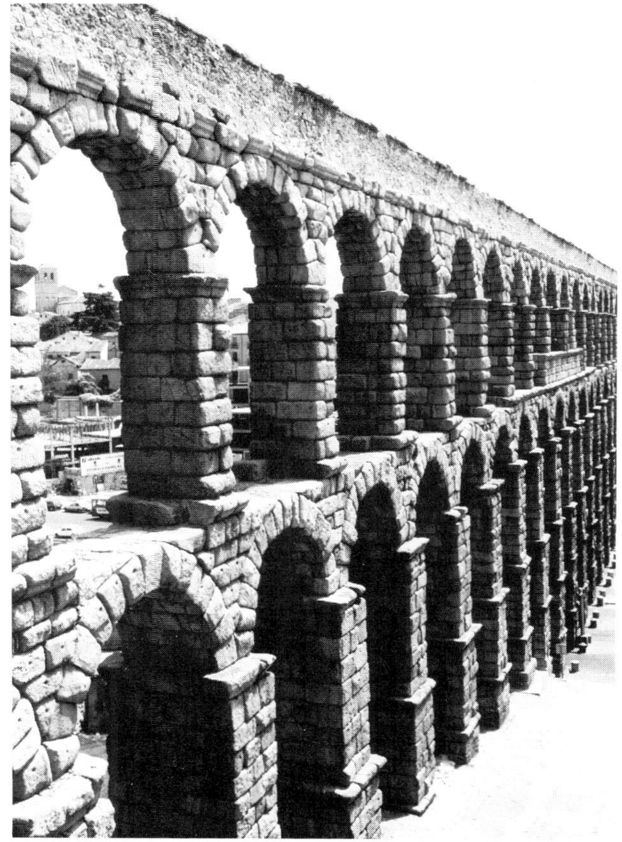

Fig. 127 The long and high aqueduct at Segovia, Spain (AQ60).

The aqueduct at Segovia (AQ60; Casado, 1972: 33–90; Wiseman, 1956: 82; Robins, 1946: 66–7), even finer, is still in use. It carries water 18 km from the Río Frio and crosses the deep valley outside the old hilltop town by a slender and lofty bridge, over the modern Plaza del Azoguejo to the walls of the ancient town above. Part of the bridge is shown in Figure 127, looking from the town upstream.

The alignment is not straight. The initial approach, from the south east, is along a wall beside the San Ildefonso Road. It reaches a roofed chamber and then veers to the left, through an angle of 15°, to the first alignment. This is carried initially on a further 65 m of wall, and then on a series of six small arches, with spans of 2.2 to 2.4 m clear. It turns again to the left, through an angle of 13°, to the second alignment, carried on twenty-five arches, from 4.9 to 5.1 m in span. A further turn of 19° to the left brings it to the long third alignment, with forty-four spans from 4.9 to 5.1 m, rising to a height of about 14 m. Here it turns sharply right, through 53°, to the fourth or final alignment. To this point, the arches have been simple in form, with only a single tier. The corner pier, although still slender, is more substantial than the others, with a plan dimension around the outside of the corner of 5.5 m or, around the inside, 3.8 m. Immediately around the corner, the first span is still full height, but then a double tier of forty-two arch spans continues in a straight line across the Plaza del Azoguejo, finishing with a final full height span against the city wall. In this section, the distance centre to centre of piers is 6 m, the upper span is 4.8 m clear, and the lower 4.2 m clear.

It follows, then, that the bridge has a total of 199 spans. The length is commonly quoted as 728 m. The maximum height is 28.5 m above the ground, with the foundations a further 5.5m deep. Casado (1972: 74) has a detailed sketch of a typical main span. The water channel or *specus* is small, 0.3 m × 0.3 m, and open. The upper tier is 1.8 m thick and about 8.7 m high. The lower piers are stepped, and vary in size from 2.5 m (transverse to the bridge) by 1.8 m to 3.6 m by 2.4 m. The hydraulic gradient is 1.2 per cent. The stones are of white granite, laid without mortar.

Of the forty-two double spans, numbers 32 and 33 have the lower tier surmounted by a deeper stone panel, and the central pier above this panel has statues of the Virgin Mary on one side and St Sebastian on the other. Casado attributes the aqueduct to Claudius and this central panel carried an inscription (Casado, 1972: 69), now indistinct but believed to read

TI.CLAUDIUS. PONT. MAX. VIII COS. III TRIBUNICIA POTESTATE VIII IMP.P.P.
OMNIUM FECIT.

The inscription occupies two lines, as shown, with room for additional preceding and following letters or words on each line. The adueduct is now in good condition, but it has undergone a number of restorations. It was damaged in the eleventh century, restored in the fifteenth and again in the eighteenth, particularly in the second and third alignments.

Various authors have commented on the difference in style between the Pont du Gard and the aqueduct at Segovia. Smith, for example (Smith, 1976b: 89), says,

> Early bridges such as the Pont du Gard were structurally clumsy ... During the first century AD more elegant and structurally efficient shapes evolved, of which the Segovia and Tarragona bridges are good examples. The bridge-on-a-bridge concept can be seen giving way to a more monolithic form, in which the piers are more nearly continuous from foundation to crest with a substantial degree of arched bracing between.

Although the variation in structural form is there, more truly the great contrast is in the selection of span length. The designer at Segovia has chosen clear spans of only 4.8 m compared with the 19.2 and 24.4 m spans at the Pont du Gard.

It is possible to find rough parallels to both structures. The Claudian aqueducts of Rome, for example – the Aquae Claudia and Anio Novus (Section 7.2) – generally used the smaller spans. On the other hand, the Ponte San Antonio, which was also on the Anio Novus, had a 10 m span, while the Ponte San Pietro on the earlier Aqua Marcia (144–140 BC) had a 16 m span. These and other parallels indicate that the Romans were prepared to use two kinds of aqueducts depending on the site: some with multiple short spans in long arcades or in the crossings of wide unobstructed valleys, and others with larger spans in deep valleys where a major stream had to be crossed, as at the Pont du Gard.

There are more parallels to the lofty braced piers at Segovia than to the broad spans of the Pont du Gard. One of these parallels is in the Los Milagros aqueduct bridge at Mérida (AQ63). Mérida was served by three aqueducts: Cornalvo (AQ61), San Lázaro (AQ62) and Los Milagros (Casado, 1972: 91–122; Wiseman, 1956: 161–2). All three enter the city from the north, with the oldest, the Cornalvo, rising to the north east and terminating near the Roman theatre. It ran essentially along the ground, with no structure higher than a low wall. On the other hand, the aqueducts of San Lázaro and Los Milagros each had major bridges across the valley of the Albarregas River, a few hundred metres north of the town, about 600 m east and west respectively of the station.

Figure 128 is a photograph of the Los Milagros aqueduct (AQ63), close to the river and the Puente de Albarregas (SP16, Section 5.7). As can be seen, the aqueduct is high, with tall, slender, closely spaced pillars joined by three tiers of arches, only some of which are still intact. The principal part of the existing structure is 178 m long, with twenty-six columns spaced 7.3 to 7.6 m

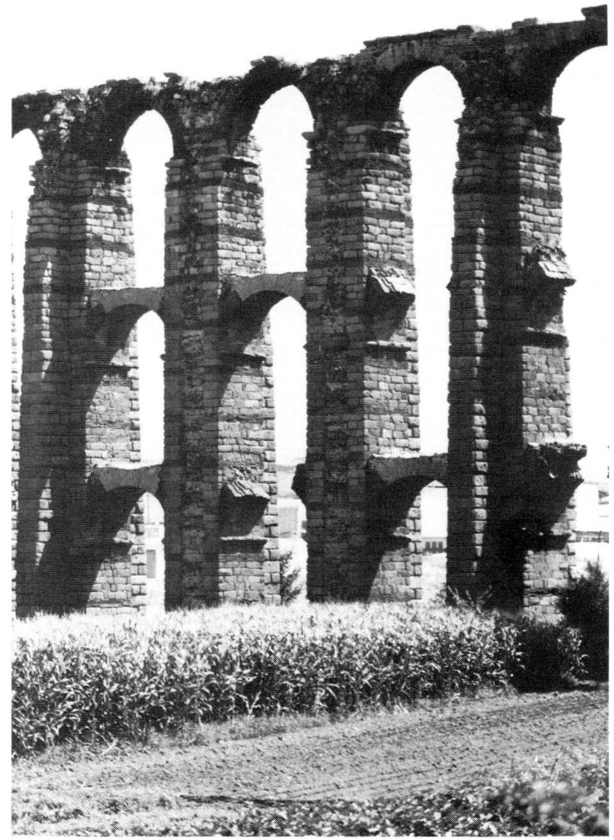

Fig. 128 Lofty piers of the Los Milagros aqueduct in Mérida (AQ63). The collapse of some of the intermediate arches between the columns is referred to in Section 9.4.

centre to centre, and a maximum height of 26 m. This portion lies north of the railway and includes a river crossing, with one of the pillars founded in the stream. South of the railway, the aqueduct turns to the south west, and here there is a further row of seven columns, including a heavy corner column, with heights from 15 to 20 m, and centre to centre spacings from 6.8 to 7.0 m.

In the major section, the clear span of the arches is normally 4.5 m, with some spans up to 4.7 or 4.9 m. The columns were originally cruciform in section, with a central portion from 2.9 to 3 m square, and engaged piers on the external faces about 1.3 m wide by 2 to 2.1 m deep, giving a total transverse dimension of about 7.3 m. The width of all three arches, and of the masonry carrying the *specus*, is about 2.7 m. The *specus* itself is open and small, about 0.4 m wide by 0.3 m deep.

Only two spans remain of the Roman aqueduct of San Lázaro (AQ62), on the same alignment as a later aqueduct bridge of the fifteenth century. These spans crossed a roadway, with a lower tier of heavier construction than that of the single tier above. The spans were 3.5 and 3.8 m clear, corresponding to about 5.8 m centre to centre of columns. The overall height is estimated to have been 15 or 16 m.

Casado (1972) dates all three aqueducts in the time of Hadrian (AD 117–38). In the case of the aqueduct bridges of Los Milagros and San Lázaro, this dating is largely on the basis of structural form.

There are substantial remains at Seville of an aqueduct (AQ64) which enters the city from the east (Casado, 1972: 123–43). Originally Roman, it was reconstructed in AD 1189, and many of the remains appear to be of a later date. The Roman town of Italica, 9 km north west of Seville, was often regarded as the birthplace of the emperor Hadrian, and here there are also aqueduct remains (AQ65), many on the ground, but with the evidence of some arcades (Casado, 1972: 143–4).

Almuñécar, on the south coast east of Málaga, has an important aqueduct (AQ66; Casado, 1972: 145–52; Smith, 1976a: 54). It rises a little over 4 km north of the town on the right bank of the Río Verde, goes south along the substantial aqueduct bridge of Torrecuevas, passes west through the ridge between the Río Verde and the Río Seco, and then follows the left bank of that stream south to the town. The bridge at Torrecuevas has nineteen simple spans, most of 4.9 m clear, but with two smaller spans. Along the left bank of the Rio Seco are another three substantial aqueduct bridges, the first with six simple spans, the second with two tiers and ten spans in the upper tier, and the third again with two tiers and with eleven spans. Throughout, the typical clear span is 4.9 m. Finally, the aqueduct crosses to the town by an inverted siphon, with a single 150 mm pipe in a below-ground gallery 1.65 m high by 0.7 m wide.

Water was brought into Toledo from the south, from the vicinity of Los Yébenes, on the River Guadalerzas. For the most part, the aqueduct (AQ67) ran along the ground, but at Toledo it crossed the Tagus River down-stream of the Puente de Alcántara, below the modern Puente Nuevo (Casado, 1972: 167–77). The remains are minimal and consist only of the lower part of an arch on the left bank, part of a basement wall on the other, and some foundations in the stream. However, they have given rise to considerable speculation (Smith, 1976a: 66–7). Casado reproduces a proposed reconstruction by Alphonso Rey Pastor. This shows a massive three-tiered aqueduct bridge, reaching to a height some 80 m above water-level in the stream. The lowest tier is over 40 m high – possibly about 41 m above water-level – with three spans, each of the order of 25 m clear. This is shown surmounted by tiers with heights of 22 and 17 m, made up of smaller arches, three to each main span. The length of the upper tier is shown as about 300 m.

The height of Rey Pastor's bridge was determined by the known water-levels in the aqueduct on either side of the gorge. If it existed, it would have been by far the highest bridge ever built by the Romans, about one and two-thirds the height of the Pont du Gard. The lowest tier by itself is about 85 per cent of that height.

Smith (1976a: 67) believes that Rey Pastor's bridge never existed. He argues that a height so much in excess of its nearest rival is difficult to accept, and goes on to say, 'Then there is the question of the bridge's fate. Where is it? What has happened to the huge quantity of material it must have contained?' He suggests instead, 'The inference is obvious. Only one tier of Rey Pastor's bridge in reality ever existed and it was the *venter* bridge of a siphon some 50 metres (or possibly 40 m) deep.'

There are remains of a Roman aqueduct in Barcelona (AQ68; Casado, 1972: 178–86). Chevallier (1976: 102) refers also to 'the bridge-aqueduct of Lladenet' (or Lladoner). This crossed a gorge 21 m deep, 1.5 km on the Barcelona side of the Cruz de Ordal, a pass 32 km west of the town. The aqueduct of Alcanadre (AQ69; Casado, 1972: 189–95) lies in the Ebro valley, west of Calahorra. The extensive arcade has clear spans of 4.5 m. There are also Roman arches included in a later aqueduct at Valencia de Alcántara (AQ70) west of Cáceres, close to the Portuguese border.

No records have been found of Roman aqueducts in Portugal itself. There are later aqueduct bridges at Elvas (1498–1622) and Lisbon. Both are substantial, but the latter is particularly large – 69 m high and 900 m long. Called Alcántara, it forms part of the Aqueduto das Aguas Livres (Smith, 1976b: 125).

7.5 Africa

Section 5.8 described bridge-aqueducts at Sbeïtla (A14, AQ71) and Khémissa (A17, AQ72) in Tunisia. These were either dual-purpose structures, carrying both an aqueduct and a road, or alternatively carried an aqueduct only. That section also described bridges at Tébessa (A15) and Chemtou (A26), and there were aqueducts at these centres also (AQ73, AQ74; Smith, 1976b: 88; Robins, 1946: 67). The one at Chemtou was long, and located to the north of the bridge (Babelon et al., 1893: 31.70). Musti is

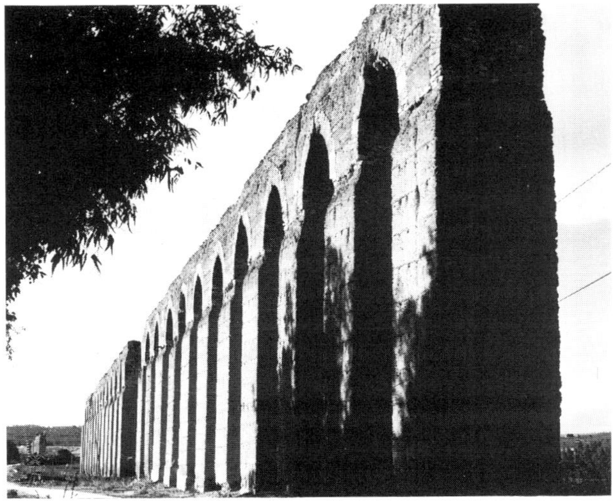

Fig. 129 Part of one of the largest of all Roman aqueducts, from Zaghouan to Carthage. This part lies south of the Tunis to Jedeida road, west from Bardo (AQ76).

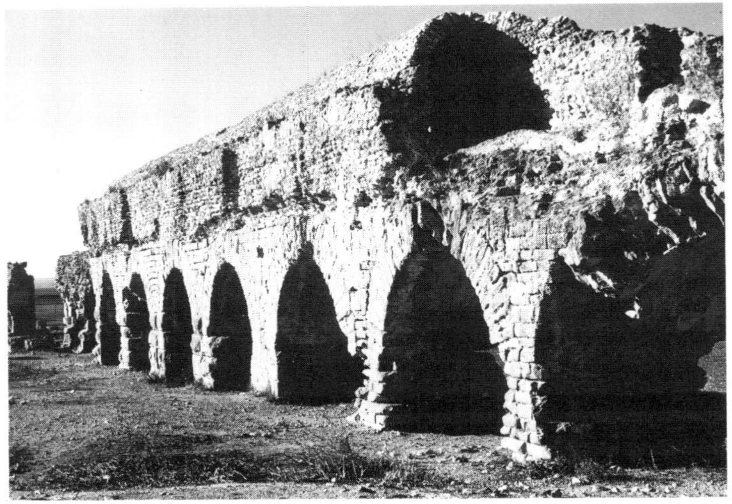

Fig. 130 The Zaghouan–Carthage aqueduct near Mohammedia showing the arch-roofed *specus* (AQ76).

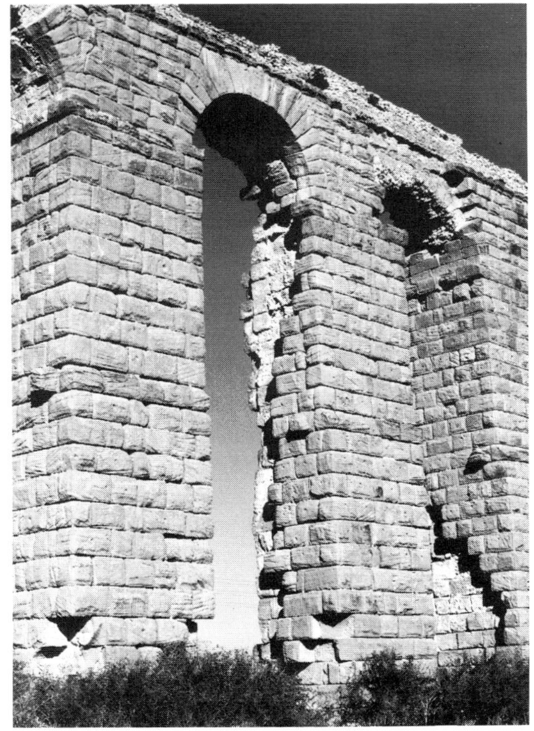

Fig. 131 A more southerly portion of the Zaghouan–Carthage aqueduct, near Oudna (AQ76).

another Tunisian town with a Roman aqueduct (AQ75), west south west of Dougga. But the outstanding aqueduct bridges in this part of Africa were along the Zaghouan–Carthage aqueduct (AQ76); near Constantine (AQ77); and at Cherchell (AQ78).

Carthage lay on the north western coast of the Gulf of Tunis, to the south west of which is the Bay of Tunis, a land-locked salt lake which opens into the gulf by a narrow entrance (see Section 5.8). Further to the south west is another salt lake, the Sebkhet Sejoumi, this time completely surrounded by land, with the old town of Tunis lying between it and the bay. Zaghouan lies 45 km south of Tunis (BNI, 1945a: 129, 433), about 52 km from Carthage; the two are joined by an aqueduct some 80 km long. The route of the aqueduct is initially north west from Zaghouan and then north east to the vicinity of Oudna. It crosses the major road P3, which joins Tunis to Thuburbo Maius and el Fahs, about 17 km from Tunis, and then turns north west, to run well to the west of the Sebkhet Sejoumi and join the Tunis to Jedeida road, P7, about 12 km beyond Manouba. Finally, it turns east to Carthage (Babelon et al., 1893: 28.48).

There are fine and extensive sections of aqueduct bridge still to be seen south of the Jedeida road (Fig. 129), and along route P3 south of Mohammedia (Figs. 130 and 131). Parts of the structure are faced with stone; others are finished in concrete; and there is evidence of late repairs. There are said to be other substantial sections near the old tram route from Tunis to Bardo (BNI, 1945a: 433) and near Zaghouan (Raven, 1969: 49). Smith (1976b: 88) describes it as 'one of the longest of all Roman aqueducts'.

Van der Heyden and Scullard (1963: 173, fig. 458) have a most extraordinary photograph of an aqueduct (AQ77), which served the Algerian city of Constantine, Roman Cirta. Four stone arches are shown, on the Oued Rhumel where both the stream and the aqueduct pass through an immense natural arch of rock. If it is assumed that the aqueduct is about 10 m high – and it may well be higher – then the rock arch has an overall height of about 130 m, a clear height of at least 100 m and a span almost as large. The Oued Rhumel rises to the south and west of Constantine and the rock arch and this portion of the aqueduct must lie in this direction. Robins (1946: 67) refers to 'the comparatively late aqueduct of Koudiat-Ati, feeding a reservoir at Constantine', built by Justinian with five arches remaining 'about sixty feet high and composed of huge blocks of limestone without mortar'. Constantine has a suburb of this name to the south west of one of the two gorges that surrounded the old town (BNI, 1944a: 71).

Cherchell is the modern name for Roman Caesarea Iol, on the coast about 100 km west of Algiers. Smith (1976b: 88) describes an aqueduct bridge here (AQ78) built by Hadrian, 35 m high. Volubilis in Morocco, south of Tangier, also has the remains of an extensive water supply system (AQ79; Raven, 1969: 92), with an aqueduct.

7.6 The east

Split, on the Adriatic coast of Yugoslavia, is famous for Diocletian's Palace. Roman Salona lay 5 km to the north east and the remains of Justinian's aqueduct (AQ80) may be seen beside the road between the two towns.

Kavala, in north eastern Greece, is the port of the ancient town of Philippi. A two-tiered Roman aqueduct here (AQ81) was still used this century to bring water from Mount Pangeus. The construction is unusual (Sitwell, 1981: 130), with very heavy piers, relieved in the upper tier by double arches, one above the other. This upper tier has alternating large and small spans, with the small spans subdivided vertically so as to form, in effect, a third tier.

Modern Turkey has a significant number of Roman aqueducts. The first is at Istanbul (AQ82) in European Turkey, 'built by Hadrian, rebuilt by Valens and restored by Suleiman the Great' (Robins, 1946: 61). Erythrai, on the coast near Çesme west of Izmir or Smyrna, had a Byzantine aqueduct (AQ83; Akurgal, 1978: 233; Bean, 1966: 155, 157). Further south, Ephesus still has the remains of a fine aqueduct (AQ84): to quote Akurgal (1978: 34), 'One of the finest and oldest aqueducts in Anatolia is to be found just outside Ephesus on the Söke road'; that is, to the south. Bean (1966: 178, pl. 41) has a photograph of a two-tiered structure, with three major arches still standing in the lower tier and five above. He dates it in the time of Augustus (31 BC–AD 14).

Laodicea also had a major aqueduct (AQ85; Bean, 1971: 252, 255–7; Akurgal, 1978: 237). It rose to the south of the town and was about 8 km long. Initially, it was carried along the ground, but then crossed a valley some 140 m wide by an aqueduct bridge, with arches up to 3 m high. Closer to the town, the water was brought across the present railway by pipes under pressure in an inverted siphon. South of Ephesus, the neighbouring towns of Alabanda and Alinda each had Roman aqueducts. In the case of Alabanda (AQ86; Bean, 1971: 185, 187, pl. 45). one stone span remains across the stream called the Kemerderesi, south of the town, but 19 km to the south west over the Marsyas at Incekemer (Bean, 1971, pl. 54) are more substantial remains, with three spans of a bridge (see also E20, Section 5.9) that probably forms part of the same aqueduct. The remains at Alinda (AQ87) are more impressive again (Bean, 1971: 197–8, pl. 51). To quote Bean, 'Beyond the second acropolis . . . runs an aqueduct still in good preservation. It is well and strongly built, though without much pretension to elegant finish. There remain four arches and a stretch of solid wall pierced by a gate . . .'

Phaselis was on the southern coast, south of Antalya. Here was another aqueduct (AQ88; Bean, 1968: 162, 164, pl. 71; Akurgal, 1978: 266) which, to quote Bean again, has 'high arches which are still quite well preserved'. However, east of Antalya at Aspendos (AQ89) is an aqueduct described by Akurgal (1978: 334) as 'one of the best preserved and most impressive Roman aqueducts to be found in Turkey'. Bean (1968: 73–6, pl. 26) goes even further: 'it provides probably the best surviving example of a Roman aqueduct (others, like the Pont du Gard or the aqueduct at Volubilis, are rather better preserved, but less interesting in point of construction).' The extensive remains include at least twenty arches and many additional columns. At one point, the structure rose to a height of 30 m, possibly with three tiers, the water then dropping in pipes through an inverted siphon to a substantial lower arcade. The length of this crossing was about 800 m. It is believed that it was built in the second century AD.

At nearby Side, on the coast, was another major aqueduct (AQ90; Bean, 1968: 83, 86–7, pl. 29; Akurgal, 1978: 337, 339), this time dating from perhaps the latter half of the second century BC. The remains are of substantial construction, with two tiers, at least three arches in the lower tier, and one in the upper tier remaining.

Further to the east, Brice (1966, pl. 9) has an illustration of an aqueduct bridge in the district of Cilicia Tracheia or Stony Cilicia. This may well be on one of two aqueducts (AQ91) which served the ancient town of Elaiussa-Sebaste (Akurgal, 1978: 344), north east of modern Silifke. Robins (1946 68) speaks of a major aqueduct bridge at Antioch (AQ92), either Antakya or Pisidian Antioch, possibly the latter. He quotes a length for one bridge of 210 m and a height of 60 m, 'the lower part consisting mainly of solid wall and the upper part of a series of arches with massive pillars'. If this height were confirmed, it would be the highest of all Roman aqueducts.

Finally in Turkey, there was a Roman aqueduct which brought water from the Kâhta Çay to Samosata along the north or right bank of the Euphrates River (AQ93; Yorke, 1896: 323; Mitford, 1974: 160, pl. 4). To quote Yorke,

> The water was carried mainly through tunnels and rock-cut conduits, but by bridges across the mouths of the tributary valleys. Near Alakeupri (four hours from Samosata) these bridges assume very large proportions. Originally, their arches were wide, and built of squared stones of considerable size – work of the third century AD – but later, in order to strengthen the waterway, the arches were almost wholly closed in with coarse masonry of small stones.

In Palestine, there is a long stone-arched aqueduct bringing water from the north to Acre, Roman Ptolemais, but this is late, having been built by Ahmed el Jazzar in the late eighteenth century (BNI, 1943: 36, 297–9, 388).

Further south, Herod (37 BC–AD 4) built his port at Caesarea in 22 BC and this was supplied by three aqueducts (AQ94). The first is presumed to have been built by Herod and rises about 7 km north west of the town, beyond the Nahr Zerka or Crocodile River. It ran west, passed through a sandstone ridge near the coast, and then turned south along the coast to the town. This was supplemented in the time of Hadrian (AD 117–38) by a second stage which collected water from springs a further 7 km to the north west. These two channels were carried partly through tunnels and partly on the ground, but also on substantial arcades: Peleg (1984) has a photograph showing about twenty arches near the town. He describes them as having 'a distinct architectural design, characterized by a continuous moulded cornice above the top of the arches. The arches are built of sandstone ashlars and the channels are of Roman concrete.' These two aqueducts form the so-called 'high-level aqueduct'. The 'low-level aqueduct' was a separate later structure which brought water from a dam on the Crocodile River by a 5 km long channel, cut initially in rock, and then formed in the sand dunes with a continuous roof of barrel-vaulted stone masonry.

Braunschweig (1984) describes other water supply systems in towns of the eastern Mediterranean, but it appears that none used bridges. There is no record of Roman aqueduct bridges in Mesopotamia, although the massive dam bridge of Band-i-Kaisar described in Section 5.9 fed water for irrigation through a system of channels and tunnels.

It is appropriate, however, to describe here the remarkable pre-Roman aqueduct built by Sennacherib in 690 BC to provide water for Nineveh. It was not the city's first water supply, for in 703 BC he had built a dam across the River Khosr and directed water into a 16 km long canal 'dug with iron pick-axes' and in 694 BC augmented this by the construction of additional diversion channels and dams (Smith, 1976b: 78; de Camp, 1977: 68). But these were not sufficient, and in 690 BC he dammed the Gomel Su or Afrush River at Bavian by a masonry dam in a deep gorge, feeding water through stone sluice gates to a stone-paved canal at least 58 km long, which joined the River Khosr near Nineveh. At the modern village of Jerwan it crossed a tributary of the Gomel Su and here 'the canal, 72 feet wide, crossed a ravine by a stone aqueduct 918 feet long with five pointed arches' (BNI, 1944b: 633). Smith (1976b: 76; based on the work of Jacobsen and Seton Lloyd, 1935) has a drawing of this aqueduct bridge. The arches are corbelled, with large flat stones laid on horizontal beds, shaped and brought forward progressively to form each narrow pointed opening. The construction of the work took Sennacherib only 15 months and is commemorated by a great relief and inscription carved in the cliff at Bavian.

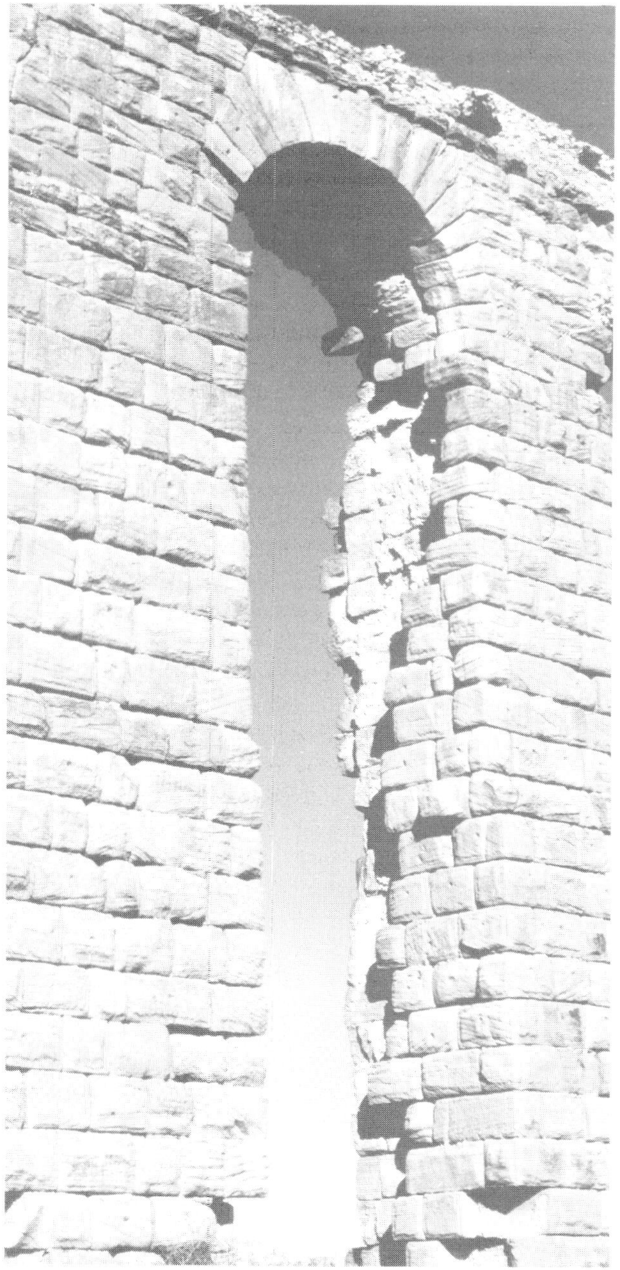

8

DESIGN AND CONSTRUCTION OF ROMAN ARCHES

8.1 Introduction

The Romans built stone arch bridges for at least 400 years. Blake (1947: 209) refers to the systematisation of roads and bridges undertaken by the Romans in Etruria in 174 BC and cites the Ponte San Lorenzo near Bulicame (I72, Section 5.4) as an old stone bridge. The small Ponte di Nona (I22, Section 5.2) near Rome may date from 173 BC. The Pons Aemilius (I5, Section 5.2) in Rome, now called the Ponte Rotto, was originally built in 179 BC with stone piers and a timber deck, but rebuilt in 142 BC in stone, to become the first of Rome's stone bridges. At the other end of the scale, Section 5.9 describes the dam bridge of Band-i-Kaisar at Shushtar (E42) in Iran built, under compulsion, by Valerian and his troops about AD 260, with possibly as many as thirty-five arch spans. These dates cover 434 years for stone arches. The actual period would have been longer – perhaps back to the start of the Via Appia in 312 BC, and on through the reigns of emperors such as Diocletian (AD 284–305) and Constantine (AD 306–37) to the fall of the empire in the west in AD 476, covering some 600 to 800 years.

Chapter 3 argues that there were appropriate structures in Roman life for the transmission of technical information and for the development of experience. In these circumstances, it is only reasonable to suppose that the Romans had rules for bridge design: indeed, the alternative, to suppose that each bridge was built as if it were alone, starting *ab initio* as a pioneering endeavour, is hardly worthy of consideration. The difficulty is that no written record of these rules has been passed down; they may be guessed at only by a study of the bridges themselves and this is one of the aims of this chapter (see also Mainstone, 1959 and 1965–6; Cowan, 1977). The chapter also discusses construction procedures for stone arch bridges. Their analysis by modern methods is considered in Chapter 9.

It is appropriate to start by illustrating some arch terminology (Fig. 132; see also Appendix 1), which may conveniently be summarised by describing briefly the erection of a bridge.

Construction would begin with *abutments* at the sides of the opening, and the *foundations* and *piers* at the intermediate supports. The *arch rib*, semicircular in profile in most but not all Roman bridges, rises from the *springings* on the piers or abutments, being composed of individual stones called *voussoirs*, with radial joints, and finished or closed by a final stone at mid-span called the *keystone*. The underside of the arch rib is the *intrados* or *soffit*. The *extrados* or upper face is curved and unsuitable for traffic. It is necessary to find some means of filling the space, called the *spandrel*, between the *deck* or road surface and the curved arch rib. In modern bridges, this may be done in such a way that the spandrel is partly open. In Roman bridges the spandrel is said to be solid or closed, and was filled – occasionally with soil, but usually with concrete,

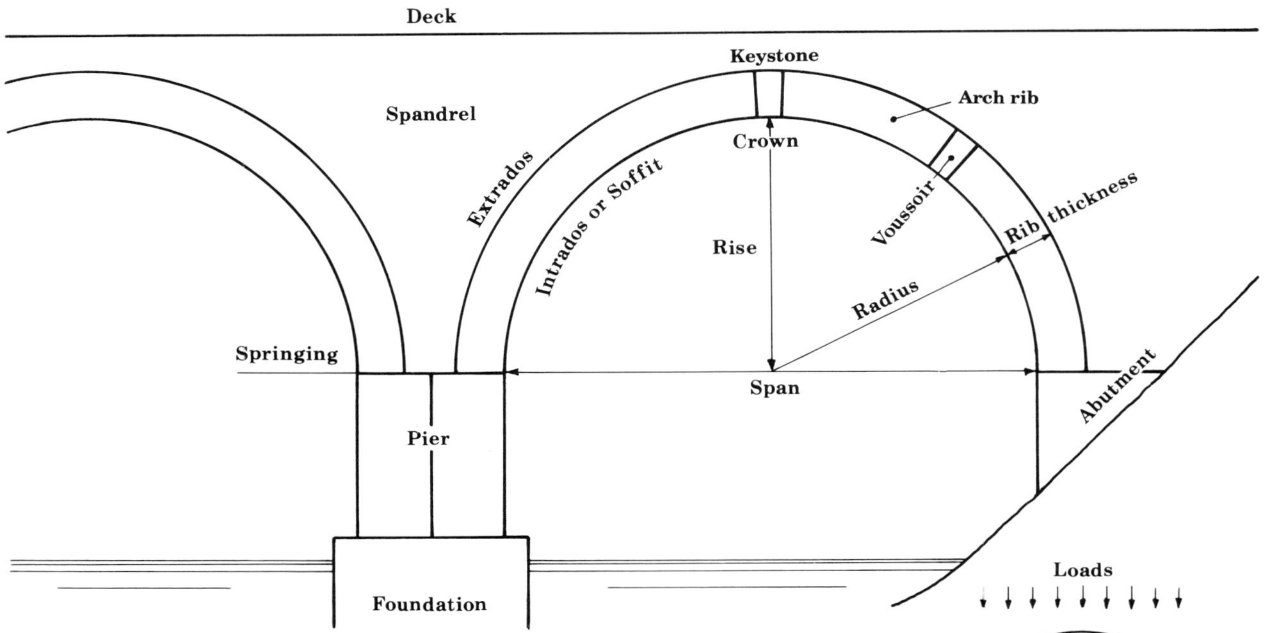

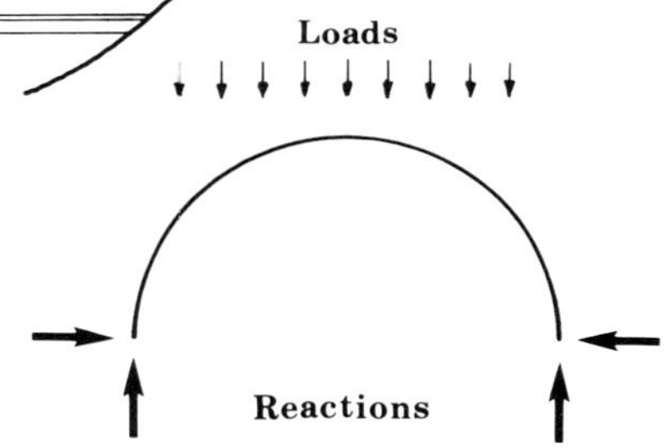

Fig. 132 Technical terms used in arch construction.

retained at the side of the bridge by vertical stone retaining walls – the *spandrel walls*. The *rise* and *span* of the bridge are as shown.

One further point should be made. For such a structure to merit truly its designation as an arch, the arch rib must not only be curved, but supported at its end in such a way that it cannot move outwards. There must be inward acting reactions, as shown also in Figure 132. This condition was achieved in the Roman arches. Vitruvius (6.8.3–4) describes the voussoirs of an arch as wedges, with their joints directed towards a centre. He clearly recognises the horizontal thrust produced by the arch – the wedges press towards the centre, and have a tendency to push out the abutments. At intermediate piers, these thrusts are balanced, but the outer piers should be wider, to give stability to the work.

It must be remembered also that the arch has a third dimension, in the direction of the width of the bridge. The arch rib becomes a curved wall or *vault*, sometimes called a *barrel vault*.

8.2 Foundations and pier construction

Vitruvius describes three types of foundations: the spread footing (to use the modern name), built directly on the surface, or after excavation down to firm material (3.4.1); the piled foundation, supported on timber piles driven down through soft material (3.4.2); and the foundation in water, constructed within sheet pile cofferdams, dewatered, and then built either directly on the exposed bed, excavated down to firmer material, or on piles (5.12.5–6). A more detailed account is given in Section 4.3. The foundation structure itself was built of 'stone, lime and sand'; that is, either totally of squared masonry, or with a stone exterior filled with a lime mortar, or concrete (Section 4.5). These foundation types could be used both for the abutments and for the intermediate piers.

The piers themselves had a cross-section in plan which could be rectangular, pointed by triangular additions upstream and possibly also downstream, or with rounded ends (see for example Casado, n.d.: 146). They could be of solid stonework, but more often had stone faces filled with concrete. In elevation also they may be solid, or pierced with floodways, such as in the Pons Fabricius (Fig. 21) or at Sommières (Fig. 67).

As is evident from Figure 132, the pier thickness must be at least sufficient to accommodate two arch ribs. Figure 133 presents actual pier thicknesses for fifteen bridges, plotted against the larger of the adjacent spans. It should be recognised that the data are approximate, being taken largely from photographs, and the selection of bridges is fortuitous, depending on the availability of suitable photographs. Nevertheless, certain trends emerge.

A common thickness is about one third of the span – bridges 3–11 (see reference list on Fig. 133). This ratio is common for smaller bridges (less than 10 m span), but also for the bridges of larger span – Fabricius (9), Alcántara (10) and Narni (11). The thickest piers reached about one-half of the span – at Rimini (1) and Mérida (2). On the other hand, some piers were as slender as one-fifth of the span – Cestius (12), Julien (13), Bibey (14) and the aqueduct bridge, the Pont du Gard (15).

The figure also identifies bridges with flood openings in the piers. Surprisingly, these occur over the full range of thickness rather than, as would be expected, only the

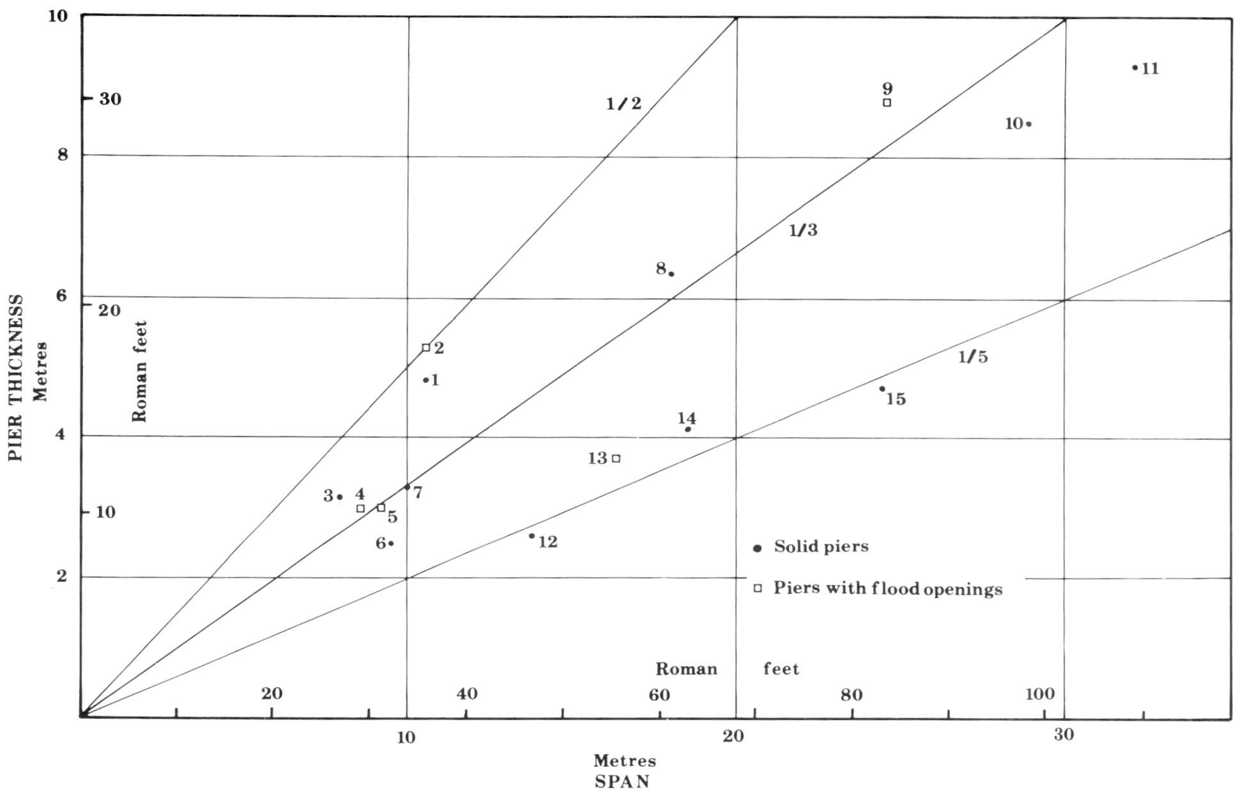

Fig. 133 Graph of pier thickness against span for fifteen bridges. The numbered bridges are identified below:

1 Ponte d'Augusto, Rimini (I117)
2 Puente Romano, Mérida (SP15)
3 Ponte Cardaro, Narni (I89)
4 Pont de Sommières (G9)
5 Puente de Villa del Rio (SP4)
6 Puente Romano, Salamanca (SP40)
7 Puente de Segura (SP22)
8 Pons Aelius or San Angelo, Rome (I2)
9 Pons Fabricius, Rome (I3)
10 Puente de Alcántara (SP21)
11 Ponte d'Augusto, Narni (I87)
12 Pons Cestius, Rome (I4)
13 Pont Julien, Apt (G5)
14 Puente Bibey (SP36)
15 Pont du Gard, Nîmes (AQ30)

thickest piers. In both cases, the practical effect would generally have been small.

The large pier thickness of all Roman arch bridges (for even 1/5 is large) was their greatest weakness in many sites, for the restricted waterway often caused high stream velocities which tended to undermine the foundations and cause collapse. The bridge at Narni (I87, Section 5.4, Figs. 37, 38) was a prime example, and one has only to see the piles of rock around the central pier of the Pons Fabricius (Fig. 21) to appreciate the measures that have been necessary for its preservation.

This point can be made in a different way. Section 5.7 quotes ratios of total opening to total length between abutments for four long Spanish bridges (Table 3).

Table 3. *Clear openings for four long Spanish bridges*

Bridge	Distance between abutments (m)	Clear opening (m)	Ratio
Córdoba (SP7)	274	169	0.62
Mérida (SP15)	721	461	0.64
Alconétar (SP20)	244	178	0.73
Salamanca (SP40)	179	142	0.80

The bridge at Alconétar no longer stands, having been destroyed no doubt by flood. That at Mérida consists of three bridges in line – the presence of the first bridge caused the stream to bypass it to the north; this second opening was crossed by a second bridge, so the river now bypassed the combined structure to the south; and this was then crossed by bridge 3. Presumably the same happened at Salamanca, where the original 179 m long fifteen-span Roman bridge was increased to a length of 356 m with twenty-six spans in the seventeenth century.

Casado (n.d.: 153) has an aerial photograph of Córdoba showing the river more than doubling its width as it negotiated the bridge.

8.3 Design of Roman arches

The arch itself was made up of the arch rib, the spandrel walls and the material in the spandrel.

Later bridges, such as early concrete arch bridges, used soil to fill the spandrel. The resulting pressure on the spandrel walls often caused them to move outwards, with distress showing at the junction between the walls and the rib. This defect is not commonly visible in Roman bridges, possibly because many of them had their spandrels filled with concrete, as a study of failed bridges indicates – see, for example, Narni (I87) and the Ponte Rotto (I5). It is also true that the Romans kept the depth of the spandrel to a minimum, with a depth from the roadway surface to the arch rib at mid-span normally less than the rib thickness. A study of twenty bridges gave an average ratio of about 0.7, with a range from about 0 to 1.3. The bridge at Pondel (I159, Section 5.5) is exceptional, with a ratio approaching 3.

Figures 134 to 138 show the elevations of five bridges, drawn from photographs and scaled using span dimensions known from the literature. A Roman bridge is essentially rectilinear, with the front face lying in a plane and the stonework of the spandrel walls and the piers laid with horizontal courses. The typical photograph is oblique – the line of sight is not normally perpendicular to the face of the bridge and the lines of the coursework are not parallel but proceed to a vanishing point. An oblique photograph of the front plane of a bridge can be rectified by back projection provided a simple rule is followed. If the image is taken from a transparency of the same size as the film used in the camera (as it normally will be) then the projector lens must have the same focal length as that used in the camera. If this is the case, then the image of the front plane may be corrected by rotating the screen until the verticals are vertical, and the horizontals are parallel and horizontal. This image is now a true elevation and may be traced and scaled. In most cases, part of the bridge will be obscured by foliage, shown by fine linework in the figures.

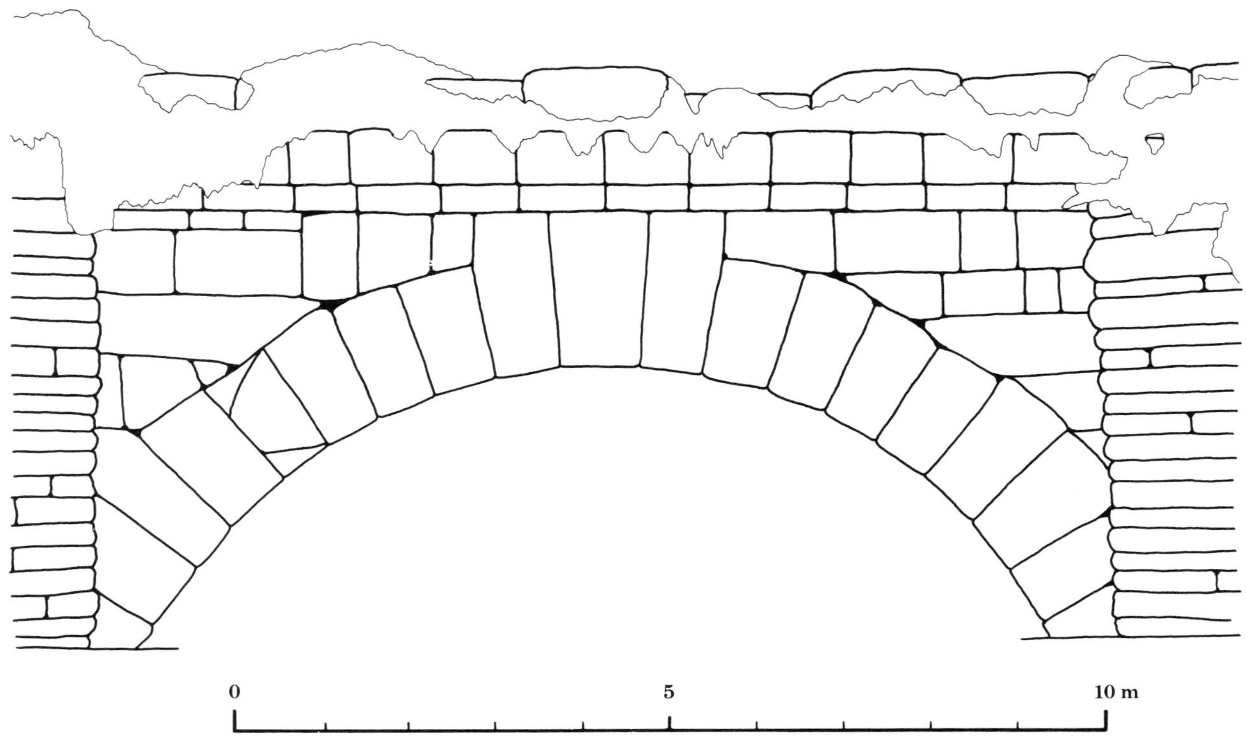

Fig. 134 Elevation of Ponte Manlio.

Figure 134 shows the Ponte Manlio (I109, Section 5.4). The massive construction is evident. The three keystones each have a height of about 1.85 m and one of the spandrel stones at the left has a length of about 2.4 m.

The next three drawings do not attempt to show the stonework of the spandrel walls. The bridge at Pondel (I159, Section 5.5) is remarkable for its great height (from the parapet to the water is about 43 m); the covered walkway with a second path on top; and the depth of cover provided over the arch rib (see also Fig. 59). The dots shown on some arch stones are holes for lifting toggles. Narni (I87, Section 5.4) is one of the great Roman bridges. Figure 136 shows the 19.6 m south span and the remnants of the 32.1 m main span. The arch rib has a projecting cornice (Fig. 38). It appears likely that this cornice continued through the spandrel. The effective rib thickness would be, then, the sum of the rib and cornice thicknesses.

The bridge at Bibey in Spain (SP36, Section 5.7) shown in Figure 137 has, like Narni, a relatively large span, in this case 18.5 m; its height is 22 m clear. The smaller span

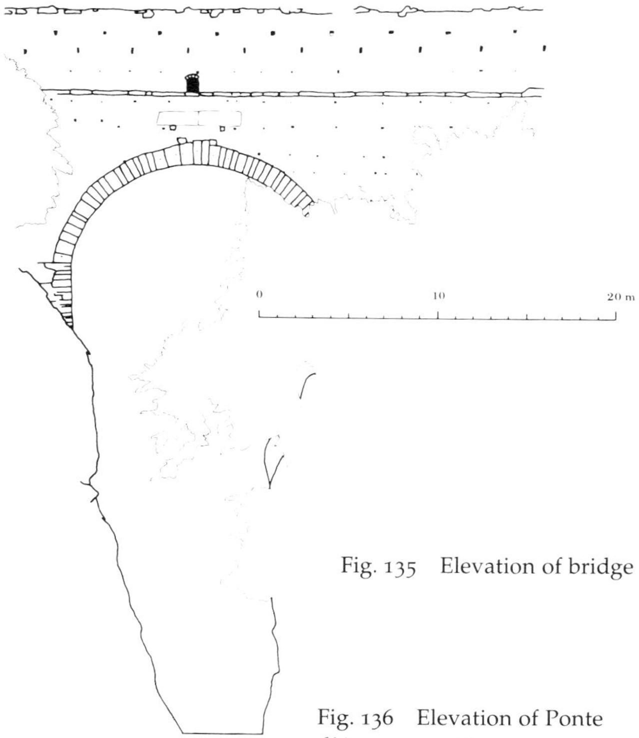

Fig. 135 Elevation of bridge at Pondel.

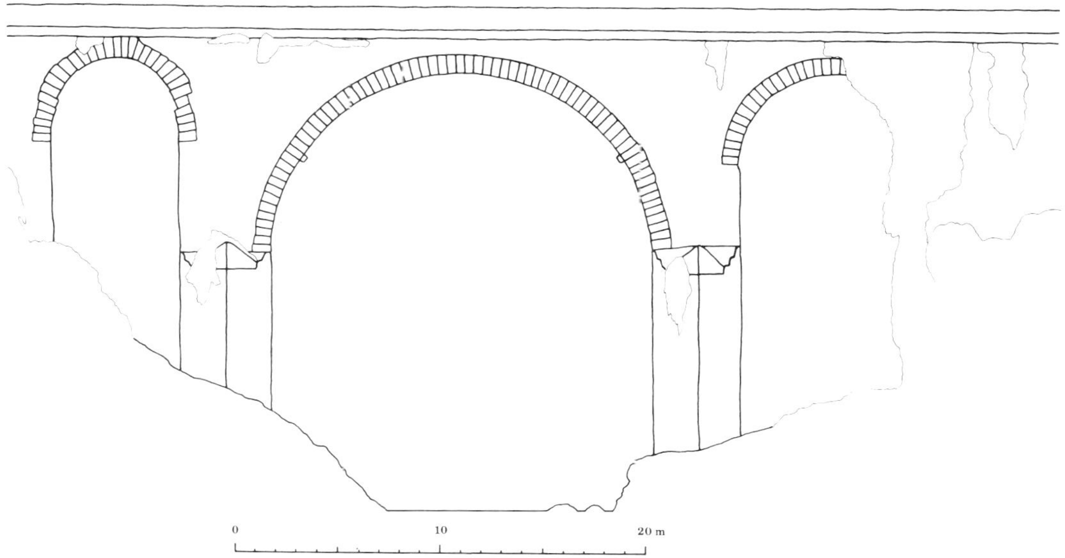

Fig. 137 Elevation of bridge at Bibey.

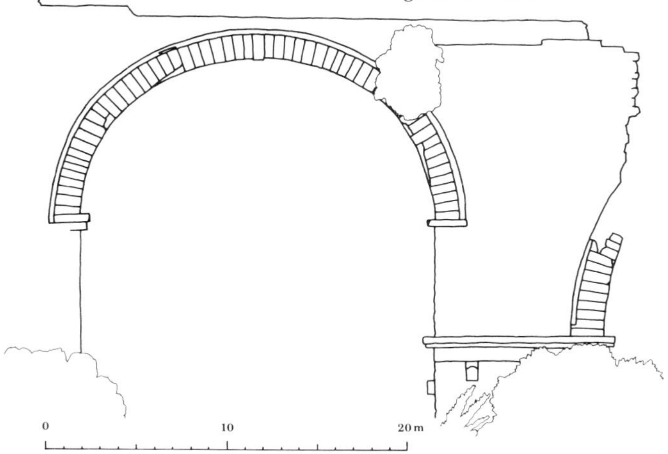

Fig. 136 Elevation of Ponte d'Augusto at Narni.

at the left shows clear signs of distress due to movement of the abutment. Soil pressure behind this high abutment has caused it to move forwards, disrupting the shape of the arch rib and causing its centre to rise (see also Fig. 93).

The greatest of all Roman spans is the Pont St Martin (I155, Section 5.5) shown in Figure 138, for its 35.6 m clear opening exceeds that of all other Roman bridges. In this case, some of the spandrel stones are also shown: the intervening spandrel spaces and the balustrade are built with stones of much smaller size (Figs. 52, 53). The arch profile is segmental – the inside radius is 18.7 m, causing the arch centre to lie about 5.8 m below the springings, with a total included angle of about 144°.

These bridges have in common a much greater irregularity of the stonework than one would normally expect. Although this may have been exaggerated to some extent in the drawing process, there is no doubt that it is real. It is unrealistic, for example, to expect the rib thickness to be constant over the span. This variation may be even more marked in the stones whose upper faces are hidden by the spandrel. There may be, in addition, deliberate changes in the rib thickness. Figure 139 is an oblique view of the Ponte di Cecco at Ascoli Piceno in Italy (I130, Section 5.4). It is apparent that, at mid-span, the top voussoirs on the left-hand face rise above the roadway within the chapel. Estimates of rib thickness taken from external photographs must then be treated with caution.

The ratio of rib thickness to clear span is the most important parameter in the design of a stone arch. Figure 140 plots the rib thickness against the span for forty-two spans, with scales in metres and Roman feet, assuming one Roman foot equals 0.294 m (see Section 4.7). In most cases, the data are for the maximum span of a bridge. In two cases, the data for two spans of the same bridge have been presented.

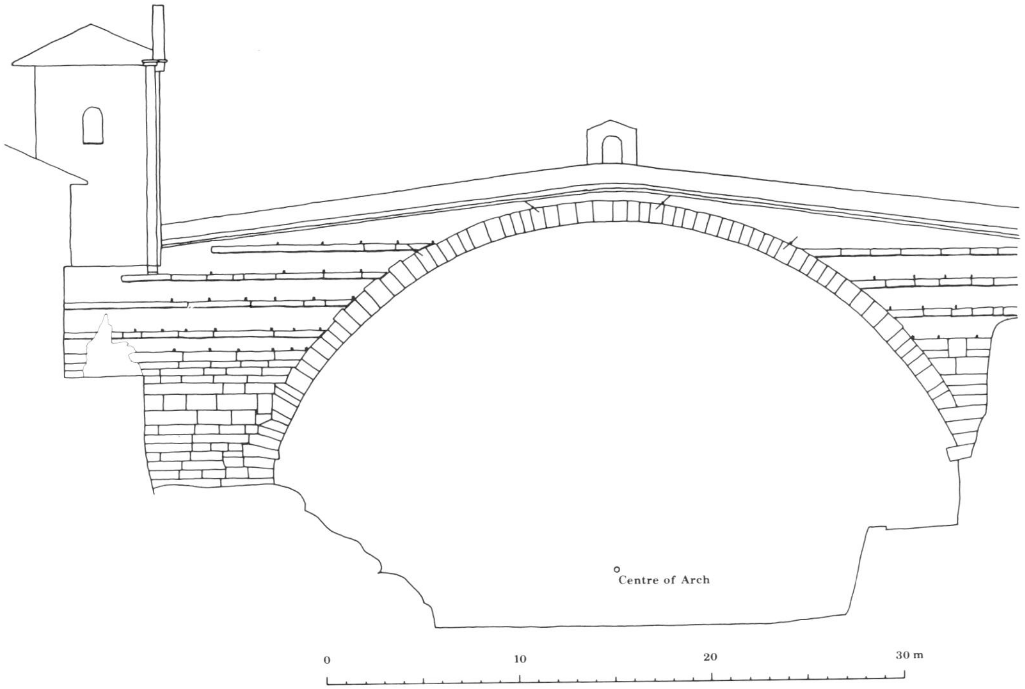

Fig. 138 Elevation of Pont St Martin.

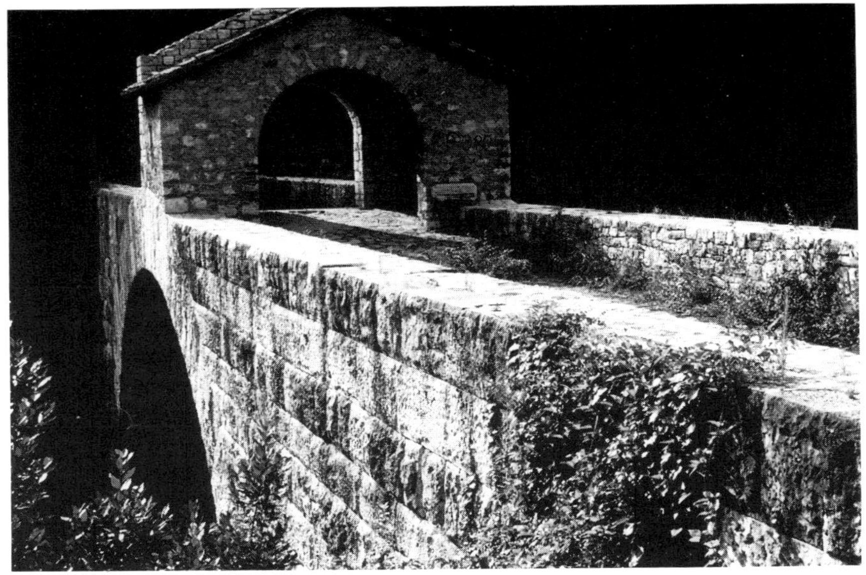

Fig. 139 Oblique view of Ponte di Cecco at Ascoli Piceno. The top of the pavement beneath the chapel and at the right lies clearly below the top of the balustrade, which in turn lies below the top of the exposed face of the arch rib.

For a few bridges the thickness of the rib is stated in the literature and has been used here. Most of the data, however, for Figure 140 have been taken from photographs – in many cases from photographs taken perpendicular to the bridge, in some from photographs corrected by oblique projection, and in a few by measuring at mid-span the ratio of thickness to rise. Figure 140 states that all points should be treated as approximate. In two cases, the approximation is emphasised by arrows – in no. 9, the Ponte dell'Abadia, because of uncertainty as to the span; in no. 12, the bridge at Kâhta, because it is such an important case and the data are uncertain. Where a range of values is known, this has been shown. Preference has been given to the smallest thickness of the rib. The figure also identifies some bridges with double ribs, such as with the cornice at Narni. This will be discussed further in Section 8.4.

Lines radiating from the origin represent a particular ratio of rib thickness to span; the figure shows the ratios 1/10 and 1/20. It is obvious that there is a considerable scatter, but when the figure is examined with care, some trends emerge.

For spans less than 50 Roman feet, or about 15 m, many bridges have a rib thickness of the order of 1/10 of the span. At the limit of 15 m span, this rule gives a rib thickness of 1.5 m or about 5 Roman feet. The figure gives the impression that the Romans were not concerned with large rib thicknesses, provided that they were smaller than this value. For the smaller spans, therefore, they often used thicknesses in excess of 1/10 of the span.

As the span exceeded 15 m, or 50 Roman feet, to avoid stone depths greater than 5 feet, the Romans could adopt one of two options: they could either use a smaller ratio, or they could subdivide the rib by building it of two thicknesses of stone. Their use of the first option showed that they knew that they could use ratios less than 1/10; Figure 140 suggests that they would commonly go down as low as 1/20, and the validity of these lower ratios for the longer spans told them that they

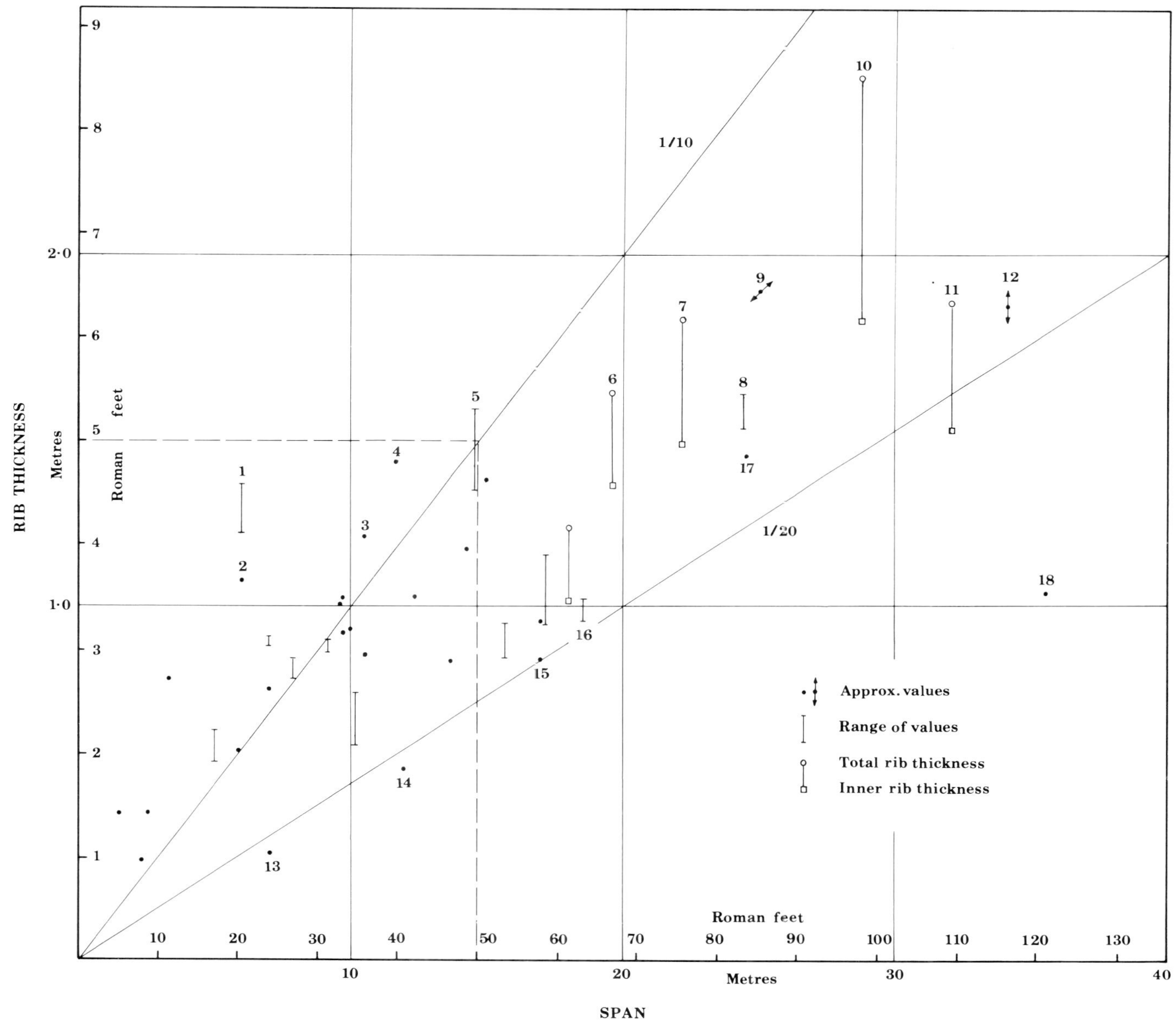

Fig. 140 Graph of rib thickness against span for forty-two arch spans. The numbered bridges are identified below:

1 Large Ponte di Nona, Rome (I23)
2 Pont de Trajan, Béja (A24)
3 Ponte d'Augusto, Rimini (I117)
4 Ponte Manlio, Cagli (I109)
5 Ponte di Cecco, Ascoli Piceno (I130), outside face
6 Ponte d'Augusto, Narni (I87) – 19.6 m span
7 Ponte di Porta Cappuccina, Ascoli Piceno (I128)
8 Pont du Gard, Nîmes (AQ30)
9 Ponte dell'Abadia, Vulci (I65)
10 Puente de Alcántara (SP21)
11 Ponte d'Augusto, Narni (I87) – main span
12 Severus bridge, Kâhta (E36)
13 Pont sur Laye, Mane (G4) – 7 m span
14 Pont sur Laye, Mane (G4) – main span
15 Ponte d'Arli (I127)
16 Puente Bibey (SP36)
17 Pons Fabricius, Rome (I3)
18 Pont St Martin (I155)

could go below 1/10 also for the short spans.

It is suggested, then, that the Roman rules for the design of arch bridges were as follows:

(1) Use as a guide the ratio for rib thickness to span of 1/10, knowing that this could be reduced to as low as 1/20;
(2) avoid stone depths greater than 5 Roman feet.

These rules are in a form consistent with Roman mathematical skills (Section 4.8). Most of the cases shown in Figure 140 will satisfy these rules. Bridges which are close to these limits, or which violate them, are numbered on the figure and named in the caption.

The rule concerning the maximum stone depth is violated in only a few cases: if the span of no. 9, the Ponte dell'Abadia, is 25 m (and this is uncertain), then the rib thickness is about 1.9 m or 6.4 Roman feet; the bridge at Alcántara in Spain (10), although built with a double rib, has an inner rib thickness of about 1.8 m or 6.1 Roman feet; the 34.2 m span bridge at Kâhta (12) has a thickness of about 1.85 m or 6.3 Roman feet.

Only a few of the shorter span bridges have thin ribs – points 13 and 14 represent two spans of the bridge near Mane in France; and the bar above these points at about 10 m span is for the bridge at St Thibéry, also in the south of France. The proximity of the points may represent local practice, or may alternatively question the authenticity of the bridges. Both are segmental, as in the Pont St Martin. Bridges with longer spans and slender ribs include the Ponte d'Arli (15), Bibey (16), the Pons Fabricius (17), Narni (11), Kâhta (12) and most of all, the Pont St Martin (18), with a ratio of rib thickness to span of about 0.029.

The rules stated above suggest that the Romans experienced difficulty with their arch proportions only when the span exceeded 15 m, or about 50 Roman feet. There is indirect evidence which supports this. Figure 141 is a cumulative frequency distribution curve for the maximum span lengths of 146 bridges. For those not experienced in statistics, the procedure followed to produce this curve was to divide possible span lengths into the ranges 0–2.49 m, 2.5–4.99 m, 5.0–7.49 m, etc., and count the number of bridges in each of the ranges; for example, there were 3 in the lowest range, 26 in the next, then 29, etc. To express this in a different way, there were 3 with a maximum span less than 2.5 m, 29 less than 5.0 m, and 58 less than 7.5 m. Dividing by 146, the total number of bridges, these figures yield the percentages 2.1, 19.9, and 39.7, as plotted in Figure 141. It will be noted that as a corollary of this method, the largest span, 35.6 m, falls in the range 35.0–37.49 m, and is plotted at 37.5 m; i.e. all bridges had spans less than 37.5 m.

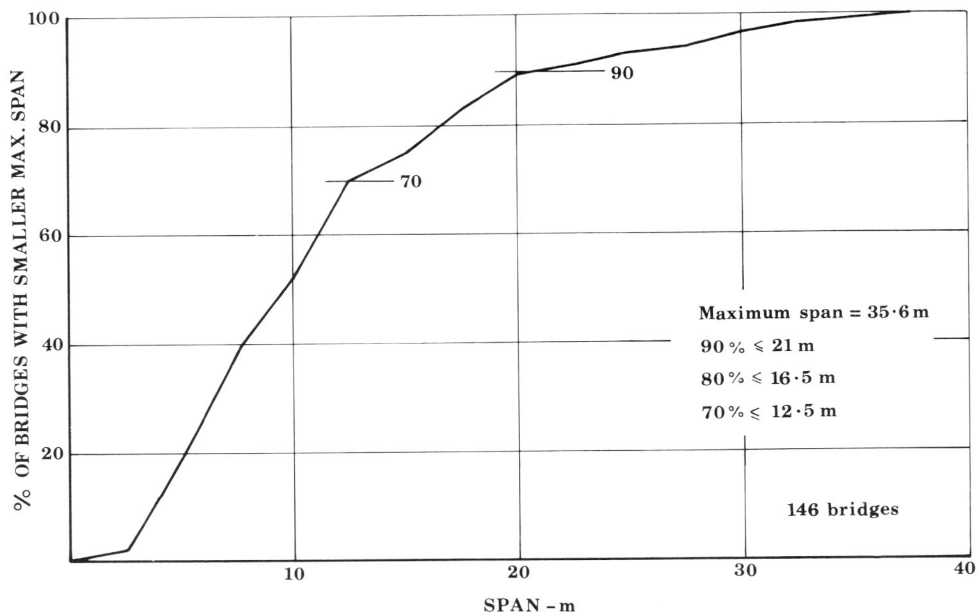

Fig. 141 Cumulative frequency distribution of maximum span for 146 Roman stone arch bridges.

The slope of the curve changes at spans of 12.5 m (70 per cent) and about 21.0 m (90 per cent); that is, 70 per cent of these bridges had a maximum span less than 12.5 m, and 90 per cent less than 21.0 m. If all spans had been counted, then the percentage less than 12.5 m would have been much higher. On the evidence, then, new difficulty arose as the span exceeded 12.5 m, and this is consistent with Figure 140 and the previous rules.

The rule for stone depth results, presumably, from the limits of Roman technology – could they conveniently cut and lift a stone of this size (see Sections 4.3 and 4.4)?

The dimensions of a voussoir will be defined here as its depth, corresponding to the thickness of the arch rib; its width, the dimension of its other face; and its length, perpendicular to the face. Figures 134 to 138 show that widths of the voussoir often vary markedly in the same span. Average figures for the width and depth in metres, and the resulting ratios of width to depth are:

Manlio (keystone)	1.0/1.85 m, giving 0.54
Pondel (main span)	0.65/1.16 m, giving 0.56
Narni (19.6 m span)	0.45/1.33 m, giving 0.34
Bibey (main span)	0.54/1.16 m, giving 0.46
St Martin	0.74/1.03 m, giving 0.71.

An average ratio is about 0.50. Assume also that the voussoir length equals the depth; then a limiting depth of 5 Roman feet would imply voussoir dimensions of 1.47 × 0.735 × 1.47 m, with a volume of 1.59 m^3. The density of a typical rock would be about 2700 kg/m^3. The mass of this stone would thus be about 4290 kg.

The keystone at Manlio is somewhat heavier. If its length was equal to its depth, its volume would be 3.42 m^3, with a mass of 9240 kg. A length of 1.2 m (instead of 1.85 m) would give a mass of 5990 kg.

These figures may be compared with the estimates made in Section 4.3 of Roman lifting capacity – 2.3 m^3 of stone, or 6200 kg, for five men in a treadmill; 1.3 m^3 or 3600 kg for three men in a treadmill; and 0.3 m^3 or 800 kg for three men at the end of a six-part pulley system suspended from a lifting pole. There is a reasonable agreement. It appears, then, that the Romans probably aimed at a maximum voussoir mass of about 5000 kg. They could lift more, say up to about 6200 kg, or alternatively could increase the rib thickness by changing the width and length of the voussoirs, or by using a double rib.

There is an important corollary: to achieve these lifts in the larger bridges, the Romans needed to use special lifting equipment, such as the five-man treadmill.

The remaining rules concern the ratio of rib thickness to span – 1/10 as a guide, and 1/20 as a normal lower bound, but extended to the remarkable ratio of about 0.029 or 1/34 in the Pont St Martin. The validity of these rules will be discussed in Chapter 9.

8.4 Segmental arches

A segmental arch is circular, but with a total included angle less than 180°. Whereas the semicircular arch is vertical at its ends, the segmental arch is inclined.

Most Roman arch bridges are semicircular. It has been said further that the Romans never built segmental arches. Needham et al. (1971: 184), for example, in volume 4 of Needham's fine series of books *Science and Civilisation in China*, suggests that this type of construction was transmitted from China to Europe by Marco Polo about 1295. The Zhao-Zhou bridge, known also as the Chaochow or An-ji (or Anchi, meaning safe crossing) bridge, is a remarkable Chinese segmental arch with open spandrels and a span of 37–37.4 m, built about AD 600 (Qian Ling-Xi, 1987; Mao Yi-Sheng, 1978: 4).

Although it is true that the segmental arch became popular in Europe after the thirteenth century (O'Connor, 1975), it is not true to say that the Romans did not build segmental arches. A study of Roman bridges suggests that three different classes of structure exist.

First there are cases where the departures from the semicircle are small, and could have resulted from errors during manufacture.

Second are bridges with a deliberate but small departure from an included angle (2θ) of 180°. The best example is the Pont St Martin, with 2θ equal to 144° (Fig. 138). Similar examples are the Pont de Pierre (I158, Section 5.5) at Aosta, and the bridge at Mane (G4, Section 5.6), whose included angles are difficult to assess properly from present data, but appear to be about the same as for the Pont St Martin. It may be significant that all three lie on the same section of road.

Third are bridges which are markedly segmental. The bridge of Nahr Aboul-el Assouad (E52, Section 5.9) near Tyre has a value, 2θ, of about 130°. The side span of the Puente de Alconétar (SP20, Section 5.7) shown in Figure 84 has 2θ equal to about 120°; at St Thibéry (G11, Section 5.6), it is about 115°; and in the Ponte San Lorenzo at Padua (I161, Section 5.5), about 113°. The bridge at Limyra (E23, Section 5.9) and the small side spans of the adjacent bridge at Kemer (E22) have 2θ about 83°.

To express these differently, a value of 2θ of 130° corresponds to a rise-to-span ratio of 0.32; 2θ equal to 83° gives a rise to span ratio of 0.19.

There are also other special cases. Gazzola's drawing of the small Ponte del Gran Caso near Ascoli Piceno (I131, Section 5.4; Gazzola, 1963b, no. 74) shows a segmental side span resting on the quarter-point of the main arch rib. A similar detail is visible in the Ponte di Pioraco (I119, Section 5.4).

In all of these cases, there is a temptation to dismiss the bridges as post-Roman, and in some of them there are other reasons to believe that this may be so. Gazzola's drawing of the Ponte San Lorenzo (Gazzola, 1963b, no. 54) in particular shows piers whose thickness is only 0.11 of the span, well outside the range of values shown in Figure 133. But it is believed that this cannot be done in every case. The credentials of the Pont St Martin and the bridges at Alconétar and Limyra appear to be very strong. Further, there is no doubt at all that the timber bridge of Apollodorus (T13, Section 6.3) was a segmental arch. Indeed, if Gazzola's representation were correct in this regard (Gazzola, 1963b, no. 184), it had a rise-to-span ratio of only 1 in 7, with a total included angle of about 64°. If the Romans, through Apollodorus, could build segmental arches in timber, there seems no reason why they could not have built them also in stone.

In Section 8.6, it will be seen that the Romans appear to have constructed the initial portions of their semicircular arches up from the springings to an angle with the horizontal of about 20°, without support from falsework. It is suggested, further, that they provided strong infill in the spandrel in this region to support the arch rib. It is only a small step from this point for them to realise that the lower part of the arch could as well be vertical as circular. The concept of the segmental arch may then have arisen directly from construction experience with the semicircular arch.

8.5 The coursing of Roman arch ribs

The arch rib is a three-dimensional structure. A variety of methods may be used to form it from individual stones (see also Section 4.4). Figure 142 shows the four most common.

In Type A, if the arch is viewed from the underside, it will be seen to be formed of separate unbonded ribs. This is the simplest form of construction, but it reduces the integrity of the structure. It may simplify the falsework, as one rib may be completed at a time.

Alternatively, the joints may be staggered across the width of the structure, as in Type B, so that the underside of the vault exhibits the so-called stretcher bond, common in walls and modern brickwork. The joints are staggered, so as to cause interlocking between successive courses, although the actual arrangement may be less regular than that shown.

The use of stretcher bond means that, at the outside faces of the bridge, adjacent voussoirs extend different distances into the vault, and the alternate stones are necessarily long. In Type C, the arch rib is constructed so that its face at the side of the bridge shows alternate long and short stones. The long stones of the face form the headers of the underside, the short stones in the face are the stretchers below, and vice versa. If this construction is continued over the whole width of the bridge, then the arch rib is as in C, but other possibilities also exist.

For the thicker arch ribs, it may be necessary to use two unbonded ribs on the face, to form two vaults, one lying outside the other, as in Type D. The outer vault is shown extended in the figure, to illustrate the coursing, but in fact the stones of the inner vault would be laid first.

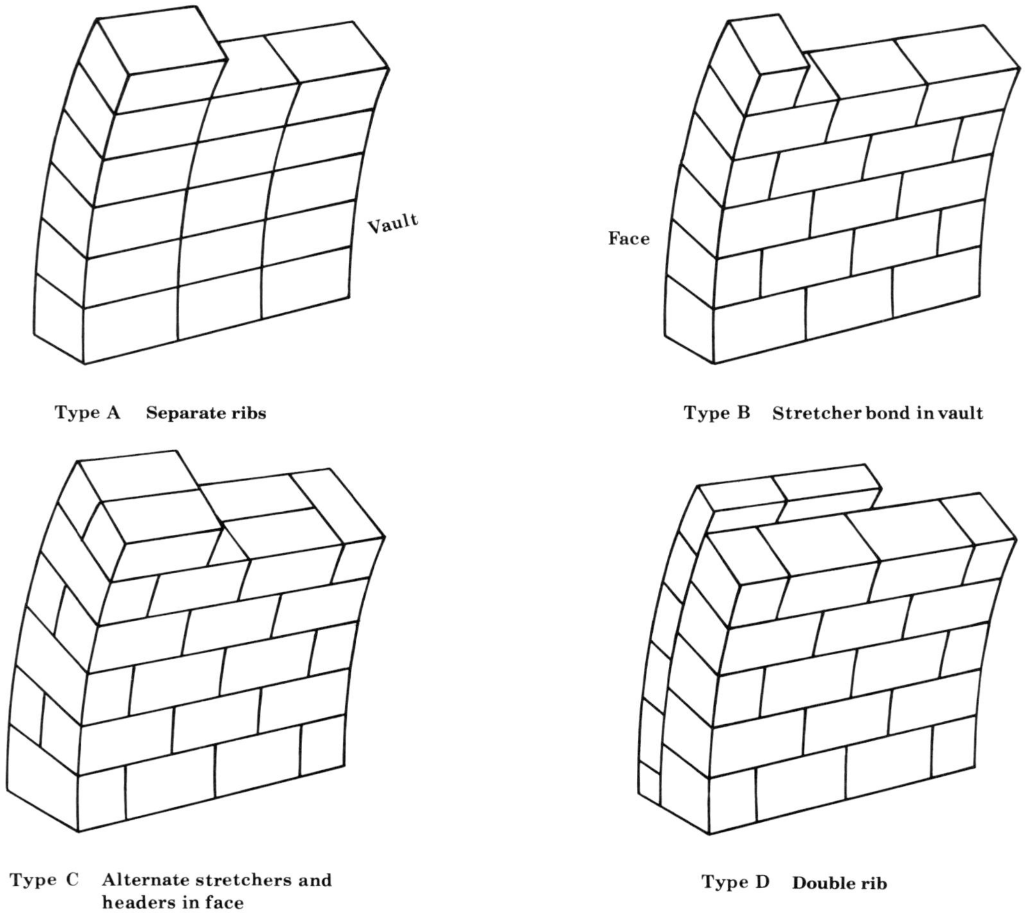

Fig. 142 Coursing of Roman arch ribs.

The second of these forms of construction, Type B, is by far the most common. A study of forty-seven stone bridges and aqueducts showed that twenty-nine, or 62 per cent, were of this type. Five, or 11 per cent, were of Type A; four, or about 8 per cent, may have been of Type C; six, or 13 per cent, were of Type D; and another three, or 6 per cent, were of a form not included in Figure 142.

The five of Type A are geographically linked. The Pont St Martin (I155, Section 5.5, Fig. 53) is of this form, with five ribs separated by narrow bands of concrete infill in the upper part of the span, but not extending to the abutments. Only one rib of the bridge at Chatillon (I157, Section 5.5, Fig. 55) still stands, but it also may have been of this form. The Pont du Gard (AQ30, Section 7.3, Fig. 122) has three separate ribs; there are four at Sommières (G9, Section 5.6, Fig. 67); and two in the Roman part of the bridge at Boissezon (G10, Section 5.6, Fig. 68).

It is difficult to identify with assurance bridges of Type

C, for it is not uncommon for bridges of Type B to have some divided voussoirs. The four counted above include three in or near Ascoli Piceno: at Quintodecimo (I126, Section 5.4, Fig. 46); at Arli (I127, Section 5.4, Fig. 47); and the Ponte di Cecco in Ascoli (I130, Section 5.4, Fig. 50). The identification may be certain only in the last case.

The bridges with double ribs, Type D, include the bridge at Narni (I87, Section 5.4, Figs 37, 38) and the somewhat similar Ponte di Porta Cappucina (I128, Section 5.4, Fig. 48). In both cases, there is a projecting cornice believed to extend through the spandrel. Alcántara (SP21, Section 5.7, Fig. 86), the much smaller bridges at Montignies-St Christophe (G25, Section 5.6, Figs. 71, 72), and the Ponte della Rocca (I69, Section 5.4, Fig. 35) certainly have a double rib. The Ponte San Angelo (I2, Section 5.2, Fig. 20) has also been placed in this category, although in this case the classification is uncertain. The face of the arch rib is decorated with a cornice above, and a stepped profile below. Here and in Figure 140 it has been assumed that the cornice is separate. There are also other difficulties, caused no doubt by the application of a decorative facing to the bridge. In particular, voussoirs appear to extend into the spandrels, as is also the case in the Pons Cestius (I4, Section 5.2, Fig. 22).

It should be understood that this sample is incomplete – it consists of those bridges for which an identification could be made, usually from photographs, and although there is a substantial overlap, it is not the same as that used in Figure 140.

8.6 Arch construction

As stated in Section 4.1, the matter of bridge construction is closely linked with the question of available technology. It is appropriate, then, to review the content of Chapter 4 before proceeding, in particular the passages in Vitruvius where he discusses the construction of foundations (see Section 4.3). In the same section, an illustration is given of a Roman crane (Fig. 8), and it is estimated that it could lift 6200 kg at a radius of 13 m, or 11 5 m clear of its own supports. This radius is reviewed further here. Section 8.3 compares this lifting capacity with the weight of stones in Roman arch bridges. As described in Section 4.3, the Romans also used simpler lifting devices with a smaller capacity.

It is also appropriate to remember the conclusions given in Section 6.1, that many early Roman bridges were of timber, and that 'stone bridges were likely to be second or third generation structures, built to replace earlier crossings, constructed in a stable environment, with access to adequate and known resources and a developed infrastructure'. A corollary of this conclusion would be that, when the stone arch was constructed, there was often an existing bridge providing access from one side of the crossing to the other, and that, if necessary, this could be used to aid the construction by providing transport for materials and equipment.

No record has been left of the construction procedures used by the Romans for the arch. However, the structures themselves provide some indications because of the features built in to aid construction. Of these, the most common are protruding stones and holes left in the arch. It is possible also that Roman methods for arch construction were handed down over the centuries with little change, in which case records of medieval arch construction may have some relevance. In this regard, the work by Fitchen (1961) is of the highest value: he not only describes methods used for Gothic arch construction, but also speculates on the falsework (or centering) used for the Pont du Gard (p. 10, etc.). Hauck (1986: 111) reproduces an alternative suggestion by Léger (1875) for the same structure. Details of arch falsework may also be found in Parsons (1968, figs. 125, 129, 155, 156, 189, etc.), Melan (1924, vol. 2, figs. 140, 141, 147, 152, 153, 159, etc.), Smiles (1862, vol. 2: 182), Jackson (1971: 85, 91, etc.), Pannell (1964: 353), and Bracegirdle and Miles (1973: 54). A characteristic of many of these examples is the use of radial struts, either generally or locally, to support the outer, circumferential falsework.

It is clear that the Romans supported the falsework at least partly on the bridge itself. In the case of the Pont du Gard (Fig. 122), there are cornices at springing level, stones extending from the faces of the piers, and protruding ribs extending within the intrados with their upper faces about 20° above the springing. The Ponte Rotto in Rome (Fig. 23) has similar protruding stones above the springing, and also at a similar height in the intrados of the arch rib of the broken span. The nearby Ponte Cestio (Fig. 22) also has these stones, but not in all spans. It would appear that the Romans sometimes cut them off after completion, although in many cases they were left. There are many such examples, and also of bridges with large holes in the intrados. The bridge at Quintodecimo (Fig. 46) is one such case, and there are many more. The use of a substantial cornice at springing level appears to have served also this practical purpose.

The nature of these supports can be used to form some idea of the falsework itself. First it must be observed that it varied greatly from bridge to bridge; but there were often some common features, which will be identified here.

It is noticeable that the first indication of falsework supports often occurs some distance above the springing. The importance of this observation is that it serves as a reminder that the arch rib could be cantilevered upwards some distance above the springing without support. Section 9.5 suggests that this may be as high as about 24° for a semicircular rib whose thickness is one-twentieth of the span. Construction of the arch rib would then be in two steps – first without and then with falsework. This separation of the early construction of the lower part of the rib into a distinct stage suggests that the Romans may also have built it in other ways. In particular, they may well have filled the spandrel in this space first, and may also have filled it in a better way – either by the use of internal squared stones, or with concrete. A concrete infill may also have been used for the higher levels, but then again it may not, and the analytical work of Chapter 9 considers this possibility, with strong infill in the lower parts, and loose fill above.

The other point of interest is that, where holes are used, they are often relatively widely spaced in the direction parallel to the stream flow. Three holes are common. This suggests that the falsework may have consisted of three parallel frames. Their spacing, then, would be so great that it would be unlikely that they were covered simply by planking to provide continuous support for the arch rib. Rather, it would be more likely that the decking of the falsework consisted of a series of many parallel members on edge, arranged so as to support the individual stones along discrete lines parallel to the flow.

The falsework frame at the level above the springing would have had to support the cantilevered rib against inward movement. It would also have had to be such that, if it was inserted in holes in the rib, it was capable of being removed after the span was complete. These observations lead to the conjectured falsework arrangement shown in Figure 143.

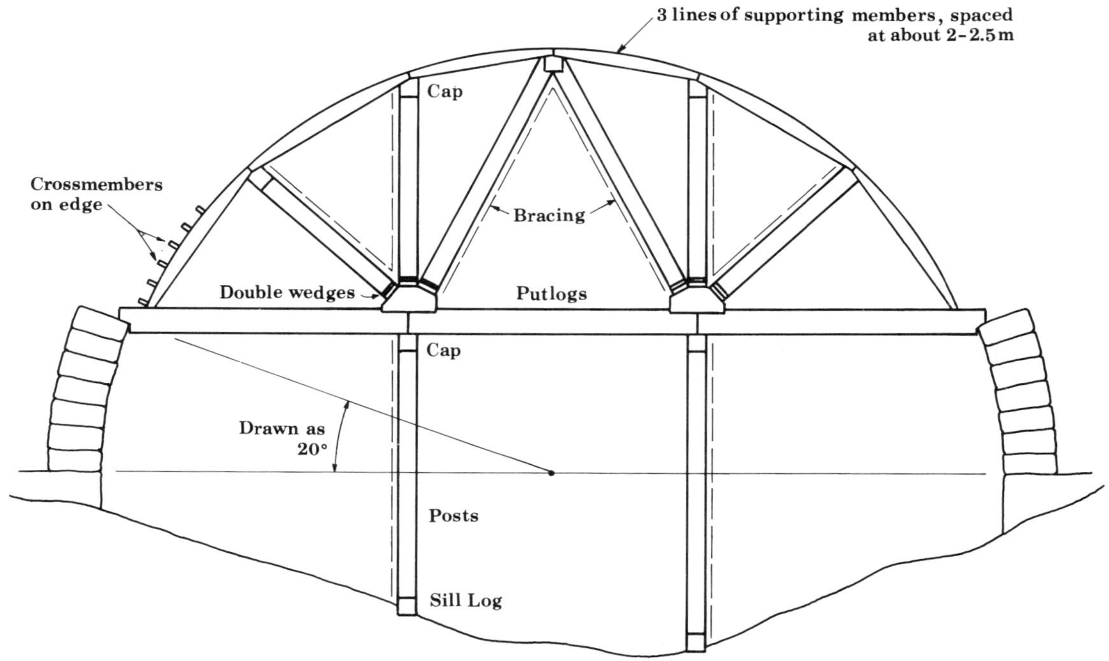

Fig. 143 One of a number of possible falsework arrangements for the construction of Roman arches.

A standard problem with all arches arises when it is desired to remove the falsework. At this stage much of the weight of the arch rib is taken by the falsework members. They are compressed, in the sense that they are shorter than they would be in the unloaded condition. If unloaded, the falsework would move outwards in the radial direction.

In modern concrete arches, this problem is overcome by building hydraulic jacks into the arch rib at closure, and these are operated so as to cause the arch to rise from its falsework. The only means available to the Romans was to build wedges (or something of this kind) into the falsework. On completion, these would have been driven free, so as to lower the falsework away from the arch. The force necessary to remove a wedge is proportional to the load acting on it. If this load is large, then the removal force is also large. If the wedges are built into the falsework at a low level, as shown in Figure 143, there will be few of them, and the load on each will be large. Alternatively, there may be more of them, built in at a higher level, such as at the ends of the circumferential members.

The critical stage of arch erection is when the falsework is being removed. If the arch thickness is too small, the arch will buckle, with one part collapsing back onto the falsework and other parts moving away. Alternatively, the increased horizontal thrust on the abutments or piers may cause them to move outwards, allowing the arch to fall. Heyman and Padfield (1972) have described how Telford's Over bridge near Gloucester came close to collapse in this way.

In the case shown in Figure 143, the ends of the horizontal falsework members, called putlogs, are notched so as to provide horizontal support to the lower sections of the arch. There may be also vertical support provided at these points, as shown by Fitchen (1961: 10). He suggests also that the upper part of the falsework could be removed intact and moved to another span. This cannot be done with the system shown here; rather the falsework must be removed member by member, and then re-erected. It is significant that the span lengths of the Pont du Gard vary by about ±10 per cent, and this fact argues against the reuse of intact falsework.

In Figure 143, the putlogs are supported by six vertical posts. These may be driven into the stream bed, or alternatively, in a rocky site, they may rest on sill logs, placed directly on the rock. It is essential that adequate bracing be placed in the direction parallel to the stream flow, both between these posts, and also between the upper radial members. If the bridge is high, then these posts will become long, and it may be better to replace them by inclined members resting on the lower parts of the piers.

For a multi-span structure, the construction sequence must be designed to keep the temporary out-of-balance horizontal thrusts on the pier tops within acceptable bounds, and also to observe the limits imposed by the available equipment. The crane shown in Figure 8 has a limited reach, both horizontally and vertically. It may be mounted either on the bridge itself, or on temporary timber platforms constructed beside the bridge, on piles driven into the stream bed. Figure 144 shows a possible construction sequence which does not require the use of separate, temporary crane platforms. At the stage shown, the crane is working from the top of pier 3. At the close of the previous stage, the crane was at pier 2, with the arch ribs of spans 1–2 and 2–3 complete, falsework in spans 1–2 and 2–3, and the spandrel fill brought forward to A.

After the crane has been moved, the first step is to bring the spandrel fill forward to B, allowing the falsework to be removed from span 1–2.

With the crane on pier 3, the following operations are carried out:

(a) construction of pier 4, the lower level of the arch rib of span 3–4, and a small extension of the pier, ready to receive the crane in the following stage;
(b) erection of the falsework of span 3–4;
(c) construction of the arch rib in span 3–4, working symmetrically up from the piers.

It can be seen that there will be unbalanced horizontal thrusts applied from the arches to the pier at various stages. Although there are alternative procedures that would reduce these forces, these would require either a longer crane, or separate crane platforms.

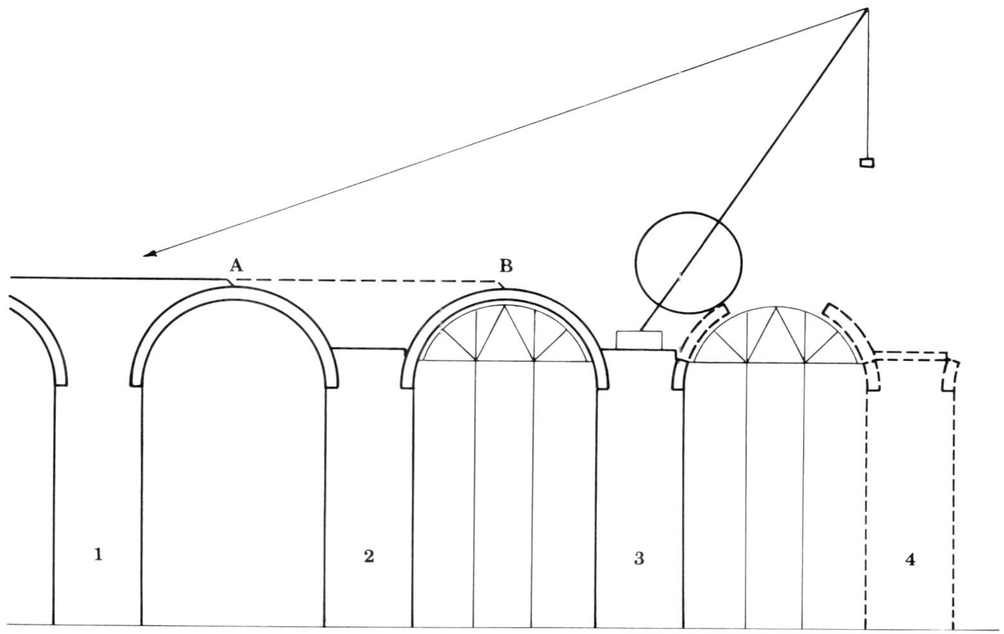

Fig. 144 Possible construction sequence for an arch bridge with multiple spans.

The clear reach required of the crane in the sequence shown in Figure 144 is one span length plus one pier width. For example, with a span length of 10 m and a pier width of 5 m, the clear reach required would be 15 m, or possibly a little less, for in practice the lift can be pushed horizontally so that the lifting ropes are inclined from the vertical.

Suppose the process of Figure 144 is carried out simultaneously from both abutments with two cranes. Then the final span can be longer, with the clear reach required of the crane at this stage being one-half of the span. The same condition applies to a single-span bridge. The Romans' largest such span was 35.6 m at Pont St Martin (I155, Section 5.5). It follows that this would have required a clear reach of 18 m, if it was constructed in this way. These figures are not greatly in excess of the 11.5 m clear reach deduced from the approximate representation of a crane shown in Figure 8. It is argued, then, that this estimate is a little small and that the Romans commonly used cranes with a clear reach of 15 m and in exceptional cases up to 18 m.

This study also provides an added reason for the Romans' preference for the use of smaller spans (Section 8.3).

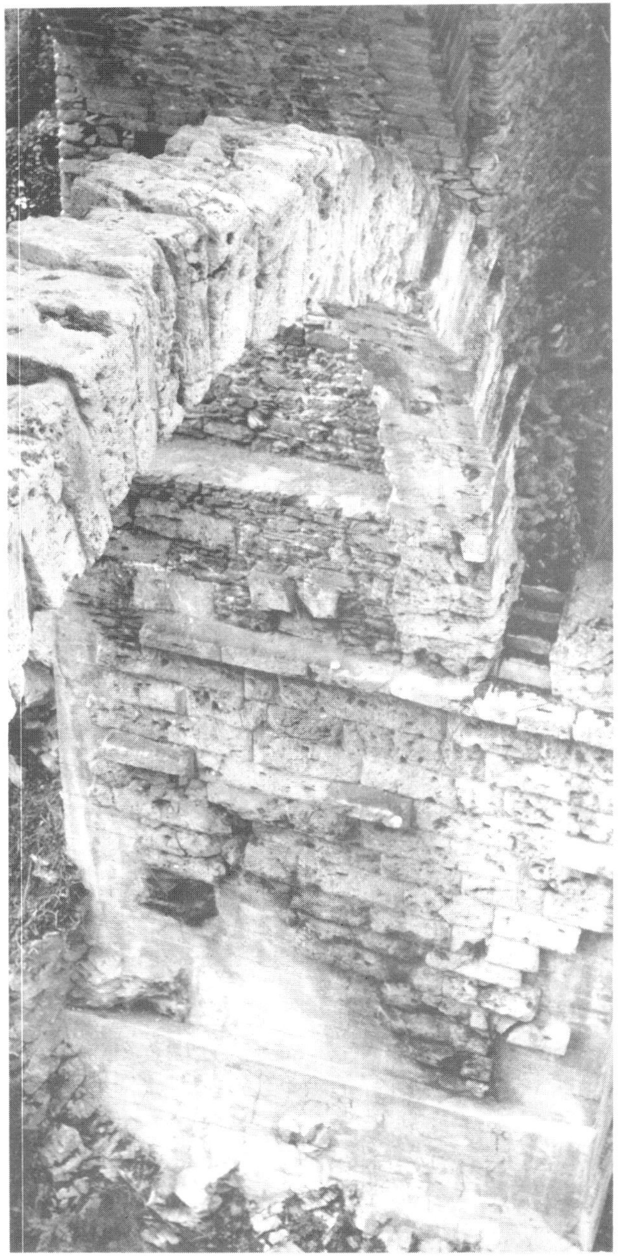

9

ANALYSIS OF ROMAN ARCHES

9.1 Introduction

The essential purpose of this chapter is to subject to modern analysis the Roman rules for arch design identified in Section 8.3. There is a sense in which these rules must necessarily be correct, for they have led to arches which have stood for centuries. However, they may conceivably have been conservative, and, in any case, a study of their validity will lead to an improved understanding of arch behaviour, and this is of some importance.

The description given here is expressed in non-technical language. It owes much to the work of Heyman (1966, 1969, 1971, 1980, 1982, etc.; and with Padfield, 1972), who also has given full accounts of the history of arch analysis in recent centuries (1972, 1976). This work has led to a renewed interest in the analysis of masonry arches, supported by the realisation that many of them, particularly those built in the last few centuries as road–rail overpasses in Europe, Britain and elsewhere, represent a valuable resource which should as far as possible, for economic reasons, be kept in service. There has been in recent years an expansion of the technical literature on masonry arches, although this is still incomplete. Some of this may be found in the publications of Chettoe, Cooke, Crisfield, Harvey, Hauck, Jennings, Page, Pippard, Selberg, Vilnay, Walklate and others. There has also been an interesting recent study by Qian Ling-Xi (1987) of the Zhao-Zhou Chinese bridge of about AD 600.

An account of the history of arch analysis will require the use of some technical language that has yet to be defined (see Section 9.2 and Appendix 1). It will therefore be reserved for the closing section of this chapter (Section 9.6).

9.2 Lines of pressure and form load

A *force* cannot be seen but can be recognized only by its effect – it tends to accelerate a body. The most common force is that due to gravity. It is downwards and vertical and has magnitude. A person, for example, may be said to 'weigh' 62 kg, although more truly this is his or her mass. If a person of this mass is allowed to fall freely, he or she will accelerate at 9.81 m/sec/sec and this demonstrates not only that there is a force present, but that its magnitude is 608 Newtons; that is, 62 × 9.81. The earth's gravitational force, although distributed through the mass, is equivalent to a single resultant force, acting through the centre of gravity. The force, then, has magnitude, direction and location. It can be represented by an arrow on a fixed line; that is, one whose direction and location are both defined. A force need not necessarily be vertical, even though caused initially by gravity. For example, if a person hangs from a rope, and the rope is passed over a pulley, the final force will act along the line of the final rope, and may be in any direction.

Consider, in Figure 145a, the vertical force V_1 and the horizontal force H. They combine to form a single resultant force R_1, passing through their point of intersection,

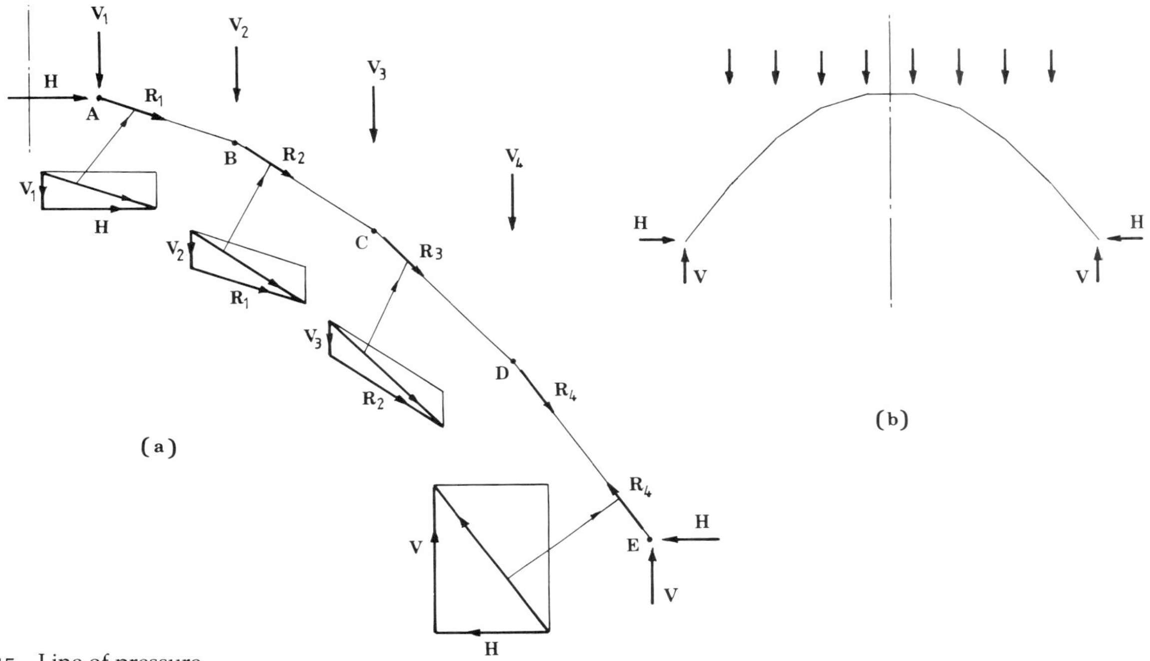

Fig. 145 Line of pressure.

A, where the magnitude and direction of R_1 may be determined from the diagonal of a parallelogram (in this case a rectangle) drawn with sides H and V_1. Similarly R_1 and V_2 may be combined to form R_2; $R_2 + V_3 = R_3$, etc. until R_4 is formed. Suppose that at E a pair of forces H and V are applied, so that their resultant is equal and opposite to R_4. Then the whole force system is in *equilibrium*. If the final forces H and V are supplied by suitable supports, such as the ground, then they are called *reactions*.

The line ABCDE has a shape not unlike the right-hand half of an arch. If it is combined with its mirror image, the diagram shown in Figure 145b is derived. Line ABCDE in (a) and the full line in (b) are each examples of a *line of pressure*, defined as a line whose tangent at any point is the line of action of the forces to the left (or to the right) of that point. If the applied loads are distributed, so that they correspond to a sequence of small, closely spaced loads, such as the gravity forces on a body, then the line of pressure will be curved, instead of segmented as in Figure 145. The shape of the line of pressure depends not only on the loads, V_1, etc., but also on the horizontal reaction, H. If H is increased, for example, then the line of pressure will be flatter, with a smaller rise at mid-span.

The line of pressure shown in Figure 145 represents the shape of an ideal arch, built so that the resultant force passes always along its centreline. Conversely, for a given arch, the loads F_1, etc., can be derived so that their line of pressure passes along the arch centreline. This system of loads is called the *form load*.

Consider instead the pair of parallel forces shown in Figure 146a; they also can be combined into a single

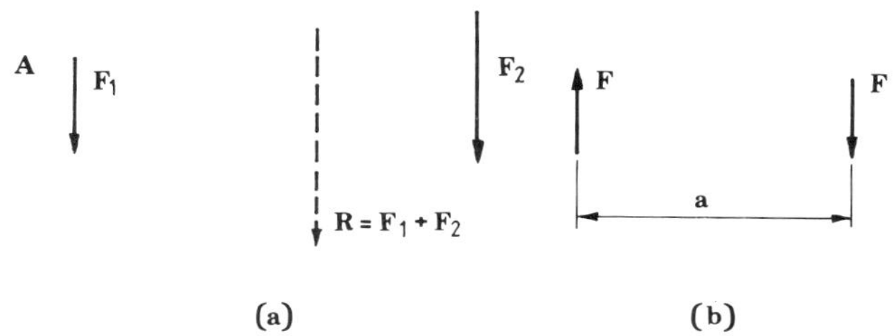

Fig. 146 The concepts of moment and couple.

resultant force. The *moment* of force F_2 about A is defined as equal to the magnitude of F_2 multiplied by the perpendicular distance from A. Then the resultant force is determined by the following conditions: (1) it must be parallel to F_1 and F_2; (2) its magnitude must equal the sum of F_1 and F_2; and (3) its moment about any point, such as A, must equal the moment about A of the original forces.

This process breaks down in (b), where the forces are equal and opposite – they cannot be replaced by a resultant force. Instead, they are called a *couple*, and the magnitude of the couple equals the magnitude of each force times the distance between them. Any force system may, in the limit, be reduced to a single force and a single couple, although in some cases either the resulting force or the resulting couple may be zero. If such a couple acts internally at a point within a member, between one part of the member and the other, then it is called a *bending moment*, which may be calculated as the moment about the point in the member of the forces taken either to the left or to the right of that point.

Figure 147a shows the line of pressure for a vertical stone wall, subject to the action of its own weight and a sequence of increasing horizontal loads, H_1 etc. Consider instead Figure 147b. If the weight of the top stone passes outside its point of contact with the stone below, then the upper stone will fall. In this case, the weight of the upper stone is, at this point, the only force in the line of

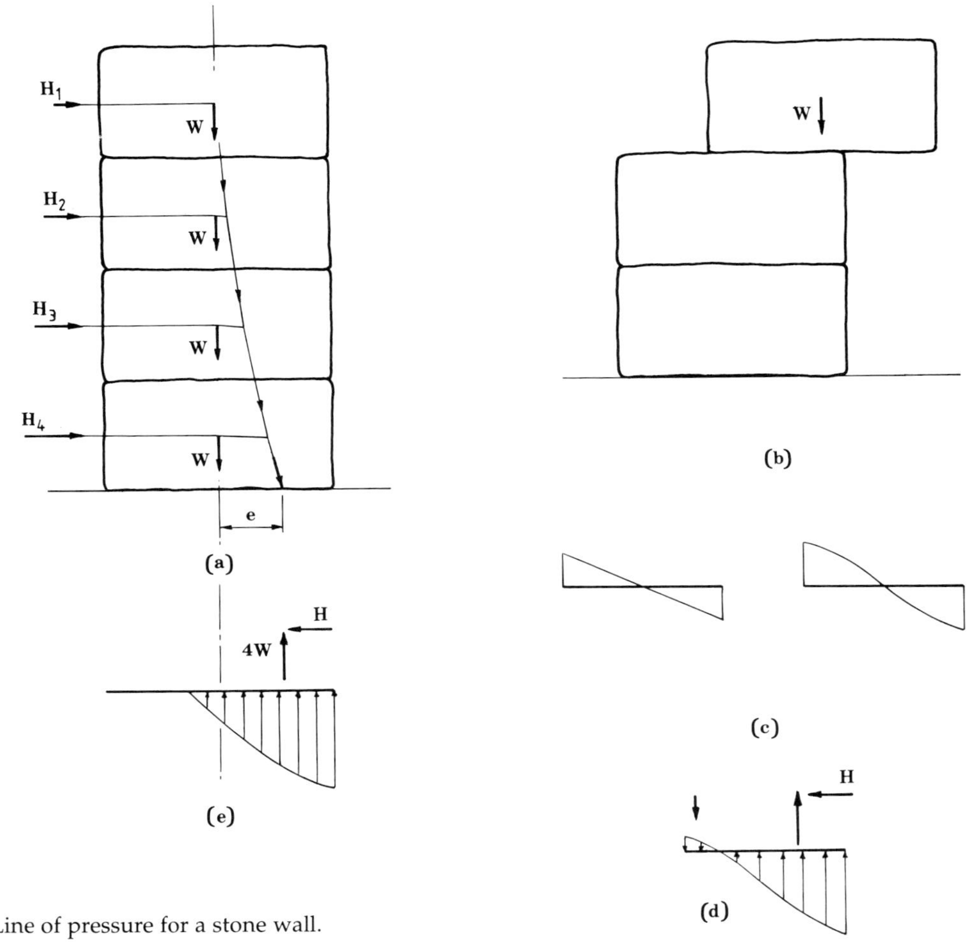

Fig. 147 Line of pressure for a stone wall.

pressure. Similarly in (a), if the line of pressure at any joint passes outside the boundary of the wall, then collapse will also occur, with the upper part rotating about the corner of the lower stone; the corner is said to act as a hinge.

The situation is different if, instead of being composed of a series of stones, the wall is integral, carved out of some continuous solid material. As shown in (a), the resultant force at the base of the wall lies at an eccentricity e from the centre of the wall. There is here a bending moment of magnitude 4W × e.

Stress is defined as force per unit area. If the total weight, 4W, at the base of the wall acted alone, it would produce a uniform *compressive* stress equal to 4W divided by the area. If, instead, the bending moment 4We acted alone on a solid wall, then it would produce stresses something like one of those shown in Figure 147c. The combined stresses may be as in (d) – those at the right are compressive; those at the left are said to be *tensile*. These stresses may be replaced by the resultant reactive forces also shown in (d). There is also a reactive force parallel to the base, called a *shearing force*, H. The combination of these reactive forces would hold the wall in equilibrium.

The stone wall, however, is incapable of transmitting tensile stresses across its joints. Its stresses would, then, be something like those shown in (e).

The loads applied to an arch rib are its own weight and the forces from the spandrel. The latter derive from the weight of the spandrel material and also from any loads. The self weight of the arch rib is vertical, and if the arch thickness is constant, the resulting distributed loads are constant per unit length around the rib. More generally, these loads will be expressed here as p′ per unit circumferential length, where p′ is constant for a rib of constant thickness. Alternatively, if a uniform load is applied to the deck, this will be constant per unit horizontal distance. Such loads will be expressed here as p. Then p and p′ are related at a point by the slope of the arch rib. If the rib is horizontal, p equals p′.

Concerning the spandrel weight, it has been customary, for the purposes of analysis, to divide the spandrel into sections by taking vertical planes parallel to the stream flow, and applying the weight of the material in each of these sections as a vertical load to the arch rib. This is clearly incorrect. If the spandrel material is soil, then it may apply also horizontal components and these may vary between wide limits, from a lower bound called active soil pressure to an upper bound called passive pressure.

If the form load of an arch is specified by setting p constant, then it may be shown that the correct shape of the arch is *parabolic*. If, instead, p′ is constant, then the correct shape is a *catenary*; that is, the inverse of the shape adopted by a uniform chain loaded by its own weight. In both cases, the rise of the arch may be varied by adjusting the horizontal thrust, H.

It has been seen that it is possible to derive the form load for any arch shape, and this may be expressed alternatively by p or p′, where both will in general vary with the distance, x, from mid-span. Figure 148 plots form loads for circular, parabolic and catenary arches. The shapes themselves are also plotted, for arches scaled so that each has the same radius of curvature, R_0, at mid-span. For the case shown, with a span length equal to $2R_0$, the rises are R for a circular arch, 0.543R for a catenary, and 0.5R for a parabola.

The values of p and p′ are shown in the lower part of the figure, and must be scaled by the multiplier H/R_0. It will be recalled that Figure 145 showed that, beginning with a horizontal force H, the line of pressure may be curved downwards to the right by a succession of vertical loads. In view of the shapes of the arches, it is only to be expected that p and p′ will increase in the order parabola, catenary, and circle. The circular shape is vertical at the right hand end: to achieve this it is necessary that both p and p′ become infinitely large as the end is approached. It should be observed that p and p′ are simply alternative ways of describing the same form load. If a load is distributed around the arch, p′, then a part of this load is expressed by multiplying the intensity by a curved circumferential length. As this length must be greater than its horizontal projection, so, for the same load, p′ must be less than p, and this appears also on the diagram.

It is not uncommon for the Romans to be criticised for their use of the semicircular shape. Some conclusions are obvious – if p were constant, then the shape should be parabolic; if p′ is constant, then the catenary is the correct shape. However, in a Roman arch, neither p nor p′ is constant, but rather both increase as the end is approached. Further, the loads applied to the arch rib from the spandrel fill will not be vertical. When these facts are remembered, it may be concluded that a circular shape is not as faulty as was first thought, although a segmental arch, using only the central part of the circle, would be better.

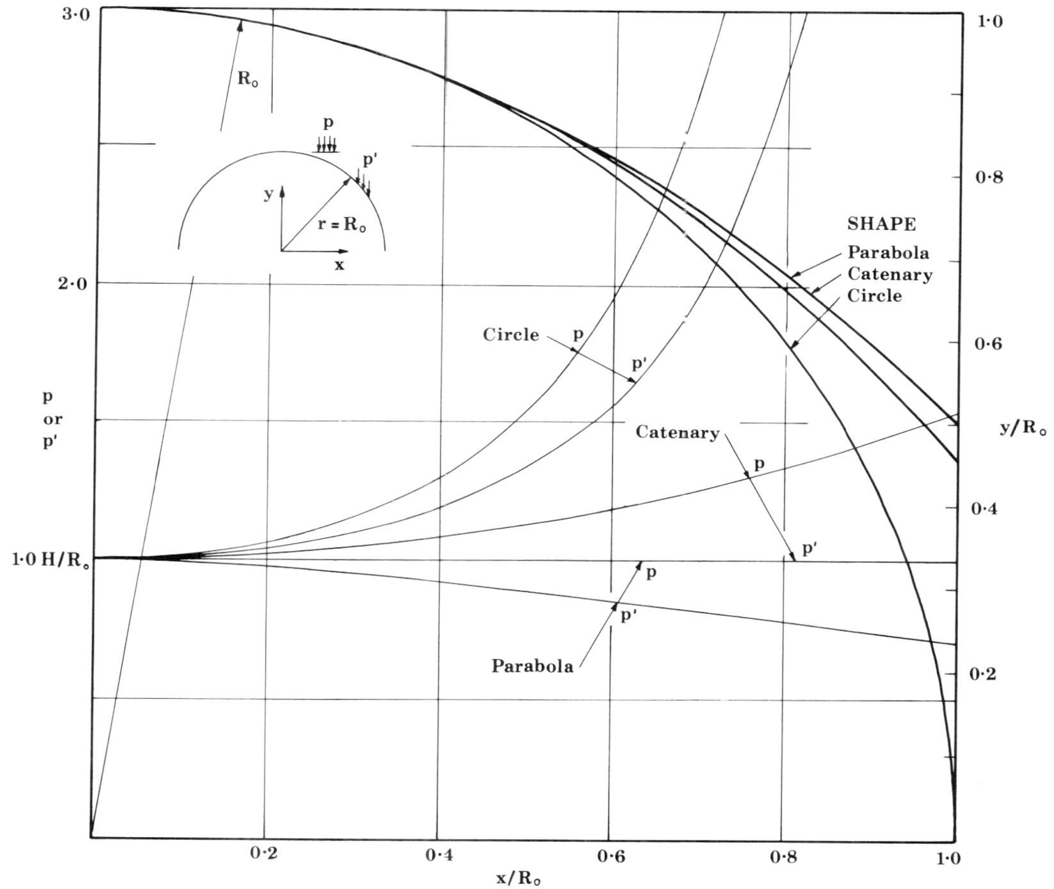

Fig. 148 Form loads.

9.3 Collapse of masonry arches

Heyman (1966, 1969, etc.) restated the conclusions of Couplet, Coulomb and others (Section 9.6) when he pointed out that, for a masonry arch to be stable, it must be possible to construct at least one line of pressure that is consistent with the applied loads, and which lies fully within the arch rib. This conclusion is based on the following assumptions: (1) the joints between voussoirs possess no tensile strength; (2) the compressive strength of the masonry is sufficiently large that it may be considered to be infinite; and (3) sliding failure between voussoirs cannot occur. All three are reasonable.

Figure 149a shows a limiting case, where the rib thickness has been reduced until the condition necessary for stability has just been satisfied. The line of pressure shown is of the shape arising from the self weight of the arch rib. It touches the face of the arch at the five points A, B, C, D and E, alternately on the outer and inner sides. At each of these points a hinge may form, with one part of the structure rotating with respect to the other. One mode of collapse is shown in (b): the crown of the arch falls, while the intermediate points B and D move outwards. The arch with these five hinges forms a mechanism, and is unable to bear load. In fact, only four such hinges are generally required, but in this case five form because of symmetry. An arch bridge with three hinges is stable, and in modern steel and concrete structures, three hinges may be deliberately inserted, such as at A, C and E.

Heyman's (1969: 367–8 and elsewhere) formulae may be used to calculate the minimum thicknesses of circular arches of constant thickness under the action of their own weight.

Hinges B and D form at an angle β from the crown (C), where β satisfies the equation

$$\beta \cos \beta \quad \frac{2\beta \cos \beta + \sin \beta \cos^2 \beta + \sin \beta}{2\beta \cos \beta + \sin \beta \cos^2 \beta - \sin \beta \cos \beta} = \theta \cos \tfrac{1}{2}\theta \qquad (1)$$

θ is the half-angle shown in Figure 149a. The equation may be solved by trial and error. For θ equal to 90° (the semicircular arch), β is 58.83°; and generally β is of the order of 0.65–0.7θ. Then the minimum thickness is given by

$$\frac{t}{R} = 2\,\frac{(\beta - \sin \beta)(1 - \cos \beta)}{\beta\,(1 + \cos \beta)} \qquad (2)$$

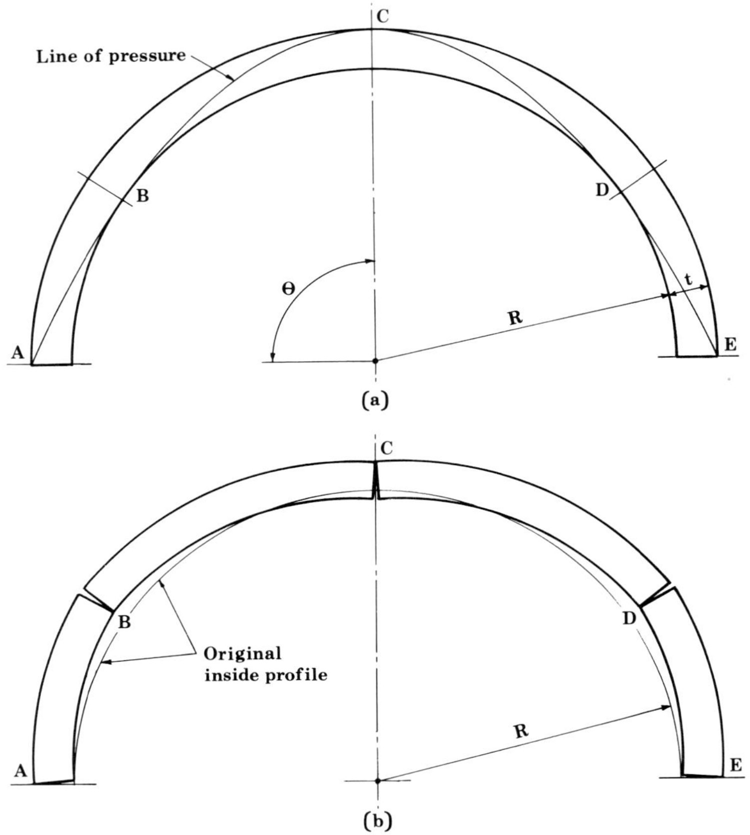

Fig. 149 Collapse of masonry arch.

Also, for a segmental arch, the clear span L is 2R sin θ, and this may be used to devise limiting values of t/L from t/R.

Both of these ratios are plotted in Figure 150. For a semicircular arch, with θ = 90°, the minimum t/R = 0.1060. As shown in the graph, the necessary thickness reduces rapidly as θ is reduced. For example, when θ = 80°, t/R = 0.0680; θ = 70°, t/R = 0.0410.

Section 8.3 suggested that the Roman rules/or design were:

(1) Use as a guide the ratio for rib thickness to span (t/L) of 1/10, knowing that this could be reduced to as low as 1/20;
(2) avoid stone depths greater than 5 Roman feet (1.48m)

It also plotted rib thicknesses for many bridges in Figure 140 and identified the Pont St Martin as a segmental bridge with a particularly slender rib, with a rib thickness

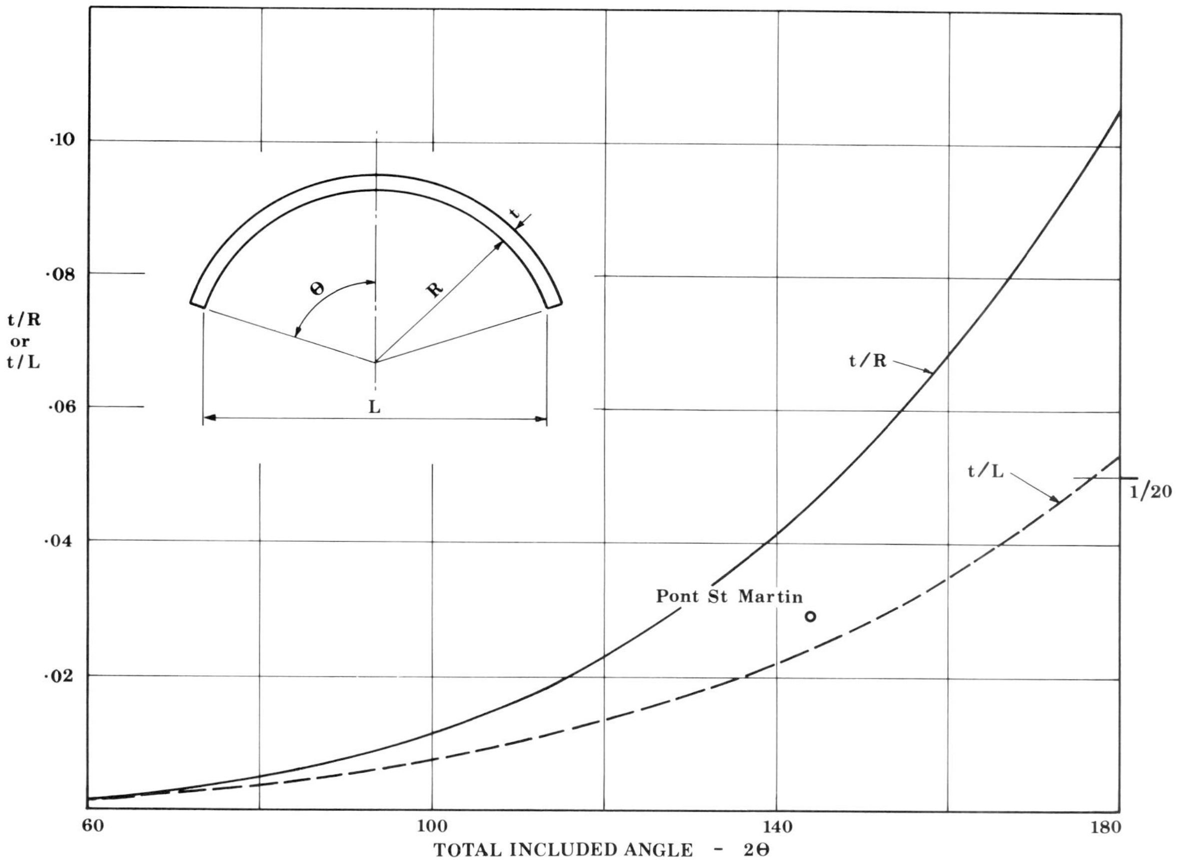

Fig. 150 Theoretical minimum rib thicknesses for perfect arches of semicircular and segmental profile.

equal to 0.029 times the 35.6 m span, and a total included angle of about 144° (i.e. $\theta = 72°$).

The present analysis suggests that the value t/L equal to 1/10 is comfortably safe, but that 1/20 (t/R = 0.10) is close to the theoretical limiting value of t/R = 0.106 for a semicircular arch. It is on the unsafe side but, if 2θ were reduced from 180° to about 177°, a very small adjustment, then the arch would become stable.

It can also be seen that, for an included angle of 144°, the rib of the Pont St Martin is a little thicker than required, with t/L = 0.029 as compared with the required ratio of 0.024.

It is tempting to conclude, therefore, that the Romans were about right in their design rules. Before reaching this conclusion, however, it is necessary to address some other questions.

(1) The above analysis considers only the arch weight; that is, for the case at decentering with no spandrel fill.

(2) It assumes that the arch is built precisely to the required geometry.

One of the difficulties which arise in the analysis of Roman arches is that it is not clear how the spandrel fill should be treated. At one extreme, it may be a loose mass applying non-uniform loads to the arch, and this situation may be worsened if the fill is not placed symmetrically. At the other extreme, if the spandrel is filled with concrete, and this appears likely in many cases, then it will have a substantial strength and will aid the arch rib in supporting its loads. At the least it appears likely that the lower parts of the spandrel fill were strong, as argued in Section 8.6.

The service loads applied to Roman masonry bridges would have been small, compared with the self weight of the structures themselves. Section 4.3 describes vehicles that could carry up to one half of a tonne. Heavier loads than this could arise during construction: Sections 4.1 and 8.3 suggested that a single stone might weigh up to 6.2 tonnes; and a load may be drawn across the bridge during construction. Even this load is relatively small. An arch with a rib thickness of one-tenth of the span, and with voussoirs whose width is one half of the rib thickness, will require thirty-five such stones to form a span. When the width of the bridge is considered, this number could readily rise to 150 or so in the vault as a whole. To the weight of these stones must be added the weight of the spandrel itself.

It is not argued that the weight of this service load is negligible. Rather it is suggested that the combination of spandrel weight and strength with service loads is unlikely to achieve an effect that is worse than that which occurs before the upper part of the spandrel is placed.

The question of defective arch geometry cannot be dismissed so readily. An imperfect shape is visible in many existing arches and may arise in a number of ways:

(a) incorrect geometry of the falsework;
(b) inadequate rigidity of the falsework;
(c) rotations at the joints between voussoirs when the load is first applied (and possibly later);
(d) imperfect mating surfaces between voussoirs, such as rounded edges, so that when collapse occurs, rotation may take place at a point some distance within the boundary of the arch rib;
(e) variation in the thickness of the arch rib; and
(f) foundation movements.

Estimation of the magnitudes of these effects is, at this stage, purely a matter for judgement. Nevertheless, an analysis has been made to adjust the minimum thicknesses given in Figure 150 for three kinds of imperfections.

It is assumed first that, as a result of falsework errors or deflections, the distance from the centre point to the inside of the arch rib may vary anywhere over the range $(1 \pm 0.005)R$. For a semicircular arch of span 20 m, the radius R is 10 m, and the resulting error is ± 50 mm. Referring to Figure 149, it is assumed further that these errors may occur at B, C and D, so that the arch is built 0.005R outside its correct position at B, and 0.005R below at C. The necessary rib thickness, t, must then be increased by 0.010R.

If the arch rib is constructed without mortar between the voussoirs, there will be contact only at certain points in each joint. The joint may be capable of rocking; or, alternatively, crushing at these points under load may allow rocking. It is assumed here that this rotation will be equivalent to ± 1 mm over one half of the rib thickness. The number of such rotations will depend on the width of the voussoirs, being greater if they are thin. It will, therefore, be assumed further that the voussoir width will be 0.35 times its depth, an assumption which is not unreasonable for the larger spans. These additional curvatures will be arranged so as to be compatible with fixity at the springings and also to maximise the resulting imperfections. The resulting increases in rib thickness may be shown to be:

Total included angle (2θ)	Increase in t
180°	$0.00129\,(R/t)^2$
160°	$0.00103\,(R/t)^2$
140°	$0.00080\,(R/t)^2$

Finally, it will be assumed that gaps at the joints between voussoirs may extend up to 50 mm in from any face.

The resulting changes in the minimum thickness do not vary linearly with either R or L. They are plotted in Figure 151 for arches of various spans, L, for included angles, 2θ, of 180, 160 and 140°. For the case of the semicircular arch, this figure also shows the minimum thickness, $t = 0.106R$, or $0.053L$, for the arch without imperfections, and the progressive movement of the minimum thickness line as the imperfections are added. Also shown are the rules, $t = L/10$ and $t = L/20$, and the thicknesses of the more slender Roman arches, taken from Figure 140.

It can be seen that, for all of these Roman arches, the actual rib thickness is less than that derived for $2\theta = 180°$, and the same is the case for rib thicknesses derived from the rule $t = L/20$. The rule $t = L/10$ is conservative for all practical cases.

Section 8.6 mentioned that in Roman construction the arch rib was often cantilevered upwards for some distance above the springing lines without the use of falsework. This will be discussed further in the following section. It also suggested that in these lower regions there may have been a strong backfill. In these circumstances, the hinges A and E of Figure 149 could not form at the springings, but would be forced to act at a higher level, possibly at an angle of about 20° above the springings. The central portion of the rib would then act as a segmental arch with an included angle, 2θ, of 140°.

It is appropriate, then, to compare the span/20 rule, and the thicknesses of actual arches, with the minimum thicknesses required for segmental arches in this range.

Figure 151 shows that the span/20 rule is safe, provided that the effective included angle, 2θ, is not more than 166° for a 40 m span, 160° for a 20 m span, and 147° for a 10 m span.

As far as the individual cases are concerned, Figure 151 shows that if there is firm backfill to 20° above the springings ($2\theta = 140°$), all of the Roman arches plotted in Figure 140 and 151 are safe, excepting for the 7 m span at Mane (13) and the Pont St Martin (18). The first is a small bridge: the data may be incorrect to this extent; and the arch may receive some further spandrel support.

In the case of the Pont St Martin, the disparity is greater, particularly when it is remembered that it is deliberately segmental, with an included angle of about 144°, as shown in Figure 138. However, this bridge differs from all other Roman bridges in certain respects. Both spandrel walls and the underside show solid stonework to a further angle above the springings. Above this level, the arch is built with four separate rings, with bands of concrete showing between them (see Fig. 53). The spandrel walls have small stones except for prominent horizontal rows of larger stones at intervals in height of about 1.5 m. Section 5.5 quotes Blake (1947: 217), who also refers to this change in construction. It is likely that this form of spandrel wall construction also indicates the use of a concrete infill placed in 1.5 m lifts, with the larger stones used to tie the wall back into the infill. There is some evidence that these tie-backs have not been entirely effective. Figures 52 and 53 show four inclined metal bars near the crest of the arch, overlapping the spandrel and the arch rib. It is probable that these form part of a modern anchorage system added to control outward movements of the spandrel, with four metal rods passed horizontally through the entire bridge, to join a bar at one side to a bar at the other. However, the concrete between the arch ribs extends fully to the crest. It is suggested, then, that the Pont St Martin represented a new development in Roman bridge design which relied

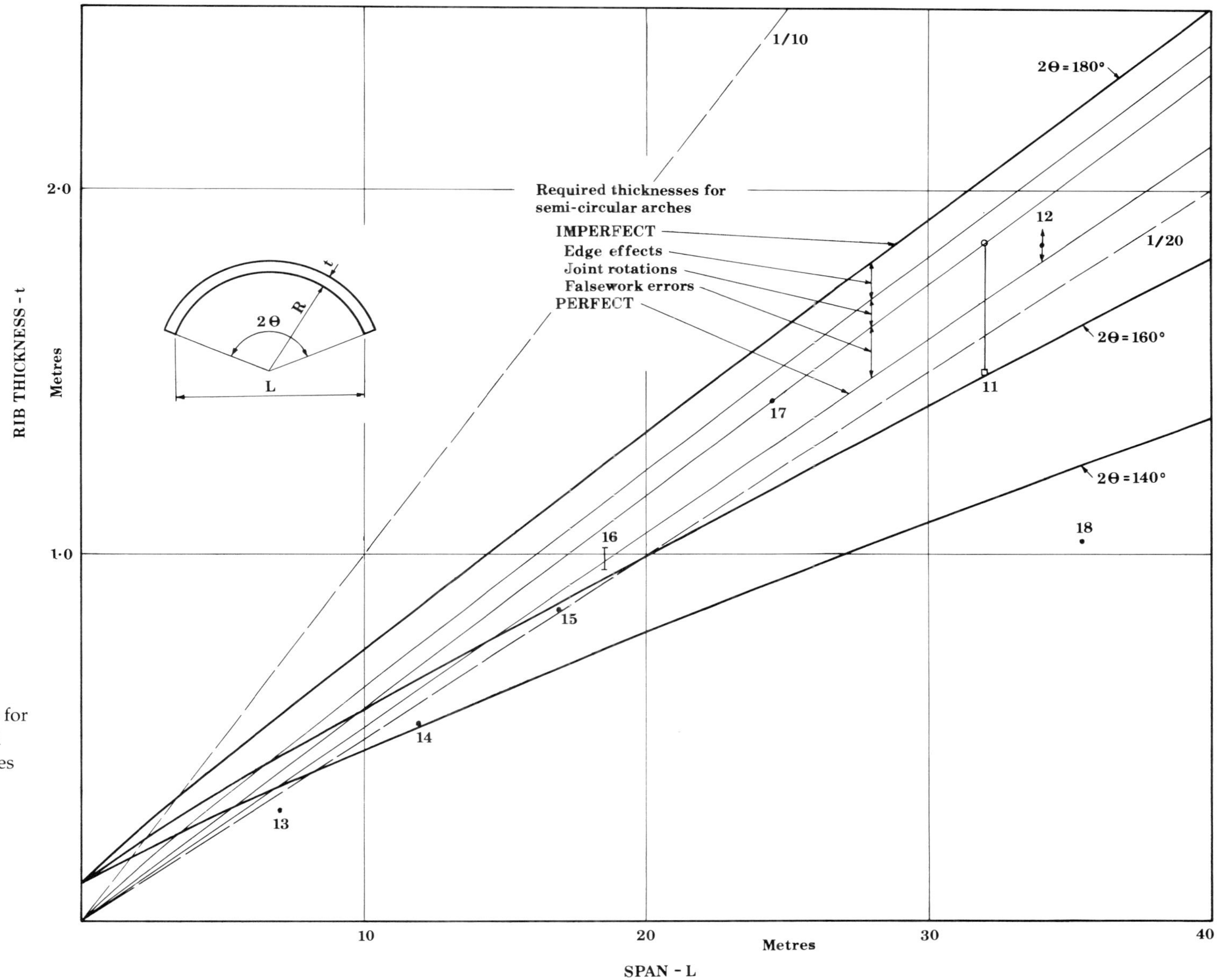

Fig. 151 Theoretical minimum rib thicknesses for imperfect arches of semicircular and segmental profile, compared with the actual rib thicknesses of the more slender Roman arches shown in Figure 140:

11 Ponte d'Augusto, Narni (I87) – main span
12 Severus bridge, Kâhta (E36)
13 Pont sur Laye, Mane (G4) – 7 m span
14 Pont sur Laye, Mane (G4) – main span
15 Ponte d'Arli (I127)
16 Puente Bibey (SP36)
17 Pons Fabricius, Rome (I3)
18 Pont St Martin (I155)

more explicitly on the use of concrete in the spandrels.

To return to the question 'Were the Roman rules for arch design correct?' the answer is, 'Yes, provided it is recognised that they relied on support from the spandrel in the lower regions of the arch', and this, no doubt, was a principle that any experienced builder of semicircular arches would soon learn.

9.4 Delayed collapse

The previous analysis assumes that the most critical stage for the arch was when it was removed from its centering. It does not explain why some arches collapsed at a later stage. One puzzling example lies in the arched supports between the columns of the Los Milagros aqueduct in Mérida (Fig. 128).

This question of delayed collapse is not easy to address in general terms. Rather, it is suggested that any proper account must be based on a detailed study of particular bridges that have collapsed. It is, however, intrinsic to the present study, for, if there is some general cause of delayed collapse that has not been recognised, then the previous analysis is incorrect.

The most common cause is likely to be foundation movement. The wide Roman bridge piers (Section 8.2) caused a major obstruction to stream flow. In floods, large stream velocities would occur around the piers, undermining them, eroding the foundation material and causing the piers to collapse. However, this scenario implies that the pier fell, and that this caused the bridge to collapse. It does not explain cases, such as at Narni (I87, Section 5.4) and others, where the span has fallen while the piers have remained intact. At Narni, all three piers stood until 1885, even though three of the four spans had collapsed.

An explanation may lie in the following. At Narni, when piers 2 and 3 collapsed, they fell upstream, clearly because they had been undermined on the upstream side, and this would be expected. Consider, then, the stage when they had been undermined, but when the spans were still intact. The tendency of the pier tops to move upstream would have been counteracted by shear in the arch ribs. If these shear forces became excessive, they could have fractured the arch ribs, dislodging stones to form a characteristic diagonal fracture, and causing the arch to fail. Such a fracture is visible in the remnants of the eastern arch of the Pons Aemilius or Rotto in Rome (Fig. 23, Section 5.2). Smith (1990–1), in a recent article on the Pont du Gard, has a photograph (fig. 7) showing a perceptible upstream curvature of its upper levels which, in his words, 'might have been due to settlement'. The shape is consistent with that which would result from settlement of the upstream ends of the piers, and this is possible even though they appear to be founded on rock above the normal water-level. Another reason suggested by Smith is that the piers were cut away in the fifth and seventeenth centuries to provide room for a roadway above the first tier. This seems less likely as it is believed that the roadway was on the downstream side, and the resulting pier deformations should have caused horizontal movements of the upper structure in the downstream sense. Whatever the cause, the curvature indicates the presence of shear stresses in the arch rib and forms a clear example of the mechanism which may lead to delayed collapse, as described here.

There are, however, other general explanations of delayed collapse. Three may be mentioned here. Foundation movements and excessive stresses in the masonry may cause changes to the arch shape that cause it to become unstable. Fretting of the corners of the highly stressed stones at the hinges shown in Figure 149 may reduce the effective thickness of the arch in such a way that it can no longer contain the line of pressure. And finally, there may have been excessive service loads, although it appears unlikely that these would have occurred in Roman times.

9.5 Construction analyses

Analyses are given here for two construction stages.

Figure 152 shows the lower part of the arch, built upwards from the springing without support from falsework. Such a free standing member is called a cantilever. It may fail in two ways. If it is taken up too high, its centre of gravity will fall outside its base and it will fall inwards. Alternatively, the top stone may slide off.

The second case is the more easily analysed. If the limiting coefficient of friction were 0.5, then the angle below the top stone should not exceed 26°. The total cantilever angle to the top of the top stone would be a little greater – about 29° if there were thirty-three stones in the half-arch.

Overturning analyses have been carried out for arches of various thicknesses, with the ratio t/R equal to 0.10, 0.15 and 0.20. At the limit, the resultant weight passes through the inner corner of the base, O, with the distance e equal to zero, for cantilever angles of 31.1°, 37.9° and 43.5° respectively.

However, this analysis is for an extreme case, as it would imply that, at the limit, any disturbing force, such as that due to wind, or from impacts during construction, would cause the rib to collapse. One method of safeguarding the arch would be to ensure that the eccentricity, e, is never less than one tenth of the height, h, from the base to the centre of gravity. Then the limiting angles become 23.5°, 30.21° and 35.8°. At these angles, the values e/t are 0.21, 0.18 and 0.16. However, this ratio, e/h equal to 0.1, is still not entirely comfortable. For e/h equal to 0.15, the limiting cantilever angles are 20.9°, 27.2° and 32.5°. When these are modified by the limit given above for sliding of the top stone, suitable values are 21°, 27° and 29° for t/R equal to 0.1, 0.15 and 0.20.

Another analysis of some interest is for the horizontal thrust, H, and the corresponding falsework loads during erection. Figure 153 is for an arch with t/R equal to 0.1, a

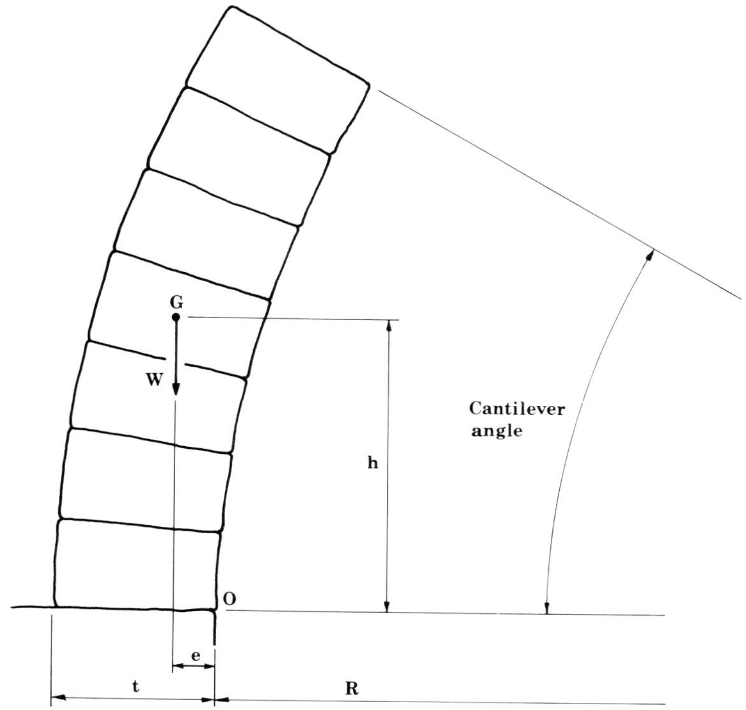

Fig. 152 Lower portion of a semicircular arch, constructed without falsework.

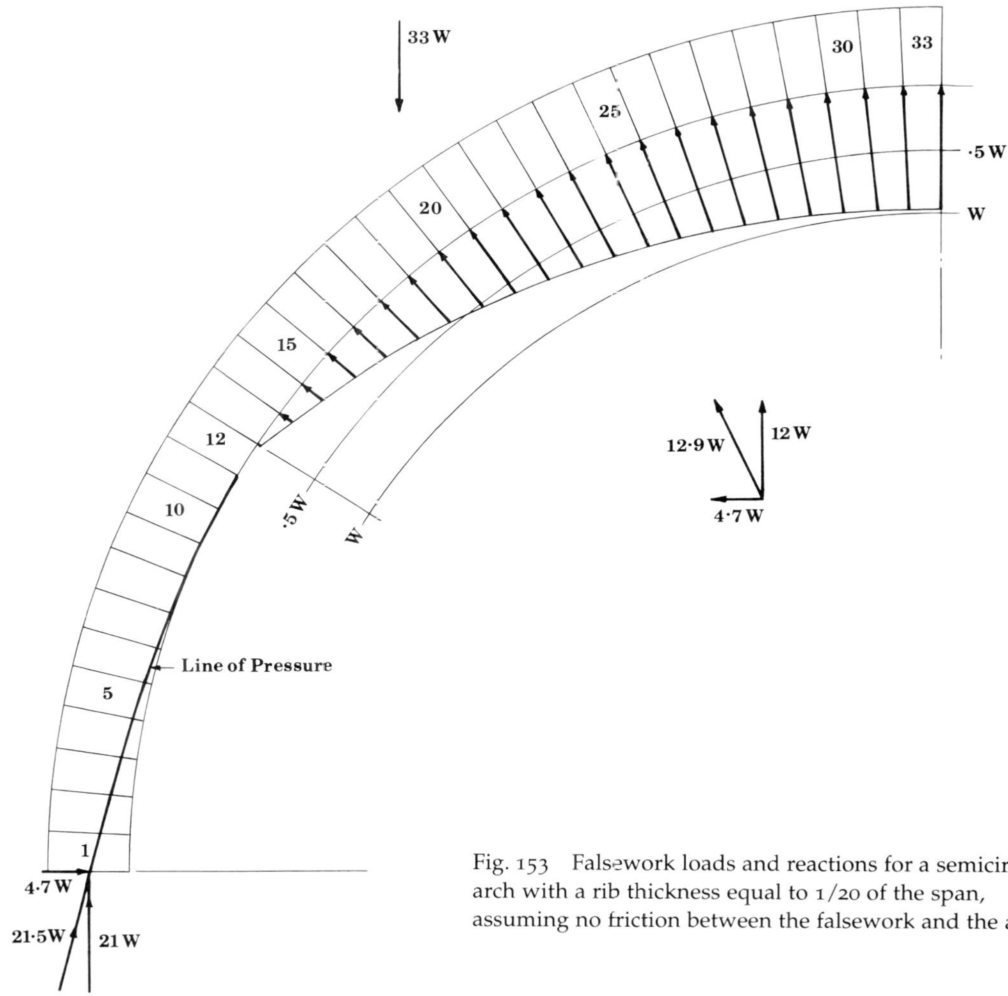

Fig. 153 Falsework loads and reactions for a semicircular arch with a rib thickness equal to 1/20 of the span, assuming no friction between the falsework and the arch.

little thinner than the limits given in either Figure 150 or 151 for a semicircular arch. If the thickness of each voussoir is one-half of the rib depth, then very nearly thirty-three voussoirs are required for each half of the arch. This figure (33) has been taken here, although more normally there would be an odd number of voussoirs in the total arch, to provide for a keystone at the crown. The following assumptions have been made:

(a) there is no friction between the arch and the falsework; and
(b) as each voussoir is laid, it is supported by bearing on the previous stone at a point near the intrados, and at its upper inside corner by a radial force from the falsework.

These assumptions correspond to only one of a number of conditions which may arise, but are reasonable for the present purpose. Results are presented in Figure 153.

At mid-span there is no contact between adjacent stones until the arch is decentered. The reaction from each stone on the falsework is 0.474 times the weight W of a single voussoir, with the two totalling 0.95W. This is plotted beneath the arch, as are also the successive reactions at the joints. It can be seen that these reactions reduce, and become zero below stone 12; that is, below an angle from the springing of 32.7°. From this point to the base, the line of pressure moves away from the intrados, to reach a distance of 0.505t within the arch at the springing.

The resultant forces acting on the half-arch are shown on the diagram. The horizontal thrust at the base is 4.728W, or 0.0716 times the complete weight. This arch would collapse as the falsework is removed, as its thickness of 0.1R is less than the required value of 0.106R. However, for this slightly thicker but stable arch (t = 0.106R), the horizontal thrust is 0.197 times the total weight. It may be expected, then, that for a semicircular arch whose thickness is close to the minimum value, the horizontal thrust, H, will increase by about 2.7 times during the decentering operation.

Decentering is therefore a crucial test, not only of the stability of the arch rib, but also of the strength of the abutment and piers under horizontal load.

9.6 Historical review

As stated in the introduction to this chapter, much of the work of this historical review is taken from the publications of Heyman, in particular his book *Coulomb's Memoir on Statics: an Essay in the History of Civil Engineering* (1972) and his later paper 'Couplet's engineering memoirs 1726–33' (1976).

Robert Hooke is famous not only for his Hooke's Law, which forms the basis of modern elastic structural theory, but also for the form he chose for its publication. In 1675, he published a series of anagrams. Number 3 read 'ceiiinosssttuu' (Pannell, 1964: 319), or, when rearranged, 'ut tensio sic vis' – 'as the extension, so the force', or in modern terms, stress is proportional to strain. The third was 'Ut pendet continuum flexile, sic stabit contiguum rigidum inversum.' Truesdell's translation is, 'As hangs the flexible line, so but inverted will stand the rigid arch' (Truesdell, 1960).

It appears that, at that time, the catenary problem also had not been solved, but this does take away from the importance of Hooke's observation, for, at the least, it would have been possible then to determine experimentally the form shape of an arch by experiments on a loaded cord. By 1695, La Hire had developed the concept of the funicular polygon, the graphical method of producing the line of pressure discussed in Section 9.2 (Heyman, 1972: 82), and used this to consider the reverse problem, to find what loads would be necessary to achieve an arch of specified shape – the concept of the form load. He deduced correctly that, for a semicircular arch under vertical loads, the intensity of these loads would become infinite at the abutments. The solution of the catenary problem, with loads attached to a flexible chain supported at its ends, appears to be due to Euler in 1744.

To return to the arch, in 1697 Gregory (Heyman, 1972: 75f.) had restated Hooke's proposition, 'In a vertical plane, but in an inverted situation, the chain (or catenaria) will preserve its figure without falling, and therefore will constitute a very thin arch.' He goes on to consider an arch of different shape and with real thickness: 'And when an arch of any other figure is supported, it is because in its thickness some catenaria is included.' This concept of the limiting thickness of an arch rib was taken up by Couplet (Heyman, 1976) in 1730. Heyman considers his contribution to arch theory to be of outstanding importance: 'He had clear notions of lines of thrust and of mechanisms of collapse caused by the formation of hinges, he stated explicitly the simplifying assumptions necessary for his analysis, and he used these ideas to obtain an essentially complete and correct solution to the problem of arch design' (Heyman, 1972: 170f.; 1976: 35). His simplifying assumption was that hinges formed half-way between the springings and the crown. For a semicircular arch, he deduced that the minimum value of the thickness was 0.101 times the radius (the true figure is 0.106). For a segmental arch with a total included angle of 120°, he gave 0.0195, instead of the true 0.0226.

The work of Gregory and Couplet was known in 1743 to Poleni (Heyman, 1972: 177–8), who was one of a number called in to investigate the stability of St Peter's cathedral in Rome. Couplet's work was largely forgotten, and it was not until 1773 that Coulomb solved again the problem of determining the minimum thickness of the arch rib, in his case precisely, for he used a variational process to determine the exact locations of the intermediate hinges (Heyman, 1972: 181).

Heyman (1972: 183) has also an account of tests made on model arches by Boistard in 1800. He tested twenty-two, all of 2.44 m span, of which four were semicircular and five segmental (the others were oval). His first test was of a semicircular arch with a thickness equal to 0.08 times the radius. As predicted by Couplet, it collapsed when the falsework was removed. The complete span was formed of forty-eight voussoirs. Boistard noted that, during construction, the first ten voussoirs stood without support from the falsework; that is, up to an angle of 37.5° from the springing. This angle is larger than those estimated in Section 9.5, but it should be remembered that it would be dependent on the exact mode of construction. The important point is that the ability of the lower portion of the arch to stand without support from the falsework is recognised here, at it was by the Romans so many years before.

There is a lesson in this final observation: it is much harder to anticipate behaviour than it is to observe it in practice. The Romans were practitioners. Although they were no doubt incapable of describing behaviour in the terms mentioned here, yet one should not be hasty in underestimating their ability to judge the real behaviour of the arches they built for so many years. If Boistard could successfully observe the behaviour of a model arch, then the Romans could surely observe that of one which was full size.

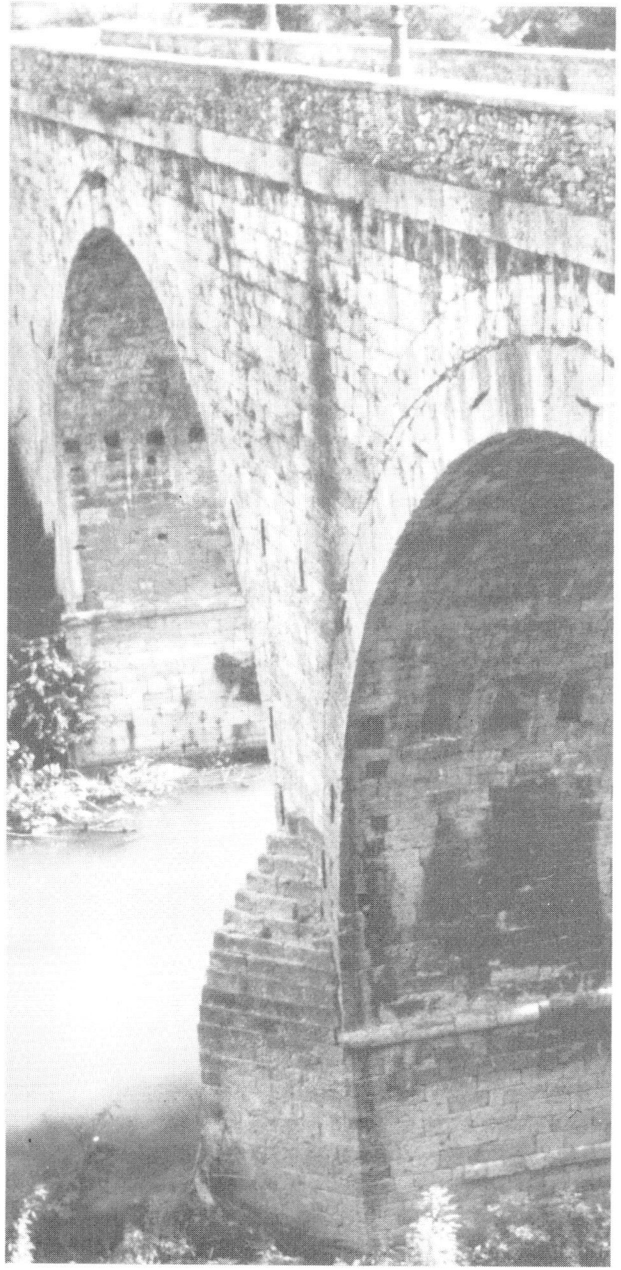

10

THE ROMAN ACHIEVEMENT

Roman achievement in bridge construction may be measured in a variety of ways. It may be measured first by their *number*. Appendix tables A, B and C list 330 stone traffic bridges, 34 timber bridges and 94 aqueducts, of which 54 had bridges. Many of these bridges are still standing, and a surprising number are still used to carry vehicles. For example, the bridges at Mérida (SP15, SP16), Rimini (I117), Sommières (G9), Ascoli Piceno (I128), Chaves (SP39) and Salamanca (SP40) still serve as city bridges, while those at Quintodecimo (I126), d'Arli (I127) and Pont St Martin (I155) perform similar functions in smaller towns. One of the main roads from Portugal to Spain is still carried over the Roman bridges at Segura (SP22) and Alcántara (SP21). The bridges at Villa Formosa (SP17), Bibey (SP36), el Kantara (A19), and the large Ponte di Nona (I23) are also main road bridges; the Ponte Grosso (I105), Pont Julien (G5), the bridge over the Oued Nebana (A6) and that at Kâhta (E36) all carry country roads. Some have had a change in function, to serve as footbridges, like the old bridges of Rome, set aside no doubt partly to preserve them. Others still stand but are out of use.

These lists are by no means exhaustive. Any road, except in a desert, will be found to cross a succession of major and minor streams and it must do so by ferries and fords, or by bridges and culverts. It is to be expected that Roman timber bridges will have long since disappeared (but see Section 6.4, T19, for the bridge at Wallasey). The smaller bridges or culverts may also have gone, either buried under silt carried down by the stream, or by having their stones taken for adjacent building. When all things are considered, it is likely that the Romans built thousands of bridges.

Their achievement may be measured secondly by the geographical *extent* of their bridges: from Hadrian's Wall in the far north west, to the bridges of Africa in the south, and across to Turkey, Palestine and Shushtar (E42) in the east. It is an emotional experience to find the Pont de Trajan (A24) in the distant fields of Tunisia; to walk down the Via Appia to the Pons Auruncus (I45), almost hidden in an Italian farm; or to see the lofty bridge at Pondel (I159), in the foothills of the Italian Alps. All are impressive reminders of the spread of Roman authority and technology.

The third measure of Roman bridges is their *size*. The longest span of existing Roman stone arch bridges is 35.6 m at Pont St Martin (I155); the greatest length, 755 m, in the original 62-span Puente Romano (SP15) at Mérida; and the greatest heights, 40–5 m, at Pondel (I159) and 47.8 m at the Pont du Gard (AQ30). The timber arch bridge of Apollodorus (T13) across the Danube was even longer, with, perhaps, twenty-one 32.3 m spans between piers 17.6 m wide and a total length of 1032 m.

In 1975, I published a study (O'Connor, 1975) of bridge-building achievement for all kinds of bridges up to that date, using maximum span length as the criterion. It may be noted first that this study used existing data, which were by no means reliable, and would have omitted some major bridges. Nevertheless, its major conclusions were correct. Until 1755, all major bridges recorded in that study were of masonry. In that year, and in 1758, three timber arch bridges were built by the Grubenmann brothers (Beckett, 1969: 88), with spans of 59, 73 and 119 m. The first metal bridge was built at Coalbrookdale, England, in 1779 as an arch with a 30 m span (Cossons and Trinder, 1979; Bracegirdle and Miles, 1974); the Wear bridge at Sunderland of 1796 (Smiles, 1862) had a 72 m span, and following that date there was a rapid proliferation of bridge types, with the stone arch eclipsed but not extinct.

The stone or brick arch was therefore the major structural form for bridge construction until 1755. There would also have been, of course, timber bridges and other primitive bridges, such as suspension bridges built with vines. In that study (O'Connor, 1975), the only pre–1755 masonry arches that were identified with spans greater than the Roman bridges were five bridges built between 1204 and 1377 – at Bern, 46 m; Toledo, 40 m; Ceret, 45 m; Verona 49 m; and over the River Adda at Trezzo, 72 or 76 m (Pannell, 1964: 220; Hosking, 1839; Degrand and Resal, 1887–8). The first three still stand, while the fourth has been rebuilt. The last was destroyed in 1416. Although the dimensions quoted here should be treated with caution, there is little doubt that this group of large spans was built about that time. By contrast, it should be noted that it was not until the mid-eighteenth century that similar spans were repeated. It was concluded, therefore, that apart from these few exceptions, the maximum span achieved by the Romans was not exceeded until 1755 (at Pontypridd, 43 m).

More accurate information suggests that these conclusions were not entirely correct. For example, the Chinese Zhao-Zhou bridge at An-ji or Anchi was built about AD 600 with a span of about 37 m. The bridge at Martorell (SP2), with a span of 37.3 m, was attributed in that publication to the Romans, whereas it is now believed that this span is for a bridge built by the Moors about AD 1283 on an earlier Roman bridge. A similar structure is the Malabadi bridge of south east Turkey (Fig. 103), built about AD 1150 with a span of 38 to 41 m. More truly, it appears that no accurate review has ever been published of bridge construction in the post-Roman period and it appears only reasonable to suggest that such a review should pay closer regard to regions with civilisations which continued in the post-Roman period, such as in Turkey or Moorish Spain.

Despite these reservations, the Roman achievement measured by bridge size is most impressive, for even the data given above suggest that, apart from the Zhao-Zhou bridge in AD 600, it was not until AD 1150 that the Roman maximum span was exceeded.

The final measure of Roman achievement lies in their *technology*. It has been customary to belittle the Romans as not interested in technology, or capable only of using the technology of others. Now it is true that the Romans learned from others; that they employed, for example, Greeks in their service; and that they did not invent the arch. The arch and the vault appear to have originated in Egypt and Mesopotamia centuries before the Romans. However, this is hardly the point, for, to quote Robertson, 'The Romans were the first builders in Europe, perhaps the first in the world, fully to appreciate the advantages of the arch, the vault and the dome' (1943: 231). If it is suggested that others before the Romans saw the advantages of the arch for bridge construction, one has only to ask, 'Where, then, are their bridges?'

To build an arch bridge required much more than a concept, even though this concept was one which was so valuable. It required skills in all those matters raised in Chapter 4: to lay out the site (Section 4.7); construct the foundations (Section 4.3); select the materials – stone, timber or concrete, remembering that concrete was a Roman discovery (Sections 4.4, 4.5, 4.6); make heavy lifts (Section 4.3); provide fasteners for falsework and other purposes (Section 4.2); and use the necessary tools (Section 4.2). Not only so, but it required the ability to plan, and to develop the design rules identified in Chapter 8 and justified in Chapter 9.

To build a bridge is not merely an academic exercise – rather it is a practical task whose magnitude will vary with the size of the bridge. It is an exercise in management and logistics and, in a sense, it is also a conflict, as a bridge builder must contend with certain major forces of nature. It requires him to overcome the force of gravity – to handle major lifts can be a frightening experience – and in many structures to counter the savage effects caused by a stream in flood. It requires not only immediate success, but a success that will last – as Lacer said in Spain some 1900 years ago, with a correct purpose but some exaggeration, 'I have built a bridge which will remain for ever.'

How then should Roman bridge construction be assessed? As one of the most successful, extensive, and lasting of all human, material achievements.

APPENDICES

1 GLOSSARY OF TECHNICAL TERMS

(This glossary includes only engineering or unusual terms)

abutment — the end of a bridge

arch — a curved member, shaped and supported in such a way that applied loads are transferred to the supports primarily by axial compression in the arch rib (see Fig. 132)

arch rib — the primary, curved structural member of an arch (see Fig. 132)

Archimedean screw — a pumping device, consisting of an inclined tube containing a large, coaxial, closely fitting screw, rotation of which causes water to rise up the tube

axial load — a load applied parallel to the length of a member

beam — a member loaded with transverse loads which it supports by virtue of its bending strength; that is, its capacity to transmit bending moments

bed — one of a horizontal system of jointing planes in natural rock or masonry construction

bending moment — an action at a cross-section of a member, calculated as the moment about an axis in the cross-section, of the forces applied to the member on one side *or* the other of that cross-section

block and tackle — a pair of pulley blocks with their ropes and attachments

brittle — the adjective given to a material which fails suddenly

caisson — a bridge foundation, having the form of a hollow pier emptied of water (or dewatered)

calcin — a hard layer formed on the surface of limestone by exposure to the atmosphere

cantilever — a beam that is fixed at one end, free elsewhere

capstan — a windlass with its axle vertical

catenary — the shape of a uniform flexible string, hanging between two supports that are separated horizontally, under the action of its own weight

cement — in masonry construction, a fine powder which hardens by combination with water

cement, hydraulic — a cement which hardens under water

centering — another name for falsework (q.v.), referring particularly to the temporary members used to support an arch during construction (see Figs. 143, 144)

chain dogs — a pair of metal hooks linked by a chain, used to lift a stone by engaging holes cut in two opposing faces of the stone (see Fig. 13)

cofferdam — a temporary watertight wall, often of sheet piles, built to allow a site to be dewatered, so that the permanent construction may proceed in the dry

column — an axially loaded member in compression

compression — a force or stress whose sense is such that it tends to cause shortening of a member

concrete — the combination of pieces of stone in a hardening mortar

cornice — an ornamental projecting course, either horizontal or around the upper edge of an arch rib

couple — a pair of coplanar, equal and opposite forces – the magnitude of a couple is equal to the product of the magnitude of each force by the perpendicular distance between them (see Fig. 146b)

cramp — an article, usually of metal but sometimes of wood, used to join stones in masonry construction

density — mass per unit volume

design — the process of inventing a structure or system to achieve some desired purpose

design drawing — a drawing prepared to describe a structure before it is built

durability — the ability of a material to resist corrosion, such as atmospheric corrosion or wear

extrados	the upper surface of an arch rib (see Fig. 132)
falsework	temporary members used to support a structure during construction (see Figs. 143, 144)
floodway	a channel used to carry water in periods of high flow, such as an opening through a pier top
foot (Roman)	about 0.294 m (or 0.963 imperial ft)
force	that which, when applied to a body, causes it to accelerate, the acceleration being directly proportional to the magnitude of the force and inversely proportional to the mass of the body
form load	the unique system of loads which causes the axial force in an arch rib to act everywhere along its centreline (see Fig. 145)
formwork	temporary supports for concrete when it is first placed, before it gains its strength
foundation	the base of a structure, used to transmit forces from the structure to the ground
fracture	the process whereby a material fails by direct separation across a crack
fretting	the type of material failure characterised by the loss of small particles, due either to reaction with the environment, or high local stresses
frig-bob saw	a handsaw used to cut soft stone
funicular polygon	the shape of a flexible string when it is anchored at two points and loaded with a system of point loads
girder	alternative name for a beam
grain	a vertical plane along which hard rocks such as granite may be split; or the longitudinal fibrous structure of timber
gravitational force	the attractive force exerted between the earth and any body, whose magnitude is such that for a body near the earth's surface, the resulting acceleration 'g' is 9.81 m/sec/sec (or 32.2 ft/sec/sec)
head tank	a tank used to feed water into a pipe or conduit
hodometer	a distance-measuring device
intrados	the lower face of an arch (see Fig. 132)
keystone	the central stone of a masonry arch (see Fig. 132)
kilogram (kg)	the mass of a standard object called the kilogram (1 kg = 2.205 lbm)
kilo-Pascal (kPa)	a stress equal to 1000 Pa (1 kPa = 0.1450 lbf/sq. in.)
levelling	the process of determining the heights of points, or of ensuring that they are at the same height
lever	a beam resting on a pivot, with one end placed under the object to be lifted and the other loaded; if the pivot is closer to the object than it is to the applied load, then a mechanical advantage is achieved
Lewis	a device used to lift a stone by engaging a specially cut hole in its upper face (see Fig. 12)
lift	a term used for the object lifted by a crane, or for its weight
lime	the material produced by heating limestone in a kiln to form calcium oxide, or quicklime; or by the subsequent addition of water to form calcium hydroxide, or slaked lime
line of pressure	the line representing the location of the resultant axial force in an arch rib (see Fig. 145)
load	another name for force – a load may be applied locally (a point load) or distributed over an area or volume
mass	the amount of matter in a body
mechanical advantage	the ratio of the force output of a mechanical system to the input force
mega-Pascal (MPa)	a stress equal to 1, 000, 000 Pa (1 MPa = 145.04 lbf/sq. in.)
member	part of a structure, long compared with its depth, and connected to the remainder of the structure either at its ends, or at discrete points along its length
metre	the basic unit of length in the S. I. system (1 m = 3.281 ft)
moment	the product of the magnitude of a force by the perpendicular distance to a point or axis
mortar	a hardening mixture of a cement, sand and water
Newton (N)	the force which will give to a mass of one kilogram an acceleration of one m/sec/sec (1N = 0.2248 lbf)
Pascal (Pa)	a unit of stress, where one Pascal equals one Newton/m^2 (1 Pa = 0.145×10^{-3} lbf/sq. in.)
pier	a vertical supporting member; in a bridge, the intermediate supports
pile	a member pointed at one end and driven into the ground to provide support to a structure
piledriver	a machine used to drive a pile
plug and feathers	a tapered rod (the plug) inserted into a split sleeve (the feathers); after the sleeve has been placed in a hole drilled in rock, the plug is driven and acts as a wedge to split the rock (see Fig. 9)
plumb bob	a surveying device, consisting of a weight pointed at its base and hung by a vertical cord
pound force (lbf)	the force exerted by gravity on a mass of one pound; it follows that a force of 1 lbf will give to a mass of 1 lbm an acceleration of 32.2 ft/sec/sec
pound mass (lbm)	the mass of a standard object called the pound
pounder	a stone attached to the end of a vertical rod or handle, operated by lifting the handle and dropping the stone onto the surface to be worked

pozzolan	a fine material of volcanic origin which hardens in combination with lime and water
pulley	a grooved wheel mounted on an axle, used to change the direction of a rope passed around the wheel
pulley block	a system of pulleys mounted on a single axle; or a pair of such blocks joined by a single rope passed around the pulleys in such a way that a tension force applied to the rope end causes an increased force to be applied by the blocks
putlog	a horizontal falsework member
reach	as applied to a crane, the maximum horizontal distance that can be achieved between the object that is lifted and the base of the crane
reaction	a force or moment at a support, called into play to oppose the applied loads
rift	a vertical plane along which hard rocks such as granite may be split
rise	the height of the crown of an arch above its springings (see Fig. 132)
sclerometer	a device for measuring surface hardness
screw	in principle, a rod with a spiral groove or thread cut into its sides, so that rotation of the screw will cause a longitudinal movement of an object engaging the thread
segmental	the adjective applied to an arch whose shape is less than a full semicircle
shearing force	a force component perpendicular to the length of a member
sheerlegs	a pair of posts resting on the ground and tied together at the top, held in position by ropes attached to the top, and used to support a lifting device, such as a pair of pulley blocks
sheet pile	a wide pile, which may be driven as one in a sequence of such piles in contact to form a wall or sheet
sill	a horizontal ledge or surface, such as below a window
siphon	a sealed tube carrying a stream of liquid, with the outlet lower than the level at entry, and with intermediate levels *higher* than both
siphon, inverted	a sealed tube carrying a stream of liquid, with the outlet lower than the level at entry, and with all intermediate levels *lower* than both
soffit	another word for intrados; that is, the lower face of an arch (see Fig. 132)
span	the distance between adjacent supports
spandrel	the space above the rib of an arch (see Fig. 132)
Spanish windlass	a device which applies a tension force through the twisting of a multiple set of ropes
specification	a description of a structure prepared prior to construction, which may state also the required standards of construction
specus	the water channel of an aqueduct
spread footing	a foundation placed upon the ground or rock and which spreads the loads to it
springing	the end of an arch rib (see Fig. 132)
stress	force per unit area
string course	an ornamental, projecting horizontal course in a wall
tension	an axial force, or stress, acting on a member in such a sense that, if the member fractures, the fractured surfaces will separate
ton (T)	a mass equal to 2240 lbm; or a force equal to 2240 lbf
tonne (t)	a mass equal to 1000 kg (1t = 0.984T)
torque	a moment about the longitudinal axis of a member
treadmill	a large wheel, mounted on a horizontal axle, and of such size that men walking within the wheel cause it to turn
units, imperial	the older British system, which includes as basic units the yard for length, the pound mass (lbm) for mass, and the second (sec) for time; the foot (ft) is one third of a yard
units, S.I	the later Système International, adopted as a unified standard by many countries, which includes as basic units the metre (m) for length, the kilogram (kg) for mass, and the second (sec) for time
vault	an arch rib viewed in three dimensions, with a width large compared with its thickness
venter	an aqueduct bridge used to support the lower part of an inverted siphon
voussoir	a wedge-shaped stone, used with others to form an arch rib (see Fig. 132)
waterwheel	a wheel with paddles, turned by the action of water
wedge	a tapered piece, formed between two planes with a small included angle, which when driven into a gap applies a large separating force
weight	the force applied to an object by the earth's gravitational field
windlass	a drum mounted on an axle – a rope is wound on the drum and tensioned by rotating the drum

2 USE OF THE JAPANESE ABACUS OR *SOROBAN*

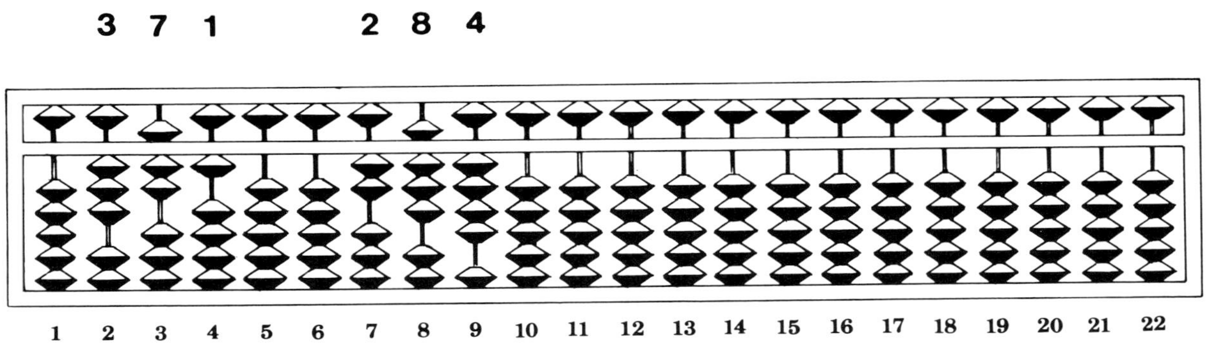

Fig. 17 Japanese abacus or *soroban* (based on Mikami, 1913, fig. 1).

Figure 17 shows the Japanese *soroban*. The counters on each wire can represent a number in the range 0–10. If no counters are touching the horizontal partition strip, the number is 0. By moving the lower counters up – the one nearest the strip, then the next, etc. – the numbers 1 to 5 may be represented. When 5 is reached, the lower counters are moved down, and the upper counter is moved to touch the partition strip.

To represent a three digit number – say 371 – three wires are used. Figure 17 numbers the wires or columns for reference. The number 371 is shown in columns 2–4. To represent two such numbers, any two groups of three wires may be used, but spaced apart by wires that are not in use. Figure 17 shows also the number 284.

Suppose it is required to add 371 and 284. Then 371 is set on the abacus, but not 284. The 284, when verbalised, is two hundred and eighty-four. When the two hundred is read or heard, two counters are added to the three existing counters touching the partition strip in column 2. When eight is heard, then 8 is to be added to the 7 in column 3. It is immediately recognised that 8 + 7 will not fit. A child on the Arabic system will have learnt by heart, 'eight plus seven equals fifteen'. A Japanese child will remember instead the simpler rule that the new number 8 has a complement of 2 – or eight plus two equals ten. He or she will then subtract 2 from 7 in column 3 – to give 5 – and add one to column 2.

Similarly, in subtracting a number 8, the Japanese child remembers 'eight away, two remains', adds 2 to the appropriate digit and subtracts 1 from the digit to the left.

The abacus may also be used for multiplication and division. Multiplication, being easier, will be described here; the division process is fully described by Mikami (1913). The Japanese child learns mental multiplication for numbers from 1 to 9 is a manner similar to that for the Arabic system, but with some minor variations – the word 'times' is omitted, and the smaller number is always taken first: 'Four times three equals twelve', for example, becomes 'three–four–twelve'.

To multiply 371 by 284, the numbers are set on the abacus as shown in Figure 17. The order of operations is

1 371 × 4 gives 1484
2 Add on 371 × 80, or 29, 680, to give 31,164
3 Add on 371 × 200, or 74,200 to give the answer 105,364.

Note that the second step adds 0 to the units figure and the third adds 0 to both the units and the tens figures.

The left-hand number, or multiplier, has three digits. The result is set with its units in column 12, or three columns to the right of the units column of 284. This number, 284, is progressively overwritten by the answer.

The abacus will show the following results at the close of each step:

Column number	1	2	3	4	5	6	7	8	9	10	11	12
Initial		3	7	1			2	8	4			
After step 1		3	7	1			(2	8)	1	4	8	4
After step 2		3	7	1			(2)	3	1	1	6	4
After step 3		3	7	1			1	0	5	3	6	4

To return to step 1, the multiplication 371 × 4 is carried out, as with Arabic notation, from right to left, commencing with 1 × 4 = 4, placed in column 12. To this is added 7 × 4 = 28, with the 8 in column 11 and the 2 in column 10. Finally 3 × 4 = 12. The digit 2 is added to 2 in column 10, to give 4, and the 1 is placed in column 9, overwriting the 4 of the original number.

Similarly, in step 2, the multiplication 371 × 8 is carried out, with the result of the first stage, 1 × 8 = 8, added to column 11. There is already 8 in this column: 8 + 8 = 16, so 6 is placed in column 11 and 1 *added* to column 10 (to give, at this stage, 5). These stages are:

Column number	7	8	9	10	11	12
After Step 1	(2	8)	1	4	8	4
Add on 1 × 8 = 8 in col. 11	(2	8)	1	5	6	4
Add on the 6 of 7 × 8 = 56 in col. 10	(2	8)	2	1	6	4
Add on the 5 of this 56 in col. 9	(2	8)	7	1	6	4
Add on the 4 of 3 × 8 = 24 in col. 9	(2)	1	1	1	6	4
Add on the 2 of this 24 in col. 8	(2)	3	1	1	6	4

The third step proceeds similarly.

There are, of course, many similarities between this process and Arabic multiplication. Surprisingly, for an expert operator, the abacus is faster, presumably because no writing is required and the associated mental processes are designed for use with the instrument.

APPENDIX TABLE A MASONRY BRIDGES

No.	Name	Stream/ river	Town	Location	Spans[1] (m)	Height[2]	Gazzola no.
				1 Rome and vicinity			
I1	Pons Mulvius	Tiber R.	Rome	5 km north	9, 4 × 18, 9	M	29
I2	Pons Aelius or San Angelo	Tiber R.	Rome	In Rome	6 × 18	M	179
I3	Pons Fabricius	Tiber R.	Rome	In Rome	24.2, 24.5	M	40
I4	Pons Cestius	Tiber R.	Rome	In Rome	3 × 13.7	M	41
I5	Pons Aemilius or Rotto	Tiber R.	Rome	In Rome	6, incl. 4 × 24.3	M	28
I6	Viadotta sulla Via Aurelia	–	Rome	P. Cestius–Janiculum	(1S)	–	10
I7	–	Fosso di Palidoro	Palidoro	Beneath farmhouse	(2S)	–	–
I8	Primo Ponte	Fosso di Sta Maria Morgana	Sta Marinella	V. Aurelia	(1S)	L	12
I9	Secondo Ponte	"	"	V. Aurelia, at 60.9 km	(1S)	L	13
I10	–	Fosso di Vallemplice	"	V. Aurelia	(1S)	L	14
I11	Ponte del Diavolo	Torrente Fiumaretta	"	V. Aurelia	7	L	15
I12	Ponte Sodo	Valchetta	Veii	On N boundary			1
I13	–	Fosso della Valchetta	Castel Giubileo	0.25 km N	2 × 7	L	88
I14	Ponte Salario	Anio R.	Rome	5 km N on Via Salaria	8.5, 26.7, 8.5	M	283
I15	Ponte Nomentano	Anio R.	Rome	6 km NE on Via Nomentana	(S, L, S)	M	282
I16	Ponte Mammolo	Anio R.	Rome	8 km ENE on Via Tiburtina	(S), 16, (S)	M	59
I17	Ponte Lucano	Anio R.	Tivoli	3.5 km W on Via Tiburtina	(5S)	L	51
I18	Ponte dell'Acquoria	Anio R.	Tivoli	1.2 km W	(7S)	–	52
I19	Ponte Scutonico	–	Arsoli	2 km S on Via Valeria	(1S)	L	49
I20	Ponte San Giorgio	–	Arsoli	3 km N on Via Valeria	(1S)	L	48
I21	Ponte di Agosta	Anio R.	Agosta	8 km S of Arsoli	(7S)	L	134
I22	Small Ponte di Nona	T. Marrana	Rome	12 km E on V. Praenestina	6	L	16
I23	Large Ponte di Nona	T. Marrana	Rome	12 km E on V. Praenestina	(S), 7 × 6, (S)	H	43
I24	Ponte di Terra	F. di Terra	Rome	21 km E on V. Praenestina	4.9	L	20
I25	Viaduct over the F. di Ponte Terra	Fosso di Ponte Terra	San Vittorino	1 km NE	(4M)	M	–
I26	Ponte Amato	Fosso Collafri	Gallicano	0.4 km SW on V. Praenestina	5	M	17
I27	Ponte Sardone	–	Palestrina	0.7 km SW on V. Praenestina	–	–	18
I28	–	Fosso del Giardinetto	Rome	11 km E on V. Labicana	5.7	L	19
I29	Viadotta di Valle Ariccia	Valle Ariccia	Rome	27 km SE on V. Appia	(3S)	L	2
I30	Ponte di Vigna Capoccio	Fosso di Carano	Velletri	4 km SW on V. Appia	(1S)	–	33
I31	Ponte Catena	–	Cori	Outside Porta Ninfina	(1S)	H	99
I32	–	Fosso Malafede	Ostia	9 km NE on V. Ostiensis	(2S)	L	3
I33	Ponte presso Tor di Valle	–	Rome	10 km SW on V. Ostiensis	(1S)	L	4
I34	Ponte di Mezzocamino	–	Rome	–	(1S)	L	5
I35	–	Numicus R.	Ostia	2 km SE on V. Severiana	(3S)	L	–

No.	Name	Stream/ river	Town	Location	Spans[1] (m)	Height[2]	Gazzola no.
			2 Southern Italy and the islands				
I36	Ponte del Diavolo	R. Spalla Bassa	Cassino	0–12 km SW	5.5	L	37
I37	Ponte di Cassino	Valle del Rapido	Cassino	About 3 km N	S	L	38
I38	Ponte di S. Giuliano	R. Rapido	Atina	In Atina	(2M)	M	87
I39	Ponticello di Isernia	–	Isernia	–	(1M)	L	32
I40	Ponte del Diavolo	R. Titerno	Campobasso	–	(1M)	L	216
I41	Ponte di Tre Ponti	Pontine Marshes	Terracina	About 15 km NW on V. Appia	(2S)	L	11
I42	Ponte Alto	–	Terracina	On V. Appia	(1S)	L	100
I43	Ponte Antico	Fosso dell'Acqua Traversa	Terracina	On V. Appia	(1S)	L	101
I44	Ponte di Sperlonga	–	Sperlonga	Near Villa of Tiberius	4.7	L	117
I45	Pons Auruncus	–	Sessa Aurunca	1 km W	(21M)	L	–
I46	Ponte Tufaro	–	Beneventum	About 10 km SW	(3S)	L	102
I47	Ponte Apollosa	–	Beneventum	About 8 km SW	(S,M,S)	M	103
I48	Ponte Corvo	–	Beneventum	About 5 km SW	(2S)	L	109
I49	Ponte Leproso	–	Beneventum	About 1 km SW	(5S)	L	108
I50	Ponticello	–	Beneventum	In Beneventum, to E	(1M)	M	107
I51	Ponte Valentino	R. Calore	Beneventum	5 km E on V. Traiana	15, 18, 15	M	130
I52	Ponte dei Ladroni	V. d. Ferrara	Beneventum	17 km ENE on V. Traiana	3, 10, 14	–	–
I53	Ponte delle Chianche	T. d. Buonalbergo	Montecalvo	3.5 km NW on V. Traiana	(6M)	L	–
I54	Ponte Rotto	Cervaro R.	Foggia	10 km S on V. Traiana	(17, S–M)	L	–
I55	Ponte Rotto	Carapelle R.	Foggia	15 km SE on V. Traiana	(10, S–M)	L	–
I56	Ponte sul Ofanto	Ofanto R.	Canosa d. P.	5 km W on V. Traiana	(7, S–M)	L	133
I57	Ponte Pietra	Ofanto R.	Canosa d. P.	5 km W on V. Traiana	(2)	–	132
I58	Ponte sul Savuto	Savuto R.	Scigliano	V. Popillia (or Annia)	20.5	M	226
I59	Ponticello sul F. Selino	R. Selino	Selinunte	Sicily, near Castelvetrano	(1S)	L	47
I60	Ponte sul F. Mannu	R. Mannu	Porto Torres	Sardinia, to W of town	5.4, 5.4, 5, 5.3, 5.2, 12.5, 19.5	M	152
I61	Ponte di Gavoi	–	Gavoi	Sardinia, SW of Nuoro	(5, S–M)	L	154
I62	Ponte di S'Antioco	–	S'Antioco	SW Sardinia	(2S)	L	153
I63	Ponte del Tirso	R. Tirso	Oristano	Sardinia	Ruins	–	155
			3 Central Italy				
I64	Cloaca di Porta San Clementino	R. Marta	Graviscae	Near Tarquinia on V. Aurelia	(1S)	L	21
I65	Ponte dell'Abadia	R. Fiora	Vulci	8 km N of Montalto d. Castro	(S), c.25, (S)	H	30
I66	Ponte del Diavolo	–	Manziana	W of L d. Bracciano	7	L	22
I67	Ponte Piro	–	Barbarano	4 km SE of Bieda on V. Clodia	–	–	36
I68	Ponte del Diavolo	–	Bieda	In gorge below town	(3S)	L	35
I69	Ponte della Rocca	–	Bieda	1 km W on V. Clodia	(1S)	M	34
I70	–	–	Sette Vene	15 km N of Veii on V. Cassia	(1S)	–	–
I71	Ponte S. Nicolao	–	Viterbo	Via Cassia	(1M)	L	156
I72	Ponte S. Lorenzo	–	Bulicame	1 km NE of Viterbo on V. Cassia	(1S)	L	39

I73	Ponte sul Tevere	Tiber R.	Sigliano	–	–	–	135
I74	Ponte Nepesino	Fosso Cerreto	Nepi	3 km S on V. Amerina	(4S)	L	–
I75	Ponte di Nepi	–	Nepi	–	(1S)	L	25
I76	Ponte sul Fosso Tre Ponti	Fosso Tre Ponti	Nepi	4 km N on V. Amerina	4.6	M	23
I77	Ponte sul Fosso della Mola	Fosso della Mola	Nepi	2 km E	(1M)	L	24
I78	Ponte Ritorto	Fosso di Chiavello	Civita Castellana	4 km E on V. Flaminia	5.9	L	31
I79	Ponte presso Torre Pastore	–	Civita Castellana	0.3 km N of P. Ritorto	3.1	L	6
I80	Voltarella	–	Civita Castellana	5 km NE on V. Flaminia	3.1	L	7
I81	Ponte sul Treia	Treia R.	Civita Castellana	5 km NE on V. Flaminia	–	–	8
I82	Muro del Peccato	–	Civita Castellana	N of Ponte sul Treia	–	–	9
I83	Ponte sul R. Fratta	R. Fratta	Orte	13 km SE on V. Flaminia	–	–	27
I84	Ponte Picchiato	R. Miccino	Orte	12 km SE on V. Flaminia	11.5	L	26
I85	Pons Minucius	R. Tiber	Otricoli	3 km S on V. Flaminia	–	–	44
I86	Ponte Sanguinaro	Sanguinaro	Otricoli	7 km NE on V. Flaminia	–	–	45
I87	Ponte d'Augusto	R. Nera	Narni	0.5 km N on V. Flaminia	19.6, 32.1, 18, 16	H	60
I88	Ponte Calamone	–	Narni	2.5 km N on V. Flaminia	3.3, 1.4, 3.3	L	105
I89	Ponte Cardaro	–	Narni	5 km N on V. Flaminia	3.4, 5.5, 7.9, 5.5, 3.4	L	104, 106
I90	Ponte di S. Giovanni de Butris	–	Acquasparta	1 km S on V. Flaminia	2 × 8.7	L	62
I91	Ponte Fonnaia	T. Naia	Acquasparta	3 km N on V. Flaminia	3.4	L	61
I92	Ponte del Diavolo	–	Massa Martana	13 km N on V. Flaminia	3.3	L	63
I93	Ponte Toro	–	Terni	6 km NE on V. Flaminia	9	L	97
I94	Ponte Sanguinario	T. Tessino	Spoleto	Beneath Piazza Garibaldi	(S), 6.2, 6.8, (S)	L	53
I95	Primo Ponte	–	Foligno	Via Sudena	(4S)	L	64
I96	Secondo Ponte	–	Foligno	–	(7S)	L	65
I97	Terzo Ponte	–	Foligno	Via Giovanni dell'Acqua	9, 9, 10, 9, 9	L	66
I98	Quarto Ponte	–	Foligno	Via delle Grazie	8, 10, 8	L	67
I99	Primo Ponte	–	Gualdo Tadino	In vicinity	*c*.4	M	110
I100	Ponte Spiano	R. Fonturce	Sigillo	6 km N on V. Flaminia	3.25	L	90, 111
I101	Ponte di Fossato di Vico	F. Regalino	Fossato di Vico	*c*. 1.5 km E of V. Flaminia	(1S)	L	69
I102	Ponte Etrusco	–	Sigillo	7.5 km N on V. Flaminia	4	L	222
I103	Ponte Voragine	–	Scheggia	3 km N on V. Flaminia	3	L	223
I104	Ponte Grosso a Ponterriccioli	Fosso della Scheggio	Scheggia	6 km N on V. Flaminia	[illegible] × 3.4	L	95
I105	Ponte Grosso	Burano R.	Cantiano	N on V. Flaminia	[illegible] × 7	L	98
I106	Ponte dei Ciclopi	–	Cantiano	*c*.3 km N on V. Flaminia	3.7	L	224
I107	Ponticello	–	Cantiano	0.2 km N of P. d. Ciclopi	(S)	L	225
I108	Ponte Taverna	Burano R.	Cagli	0.5 km S on V. Flaminia	4.8, 16.2	–	93
I109	Ponte Manlio	T. Bosso	Cagli	0.5 km N on V. Flaminia	11.7	L	46
I110	Ponte di Smirra	–	Acqualanga	2.0 km S on V. Flaminia	3.8	L	92
I111	Ponticello di Case Nuove	–	Acqualanga	N on V. Flaminia	1.4	L	91
I112	Ponticello dell'Abbazia	–	Acqualanga	4 km N on V. Flaminia	2.1, 2.8	L	94
I113	Ponte di Traiano	Metaurus R.	Fossombrone	5 km SW on V. Flaminia	4, 11.3, 11.3	M	178
I114	Ponte di Diocleziano	Metaurus R.	Fossombrone	2.5 km SW	*c*.20	H	230
I115	Ponte detto delle Piangole	Metaurus R.	–	Pesaro region	(4M)	M	78

No.	Name	Stream/ river	Town	Location	Spans[1] (m)	Height[2]	Gazzola no.
I116	Ponte di Porta Rimini	Foglia R.	Pesaro	–	28.7, 6.3	M	84
I117	Ponte d'Augusto	Marecchia R.	Rimini	On V. Aemilia	8.7, 8.9, 10.6, 8.9, 8.0	L	83
I118	Ponte sull'Ausa	T. Ausa	Rimini	Near Arch of Augustus	(1S, 1M)	L	82
I119	Ponte di Pioraco	–	Pioraco	*c.*25 km NE of Nocera Umbra	(1S)	L	85
I120	Ponte di Pollenza	–	Pollenza	10 km SW of Macerata	(M,S)	L	86
I121	Ponte Genga	Sentino R.	Fabriano	16 km NE	8	M	79
I122	Ponte di Caudino	T. Lorenzino	Caudino	Near Arcevia	8	L	68
I123	Ponte Sambuco	–	Rieti	S on V. Salaria	(S)	L	96
I124	Ponte Romano	R. Velino	Rieti	In Rieti	–	L	284
I125	Ponte Romano	T. Garafo	Acquasanta	On V. Salaria	10.5	M	72
I126	Ponte di Quinto-decimo	R. Tronto	Quintodecimo	24 km W of Ascoli P. on V. Salaria	17.1	M	75
I127	Ponte d'Arli	R. Tronto	Arli	14 km W of Ascoli P. on V. Salaria	17.0	M	76
I128	Ponte di Porta Cappuccina	R. Tronto	Ascoli Piceno	NW entrance to city	22.2	H	70
I129	Ponte di Porta Tufilla	R. Tronto	Ascoli Piceno	N entrance to city	(2M, 1S)	M	77
I130	Ponte di Cecco	R. Castellano	Ascoli Piceno	E entrance to city	14.5, 7.2	H	71
I131	Ponte del Gran Caso	T. Gran Caso	Ascoli Piceno	2 km E on V. Salaria	6	M	74
I132	Ponte della Scutella	F. Cavignano	Ascoli Piceno	On V. Salaria	10.8	L	73
I133	Ponte Nascosa	–	Civitamassa	*c.*9 km NW of L'Aquila on V. Caecilia	(S)	L	58
			4 Northern Italy				
I134	Ponte Vecchio	–	Rapallo	In town	(M)	L	150
I135	Secondo Ponte della Val Quazzola	–	Quiliano	Near Ricchini	6	L	136
I136	Terzo Ponte della Val Quazzola	–	Quiliano	2 km beyond I135	*c.*6	L	137
I137	Quarto Ponte della Val Quazzola	–	Quiliano	Volta locality	–	–	138
I138	Quinto Ponte della Val Quazzola	–	Quiliano	Colle sul Rio d. Scarroni	–	–	139
I139	Sesto Ponte della Val Quazzola	–	Quiliano	R. d. Gallo	–	–	140
I140	Primo Ponte di Val Ponci	–	Finale Ligure	In Val Ponci on V. Julia Augusta	(S	L	141
I141	Secondo Ponte di Val Ponci	–	Finale Ligure	V. Ponci on V. Julia Augusta	(S)	L	142
I142	Terzo Ponte di Val Ponci	T. Voze	Finale Ligure	V. Ponci on V. Julia Augusta	4–5	L	143
I143	Quarto Ponte di Val Ponci	–	Finale Ligure	V. Ponci on V. Julia Augusta	–	–	144
I144	Quinto Ponte di Val Ponci	–	Finale Ligure	V. Ponci on V. Julia Augusta	5.7, (S)	M	145
I145	Pontetto	–	Loano	On V. Julia Augusta	(S)	L	146
I146	Pontaccio	–	Loano	On V. Julia Augusta	10	L	147
I147	Ponte Lungo	–	Albenga	On V. Julia Augusta	10 × (6.6–8.4)	L	149
I148	Ponte sul R. della Torre	R. della Torre	S. Lorenzo al Mare	On V. Julia Augusta	4	L	148

I149	Ponte San Vito	T. Uso	Rimini	10 km W on V. Aemilia	(S)	L	80
I150	Ponte di Sant' Arcangelo di Romagna	T. Uso	Rimini	*c.*12 km W on V. Aemilia	*c.*4	L	81
I151	Ponte sul Rubicone	R. Rubicon	Savignano	On V. Aemilia	3 × 6.2	L	57
I152	Ponte sul Reno	R. Reno	Bologna	On V. Aemilia	–	–	277
I153	Ponte di Corso V. Emanuele	–	Milan	Beneath C. V. Emanuele	2.6	L	231
I154	Ponte sulla Dora	Dora Baltea R.	Ivrea	In Ivrea	(L, M, S)	M	118
I155	Pont St Martin	T. Lys	Pont St Martin	In Pont St Martin	35.6	M	119
I156	Ponte di St Vincent	T. Cillau	St Vincent	1 km east	9.7, 3.1	L	122
I157	Ponte di Chatillon	T. Marmore	Chatillon	In Chatillon	15	M	123
I158	Pont de Pierre	T. Buthier	Aosta	E of Arch of Augustus	17.1	L	121
I159	Ponte del Pondel	R. Cogne	Pondel	*c.*11 km SW of Aosta	14.2	H	120
I160	Ponte Petreo	–	Ravenna	*c.*3 km S	–	–	278
I161	Ponte S. Lorenzo	Bacchiglione R.	Padua	To E of town	12.8, 14.4, 12.5	L	54
I162	Ponte Corbo	Bacchiglione R.	Padua	To E of town	(3S)	L	55
I163	Ponte S. Matteo	Bacchiglione R.	Padua	To E of town	–	–	56
I164	Ponte Altinate	Bacchiglione R.	Padua	To NE of town	(3S)	L	129
I165	Ponte di. Fiumicinetto	T. Gressaga	S. Dona d. Piave	N	3 × *c.*7	L	219
I166	Ponte sul Canalat	T. Canalat	Ceggia	–	8.3, 7.4, 6.8	L	220
I167	Ponte sul Lemene	R. Lemene	Concordia Sag.	–	–	L	218
I168	Ponte sul Nafissa	R. Nafissa	Aquileia	–	(1S)	L	221
I169	Ponte Pietra	Adige R.	Verona	NE corner of town	(5M)	L	–
I170	Ponte Pusterla	Bacchiglione R.	Vicenza	On V. Claudia Augusta	(3M)	L	126
I171	Ponte degli Angeli	Bacchiglione R.	Vicenza	Near Chiesa d. Angeli	(5M)	L	124
I172	Ponte Furo	Retrone R.	Vicenza	Near Campo Marzio	2 × 4.6	L	125
I173	Ponte di Susegana	–	Susegana	–	5.3	L	127
I174	Ponte de Arena	–	Monza	–	5.1	L	232
I175	Ponte sul Adda	Adda R.	Olginate	S of Lecco	4.8	L	280
I176	Ponte della Regina	Brembo R.	Almenno	Lecco–Bergamo	(8M)	M–H	180
I177	Ponte Marzio	Serio R.	Gorle	Near Bergamo	(3M)	L	279
I178	Ponte di Promontogno	–	Promontogno	Near Chiavenna	(1S)	M	290
I179	Ponte Romano sul Passirio	Passirio R.	Merano	–	(M, S)	L	281
I180	Ponte Ladricio	Isarco R.	Franzenfeste	–	–	–	128

5 Gaul

G1	Pont des Esclapes	Le Béal	Fréjus	2.5 km W on V. J. Augusta	(3S)	L	–
G2	Pont Flavien	R. Touloubre	St Chamas	E of Arles on V. J. Augusta	12.3	L	176
G3	–	R. Buès	Ganagobie	W bank of R. Durance, S of Sisteron	–	–	–
G4	Pont sur Laye	Laye T.	Mane	1 km W	3, 7, 12	L	175
G5	Pont Julien	Coulon R.	Apt	7 km W	10.3, 16.3, 10.3	L	173
G6	Pont sur Rhône	R. Rhône	Arles	–	–	–	267
G7	Pont de St Gabriel	–	Tarascon	–	3	L	265
G8	Pont d'Ambroix	R. Vidourle	Ambroix	W of Gallargues on V. Domitia	(5M)	L	166
G9	Pont de Sommières	R. Vidourle	Sommières	In town	(2S), 13 × 8.5, (2S)	L	171

No.	Name	Stream/river	Town	Location	Spans[1] (m)	Height[2]	Gazzola no.
G10	Pont de Boissezon	R. Bénovie	Boissezon	1 km S of Sommières	6, 8.8, 9.4, 8.8, 6	L	172
G11	Pont de St Thibéry	Hérault R.	St Thibéry	1 km SE on V. Domitia	9 × 10–12	L	169
G12	Pont Serme	–	Béziers	14 km W on V. Domitia	–	L	170
G13	–	–	Narbonne	Beneath houses	c.15	L	–
G14	Pont sur Couesnon	–	Vendelais	–	3	L	167
G15	Pont de Vaison-la-Romaine	Ouvèze R.	Vaison-la-Romaine	In town	17	L	174
G16	–	R. Escoutay	Viviers	–	–	–	–
G17	Pont Vieux	Hérault R.	Le Vigan	–	–	–	291
G18	Pont d'Ascain	Nivella R.	Ascain	7 km SE of St-Jean-de-Luz	–	–	263
G19	Pont sur Rhône	R. Rhône	Vienne	To Ste Colombe les Viennes	–	–	262
G20	Pont sur Loire	R. Loire	Lavoute	–	–	–	264
G21	Pont de César	Velaj R.	Mondouilloux	On V. Bolena	–	–	266
G22	Pont de Rolampont	Marne R.	Rolampont	1–1.5 km SE	(3S)	L	268
G23	Pont d'Outremecourt	Mouzon R.	Bourmont	–	3, 4.7, 5.8, 4.7, 3	L	261
G24	Pont de Domqueur	–	Domqueur	16 km E of Abbeville	2.3	L	227
G25	Pont de Montignies-St Christophe	R. Hantes	Montignies-St Christophe	1 km NW (Belgium)	13 × 1.5	L	165
G26	Pont d'Ombret	R. Semois	Arlon	Arlon–Trier rd (Belgium)	–	–	168
G27	Pont sur Sure	R. Sure	Echternach	(Luxembourg)	–	–	292
G28	Pont sur Sure	R. Sure	Wassabillig	(Luxembourg)	–	–	293
				6 Spain and Portugal			
SP1	Puente Viejo	R. Fluvia	Besalu	30 km NW of Gerona	–	–	287
SP2	Puente del Diablo	R. Llobregat	Martorell	E end of town	c.16, 37.3	H	187
SP3	Puente de Monistrol	R. Llobregat	Monistrol	In town	37, 27, 17, 12	M	188
SP4	Puente de Villa del Río	R. Salado de Porcuna	Villa del Río	10 km E of Montoro	3, 9.1, 3.6	L	160
SP5	Puente del Jándular	R. Jándular	Andújar	5 km W	7 × 5.1, (2S)	L	–
SP6	Puente Mocha	R. Guadal-mellato	Córdoba	10 km E	10 × (4.5–6.6)	L	–
SP7	Puente Romano	R. Guadal-quivir	Córdoba	In town	16 × (7.9–12.0)	L	159
SP8	Puente de Carmona	–	Carmona	28 km E of Seville	4, 4, 5, 4, 4	L	–
SP9	Puente de Alcantarilla	R. Salado de Moran	Utrera	32 km SE of Seville	5.4, 5.7	L	192
SP10	–	R. Guadiana	Mertola	–	–	–	–
SP11	Puente de Piedra	R. Segre	Lérida	In town	–	–	158
SP12	Puente de Piedra	R. Ebro	Zaragoza	In town	–	–	–
SP13	Puente de Alcántara	R. Tagus	Toledo	E approach to town	16, 28.3	H	286
SP14	Puente de Medellín	Guadiana R.	Medellín	35 km E of Mérida	28 × 12	L	275
SP15	Puente Romano	Guadiana R.	Mérida	S entrance to town	62 × (5.6–11.6)	L	161
SP16	Puente de Albarregas	Albarregas R.	Mérida	NW entrance to town	1.3, 1.7, 3 × 5.0	L	186
SP17	Villa Formosa	R. Seda	Alter do Chão	10 km W	6 × c.9.6	L	163
SP18	Puente del Vadillo	Tamuja R.	Vadillo	30 km E of Cáceres	(1M)	L	270
SP19	Puente de El Garro	Almonte R.	Alconétar	Now flooded	–	–	271
SP20	Puente de Alconétar	Tagus R.	Alconétar	Now 6 km N	16 × (7.3–15)	M	183
SP21	Puente de Alcántara	Tagus R.	Alcántara	–	13.8, 21.9, 28.8, 27.4, 21.9, 13.8	H	181
SP22	Puente de Segura	R. Elja	Segura	–	7, 7, 10, 7, 7	M	185
SP23	Puente de Cáparra	R. Ambroz	Plasencia	20 km N	8.7, 9.2	L	189

SP24	Puente de Guijo de la Granadilla	R. Alagón	Plasencia	*c.*20 km N	19.1	M	288
SP25	Puente de la Doncella	R. Ambroz	Plasencia	*c.*40 km N	9.3	L	274
SP26	Puente de Guallaminos	R. Guallaminos	Villanueva de la Vera	–	18.6	L	273
SP27	Puente de Alardos	Alardos R.	Madrigal de la Vera	–	22.8	M	272
SP28	Puente de Sta Coloma	R. Gran Valira	Santa Coloma	3 km S of Andorra la Vella	(1S, 1M)	L	228
SP29	Puente de Lumbier	Irati R.	Lumbier	At Hoz de Lumbier	(1M)	H	–
SP30	Puente de Tres Puentes	R. Zadorra	Iruna	9 km W of Vitoria-Gasteiz	(13S)	L	190
SP31	Puente de Villodas	R. Zadorra	Villodas	*c.*12 km W of Vitoria-Gasteiz	(10S)	L	191
SP32	Puente de Cangas de Onís	R. Sella	Cangas de Onís	–	7.7, 21.6, 9.5, 6.8, 4.3, 3.6	M	285
SP33	Puente de Lugo	R. Jiloca	Lugo	–	5.4, 11.2, 5.4	L	251
SP34	Puente de Orbigo	–	Santiago de Compostela	–	(>5S)	L	289
SP35	Puente de la Cigarrosa	R. Sil	Petín	In town	8, 10, 4, 19, 10	L	–
SP36	Puente Bibey	R. Bibey	Puebla de Trives	*c.*8 km towards Petín	6.1, 18.5, 8.8	H	–
SP37	Puente de Puebla de Trives	R. Xares	Puebla de Trives	Near Mendoya	7	L	193
SP38	Puente Linares	R. Limia	Bande	On Bande–Porquera road	7.5, 7 + (2S)	L	162
SP39	Puente de Chaves	R. Tamega	Chaves	In town	(*c.*13S)	L	157
SP40	Puente Romano	R. Tormes	Salamanca	In town	15 × (9.4–9.7)	L	182

7 Africa

A1	Pont d'Oued Cherchar	Oued Cherchar	Bou Ficha	4 km NE	(*c.*16S)	L	197
A2	Pont d'Oued Bou-Zid	Oued Bou-Zid	Ksar Hellal	–	–	–	207
A3	Pont d'Oued Ziuna	Oued Ziuna	Tolemaide	–	(2S)	L	276
A4	–	Nile R.	Aswan	–	–	–	236
A5	Pont d'Oudna	Oued Oudna	Oudna	200 m NE	(3S)	L	215
A6	Pont d'Oued Nebana	Oued Nebana	Kairouan	36 km N	(5), 6 × 2.6, (S) (now *c.*14)	L	212
A7	Pont d'Adpertusa	Oued Chafrou	Adpertusa	5 km W of Tunis	6.3	L	209
A8	Pont de Grich el Oued	Oued el Hamar	Grich el Oued	62 km NE of Mejez el Bab	(>3)	L	195
A9	Pont d'Ain Younes	Oued el Kebir	Téboursouk	9 km SE	(1S)	L	206
A10	Pont d'Oued el Ahmar	Oued el Ahmar	Dougga	6.5 km E	–	–	205
A11	Pont d'Ain Hedja	–	Agbia	–	–	–	203
A12	Pont sur l'Oued Djilf	Oued Djilf	Maktar	15 km E	6 × 5.4	M	199
A13	Culverts in Siliana valley	–	Maktar	E	(S)	L	208
A14	Pont de Sbeïtla	Oued Sbeïtla	Sbeïtla	300 m NE	(4M)	M	237
A15	Pont de Tébessa	–	Tébessa	2.5 km NE	4 × 3.1	L	202
A16	Pont de Gastal	–	Tébessa	possibly 30 km N	7.5	L	213
A17	Pont de Khémissa		Khémissa	2 km E	(S), 5.4, (S)	M	211
A18	Pont de Treviri	Oued Rhumel	Constantine	–	c3.5, 5.3 *c.*3.5	M	235

No.	Name	Stream/ river	Town	Location	Spans[1] (m)	Height[2]	Gazzola no.
A19	Pont d'el Kantara	Oued Kantara	el Kantara	65 km SW of Batna	10	M	253
A20	Pont de Jedeida	Mejerda R.	Jedeida	E end of town	(9S)	L	196
A21	Pont de Tébourba	Mejerda R.	Tébourba	SE of town	(c.16S)	L	–
A22	Pont de Mejez el Bab	Mejerda R.	Mejez el Bab	W end of town	(7S)	L	–
A23	Pont d'Oued Zarga	Oued Zarga	Oued Zarga	–	(2M)	L	214
A24	Pont de Trajan	Oued Béja	Béja	10 km S	5.8, 5.3, 6	L	151
A25	Pont de Civitas Avensis	Mejerda R.	Bou Salem (Souk el Khemis)	11 km E	(2S)	–	210
A26	Pont de Chemtou	Mejerda R.	Chemtou	20 km W of Jendouba	(5S)	M	194
A27	Pont d'Oued el Enja	Oued el Enja	Thuburnica	11 km NW of Chemtou	10	L	200
A28	Pont d'Oued Budjema	Oued Budjema	Annaba	1 km S, at Hippone	(11S)	L	201
A29	Pont de Duperré	Oued Chelif	Duperré	150 km SW of Algiers	–	–	198
				8 The east			
E1	Bridge at Oberzeiring	Blahbach Torrent	Oberzeiring	20 km NW of Judenberg, Austria	(1S or 2S)	L	269
E2	St Dionysen bridge	Murufer Torrent	St Dionysen	Between St Dionysen and Oberdorf, Austria	(1S)	L	233
E3	Cinque Ponti	–	Salona	7 km NE of split, Yugoslavia	(2S)	L	217
E4	Shkumbi R. bridge	Shkumbi R.	Topçias	Near Shenjan, Albania	(17S)	–	–
E5	Hadji Bektari bridge	Shkumbi R.	H. Bektari	10 km NE of Elbasan, Albania	(3S)	–	–
E6	Pons Servili	–	Orakë	W of Lake Ochrid, Albania	–	–	–
E7	Sakarya bridge	Sangarius R.	Izmit	4 km W of Adapazari, Turkey	(8–12M)	L	257
E8	–	R. Iris	Amasya	Turkey	–	–	–
E9	Karas bridge	–	Karas	10 km N of Erzerum, Turkey	(4–8M)	L	–
E10	Pons Sangarius	Sangarius R.	Çandir Köprü	55 km SE of Eskişehir, Turkey	–	–	–
E11	Rhyndacus R. bridge	Kemalposa R.	Aezani	Çavdarhisar, Turkey	c.3.9, 5.2, 6.5, 5.2, c.3.9	L	239
E12	Rhyndacus R. bridge	Kemalposa R.	Aezani	Çavdarhisar, Turkey	(c.4M)	L	250
E13	Roman bridge	Biga Çay	Biga	Over R. Granicus, Turkey	–	–	–
E14	Selinus R. bridge	R. Selinus	Pergamum	Turkey	2 × 11.4	L	204
E15	Selinus R. bridge	R. Selinus	Pergamum	Under Temple of Serapis or Kizil Avlu, Pergamum, Turkey	12.1, 12.5	L	204
E16	Caravan bridge	R. Meles	Izmir	Turkey	–	–	–
E17	–	Asopus R.	Laodicea	Turkey	8.0, 12.3, 7.1	M	238
E18	Roman bridge	–	Eğridir	Turkey	–	–	–
E19	–	–	Kireli	Turkey	–	–	–
E20	–	R. Marsyas	Incekemer	Turkey	(3M)	M	–
E21	Ponte di Rodi	–	–	Island of Rhodes	(2S)	L	164
E22	–	–	Kemer	N of Kalkan, Turkey	4.1, 4.5, 4.0+	L	–
E23	Bridge at Limyra	–	Limyra	7.5 km NE of Finike, Turkey	27 × c.10.7	L	–
E24	–	Arapsu R.	Olbia	c.5 km W of Antalya, Turkey	(1S)	L	–
E25	–	R. Eurymedon		c.5 km N of Beşkonak, Turkey	(1S)	L	–
E26	–	R. Göksu	Silifke	Turkey	5 × c.17, 2 × 4.3	L	–
E27	–	–	Tarsus	Near Adana, Turkey	(3S)	L	–
E28	Taşköprü	Seyhan R.	Adana	Turkey	(21S)	L	–
E29	Misis bridge	Ceyhan R.	Adana	E of Adana, Turkey	(9M)	L	–

E30	–	–	Antakya	N of Antakya, Turkey	(1S)	H	–
E31	Asi bridge	–	Antakya	Turkey	4 × 10	L	–
E32	Afrin bridge	R. Afrin	Aleppo	Aleppo–Cyrrhus, Syria	11.5, 12.5, 12	M	258
E33	–	Saboun-Souyou	Aleppo	70 km N, Syria	3.8, 8.4, 10, 9.8, 8.2, 3.8	M	259
E34	–	Saboun-Souyou	Aleppo	Cyrrhus–Doliche, Syria	4.2, 9.3, 4.7	L	260
E35	Göksu R. bridge	Göksu	Samsat	25 km W, Turkey	c.30	–	255
E36	Severus bridge	Cendere Çay	Kâhta	41 km E of Adiyaman, Turkey	34.2	H	254
E37	–	Arabkir Çay	Korpanik	3 km E, Turkey	30	–	–
E38	Karamağara bridge	Arabkir Çay	Agin	Turkey	17	H	–
E39	Sabrina R. bridge	Karabudak	Zimara	Turkey	–	–	–
E40	Zakho bridge	Khabur R.	Zakho	Iraq	(1S, 1L, 1S)	H	252
E41	Bridge at Eski Mosul	–	Eski Mosul	Iraq	c.15	M	–
E42	Band-i-Kaisar	Karun R.	Shushtar	Iran	(c.35M)	M	229
E43	–	Wadi Suweida	Suweida	N of Bostra, Syria	(1)	–	247
E44	–	Wadi Zeidi	Bostra	Syria	3 × 3.8	L	245
E45	–	Wadi Zeidi	el-Tali'a	W of Bostra, Syria	(2S)	L	246
E46	–	Wadi Zeidi	Gemarrin	Syria	(3S)	L	256
E47	–	Nahr er Rukkad	–	W of Bostra, Syria	(8)	–	244
E48	–	–	Qodana	W of Bostra, Syria	(3)	–	244
E49	Jisr el Majami	Jordan R.	Gesher	Palestine	c.15, c.18	M	243
E50	Jisr Banat Ya'qub	Jordan R.	–	N of Sea of Galilee, Palestine	(4M)	L	–
E51	Bridge at Maameltein	Maameltein R.	Beirut	18 km NE, Lebanon	12	L	240
E52	Bridge at Nahr Aboul-el-Assouad	R. Kasniré	Tyre	10 km N, Lebanon	(1M)	L	241
E53	Bridge of Jisr Jindas	–	Lod	35 km NW of Jerusalem, Palestine	–	–	242

[1] Either in m or (L) = large (M) = medium (S) = small
e.g. (2S) = 2 small spans
6 × 18 = 6 spans, each of 18 m
[2] H = high M = medium L = small

APPENDIX TABLE B TIMBER BRIDGES

No.	Name	Stream	Town	Location	Spans[1] (m)	Height	Gazzola no.
T1	Bridge of boats	R. Danube	Kostolac	Romania–Yugoslavia	–	–	–
T2	Bridge of boats	R. Danube	Turnu-Severin	Romania–Yugoslavia	–	–	–
T3	Caesar's bridge	R. Rhine	Coblenz	Germany	(*c*.26)	–	–
T4	–	R. Rhine	Mainz	Germany	–	–	177
T5	–	R. Loire	–	France	–	–	–
T6	–	R. Seine	Paris	France	–	–	–
T7	–	R. Loire	Orléans	France	–	–	–
T8	–	R. Thames	London	England	–	–	–
T9	–	R. Rhône	Geneva	Switzerland	–	–	–
T10	–	R. Saône	–	France	–	–	–
T11	–	R. Moselle	Trier	Germany	–	–	–
T12	Pons Sublicius	R. Tiber	Rome	Italy	–	–	–
T13	Bridge of Apollodorus	R. Danube	Turnu-Severin	Romania–Yugoslavia	21 × 32.3	–	184
T14	Pons Tirenus	Garigliano R.	Minturnae	Italy	–	–	–
T15	Justinian's bridge	R. Siberis	Sykeon	Turkey	8 × (5.4–9.6)	–	–
T16	–	–	London	Newgate, England	–	–	–
T17	–	R. Churn	Cirencester	England	–	–	–
T18	–	–	Wroxeter	England	–	–	–
T19	–	–	Wallasey	England	–	–	–
T20	–	–	Water Newton	England	–	–	–
T21	–	R. Trent	Cromwell	England	–	–	–
T22	–	R. Tees	Piercebridge	England	–	–	–
T23	–	R. Tees	Pounteys Bridge	England	–	–	–
T24	–	R. Wear	Binchester	England	–	–	–
T25	–	–	Hunwick Gill	England	–	–	–
T26[2]	–	R. Tyne	Corbridge	England	(6–11S)	–	–
T27	–	R. Rede	Risingham	England	–	–	–
T28	–	R. Rede	Elishaw	England	–	–	–
T29	Pons Aelius	R. Tyne	Newcastle	England	–	–	–
T30[2]	–	North Tyne R.	Chesters	England	(4S)	–	–
T31[2]	–	R. Irthing	Willowford	England	–	–	–
T32	–	R. Eden	Hyssop Holme Well	England	–	–	–
T33	–	R. Kelvin	Summerston	Scotland	–	–	–
T34	–	R. Forth	–	Scotland	–	–	–

[1] Either in m or (L) = large (M) = medium (S) = small
e.g. (2S) = 2 small spans
6 × 18 = 6 spans, each of 18 m

[2] Possible stone arches

APPENDIX TABLE C ROMAN AQUEDUCTS

No.	Country	Town	Name	Bridges
AQ1	Italy	Rome	Marcia	Yes
AQ2	"	Rome	Tepula	Yes
AQ3	"	Rome	Julia	Yes
AQ4	"	Rome	Claudia	Yes
AQ5	"	Rome	Anio Novus	Yes
AQ6	"	Rome	Alexandrina	Yes
AQ7	"	Minturno	–	Yes
AQ8	"	Syracuse	–	–
AQ9	"	Taormina	–	–
AQ10	"	Turin	–	–
AQ11	"	Venafro	–	Yes
AQ12	"	Viterbo	–	–
AQ13	"	Anzio	–	–
AQ14	"	Civitavecchia	–	Yes
AQ15	"	Termini Imerese	–	Yes
AQ16	"	Terracina	–	Yes
AQ17	"	Catania	–	–
AQ18	"	Pompeii	–	–
AQ19	"	Bologna	–	–
AQ20	"	Spoleto	Ponte delle Torri	Yes
AQ21	"	Tarquinia	–	Yes
AQ22	"	Alatri	–	–
AQ23	"	–	Ponte Funicchio	Yes
AQ24	France	Fréjus	–	Yes
AQ25	"	Cimiez	–	–
AQ26	"	Antibes	–	Yes
AQ27	"	St Rémy-de-Provence	–	–
AQ28	"	Aix-en-Provence	–	–
AQ29	"	Arles	–	Yes
AQ30	"	Nîmes	Pont du Gard	Yes
AQ31	"	Pézenas	–	–
AQ32	"	Vaison	–	–
AQ33	"	Die	–	–
AQ34	"	Vienne	–	–
AQ35	"	Lyon	Mont d'Or	–
AQ36	"	Lyon	Brevenne	–
AQ37	"	Lyon	Craponne	Yes
AQ38	"	Lyon	Gier	Yes
AQ39	"	Saintes	–	–
AQ40	"	Rodez	–	–
AQ41	"	Périgueux	–	–
AQ42	"	Poitiers	–	–
AQ43	"	Sens	–	–
AQ44	"	Paris	Lutèce	–
AQ45	"	Evreux	–	–
AQ46	"	Metz	–	Yes
AQ47	"	Bavay	–	–
AQ48	"	Bordeux	–	–
AQ49	"	Marseille	–	–
AQ50	"	Orange	–	–
AQ51	"	Luynes	–	Yes
AQ52	"	St Bertrand de Comminges	–	Yes
AQ53	Germany	Mainz	–	–

No.	Country	Town	Name	Bridges
AQ54	Spain	Granada	–	–
AQ55	"	Bellone-Claudia	–	–
AQ56	"	Cádiz	–	–
AQ57	"	Sadaba	–	–
AQ58	"	Chelva	–	–
AQ59	"	Tarragona	Puente de las Ferreras	Yes
AQ60	"	Segovia	–	Yes
AQ61	"	Mérida	Cornalvo	–
AQ62	"	Mérida	San Lázaro	Yes
AQ63	"	Mérida	Los Milagros	Yes
AQ64	"	Seville	–	Yes
AQ65	"	Italica	–	Yes
AQ66	"	Almuñécar	–	Yes
AQ67	"	Toledo	–	Yes
AQ68	"	Barcelona	–	Yes
AQ69	"	Alcanadre	–	Yes
AQ70	"	Valencia de Alcántara	–	Yes
AQ71	Tunisia	Sbeitla	–	Yes
AQ72	"	Khémissa	–	Yes
AQ73	"	Tébessa	–	Yes
AQ74	"	Chemtou	–	–
AQ75	"	Musti	–	–
AQ76	"	Carthage	Zaghouan–Carthage	Yes
AQ77	Algeria	Constantine	–	Yes
AQ78	"	Cherchell	–	Yes
AQ79	Morocco	Volubilis	–	–
AQ80	Yugoslavia	Salona	–	Yes
AQ81	Greece	Kavala	–	Yes
AQ82	Turkey	Istanbul	–	Yes
AQ83	"	Erythrai	–	–
AQ84	"	Ephesus	–	Yes
AQ85	"	Laodicea	–	Yes
AQ86	"	Alabanda	–	Yes
AQ87	"	Alinda	–	Yes
AQ88	"	Phaselis	–	Yes
AQ89	"	Aspendus	–	Yes
AQ90	"	Side	–	Yes
AQ91	"	Sebaste	–	Yes
AQ92	"	Antioch	–	Yes
AQ93	"	Samosata	–	Yes
AQ94	Israel	Caesarea	–	Yes

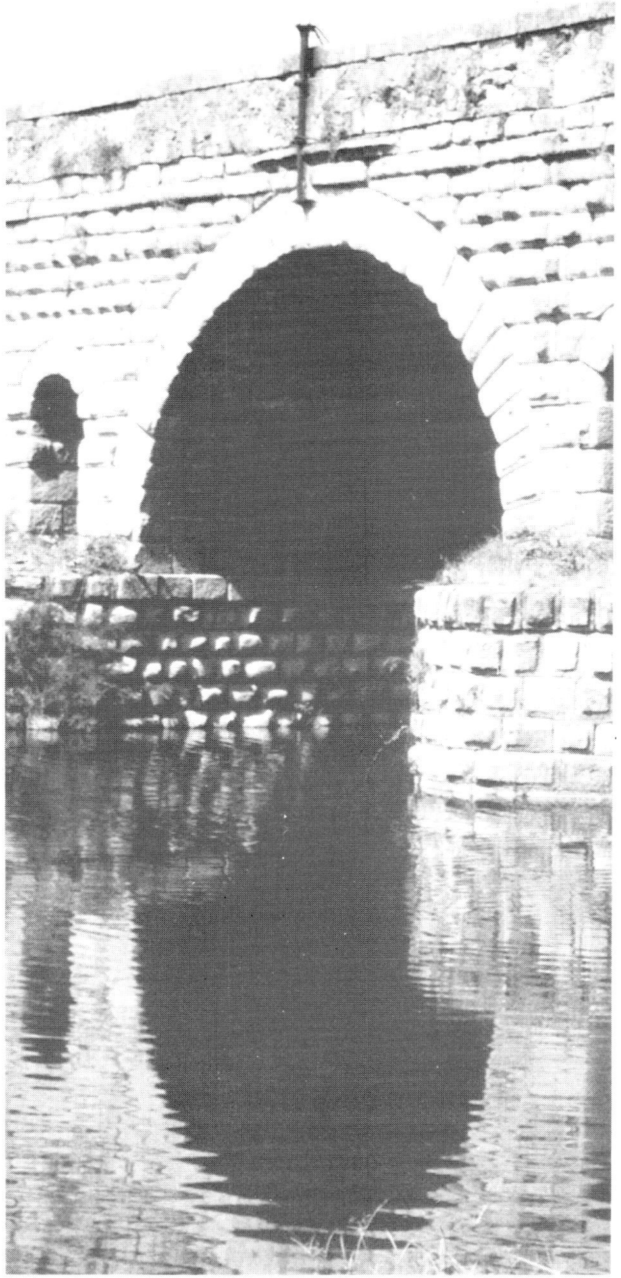

WORKS CITED

1 Ancient authors

Arrian, *Histories of Alexander and Indica: Anabasis Alexandri*, ed. and tr. P. A. Brunt, 1976–83, Cambridge, Mass. and London.

The Life of Alexander the Great, tr. A. de Sélincourt, 1958, London.

Augustine, *Confessions*, tr. W. Watts, 1912, 2 vols., London.

Concerning the City of God against the Pagans, tr. H. Bettenson, 1972, Harmondsworth.

Ausonius, tr. H. G. E. White, 1919–21, 2 vols., London.

Caesar, *The Gallic War*, tr. H. J. Edwards, 1917, London.

The Conquest of Gaul, tr. S. A. Handford, 1951, Harmondsworth.

The Battle for Gaul, tr. A. and P. Wiseman, 1980, London.

Cassius Dio, *Roman History*, tr. E. Cary, 1914–27, 9 vols., London.

Cicero, *Letters to Atticus*, tr. E. O. Winstedt, 1912, London.

Letters to Atticus, ed. and tr. D. R. Shackleton Bailey, 1965–70, 7 vols., Cambridge.

Diodorus Siculus, tr. C. H. Oldfather et al., 1933–67, 12 vols., London.

Frontinus, *De aquaeductu urbis Romae*, ed. C. Kunderewicz, 1973, Leipzig.

Herodotus, *The Histories*, tr. A. de Sélincourt, A., 1954, Harmondsworth.

Horace, *Satires and Epistles*, tr. P. Francis, 1902, London.

Livy, tr. B. O. Foster et al., 1919–59, 14 vols., London.

Pliny the Elder, *Natural History*, tr. H. Rackham et al., 1940–63, 10 vols., London.

Pliny the Younger, *The Letters of the Younger Pliny*, tr. B. Radice, 1963, Harmondsworth.

Polybius, *The Histories*, tr. W. R. Paton, 1922–7, 6 vols., London.

Procopius, *Buildings*, tr. H. B. Dewing, 1940, London.

Statius, *Silvae* IV, tr. and commentary K. M. Coleman, 1988, Oxford.

Strabo, *Geography*, tr. H. L. Jones, 8 vols., 1944– , London.

Suetonius, *The Lives of the Caesars*, tr. J. C. Rolfe, 2 vols., 1914, London.

Tacitus, *De vita Agricolae*, eds. R. M. Ogilvie and I. A. Richmond, 1967, Oxford.

Histories and Annals, tr. J. Jackson, 5 vols., 1914–37, London.

Vitruvius, *Vitruve, de l'architecture*, ed., tr. and commentary P. Fleury, 1990, Paris.

Vitruvius on Architecture, F. Granger, 1931–4, London and Cambridge, Mass.

An Inquiry into the Principles of Beauty and Grecian Architecture Including a Translation of the Ten Books of Marcus Vitruvius Pollio, J. Gwilt, 1860, London.

The Ten Books on Architecture, tr. M. H. Morgan, 1960, New York.

Vitruvius et al., *Vitruvius and Later Roman Building Manuals*, H. Plommer, H., 1973, Cambridge.

2 Other works

Akurgal, E. (1978), *Ancient Civilizations and Ruins of Turkey: from Prehistoric Times until the End of the Roman Empire*, tr. J. Whybrow and M. Emre, Istanbul.

Anderson, J. G. C. (1897), 'The road system of eastern Asia Minor, with the evidence of Byzantine campaigns', *JHS* 17: 22–44.

(1899), 'Exploration in Galatia Cis Halym Part 2: Topography, epigraphy, Galatian civilisation', *JHS* 19: 52–134, 280–318.
Ashby, T. (1902), 'The classical topography of the Roman campagna Part 1', *PBSR* 1: 209–10.
(1906), 'The classical topography of the Roman campagna Part 2', *PBSR* 3: 84–5.
(1925), 'Practical engineering in ancient Rome', *Nature* 116: 576–80.
(1927a), *The Roman Campagna in Classical Times*, London.
(1927b), *The Architecture of Ancient Rome*, London.
(1935), *The Aqueducts of Ancient Rome*, Oxford.
Ashby, T. and Fell, R. L. (1921), 'The Via Flaminia', *JRS* 11: 125–90.
Ashby, T. and Gardner, R. (1916), 'The Via Traiana', *PBSR* 8: 104–71.
Avi-Yonah, M. (1977), *The Holy Land: a Historical Geography*, Grand Rapids.
Ayerst, D. and Fisher, A. S. T. (1971), *Records of Christianity*, Oxford.
Babelon, E., Gagnat, R. and Reinach, S. (1893), *Atlas archéologique de la Tunisie*, 3 vols., Paris.
Baines, J. and Malek, J. (1980), *Atlas of Ancient Egypt*, Oxford.
Ballance, M. H. (1951), 'The Roman bridges of the Via Flaminia', *PBSR* 19: 78–117.
(1958), 'Roman roads in Lycaonia', *Anatolian Studies* 8: 223–34.
Barrow, R. H. (1968), *Slavery in the Roman Empire*, London.
Baterden, J. R. (1908), *Timber*, London.
Bates, K. L. (1900), *Spanish Highways and Byways*, London.
Bean, G. E. (1966), *Aegean Turkey; an Archaeological Guide*, London.
(1968), *Turkey's Southern Shore: an Archaeological Guide*, London.
(1971), *Turkey beyond the Meander: an Archaeological Guide*, London.
(1978), *Lycian Turkey: an Archaeological Guide*, London.
Beckett, D. (1969), *Great Buildings of the World: Bridges*, London.
Belgrand, M. (1875), *Les Aqueducs romains*, Paris.
Bergier, N. (1626), *Histoire des grands chemins de l'empire romain*, 1st edn. 1626; 2nd edn. Brussels, 1728; 3rd edn. Brussels, 1736.
Bidwell, P. T. and Holbrook, N. (1989), *Hadrian's Wall Bridges*, London.
Binnie, G. M. (1949, 50) 'Some notes on the Karun River and the Shatt el Arab', *Institution of Civil Engineers Journal* 3: 204–53.
Birley, A. R. (1963), *Hadrian's Wall: an Illustrated Guide*, London.
(1981), *The Fasti of Roman Britain*, Oxford.
Blackman, D. R. (1978), 'The volume of water delivered by the four major aqueducts of Rome', *PBSR* 46: 52–72.
(1979), 'The length of the four great aqueducts of Rome', *PBSR* 47: 12–18.
Blake, M. E. (1947), *Ancient Roman Construction in Italy from the Prehistoric Period to Augustus*, Washington.
(1959), *Roman Construction in Italy from Tiberius through the Flavians*, Washington.
Blake, M. E. and Bishop, D. T. (1973), *Roman Construction in Italy from Nerva through the Antonines*, ed. and completed D. T. Bishop, Philadelphia.
BNI (1943), *Palestine and Transjordan*, Oxford.
(1944a), *Algeria*, Oxford.
(1944b), *Iraq and the Persian Gulf*, Oxford.
(1945a), *Tunisia*, Oxford.
(1945b), *Persia*, Oxford.
Bouchier, E. S. (1914), *Spain under the Roman Empire*, Oxford.
(1916), *Syria as a Roman Province*, Oxford.
(1917), *Sardinia in Ancient Times*, Oxford.
(1921), *A Short History of Antioch 300 BC–AD 1268*, Oxford.
Boussard, J. (1957), *Atlas historique et culturel de la France*, Paris.
Bowles, O. (1939), *The Stone Industries*, New York.
Boyer, M. N. (1976), *Medieval French Bridges: a History*, Cambridge, Mass.
Bracegirdle, B. and Miles, P. H. (1973), *Great Engineers and their Works: Thomas Telford*, Newton Abbot.
(1974), *Great Engineers and their Works: the Darbys and the Ironbridge Gorge*, Newton Abbot.
Brangwyn, F. and Sparrow, W. S. (1915), *A Book of Bridges*, London.
Breeze, D. J. (1974), 'The organisation of the career structure of the *immunes* and *principales* in the Roman army', *Bonner Jahrbücher* 174: 245–92.
Breeze, D. J. and Dobson, B. (1987), *Hadrian's Wall*, Harmondsworth.
Brice, W. C. (1966), *South West Asia*, London.
British Standards Institution (1971), *Code of Practice for the Structural Use of Timber, CP112, Part 2*, London.
(1976), *BS5390: 1976 Code of Practice for Stone Masonry*, London.
(1988), *BS5268, Part 2: 1988 British Standard: Structural Use of Timber, Part 2. Code of Practice for Permissible Stress Design, Materials and Workmanship*, London.
(1989), *BS2052: 1989 British Standard Specification for Ropes made from Manila, Sisal, Hemp, Cotton and Coir*, London.
Brogan, O. (1953), *Roman Gaul*, London.
Brookes, A. C. (1974), 'Minturnae: the Via Appia bridge', *American Journal of Archaeology* 78: 41–8.
Brown, F. E. (1961), *Roman Architecture*, New York.
Browning, R. (1980), *The Byzantine Empire*, London.
Bruce, F. F. (1958), *The Spreading Flame*, London.
Bruce, J. C. (1978), *Handbook to the Roman Wall*, 13th edn, ed. C. Daniels, Newcastle upon Tyne.
Brusin, G. (1949–50), *Atti dell'Ist. Veneto di Sc. Lett. e Arti* 108: 116–20.
Buck, R. J. (1971), 'The Via Herculia', *PBSR* 39: 66–87.
(1974), 'The ancient roads of eastern Lucania', *PBSR* 42: 46–67.
(1975), 'The ancient roads of south eastern Lucania', *PBSR* 43: 98–117.
Bundgard, J. A. (1965), 'Caesar's bridges over the Rhine', *Acta Architecture* 36: 87–103.
Bunt, N. H., Jones, P. S. and Bedient, J. D. (1976), *The Historical Roots of Elementary Mathematics*, Englewood Cliffs.
Burford, A. (1960), 'Heavy transport in classical antiquity', *The Economic History Review*, 2nd series, 13: 1–18.
Burn, A. R. (1953), *Agricola and Roman Britain*, London.
Burns, R. (1968), *The Poems and Songs*, ed. J., Kinsley, 3 vols., Oxford.
Byron, R. (1964), *The Byzantine Achievement: an Historical Perspective AD 330–1453*, New York.
Calder, W. M., and Bean, G. E. (1958), *A Classical Map of Asia Minor*, London.

Caprino, C., Colini, A. M., Gatti, G., Pallottino, M. and Romanelli, P. (1955), *La colonna di Marco Aurelio*, Rome.
Cary, M. and Scullard, H. H. (1975), *A History of Rome down to the Reign of Constantine*, 3rd edn, London.
Casado, C. F. (1972), *Acueductos romanos en España*, Madrid.
(n.d.), *Historia del puente en España: puentes romanos*, Madrid.
Casson, L. (1974), *Travel in the Ancient World*, London.
Cerruti, J. (1981), 'Down the ancient Appian Way', *National Geographic* 159: 714–47.
Charlesworth, M. P., (1970), *Trade-routes and Commerce of the Roman Empire*, reprint of 2nd edn of 1926, New York.
Chettoe, C. S. and Henderson, W. (1957), 'Masonry arch bridges: a study', *ICE Proc.* 7: 723–74.
Chevallier, R. (1976), *Roman Roads*, tr. N. H. Field, London.
Cichorius, C. (1886–1900), *Die Reliefs der Trajanssäule*, Berlin.
Clifton-Taylor, A. and Treson, A. S. (1983), *English Stone Building*, London.
Coarelli, F. (1974), *Guida archeologica di Roma*, Milan.
Collingwood, R. G. and Richmond, I. (1969), *The Archaeology of Roman Britain*, revised edn, London.
Connolly, P. (1981), *Greece and Rome at War*, London.
Cooke, N. (1987), 'Instability of masonry arches', *ICE Proc.* 83 (2): 497–515.
Cossons, N. and Trinder, B. (1979), *The Iron Bridge: Symbol of the Industrial Revolution*, Bradford-on-Avon.
Coulton, J. J. (1977), *Greek Architects at Work: Problems of Structure and Design*, London.
Cowan, H. J. (1977), *The Master Builders: a History of Structural and Environmental Design from Ancient Egypt to the Nineteenth Century*, New York.
Crisfield, M. A. (1984), 'Finite element computer program for the analysis of masonry arches', *Transport and Road Research Laboratory, Laboratory Report* 1115.
Cuntz, O., ed. (1929), *Itineraria Romana, itineraria Antonini Augusti et Burdigalense*, Leipzig.
Cüppers, H. (1969), *Die Trierer Römerbrücken*, Mainz.
Curle, J. (1911), *A Roman Frontier Post and its People*, Glasgow.
Davidovitis, J. (1987), 'Ancient and modern concretes: what is the real difference?', *Concrete International* 9: 23–35.
de Camp, L. S. (1977), *Ancient Engineers*, London.
Degrand, E. and Resal, J. (1887–8), *Ponts en Maçonnerie*, 2 vols., Paris.
de Montauzan, C. G. (1908), *Les Aqueducs antiques de Lyon*, Paris.
Dijksterhuis, E. J. (1987), *Archimedes*, tr. C. Dikshoorn, with a new bibliographic essay by W. R. Knorr, Princeton.
Dilke, O. A. W. (1971), *The Roman Land Surveyors: an Introduction to the Agrimensores*, Newton Abbot.
(1985), *Greek and Roman Maps*, London.
Dorner, F. K., ed. (1981), *Vom Bosporus zum Ararat*, Mainz.
Drachman, A. G. (1963), *The Mechanical Technology of Greek and Roman Antiquity: a Study of the Literary Sources*, Copenhagen.
Dymond, D. P. (1961), 'Roman bridges on Dere Street, County Durham, with a general appendix on the evidence for bridges in Roman Britain', *Archaeological Journal* 118: 136–64.
Eck, W. (1985), *Die Statthalter der germanischen Provinzen*, Cologne/Bonn.
Emerson, W. and Gromort, G. (1925), *Old Bridges of France: a Series of Historical Examples from Roman Times to the end of the XVIIIth Century*, New York.
Ertman, P. C. (1976), *Curatores Viarum: a Study of the Superintendents of Highways in Ancient Rome*, Ann Arbor.
Fentress, E. W. B. (1979), *Numidia and the Roman Army*, Oxford.
(1984), 'Via Aurelia, Via Aemilia', *PBSR* 52: 72–6.
Finley, M. I. (1973), *The Ancient Economy*, Berkeley and Los Angeles.
(1980), *Ancient Slavery and Modern Ideology*, London.
Fitchen, J. (1961), *The Construction of Gothic Cathedrals: a Study of Medieval Vault Erection*, Oxford.
Florescu, F. B. (1969), *Die Trajanssäule – Grundfragen und Tafeln*, Bucharest.
Forbes, R. J. (1934), *Notes on the History of Roman Roads and their Construction*, 2nd edn, 1964, Amsterdam.
(1965), *Studies in Ancient Technology*, 9 vols., Leiden.
Frank, T. (1924), *Roman Buildings of the Republic: an Attempt to Date them from their Materials*, Rome.
Frederiksen, M. W. and Ward Perkins, J. B. (1957), 'The ancient road systems of the central and northern Ager Faliscus (notes on southern Etruria, 2)', *PBSR* 25: 67–209.
French, D. H. (1965), 'Prehistoric sites in the Göksu valley', *Anatolian Studies* 15: 177–201.
(1974), 'A study of Roman roads in Anatolia', *Anatolian Studies* 24: 143–150.
Frere, S. S. (1987), *Britannia: a History of Roman Britain*, London.
Galloway, G. (1914), *The Philosophy of Religion*, Edinburgh.
Ganuza, A. B. et al. (1981), *España: guía artística en color*, 4th edn, León.
Gardiner, A. H. (1920), 'The ancient military road between Egypt and Palestine', *J. of Egyptian Archaeology* 6: 99–116.
Gardner, R. (1913), 'The Via Claudia Nova', *JRS* 3: 204–30.
(1920), 'The Via Claudia Valeria', *PBSR* 9: 75–106.
Gazzola, P. (1963a), *Il ponte pietra in Verona*, Florence.
(1963b), *Ponti romani, volume II: Contributo ad un indice sistematico con studio critico bibliografico*, Florence.
González de Riancho Mazo, J. (1988), *La via romana de el Escudo*, Santander.
Gregory, J. W. (1931), *The Story of the Road: from the Beginning down to AD 1931*, London.
Grenier, A. (1931), *Manuel d'archéologie gallo-romaine*, 4 vols., reprinted 1985, Paris.
Guest, A. R. (1899), 'Itinerary from Kantara to el Arish', *Geographical J.* 13: 281–5.
Guido, M. (1972), *Southern Italy: an Archaeological Guide*, London.
Hall, W. H. B. (1898), *The Romans on the Riviera and the Rhône: a Sketch of the Conquest of Liguria and the Roman Province*, London, reprinted 1974, Chicago.
Hamey, L. A. and Hamey, J. A. (1981), *The Roman Engineers*, Cambridge.
Hammond, N. G. L. (1974), 'The western part of the Via Egnatia', *JRS* 64: 185–94.
(1988), 'The expedition of Xerxes', *CAH*, vol. 4, 2nd edn, pp. 518–91, Cambridge.
Harris, W. V. (1965), 'The Via Cassia and the Via Traiana Nova between Bolsena and Chiusi', *PBSR* 33: 113–33.
Harrison, D. (1962), *Along Hadrian's Wall*, London.
Harvey, J. H. (1972), *Conservation of Buildings*, London.
Harvey, W. J. (1987), 'Semicircular arches', *ICE Proc.* 83 (2): 845–9.

(1988), 'Application of the mechanism analysis to masonry arches', *Struct. Eng.* 66A: 77–84.
Harvey, W. J., Smith, F. W. and Vardy, A. E. (1987), 'Analysis and design of masonry arches', *Proceedings of the International Symposium on Geomechanics, Bridges and Structures*, Lanzhou, China.
Haselberger, L. (1985), 'The construction plans for the temple of Apollo at Didyma', *Sci. Am.* 253: 114–22.
Hauck, G. F. (1986), 'Structural design of the Pont du Gard', *ASCE J. Struct. Eng.* 112; 105–19.
Haverfield, F. (1912), *The Romanization of Roman Britain*, 3rd edn, revised by G. Macdonald, Oxford.
(1924), *The Roman Occupation of Britain*, Oxford.
Hemphill, P. (1955), 'The Cassia–Clodia survey', *PBSR* 43: 118–72.
Herschel, C. (1973), *The Two Books on the Water Supply of the City of Rome of Sextus Julius Frontinus, Water Commissioner of the City of Rome AD 97*, tr. C. Herschel, with explanatory chapters and intro., reissued 1973, Boston.
Heyman, J. (1966), 'The stone skeleton', *International Journal of Solids and Structures* 2: 249–79.
(1969), 'The safety of masonry arches', *Int. J. Mech. Sci.* 11: 363–85.
(1971), *Plastic Design of Frames v2 Applications*, Cambridge.
(1972), *Coulomb's Memoir on Statics: an Essay in the History of Civil Engineering*, Cambridge.
(1976), 'Couplet's engineering memoirs 1726–33', *History of Technology 1976*, ed. A. R. Hall, and N. Smith, pp. 21–44, London.
(1980), 'Estimation of the strength of masonry arches', *ICE Proc.* 69: 921–37.
(1982), *The Masonry Arch*, Chichester.
Heyman, J. and Padfield, C. J. (1972), 'Two masonry bridges', *ICE Proc.* 52: 305f.
Hill, D. (1984), *A History of Engineering in Classical and Medieval Times*, London and Sydney.
Hindley, G. (1971), *A History of Roads*, London.
Hodge, A. T. (1983), 'Siphons in Roman aqueducts', *PBSR* 51: 174–221.
(1985), 'Siphons in Roman aqueducts', *Sci. Am.* 252: 94–9.
(1989), 'Aqueducts', in I. M. Barton, ed., *Roman Public Buildings*, Exeter Studies in History 20: 127–49, Exeter.
(1992), *Roman Aqueducts and Water Supply*, London.
Hodges, H. (1970), *Technology in the Ancient World*, Harmondsworth.
Hogarth, D. G. and Yorke, V. W. (1898), 'Inscriptions from eastern Asia Minor', *JHS* 18: 306–27.
Honeyborne, D. B. (1982), 'The building limestones of France', *Building Research Establishment Report*, HMSO, London.
Hopkins, H. J. (1970), *A Span of Bridges: an Illustrated History*, Newton Abbot.
Hopkins, K. (1978), *Conquerors and Slaves: Sociological Studies in Roman History vol. 1*, Cambridge.
Hosking, W. (1839), *Bridges in Theory, Practice, and Architecture, Section II: Practice*, London.
Hyde, W. W. (1935), *Roman Alpine Passes*, Philadelphia.
Jackson, F. J. (1914), *The History of the Christian Church from the Earliest Times to AD 461*, 6th edn, London.
Jackson, P. (1971), *London Bridge*, London.
Jacobsen, T. and Lloyd, S. (1935), *Sennacherib's Aqueduct of Jerwan*, Chicago.
Jarrett, M. G. and Dobson, B., eds. (1966), *Britain and Rome*, Kendal.
Jennings, A. (1986), 'Stability fundamentals in relation to masonry arches', *Struct. Eng. Part B* 64: 10–12.
Johnson, S. M. and Scott-Giles, C. W. (1933), *British Bridges*, London.
Jones, A. H. M. (1968–70), *A History of Rome Through to the Fifth Century*, 2 vols., London.
(1971), *The Cities of the Eastern Roman Provinces*, 2nd edn, Oxford
Kahane, A. M. (1973), 'A paved Roman road east from Gabii', *PBSR* 41: 18–44.
Kahane, A., Threipland, L. M. and Ward Perkins, J. (1968), 'The Ager Veientanus, north and east of Rome', *PBSR* 36: 1–218.
Kahane, A. and Ward Perkins, J. (1972), 'The Via Gabina', *PBSR* 40: 91–126.
Kanowski, M. G., Barrett, D. S. and Bryce, T. R. (1977), *Threshold of Time*, Sydney.
Knapp, R. C. (1983), *Roman Córdoba*, Berkeley.
Knott, C. G. (1886), 'The abacus in its historic and scientific aspects', *Asiatic Society of Japan, Transactions* 14: 18–71.
Landels, J. G. (1978), *Engineering in the Ancient World*, London.
Lawrence, T. E. (1962), *Seven Pillars of Wisdom: a Triumph*, Harmondsworth.
Leary, E. (1983), 'The building limestones of the British Isles', *Building Research Establishment Report*, HMSO, London.
(1986), 'The building sandstones of the British Isles', *Building Research Establishment Report*, HMSO, London.
Léger, A. (1875), *Les Travaux publics, les mines et la métallurgie aux temps des Romains*, Paris.
Lehmann-Hartleben, K. (1926), *Die Trajanssäule, ein romischer Kunstwerk zu Beginn der Spätantike*, 2 vols., Berlin and Leipzig.
Leonhardt, F. (1982), *Brücken/Bridges*, London.
Lepper, F. and Frere, S. (1988), *Trajan's Column: a New Edition of the Cichorius Plates*, New Hampshire and Gloucester.
Levi, A. C. and Trell, B. (1964), 'An ancient tourist map', *Archaeology* 17: 227–36.
Levick, B. M. (1967), *Roman Colonies in Southern Asia Minor*, Oxford.
Lewis, M. J. T. (1984), 'Roman navigation in northern England? A review article', *Journal of the Railway and Canal Historical Society* 28: 124f.
(1984–5), 'Roman methods of transporting and erecting obelisks', *Newcomen Society Transactions* 56: 87–110.
Lloyd, S. (1980), *Foundations in the Dust: the Story of Mesopotamian Exploration*, 2nd edn, London.
Loane, H. J. (1979), *Industry and Commerce of the City of Rome (50 BC–200AD)*, New York.
Lugli, G. (1957), *La tecnica edilizia romana, con particolare riguardo a Roma e Lazio*, 2 vols., Rome.
Macaulay, T. B. (1894), *Essays and Lays of Ancient Rome*, London.
McEvedy, C. (1967), *The Penguin Atlas of Ancient History*, Harmondsworth.
MacKendrick, P. (1972), *Roman France*, New York.
MacMullen, R. (1963), *Soldier and Civilian in the Later Roman Empire*, Cambridge, Mass.
MacPherson, I. W. (1954), 'Roman roads and milestones of Galatia', *Anatolian Studies* 4: 111–20.
Magie, D. (1950), *Roman Rule in Asia Minor*, 2 vols., Princeton.

Mainstone, R. J. (1959), 'Structural theory and design, Part 1. Early proportional theories', *Architecture and Building* 34: 106–13.
(1965–6), 'The structure of the Church of St Sophia, Istanbul', *Newcomen Society Transactions* 38: 23–49.
Malinowski, R. (1979), 'Concretes and mortars in ancient aqueducts', *Concrete International* 1: 66–76.
(1982), 'Ancient mortars and concretes: aspects of their durability', *History of Technology* 7: 89–100.
Mao Yi-Sheng (1978), *Bridges in China, Old and New*, Peking.
Margary, I. D. (1973), *Roman Roads in Britain*, 3rd edn, London.
Martinori, E. (1929), *Via Flaminia*, Rome.
(1930a), *Via Nomentana*, Rome.
(1930b), *Via Cassia*, Rome.
(1931), *Via Salaria*, Rome.
Matheson, S. A. (1972), *Persia: an Archaeological Guide*, London.
Melan, J. (1924), *Der Brückenbau*, 3 vols., Leipzig and Vienna.
Meloni, M. (1975), *La Sardegna romana*, Sassari.
Mikami, Y. (1913), *The Development of Mathematics in China and Japan*, 2nd edn, 1974, New York.
Miller, K. (1887), *Die Peutingersche Tafel*, reprinted 1962, Stuttgart.
(1916), *Itineraria Romana*, Stuttgart, reprinted 1964, Rome.
Mitford, T. B. (1974), 'Some inscriptions from the Cappadocian Limes', *JRS* 64: 160–75.
Mothersole, J. (1929), *Hadrian's Wall*, London.
Munro, J. A. R. (1901), 'Roads in Pontus, Royal and Roman', *JHS* 21: 52–66.
Murray, G. W. (1925), 'The Roman roads and stations in the eastern desert of Egypt', *Journal of Egyptian Archaeology* 11: 138–50.
Naruse, K., Aoki, K. and Murakami, E. (1964), *Bridges of the World*, Tokyo.
Nash, E. (1968), *Pictorial Dictionary of Ancient Rome*, 2 vols., 2nd edn, London.
Needham, J., Wang Ling and Lu Gwei-Djen (1971), *Civil Engineering and Nautics*, vol. 4, Part 3, *Science and Civilisation in China*, ed. J. Needham, Cambridge.
New Catholic Encyclopaedia (1967), New York.
Oakley, K. P. (1950), *Man the Tool-maker*, 2nd edn, London.
O'Connor, C. (1971), *Design of Bridge Superstructures*, New York.
(1975), *The Design of Bridges: an Historical Study*, Brisbane.
(1983), *Register of Australian Historic Bridges*, Canberra.
(1985), *Spanning Two Centuries: Historic Bridges of Australia*, Brisbane.
(n.d.) *How to Look at Bridges: a Guide to the Study of Australian Historic Bridges*, Canberra.
O'Neill, H. (1965), *Stone for Building*, London.
Page, J. (1987), 'Load tests to collapse on two arch bridges at Preston, Shropshire and Prestwood, Staffordshire', *Transport and Road Research Laboratory (Crowthorne), Research Report*, no. 110.
(1988), 'Load tests to collapse on two arch bridges at Torksey and Shinafoot', *Transport and Road Research Laboratory (Crowthorne), Research Report*, no. 159.
Paget, R. F. (1973), *Central Italy: an Archaeological Guide*, London.
Pannell, J. P. M. (1964), *An Illustrated History of Civil Engineering*, London.
Parsons, W. B. (1968), *Engineers and Engineering in the Renaissance*, Cambridge, Mass.
Peleg, Y. (1984), 'The water supply of Caesarea', in *Papers at the Symposium 'Historical Water Development Projects in the Eastern Mediterranean', Jerusalem 21/22 March 1983*, vol. 82, Leichtweiss-Institut für Wasserbau, der Technischen Universität Braunschweig, Mitteilungen.
Percy, E. (1901), *Highlands of Asiatic Turkey*, London.
Pippard, A. J. S. (1948), *The Approximate Estimation of Safe Loads on Masonry Bridges*, vol. 1, *Civil Engineer in War*, London.
Pippard, A. J. S. and Baker, J. F. (1943), *The Analysis of Engineering Structures*, London.
Pippard, A. J. S., Tranter, E. and Chitty, L. (1936), 'The mechanics of the voussoir arch', *Institution of Civil Engineers Journal* 4: 281–306.
Pitts, L. F. and St Joseph, J. K. (1985), *Inchtuthil: the Roman Legionary Fortress*, Gloucester.
Platner, S. B. and Ashby, T. (1929), *A Topographical Dictionary of Ancient Rome*, Oxford.
Porter, M. W. (1907), *What Rome was Built With*, London.
Price, C. A. (1975), 'Testing porous building stone', *The Architect's Journal* 162: 337–9.
Promis, C. (1862), *Le antichità di Aosta Augusta Praetoria Salassorum*, Turin.
Purchase, W. R. (1904), *Practical Masonry*, 5th edn, London.
Qian Ling-Xi (1987), 'New insight into an ancient stone arch bridge – the Zhao-Zhou Bridge of 1400 years old', *Int. J. Mech. Sci.* 29: 831–43.
Ramsay, W. M. (1890), *The Historical Geography of Asia Minor*, Royal Geographic Society, Supplementary Papers, vol. 4, reprinted 1972, New York.
Raven, S. (1969), *Rome in Africa*, London.
Rayment, D. L. and Pettifer, K. (1987), 'Examination of durable mortar from Hadrian's Wall', *Materials Science and Technology* 3: 997–1004.
Richmond, I. A. (1963), *Roman Britain*, 2nd edn, Harmondsworth.
(1982), *Trajan's Army on Trajan's Column*, Rome.
Rivet, A. L. F. (1988), *Gallia Narbonensis, with a Chapter on Alpes Maritimae: Southern France in Roman Times*, London.
Robert, L. (1962), *Villes d'Asie Mineure: études de géographie ancienne*, 2nd edn, Paris.
Robertson, A. (1969), *River Crossings on the Antonine Wall*, in *Roman Frontier Studies*, ed. E. Birley, B. Dobson and M. Jarrett, pp. 94–101, Cardiff.
Robertson, D. S. (1943), *Greek and Roman Architecture*, 2nd edn, Cambridge.
Robins, F. W. (1946), *The Story of Water Supply*, Oxford.
(1948), *The Story of the Bridge*, Birmingham.
Rodgers, R. H. (1982), '*Curatores aquarum*', *Harvard Studies in Classical Philology* 86: 171–80.
Rogers, P. M. (1984), 'Domitian and the finances of state', *Historia* 33: 60–78.
Rose, A. C. (1935), *Via Appia in the Days When All Roads Led to Rome*, Washington.
Rossi, L. (1968), 'The representation on Trajan's column of Trajan's rock-cut road in Upper Moesia: the emperor's road to glory', *The Antiquaries Journal* 48: 41–6.
(1971), *Trajan's Column and the Dacian Wars*, English tr. revised by J. M. C. Toynbee, London.
Rostovtzeff, M. I. (1967), *The Social and Economic History of the Roman Empire*, 2nd edn, revised by P. M. Fraser, Oxford.

Ruddock, T. (1979), *Arch Bridges and their Builders 1735–1835*, Cambridge.

Ruhen, O. (1980), *Bullock Teams: the Building of a Nation*, Sydney.

Runciman, S. (1933), *Byzantine Civilization*, reprinted 1981, London.

Sadler, H. (1986), *Brücken*, Dortmund.

Salway, P. (1981), *Roman Britain*, Oxford.

Sandström, G. E. (1970), *Man the Builder*, New York.

Sasel, J. (1973), 'Trajan's canal at the Iron Gate', *JRS* 63: 80–5.

Schaffer, R. J. (1972), 'The weathering of natural building stones', *Building Research Special Report*, HMSO 3, Building Research Establishment, Garston.

Selberg, A. (1953), 'On the bearing capacity of voussoir arches', *International Association for Bridge and Structural Engineering Publication*, 13: 321–6.

Selkirk, R. (1983), *The Piercebridge Formula*, Cambridge.

Sherwin-White , A. N. (1966), *The Letters of Pliny: a Historical and Social Commentary*, Oxford.

Shirley-Smith, H. (1953), *The World's Greatest Bridges*, London.

Singer, C., Holmyard, E. J., Hall, A. R. and Williams, T. I., eds. (1954–84), *A History of Technology*, 7 vols., Oxford.

Sitwell, N. H. H. (1981), *Roman Roads of Europe*, London.

Smiles, S. (1862), *Lives of the Engineers*, 3 vols., reprinted 1968, Newton Abbot.

Smith, D. E. (1919), *Number Stories of Long Ago*, Washington.

(1925), *History of Mathematics*, 2 vols., reprinted 1958, New York.

Smith, G. A. (1935), *The Historical Geography of the Holy Land*, 26th edn, London.

Smith, N. A. F. (1976a), 'Attitudes to Roman engineering and the question of the inverted siphon', *History of Technology* 1: 45–71.

(1976b), *Man and Water*, London.

(1977–8), 'Roman Canals', *Newcomen Society Transactions* 49: 75–86.

(1990–1), 'The Pont du Gard and the aqueduct of Nîmes', *Newcomen Society Transactions* 62: 53–80.

Smith, W. (1901–14), *A Dictionary of Greek and Roman Antiquities*, 3rd edn, London.

Société Languedocienne (1905), *Géographie générale du Département de l'Hérault*, 3 vols., Montpellier.

Speidel, M. P. (1987), 'The Roman road to Dumata (Jawf in Saudi Arabia) and the frontier strategy of Praetensione Colligare', *Historia* 36: 213–21.

Spratt, T. A. B. and Forbes, E. (1847), *Travels in Lycia*, London.

Starr, S. F. (1962), *The Ancient Roads of Asia Minor: 1961 Expedition*, New Haven.

Steinman, D. B. (1959), *Songs of a Bridge Builder*, reissued 1972, Calcutta.

Steinman, D. B. and Watson, S. R. (1941), *Bridges and their Builders*, New York.

Texier, C. (1839), *Descriptions de l'Asie Mineure, vol. 1*, Paris.

Thiel, J. H. (1954), *A History of Roman Sea-Power before the Second Punic War*, Amsterdam.

Thouvenot, R. (1940), *Essai sur la province romaine de Bétique*, Paris.

Torr, C. (1885), *Rhodes in Ancient Times*, Cambridge.

Truesdell, C. (1960), *The Rational Mechanics of Flexible or Elastic Bodies 1638–1788*, in *Introduction to Leonhardi Euleri Opera Omnia II(2)*, Zurich.

Tunç, G. (1978), *Tas Köprülerimiz*, Ankara.

Tyrrell, H. G. (1911), *History of Bridge Engineering*, Chicago.

van Deman, E. B. (1934), *The Building of the Roman Aqueducts*, reprinted 1973, Washington.

van der Heyden, A. A. M. and Scullard, H. H. (1963), *Atlas of the Classical World*, London.

van Essen, C. C. (1957), 'The Via Valeria from Tivoli to Collarmele', *PBSR* 25: 22–38.

Verbrugghe, G. P. (1976), *Sicilia*, Berne.

Vermeule, C. C. (1968), *Roman Imperial Art in Greece and Asia Minor*, Cambridge, Mass.

Viatores, the (1964), *Roman Roads in the South-East Midlands*, London.

Vilnay, O. (1984), 'Buckling of masonry arches', *ICE Proc.* 77: 33–41.

(1988), 'Dynamical behaviour of three-voussoir arch', *ASCE J. Struct. Eng.* 114: 1173–86.

Vilnay, O. and Cheung, S. (1986), 'Stability of masonry arches', *ASCE J. Struct. Eng.* 112: 2185–99.

Vinson, P. (1972), 'Ancient roads between Venosa and Gravina', *PBSR* 40: 58–90.

von Hagen, V. W. (1967a), *The Roads that Led to Rome*, London.

(1967b), *Roman Roads*, London.

Walklate, R. P. (1983), 'Method for determining the permissible loading of brick and masonry arches', *ICE Proc.* 75 part 2: 585–97.

Ward Perkins, J. B. (1957), 'Etruscan and Roman roads in southern Etruria', *JRS* 57: 139–43.

(1961), 'Veii: the historical topography of the ancient city', *PBSR* 29: 1–123.

Warland, E. G. (1953), *Modern Practical Masonry*, 2nd edn, London.

Watson, G. R. (1969), *The Roman Soldier*, Bath.

Watson, W. J. and Watson, S. R. (1937), *Bridges in History and Legend*, illustrated by E. M. Watson, Cleveland.

Webster, G. (1979), *The Roman Imperial Army of the First and Second Centuries*, 2nd edn, London.

Whitby, L. M. (1985) 'Justinian's bridge over the Sangarius and the date of Procopius' *De aedificiis*', *JHS* 105: 129–48.

White, K. D. (1984), *Greek and Roman Technology*, London.

Whitting, P., ed. (1971), *Byzantium: an Introduction*, Oxford.

Winfield, D. (1977), 'The northern routes across Anatolia', *Anatolian Studies* 27: 151–66.

Winslow, E. M. (1963), *A Libation to the Gods: the Story of the Roman Aqueducts*, London.

Wiseman, F. J. (1956), *Roman Spain: an Introduction to the Roman Antiquities of Spain and Portugal*, London.

Wiseman, T. P. (1964), 'Viae Anniae', *PBSR* 32: 21–37.

(1969), 'Viae Anniae again', *PBSR* 37: 82–91.

(1970), 'Roman Republican road-building', *PBSR* 38: 122–52.

Wright, F. A. (1937), *Marcus Agrippa: Organizer of Victory*, London.

Wurster, W. W. and Ganzert, J. (1978), 'Eine Brücke bei Limyra in Lykien', *Archäologischer Anzeiger* 1978: 288–307.

Yorke, V. W. (1896), 'A journey in the valley of the upper Euphrates', *Geographical Journal* 8: 317–35.

Young, G. (1980), *Iraq, Land of Two Rivers*, London.

INDEX

(R) indicates a Roman (or ancient) place-name.
Symbols such as C3 are map co-ordinates.
Numbers in bold type are page numbers for figures or maps.